Smart Innovation, Systems and Technologies

Volume 140

The Smart Innovation, Systems and Technologies book series encompasses the topics of knowledge, intelligence, innovation and sustainability. The aim of the series is to make available a platform for the publication of books on all aspects of single and multi-disciplinary research on these themes in order to make the latest results available in a readily-accessible form. Volumes on interdisciplinary research combining two or more of these areas is particularly sought.

The series covers systems and paradigms that employ knowledge and intelligence in a broad sense. Its scope is systems having embedded knowledge and intelligence, which may be applied to the solution of world problems in industry, the environment and the community. It also focusses on the knowledge-transfer methodologies and innovation strategies employed to make this happen effectively. The combination of intelligent systems tools and a broad range of applications introduces a need for a synergy of disciplines from science, technology, business and the humanities. The series will include conference proceedings, edited collections, monographs, handbooks, reference books, and other relevant types of book in areas of science and technology where smart systems and technologies can offer innovative solutions.

High quality content is an essential feature for all book proposals accepted for the series. It is expected that editors of all accepted volumes will ensure that contributions are subjected to an appropriate level of reviewing process and adhere to KES quality principles.

**** Indexing: The books of this series are submitted to ISI Proceedings, EI-Compendex, SCOPUS, Google Scholar and Springerlink ****

More information about this series at http://www.springer.com/series/8767

Yuzo Iano · Rangel Arthur · Osamu Saotome ·
Vânia Vieira Estrela · Hermes José Loschi
Editors

Proceedings of the 4th Brazilian Technology Symposium (BTSym'18)

Emerging Trends and Challenges in Technology

Springer

Editors
Yuzo Iano
Faculty of Electrical
and Computing Engineering
University of Campinas
Campinas, São Paulo, Brazil

Osamu Saotome
Divisão de Engenharia Eletrônica
Instituto Tecnológico de Aeronáutica ITA
São José dos Campos, São Paulo, Brazil

Hermes José Loschi
Faculty of Electrical
and Computing Engineering
University of Campinas
Campinas, São Paulo, Brazil

Rangel Arthur
Faculty of Technology
University of Campinas
Campinas, São Paulo, Brazil

Vânia Vieira Estrela
Universidade Federal Fluminense (UFF)
Duque de Caxias, Rio de Janeiro, Brazil

ISSN 2190-3018 ISSN 2190-3026 (electronic)
Smart Innovation, Systems and Technologies
ISBN 978-3-030-16052-4 ISBN 978-3-030-16053-1 (eBook)
https://doi.org/10.1007/978-3-030-16053-1

Library of Congress Control Number: 2019935478

This Springer imprint is published by the registered company Springer Nature Switzerland AG
The registered company address is: Gewerbestrasse 11, 6330 Cham, Switzerland

Foreword

It is with deep satisfaction that I write this Foreword to the Book, Proceedings of the 4th Brazilian Technology Symposium (BTSym'18), held at the Mackenzie Presbyterian University, SP, Brazil, in October 2018. This event is in its fourth edition and has consolidated to become an excellent opportunity for researchers, professors, and students to present and discuss the results of their research works.

The 2018 edition of BTSym is characterized by the broad scope of the areas exposed, with papers dealing with current and priority topics for Brazilian and world technological development, including subjects related to the various branches of human, smart and sustainable future of cities.

Events such as BTSym are an essential part of the research and innovation process. Firstly, these events contribute to the promotion of research activities, which are key to a country's technological development. The dissemination of research results, as promoted by BTSym, contributes to the transformation of research findings into technological innovation. In addition, these events facilitate the sharing of findings, leading eventually to the formation of research networks, which accelerate the achievement of new results. Therefore, I would like to congratulate the BTSym General Chair, Prof. Dr. Yuzo Iano, and his group of collaborators for the important initiative of organizing the BTSym 2018 and for providing the opportunity for authors to present their work to a wide audience through this publication. Finally, I congratulate the authors for the high-quality work presented in this proceedings.

Doha, Qatar

Douglas Aguiar do Nascimento, M.Sc. Eng.
Research Associate
Texas A&M University at Qatar

Preface

This book contains the Proceedings of the 4th Brazilian Technology Symposium, an event held in Campinas, SP, Brazil, in October 2018. This book was divided into two parts. The first part on Emerging Trends in Systems Engineering, Mathematics and Physical Sciences and the second part on Emerging Trends in Human, Smart and Sustainable Future of Cities.

The Brazilian Technology Symposium is an excellent forum for presentations and discussions of the latest results of projects and development research, in several areas of knowledge, in scientific and technological scope, including smart designs, sustainability, inclusion, future technologies, architecture and urbanism, computer science, information science, industrial design, aerospace engineering, agricultural engineering, biomedical engineering, civil engineering, control and automation engineering, production engineering, electrical engineering, mechanical engineering, naval and oceanic engineering, nuclear engineering, chemical engineering, and probability and statistics.

This event seeks to bring together researchers, students, and professionals from the industrial and academic sectors, seeking to create and/or strengthen the linkages between issues of joint interest. Participants were invited to submit research papers with methodologies and results achieved in scientific level research projects, completion of course work for graduation, dissertations, and theses.

The 64 full chapters accepted for this book were selected from 152 submissions, and in each case, the authors were shepherded by an experienced researcher, with a rigorous peer-review process. Among the main topics covered in this book, we can highlight transmission line, protein-modified mortars, electromagnetic properties, clock domains, Chebyshev polynomials, satellite control systems, Hough transform, watershed transform, blood smear images, Toxoplasma gondii, operation system developments, MIMO systems, geothermal–photovoltaic energy systems, mineral flotation application, CMOS techniques, frameworks developments, physiological parameters applications, brain–computer interface, artificial neural networks, computational vision, security applications, FPGA applications, IoT, residential automation, data acquisition, Industry 4.0, cyber-physical systems, digital image processing, pattern recognition, machine learning, photocatalytic process, physical–chemical analysis, smoothing filters, frequency synthesizers, voltage-controlled ring oscillator, difference amplifier, photocatalysis, photodegradation, digital transformation, data science, hydrothermal dispatch, project knowledge transfer, immunization programs, efficiency and predictive methods, PMBOK applications, logistics process, fingerspelling recognition, cognitive ergonomics, ecosystem services, environmental, ecosystem services valuation, solid waste, university extension, and much more.

We hope you enjoy and take advantage of this book and feel motivated to submit us your papers in the future to Brazilian Technology Symposium.

Campinas, Brazil

Best wishes,
Prof. Hermes José Loschi
Technical and Finance Chair of Brazilian Technology Symposium

Acknowledgements

Our appreciation goes to a lot of colleagues and friends who assisted in the development of this book, Proceedings of the 4th Brazilian Technology Symposium (BTSym'18).

First of all, I would like to thank all the members of the Organizing and Executive Committee, for the commitment throughout the year, several meetings were held, and many challenges were overcome for the accomplishment of the BTSym 2018. Also, and with great merit, I would like to thank all the members of the Scientific and Academic Committee, and Technical Reviewers Committee, for their excellent work, which was essential to ensure the quality of our peer-review process, and collaborating with the visibility and technical quality of the BTSym 2018.

The Brazilian Technology Symposium is an event created by Laboratory of Visual Communications of the Faculty of Electrical and Computer Engineering of the University of Campinas (UNICAMP). In this way, I would like to thank the Mackenzie Presbyterian University, especially for the support and hosting of the BTSym 2018, which was fundamental for the success their accomplishment.

Beta Telecommunications played a crucial role in holding the BTSym 2018; due to financial support from it, and it was possible to consolidate with quality many BTSym 2018 organization aspects, which ensured the quality to support the authors and speakers.

Finally, I thank all the authors for their participation in the BTSym 2018, I sincerely hope to have provided an experience that was very useful and enriching in the personal and professional life of everyone, and my special thanks go to Prof. Vânia Vieira Estrela. In my almost 50 years of academic career by UNICAMP, few were the opportunity to work with a professional such her, She is definitely one of a kind, extremely efficient, hardworking and the BTSym 2018 certainly has much to thank for Prof. Vânia Vieira Estrela.

Best wishes,
Prof. Yuzo Iano
General Chair of Brazilian Technology Symposium

Contributors

Organizing and Executive Committee

- Yuzo Iano—LCV/DECOM/FEEC/UNICAMP—General Chair BTSym
- Osamu Saotome—ITA—Associate-General Chair BTSym
- Rangel Arthur—FT/UNICAMP—Vice-General Chair BTSym
- Vânia Vieira Estrela—UFF—Vice-Associate-General Chair BTSym
- Hermes José Loschi—LCV/DECOM/FEEC/UNICAMP—Technical Program and Finance Chair
- Telmo Cardoso Lustosa—LCV/DECOM/FEEC/UNICAMP—Local Arrangements Chair
- Ana Carolina Borges Monteiro—LCV/DECOM/FEEC/UNICAMP—Registration Chair
- Reinaldo Padilha—LCV/DECOM/FEEC/UNICAMP—Proceedings Chair
- Alysson Gomes de Oliveira—LCV/DECOM/FEEC/UNICAMP—Marketing Chair
- Douglas Aguiar do Nascimento—LCV/DECOM/FEEC/UNICAMP—Institutional Relationship Chair
- Gabriel Gomes de Oliveira—LCV/DECOM/FEEC/UNICAMP—Vice-Institutional Relationship Chair

Scientific and Academic Committee

- Osamu Saotome—ITA
- Vânia Vieira Estrela—UFF—Rio de Janeiro
- Luiz Cézar Martini—DECOM/FEEC/UNICAMP
- David Bianchini—PUC/Campinas
- Luis Geraldo Pedroso Meloni—DECOM/FEEC/UNICAMP
- Saravanan Chandran—Department of CSE, NIT Durgapur, WB, India
- Ana Cláudia Seixas—PUC/Campinas
- Cristiano Akamine—Universidade Presbiteriana Mackenzie

- Georgius Nikolakopoulos—University of Technology, Luleå, Sweden
- Celso Iwata Frison—PUC/Minas-Poços de Caldas
- Luiz Vicente Figueira de Mello Filho—Universidade Presbiteriana Mackenzie
- Guillermo Leopoldo Kemper Vásquez—USMP and UNI-INICTEL
- Lia Toledo Moreira Mota—PUC/Campinas
- Sergey N. Filippov—St. Petersburg Phys&Math Lyceum 239, Russia
- Lucas Heitzmann Gabrielli—DECOM/FEEC/UNICAMP
- R. Jenice Aroma—Sri Krishna College of Technology/Coimbatore
- Edgard Luciano Oliveira da Silva—EST/UEA
- Talía Simões dos Santos—FT/UNICAMP
- Janito Vaqueiro Ferreira—DMC/FEM/UNICAMP
- Vlademir de Jesus Silva Oliveira—UNEMAT/Sinop
- Hugo Enrique Hernandez Figueroa—DECOM/FEEC/UNICAMP
- Santhanam Lakshmi—Jeppiaar SRR Engineering College, Chennai, India
- Antônio José da Silva Neto—IPRJ/UERJ
- Bruno Sanches Masiero—DECOM/FEEC/UNICAMP
- Marcos Antonio do Nascimento Guimarães—UNIP/CAMPINAS—JUNDIAÍ
- Alessandra Akkari—Universidade Presbiteriana Mackenzie
- Maria Thereza de Moraes Gomes Rosa—Universidade Presbiteriana Mackenzie
- Angela del Pilar Flores Granados—FEA/UNICAMP
- Paolo Bacega—Faculdade Anhanguera
- Marcos Fernando Espindola—IFSP São Paulo
- Navid Razmjooy—Tafresh University, Tafresh, Iran
- Polyane Alves Santos—Instituto Federal Da Bahia
- Jude Hemanth—Karunya University, Coimbatore, India
- Abdeldjalil Khelassi—Abou Bakr Belkaid University of Tlemcen, Chetouane, Algeria

Technical Reviewers Committee

- Adão Boava—Universidade Federal de Santa Catarina—UFSC
- Ana Carolina Borges Monteiro—LCV/DECOM/FEEC/UNICAMP
- Amilton Lamas—PUC-Campinas
- Agord de Matos Pinto Júnior—DESIF/FEEC/UNICAMP
- Angela del Pilar Flores Granados—FEA/UNICAMP
- R. Jenice Aroma—Sri Krishna College of Technology/Coimbatore
- Joaquim Marcos Santa Rita da Silva—Instituto Nacional de Telecomunicações
- Alysson Gomes de Oliveira—LCV/DECOM/FEEC/UNICAMP
- José Alexandre Nalon—Centro Universitário Salesiano São Paulo—UNISAL
- Murilo Cesar Perin Briganti—LCV/DECOM/FEEC/UNICAMP
- Ashok Kumar—Jaypee University of Information Technology
- Luigi Ciambarella Filho—Universidade Veiga de Almeida/Develop Biotechnology

- Lucas Alves—LCV/DECOM/FEEC/UNICAMP
- Ingrid Araujo Sampaio—Universidade Estadual de Campinas
- Hermes José Loschi—LCV/DECOM/FEEC/UNICAMP
- Denise Helena Lombardo Ferreira—PUC Campinas
- Daniel Rodrigues Ferraz Izario—LCV/DECOM/FEEC/UNICAMP
- Mariana Carvalho—DCA/FEEC/UNICAMP
- Santhanam Lakshmi—Jeppiaar SRR Engineering College, Chennai, India
- Diego Pajuelo—LCV/DECOM/FEEC/UNICAMP
- Navid Razmjooy—Tafresh University, Tafresh, Iran
- Douglas Aguiar do Nascimento—LCV/DECOM/FEEC/UNICAMP
- Saravanan Chandran—Department of CSE, NIT Durgapur, WB, India
- Edson José Gonçalves—LCV/DECOM/FEEC/UNICAMP
- Alessandra Akkari—Universidade Presbiteriana Mackenzie
- Polyane Alves Santos—Instituto Federal Da Bahia
- Rangel Arthur—INOVA/FT/UNICAMP
- Reinaldo Padilha—LCV/DECOM/FEEC/UNICAMP
- Sergey N. Filippov—St. Petersburg Phys&Math Lyceum 239, Russia
- Kelem Christine Pereira Jordão—LCV/DECOM/FEEC/UNICAMP
- Josué Marcos de Moura Cardoso—LCV/DECOM/FEEC/UNICAMP
- Giovanni Moura de Holanda—FITec—Technological Innovations, Brazil
- Euclides Lourenço Chuma—LCV/DECOM/FEEC/UNICAMP
- José Yauri—DCA/FEEC/UNICAMP
- Anand Deshpande—KLS Gogte Institute of Technology, Belagavi
- Julío Humberto León Ruiz—LCV/DECOM/FEEC/UNICAMP
- Antônio José da Silva Neto—IPRJ/UERJ
- Abdeldjalil Khelassi—Abou Bakr Belkaid University of Tlemcen, Chetouane, Algeria
- Raphael Ronald—CTI (Centro de Tecnologia da Informação)
- Raquel Jahara Lobosco—UFRJ/Macaé
- Necesio Gomes Costa—UFRJ
- Claudia Maria Moraes Santos—UNIVAP
- Vitor Chaves de Oliveira—IFSP São Paulo/Mackenzie

Sponsor

- Beta Telecommunications

Contents

Part I
Emerging Trends in Systems Engineering, Mathematics and Physical Sciences

A Performance Analysis of Uprating Techniques on a 230 KV Overhead Transmission Line

Juliana Maciel Maia Beça, Jose Maurício Barros Bezerra, Jonatan Esau Mejia Quijada, Ligia Veronica Genesio Pessoa and Ronaldo Ramos da Silva

Abstract The proposal of this work is to present uprating techniques on a 230 kV transmission line to improve the transmission lines capacity, as well, comparison of uprating techniques which highlights the importance of the reconductoring technique. The methodology is validated through a computer application widely used by electric power companies to design and optimize new transmission lines, thus bringing significant gains in the heights of the conductors. Furthermore, introducing a competitive option so as to increase the transmission capacity of lines in operation. We also present some studies case conducted on a real trans-mission line in order to evaluate the benefits the techniques may offer at level voltage.

Keywords Transmission lines · Computer application · Uprating · Reconductoring

1 Introduction

Energy is a fundamental component for the economic and social development of any country. A world without electricity is practically inconceivable [1]. Furthermore, the constant increase in demand for electric power means that there is a need to expand transmission systems, such as substations and transmission lines, thus preventing poor facilities from being a limiting factor for economic development. The implementation of new transmission lines is facing difficulties due to some external factors that have a direct impact on the cost of these lines. First, the minimization of the environmental impact caused by the use of higher structures and consequently heav-

J. M. M. Beça (✉) · J. M. B. Bezerra · J. E. M. Quijada
Grupo de Pesquisa de Transmissãao e Distribuição de Energia Elétrica, Universidade Federal de Pernambuco, Pernambuco, Brazil
e-mail: julianamaciel03@hotmail.com

L. V. G. Pessoa · R. R. da Silva
Departamento de Engenharia de Linha de Transmissão Recife, Companhia Hidreletrica Do São Francisco, Pernambuco, Brazil

Y. Iano et al. (eds.), *Proceedings of the 4th Brazilian Technology Symposium (BTSym'18)*, Smart Innovation, Systems and Technologies 140,
https://doi.org/10.1007/978-3-030-16053-1_1

ier ones. And secondly, the high cost to built including the easement transmission line strips and the payment of indemnities.

As cost and environmental considerations present barriers to large investments in rebuilding existing lines and/or constructing new ones electric utilities are facing a challenge to increase the power transfer capacity of existing overhead lines [2].

The need to increase the power transfer capacity of existing overhead lines by cost-effective and environmentally friendly methods within a competitive deregulated market is directly linked with an increase in the associated conductors ampacity (i.e. current capacity) [3].

Uprating a transmission line consists of increasing its transmission capacity of electric energy to meet to a new system request in situations where the implantation of a new transmission line is economically and environmentally more impactant. That is, it consists in increasing its thermal limit or its temperature maximum operating temperature [4].

In most publications uprating has been defined as the increase in the electrical characteristic of the transmission line, while upgrading has been denied as reinforcements in the components of the line. In the recent bibliographies studied, it is verified that the definitions of uprating is an increase in transmission capacity and upgrading is improvement in the reliability. Therefore, it is verified that transmission lines uprating is composed of these two techniques [5].

This study aims to be part of an academic development, with the purpose of contributing to the academic community, to professionals in the sector, who need more energy, more efficiency and lower costs for society. Thus, the uprating techniques have a strong importance for the economy in the realization of the project, and in addition, it brings a huge advantage to the environment and to the responsible organs, due to its analysis of processes for topographic surveys and identification of techniques used nowadays.

2 Research Development

In this section, it is initiated the research by investigating several uprating techniques already applied in Brazil, and in other countries, in order to verify the best solutions for the Northeast region of Brazil. These uprating techniques were realized in the Northeast region of Brazil, in the metropolitan region of Recife city, state of Pernambuco. The transmission line project has the extension of 44 km is called Recife II-Suape and it was simulated with the computer application PLS-CADD [Power Line-Computer Aided Design and Drafting)] widely used by electric power companies to construct and optimize new transmission lines.

The techniques would need to take into account the particularities of each transmission system, adjusting the meteorological conditions of the region studied and the cost-bene t ratio of the project for society. To better characterize these advantages, a comparative study was conducted between applying conventional techniques and the new technology. It was then possible to analyze the aspects related to reusing material and minimizing the resources involved [5].

The studies were initiated by identifying the state-of-the-art uprating methods described in the literature, and their various applicabilities, and culminated with the introduction of the best solution choose to this line and it comparison to the reconductoring technology described in this paper.

3 Uprating Transmission Lines—State of Art

In this section of the article, it is described the techniques traditionally used for uprating transmission lines. During the research, the following techniques were highlighted: insulator strings exchange, increased of the strenght, improving isolation voltage, changing structures types, the relocation of steel structures and finally the reconductoring method. These techniques have been validated and compared through PLS-CADD software. Two of the most popular ways that can be employed to achieve this are re-tensioning and reconductoring [3].

The PLS-CADD is a commercial software or a tower-rental computing application that was developed and is used for the analysis and development of transmission lines. By input data, it is provided all possible structures loading and project criteria, which enables the computer to perform these efficiently. The computational application integrates in one place all the components necessary for the geometric and structural development of a line. Moreover, the uprating techniques could be simulated using the PLS-CADD after the line project simulation.

The Fig. 1 illustrates the pro le of the transmission line which in situated in Recife and has the extension of 44 km.

3.1 Uprating Transmission Lines Techniques

A. Exchange Insulator String The first method used was the exchange in the insulator string. This solution is adopted when the clearance in the catenary safety height is

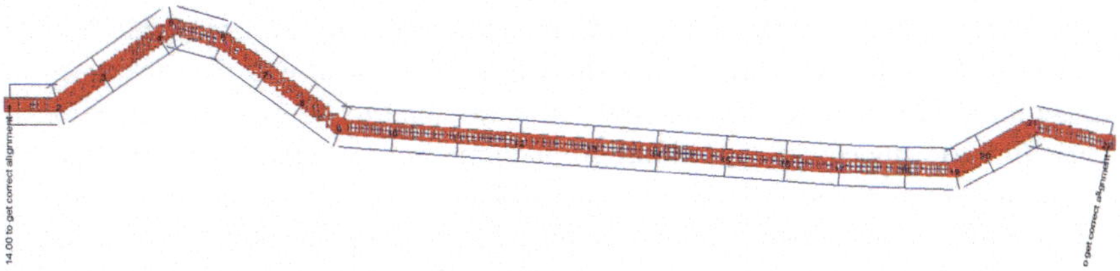

Fig. 1 Transmission Line Pro le Recife II—Suape

violated. To decrease this height the insulator string is changed around forty five centimeters and this method must be use only in polymeric insulator type. Thus, the cross-arm modification is choose between the connecting of landle to "y format" in the structure.

The Fig. 2 illustrates this technique used for uprating transmission lines. Thus, increasing the conductor ground clearance can be achieved by raising the conductors suspension point; for example by reconfiguring insulators from tangent to floating dead-end configurations. Glass disc insulators can be replaced by composite insulators with a shorter overall length and the same creepage distance [6]. A polymer insulator should have electrical ratings equivalent to its porcelain counterpart. Usually the polymer insulator will have an increased length, reduced weight, and increased leakage distance [7].

The uprating technique is illustrated in the Fig. 3 showing the reducing in three spans. This method in the insulator string should be used only in suspension structure as the string location is in the vertical direction.

B. Relocating Steel Towers On the other hand, the method relocating steel towers or changing the location of the structures, consists of redesigning the line, seeking an ideal relocation for the structures and re-adjusting their respective heights (if necessary). It was developed and tested on this 230 kV line. In adopting this method, it is possible to take advantage of the upper part of the towers, but change the feet and, often, the extensions.

Thus, using this second method could be analyzed in Fig. 3 the reducing in five spans with the change of the suspension and anchoring structures, increasing the height of these structures or changing the location of them. The challenge was modify for the next hightest structure because of the steel cost, caring about the nearest structure as the line is energized and, as well, the line pro le conditions, since the change from one to the other would be in a more viable way, in addition, the line would continue energized by making that choice even more important.

Moreover, the Figs. 4 and 5 illustrates the relocation of suspension structures. The red curve is the original curve before any modification and the green one is after the uprating modification technique.

C. Reconductoring In this topic, it is analyzed the importance of the reconductoring method giving more explanation about this technique assaying the performance of the line project whereas the conductor of the transmission line is "the key point" of the project, since from it could be analyzed all the impacts electrical, mechanical and involved cost.

The low weight of aluminum, its resistance to corrosion, low cost, high thermal and electrical conductivity, made in a short time, this metal was the most used in conductors to transport electric energy. The first metal to be used as conductor of electricity was copper. By having high conductivity compared to its mechanical strength ratio its sizing was determined by mechanical considerations. The diameter conductor turned out to be larger than required from the point of view of electrical efficiency. Due to their weight the spans were small, adding more cost for the line. An international agreement determined an annealed copper (IACS) standard for comparison with other metals [8].

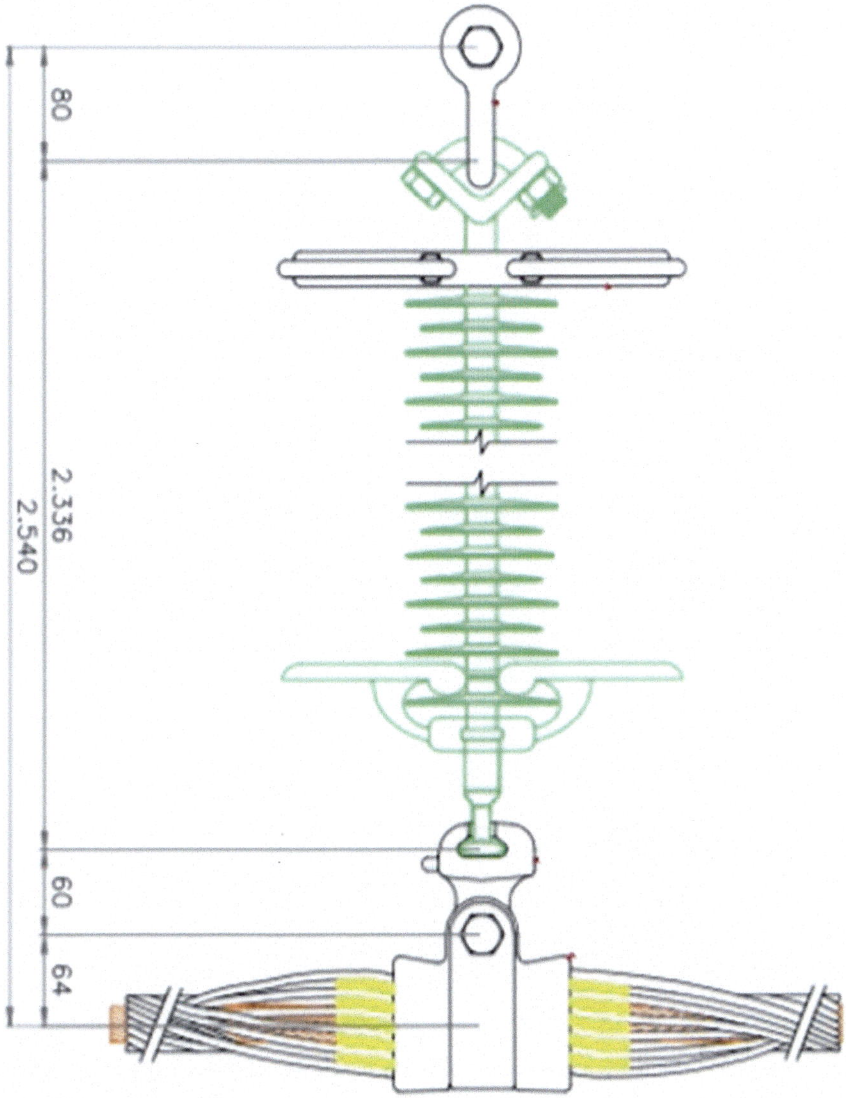

Fig. 2 Uprating technique of exchange insulator string

Back Structure Number	Ahead Structure Number	Controlling Weather Case	-Clearance--- ---Margin---- Vert. (m)	Horiz. (m)	OK	Comment
1	2	Temp max longa duracao recap	4.74	-4.71	OK	
2	3	Temp max longa duracao recap	-0.08	-1.21	NG	
3	4	Temp max longa duracao recap	-0.07	7.83	OK	
4	5	Temp max longa duracao recap	-0.50	-2.32	NG	
5	6	Temp max longa duracao recap	-0.60	-4.54	NG	
6	7	Temp max longa duracao recap	0.16	1.96	OK	
7	8	Temp max longa duracao recap	-0.03	-1.11	NG	
8	9	Temp max longa duracao recap	4.01	11.32	OK	
9	10	Temp max longa duracao recap	-0.48	-1.26	NG	Ground clear controls.
10	11	Temp max longa duracao recap	-1.28	-1.04	NG	
11	12	Temp max longa duracao recap	-1.96	-1.00	NG	
12	13	Temp max longa duracao recap	-2.13	-1.00	NG	
13	14	Temp max longa duracao recap	-1.73	-1.00	NG	
14	15	Temp max longa duracao recap	-2.13	-1.00	NG	
15	16	Temp max longa duracao recap	-1.59	-1.00	NG	
16	17	Temp max longa duracao recap	6.30	0.22	OK	
17	18	Temp max longa duracao recap	-1.13	-1.27	NG	
18	19	Temp max longa duracao recap	-0.78	-1.08	NG	
19	20	Temp max longa duracao recap	3.77	0.71	OK	
20	21	Temp max longa duracao recap	-0.03	-1.97	NG	
21	22	Temp max longa duracao recap	-0.19	-4.75	NG	

15 spans with clearance violations NG
6 spans without clearance violations

Fig. 3 Clearance in the transmission line project Recife 2—Suape

Back Structure Number	Ahead Structure Number	Controlling Weather Case	-Clearance--- ---Margin---- Vert. (m)	Horiz. (m)	OK	Comment
1	2	Temp max longa duracao recap	4.74	-4.71	OK	
2	3	Temp max longa duracao recap	0.70	0.48	OK	
3	4	Temp max longa duracao recap	0.16	7.83	OK	
4	5	Temp max longa duracao recap	-0.50	-2.32	NG	
5	6	Temp max longa duracao recap	-0.60	-4.54	NG	
6	7	Temp max longa duracao recap	0.40	1.95	OK	
7	8	Temp max longa duracao recap	0.42	-1.09	OK	
8	9	Temp max longa duracao recap	4.06	11.32	OK	
9	10	Temp max longa duracao recap	-0.48	-1.26	NG	Ground clear controls.
10	11	Temp max longa duracao recap	-1.28	-1.04	NG	
11	12	Temp max longa duracao recap	-1.96	-1.00	NG	
12	13	Temp max longa duracao recap	-2.13	-1.00	NG	
13	14	Temp max longa duracao recap	-1.73	-1.00	NG	
14	15	Temp max longa duracao recap	-1.88	-1.00	NG	
15	16	Temp max longa duracao recap	-1.33	-1.00	NG	
16	17	Temp max longa duracao recap	6.30	0.22	OK	
17	18	Temp max longa duracao recap	-1.13	-1.27	NG	
18	19	Temp max longa duracao recap	-0.78	-1.08	NG	
19	20	Temp max longa duracao recap	3.88	0.68	OK	
20	21	Temp max longa duracao recap	0.33	-1.93	OK	
21	22	Temp max longa duracao recap	-0.19	-4.75	NG	

12 spans with clearance violations NG
9 spans without clearance violations

Fig. 4 Clearance after the exchange insulator string method in the transmission line project Recife 2—Suape

The low weight of aluminum, its resistance to corrosion, low cost, high thermal and electrical conductivity, made in a short time, this metal was the most used in conductors to transport electric energy [9]. The aluminum today is dominant for the manufacture of conductors for transmission lines both in its pure form and in alloy with other elements. Beyond that, the aluminum for conductors has a conductivity of about 61% of that of copper used in cables, but because of its low specific weight the conductivity of aluminum is more than twice that of copper per unit weight [6].

Back Structure Number	Ahead Structure Number	Controlling Weather Case	-Clearance--- ---Margin---- Vert. (m)	Horiz. (m)	OK	Comment
1	2	Temp max longa duracao recap	4.74	-4.71	OK	
2	3	Temp max longa duracao recap	-0.70	0.48	OK	
3	4	Temp max longa duracao recap	0.16	7.83	OK	
4	5	Temp max longa duracao recap	-0.50	-2.32	NG	
5	6	Temp max longa duracao recap	-0.60	-4.54	NG	
6	7	Temp max longa duracao recap	0.40	1.95	OK	
7	8	Temp max longa duracao recap	0.42	-1.09	OK	
8	9	Temp max longa duracao recap	5.04	-2.26	OK	
9	10	Temp max longa duracao recap	0.43	-1.25	OK	Ground clear controls.
10	11	Temp max longa duracao recap	0.07	7.97	OK	
11	12	Temp max longa duracao recap	-1.89	-1.00	NG	
12	13	Temp max longa duracao recap	0.57	-0.00	OK	
13	14	Temp max longa duracao recap	-0.94	-1.00	NG	
14	15	Temp max longa duracao recap	3.72	0.00	OK	
15	16	Temp max longa duracao recap	0.06	8.00	OK	
16	17	Temp max longa duracao recap	6.93	0.22	OK	
17	18	Temp max longa duracao recap	-1.13	-1.27	NG	
18	19	Temp max longa duracao recap	-0.78	-1.08	NG	
19	20	Temp max longa duracao recap	3.88	0.68	OK	
20	21	Temp max longa duracao recap	0.33	-1.93	OK	
21	22	Temp max longa duracao recap	-0.19	-4.75	NG	

7 spans with clearance violations NG
14 spans without clearance violations

Fig. 5 Clearance after the relocation and alteration structure method in the transmission line project Recife 2—Suape

In order to implement the technique, the reconductoring method is the last analysis which highlights the basic system properties that in fluence its mechanical and electrical performance, with conductors of larger size or of different type that allow higher temperature operation with less sag [10].

Furthermore, for this article, the transmission line project was originally performed with an aluminum conductor steel reinforced (ACSR), the Grosbeak conductor. The analysis and results initially involve the mechanical and electrical properties of an ACSR. Internationally designated as ACSR (Aluminum Conductor Stranded Reinforced) this cable is formed by a steel core, solid or composed of several strands of galvanized steel, wrapped by one or more layers of alloyed aluminum alloy 1350. The aluminum crown provides an excellent conductivity while the steel core increases the mechanical strength of the cable [1].

The performance of these conductors is then compared with an aluminum al-loy conductors (AAAC). The aluminum alloy 6201 (aluminum-magnesium-silicon) was initially developed as an alternative to ACSR cable with high mechanical strength. Its mechanical strength is twice as high as that of alloy Al 1350. This property has allowed in some situations to replace the galvanized steel wires that form the core of the ACSR cable, thus reducing the weight per unit length of the conductors [11]. The conductor choose was the Flint which is applied in overhead circuit where greater mechanical resistance is needed than the ASC conductor and a better resistance to corrosion than the ACSR conductor.

The AAAC cable also has lower electrical losses. As the conductor has no magnetic components, the magnetic losses occurring in the case of ACSR due to the induction in the steel core, do not exist and its surface hardness, twice that of Al 3510 alloy, reduces the abrasion of the driver during launch and handling reducing

Table 1 Comparison between Grosbeak and Flint conductors

Conductor type diameter area Wc (kg/km)			
Grosbeak	ACSR 3.97	333.96	1301
Flint	AAAC 3.59	375.44	1028

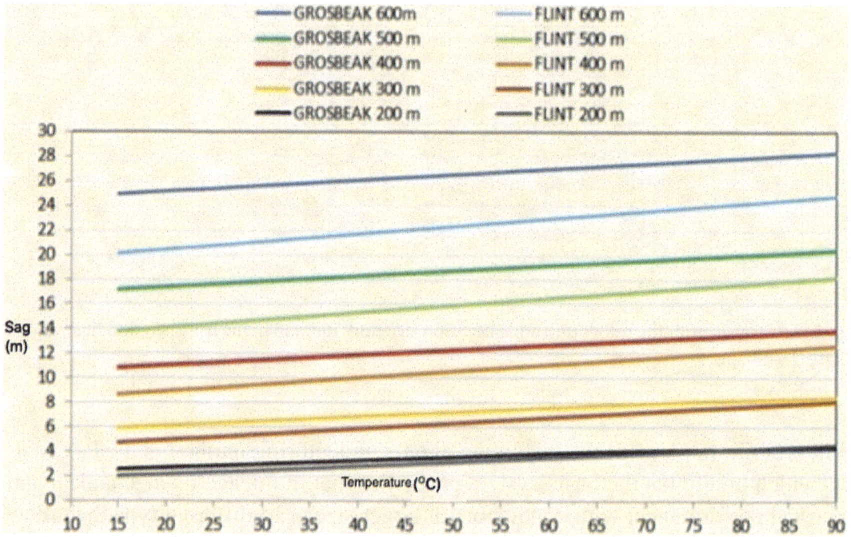

Fig. 6 Comparative evaluation between Grosbeak and Flint conductors [3]

corona losses [1]. The reconductoring method was choose between these two conductors because of their particularities and similarities properties. It is illustrated these properties in the Table 1.

The Table 1 illustrates the similarities between these two conductors because of their diameter and area. Furthermore, the decrease of the conductor size reduces the sag developed of the Flint comparing with the Grosbeak and the weight is also reduced around 300 kg/km. Nevertheless, the aluminum has greater flux of current than in steel due to the electrical conductivity of aluminum.

Thus, it can be seen in Fig. 6 that the conductor cable Flint has a smaller sag when is submitted to temperatures lower than the conductor cable Grosbeak, however, at temperatures above 70 °C the sags are equalized making the AAAC cable unattractive. Whereby for the transmission line Recife II- Suape the span was 380 m and the region is not submitted to extreme temperatures the ACSR for AAAC was required.

The Figs. 7 and 8 corresponds to the last method analyzed which was the reconductoring method, whereas it is the most expensive for that reason it was choose to be analyze after all the methods assayed before. These figures illustrates the red curve which represents the Grosbeak and the green curve the Flint.

It can be seen the difference between the catenary of these two conductors since with the change of the type of conductors the Flint is more light implying in a lower

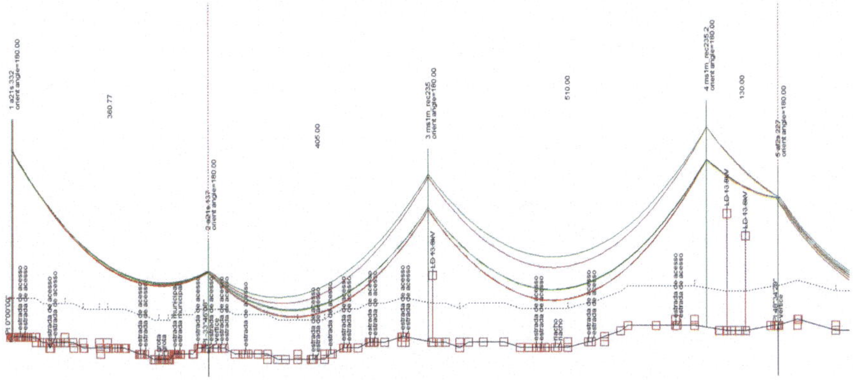

Fig. 7 Comparison between the Grosbeak and Flint catenary

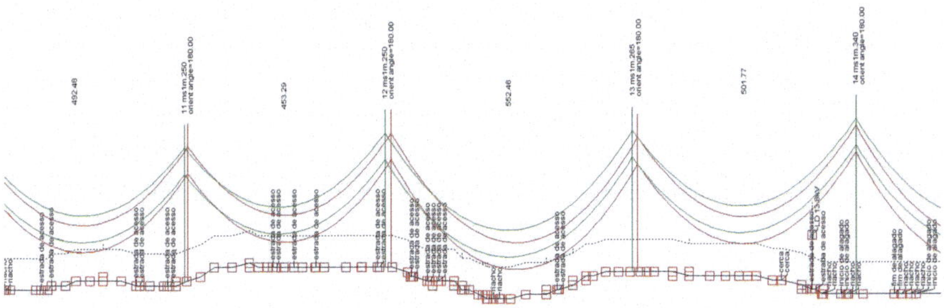

Fig. 8 Another part of the comparison between the Grosbeak and Flint cate-nary

sag and also his catenary is more traction than the Grosbeak. The Flint was choosen to modify the previous cable because of the similarities between these conductors (area and diameter) and also because of the currently use of this conductor in electric power companies to improve the transmission lines capacity.

Therefore, all previously used methods have made a great contribution to the analysis of transmission line upratings, but have not actually corrected the violation at the clearance reached by the previous conductor. For this reason, the method of reconductoring is very important, although it is the most expensive method of uprating transmission lines.

4 Conclusion

The analysis shows the importance of the uprating overhead transmission line performance whereas the electric utilities face a challenge to increase the power transfer capacity of existing overhead lines, then uprating transmission lines introduce a competitive option so as to increase the transmission capacity of lines in operation.

The article highlighted the uprating technical of transmission lines. It high-lighted its alternative characteristics to construction of new transmission lines in front of the concessionaires' willingness to meet the increasing demands of electric energy consumption.

In addition, the methods analyzed in a transmission line located in the metropolitan region of Recife illustrate the importance of increasing this capacity rather than to rebuild a new line. Thus, the methods analyzed in a transmission line situated in city of Recife illustrates the importance to increase this capacity instead of rebuilt a new line. All the methods described in this article were simulated with the PLS-CADD software which can be seen the importance in the academic field can be emphasized, as well as the choice of each method can be verified.

The case study concluded that reconductoring method should be used even it has a highly cost in its implementation, since the others did not eliminate ground heights, violators of normative criteria. It is therefore important to stress that each case of be studied specifically. Situations climatological, topographical and/or specific design criteria may lead to the adoption of different from those achieved here.

Therefore, all previously used a major contribution to the analysis of trans-mission line, but did not fully correct the safety height reached by the previous conductor. For this reason, the reconductoring method is very important, al-though it is the most expensive to increase the capacity of the transmission lines among the analyzed in this article.

Other alternatives deserve to be analyzed, for example: introduction of additional conductor, even of smaller diameter, use of technology of the special conductors. Alternatives, however, must have, in addition to the technical involved, economic constraints as final constraints to be always attended.

References

1. Da Silva, A.A.P.: Modelagem para Repotencializacao de Linhas de Transmissao atraves da Aplicacao de Condutores Especiais. M.S. thesis, Electrical Engineering Department, Universidade Federal de Pernambuco, Pernambuco, Recife, Brazil (2009)
2. Kopsidas, K., Rowland, S.M.: A performance Analysis of Reconductoring an Over- head Transmission Line Structure. In: IEEE Transactions on Power Delivery, pp. 2248 – 2256. IEEE Power and Energy Society, Manchester (2009). https://doi.org/10.1109/tpwrd.2009.2021042
3. Kopsidas, K., Rowland, S.M.: Evaluating opportunities for increasing power Ca- pacity of existing overhead line systems. In: IET Generation Transmission and Distribution, IET, vol. 5, pp. 1–10, 2011. IET Generation, Transmission and Distribution, Manchester (2011). https://doi.org/10.1049/iet-gtd.2010.0227
4. Cavassin, R.S., Fernandes, T.S.P.: Uma Abordagem de Multicriterios para Reca- pacitacao de Linhas de Transmissao. In: Revista Controle e Automacao. vol. 23 no. 6 Novembro e Dezembro 2012, Curitiba, Parana, Brazil (2012)
5. Pessoa, L.V.G.: Proposta de Tecnologia para Recapacitacao de Linhas de Transmissao baseado na Relocacao de Estruturas de Concreto. M.S. thesis, Electrical Engineering Department, Universidade Federal de Pernambuco, Pernambuco, Re-cife, Brazil (2017)
6. Labegalini, P.R, Labegalini, J.A., Fuchs, R.D., de Almeida, M.A.: Projetos Mecanicos das Linhas Aereas de Transmissao. Blucher, 2 Edic ~ ao (1992)

7. Fehr, R.: Guide for uprating rus transmission lines. Ph.D. dissertation, United Department Of Agriculture, United States of America (2005)
8. Luna, A.M. Materiais de engenharia eletrica. 2nd edn, vol. 2. Recife (1995)
9. Trash, F.R.T.: Transmission conductors—a review of the design and selection crite-ria. Southwire Company. Available at: http://www.southwire.com/processGetAr-ticle.do?commonId=6b884260ef1010VgnVCM1000002702a8c0
10. Kopsidas, K., Rowland, S.M., Boumecid, B.: A holistc method for conductor am-pacity and sag computation on an ohl structure. In: IEEE Transactions on Power Delivery, pp. 1047–1054. IEEE Power and Energy Society, Manchester (2012)
11. NEXANS. Alumnio: Condutores Nus (2008)

Mechanical and Technological Behavior of Protein-Modified Mortars

Leonardo Proença Mendes de Almeida Godoy, Diego Leandro Thomaz, André Luis Pissolati, Daniela Helena Pelegrine Guimarães and Maria Thereza de Moraes Gomes Rosa

Abstract Mortar is nowadays a very common building material at the civil construction industry. It has been employed in different ways in buildings wherein it serves for laying blocks, wall and ceiling cladding, just to name a few applications. Therefore, maximum safety for buildings is achieved when many research findings are taken into account in order to minimize possible accidents during the construction or even after the work is done. Research studies are intended to provide a better quality to mortar and the required mechanical resistance when subjected to high tensile loadings. Thus, an alternative to enhance the properties of mortar is the use of additives in its composition, which are able to improve mechanical resistance, deformation characteristics and its technological behavior. However, despite many published results on the influence of such additives, few studies have been carried out on the influence of proteins on the mortar mechanical and technological behavior. Therefore, the aim of this research is to evaluate the changes in mechanical and technological behavior of cement mortar, made by incorporating whey protein and albumin to its composition. Protein contents of 2 and 10% in weight were chosen herein. The analysis was based on the protein content and mortar aging time. Data show that contents of 2 and 10% of whey protein and albumin had a significant influence on test results.

Keywords Mortar · Protein · Additives

L. P. M. de Almeida Godoy (✉) · D. L. Thomaz · M. T. de Moraes Gomes Rosa
Center of Science and Technology, Mackenzie Presbyterian University,
Brazil Avenue, 1220, Campinas, SP 13073-148, Brazil
e-mail: leonardogodoy96@gmail.com

A. L. Pissolati
Physics Institute Geb Wataghin, State University of Campinas (UNICAMP),
Street Mendeleyev, 200, Campinas, SP 13083-860, Brazil

D. H. P. Guimarães
Lorena Engineering School (EEL-USP), São Paulo University,
Campinho Highway, Lorena, SP 12602-810, Brazil

Y. Iano et al. (eds.), *Proceedings of the 4th Brazilian Technology Symposium (BTSym'18)*, Smart Innovation, Systems and Technologies 140,
https://doi.org/10.1007/978-3-030-16053-1_2

1 Introduction

Mortar is widely used in civil construction, such as in masonry cladding, plastering walls, internal and external sealing applications. Due to its complexity, composition and building application, many factors must be considered before choosing the right materials for mortar preparation, the optimization of its viability according to each application [1].

Most pathologies and failures found in plastering mortar are cracking, detachment, cohesion loss, impermeabilization loss, efflorescence and poor thermal behavior. These problems arise from mechanical, physical and cyclic loadings, chemical reactions and the environment it is exposed to. Among the problems usually found in masonry mortar are block detachment, cracking, stains over the ceramic elements, cracking and bleaching of plastering joints, efflorescence, lack of impermeabilization and poor resistance/cohesion of the joints between blocks [2, 3].

Since many problems are usually found in mortar, many research contributions have been devoted to finding additives to solve those pathologies. Nevertheless, the use of synthetic polymers as additives to the mortar composition has been increasing in the last 40 years, and special attention has been given to influences they may have on the mortar composition. The main synthetic polymers so far are polyethylene, polypropylene, polyvinyl chloride, polyurethanes, MS polymers (modified silanes), silicones and polyolefins. The first synthetic polymers are still under scrutiny, since they are still employed in mortar compositions, sometimes with some modifications or new applications [4, 5].

However, these synthetic polymers might be hazardous substances, such as benzene, an organic compound derived from crude oils, a raw material which serves as the basis for the synthesis of many synthetic polymers, such as polystyrene, polycarbonate, epoxy resin, phenolic resins, some rubbers and plastics. Therefore, the use of biopolymers, the group from polypeptides, proteins, polynucleotides, polysaccharides, gums, resins and natural elastomers has become attractive for the preparation of mortar [6, 7].

Many researches approve the use of additives, especially fibers in cement matrices, aiming at minimizing retraction effects, reducing fissures from the cement compound, serving as a bridge for stress transfer, which also contributes to increasing resistance, deformation ability and the tenacity of the composite. Therefore, the resulting mortar will probably have significant improvements, such as a higher durability, a higher resistance against corrosion of steel reinforcements, and a lower permeability to aggressive agents (humidity, oxygen and chlorides), which could provide a high potential for applications such as laying mortar [3, 8].

The literature provides a broad overview on the influences of many additives to the composition of mortars, data from their main features, identification of changes arising from the addition of a proper additive, which may help find the best applications, according to the intended function. However, among the additives already studied, little is known about the use of proteins on the mechanical behavior of mortars and their workability is one of the limiting factors on their proper application, being considered a technological behavior of mortar.

The mechanical behavior of plaster mortars is one of the requirements for the analysis, since it allows the comprehension of their strength against loadings depending on their structural application. When masonry is exposed to a vertical loading, there will be horizontal stresses along the mortar binding the bricks. Therefore, this stress might reach the maximum shear stress, which would compromise the existing adherence at the interface between the unit and the joint, and cohesion will thus be damaged. Hence, the study of the mechanical behavior could minimize failures by forces that the plaster mortar will be exposed to, besides enabling the identification of aspects and important relationships for the proper use of the compound, such as its durability, safety and quality [9].

The main objective of this work is to evaluate the mechanical behavior and workability for cement mortar prepared by incorporating proteins to be applied as plaster mortar. Tue use of protein additives is also aimed at enhancing many of its properties regarding stresses it will support in its structural application at the civil construction industry.

2 Methodology

2.1 Mortar Preparation

Three different mixtures of Portland CP II-Z-32 cement were prepared herein. Fine aggregates were composed of and #30, #50, and #100 sieve, whereas each portion was added according to 20, 20, and 60%, respectively. Additives were whey protein (WP) and albumin (ALB) at the concentrations of 2 and 10% in relation to the cement mass.

Materials were mixed by considering their homogenization. After weighing the amount of sand #30, #50, and #100 sieve in a precision balance, model AD 2000 (resolution of 0.01 g), fine aggregates were added to a vessel. They were mixed by using a trowel for 30 s and the transferred to another similar vessel. This process was repeated three times. Mortar was cured into a humidity chamber with a relative humidity of 95%.

The cement was initially sieved with a mesh size of 1.18 mm and the weighted. When an additive was added, the mixture was performed the same way as that for sand.

After all materials were properly weighted, they were added to a vessel containing sand and cement. They were mixed with a trowel for 30 s, the transferred to another vessel. This process was repeated five times.

Mortar preparation was carried out according to NBR 16541 and the trace chosen herein was 1:3:0.7. Then 2.5 kg of the dry mixture was transferred to a tray. Tap water was measured, in weight, into two different portions: 75% of the total water and 25%

for the remaining. Once the materials were separated and the required apparatus was available, the mortar mixer was set to a slow speed for 30 s, when the water portion of 75% was added for the first 10 s. Then the speed was increased and kept for 60 s. After that the mixer was turned off. Spinning blades were removed and all the surfaces, both from the mixer and the blades, were scratched off until the mixture would adhere to the bottom of the drum. This procedure was performed within 90 s. Then the mortar mixer was turned on at a slow speed with the remaining 25% water portion at the first 10 s, and mixing was performed for 30 s. The last step was mixing for 30 s under high speed.

2.2 Density of Mortar

The analysis of the density of fresh mortar was performed according NBR 13278 technical standard. The test consisted of preparing mortar and then adding it to a cylindrical vessel of a known volume, until three layers of approximately equal heights were formed.

On each layer 20 strikes were applied by using a spatula over the perimeter of the mortar to make it uniform inside the vessel. When hitting the first layer care must be taken to avoid reaching force to penetrating the surface underneath.

After hitting the layers, the flask was released to the ground three times, at a height of 3 cm. After that no empty space should exist between the mortar and the cylinder wall. Then flattening was performed by using a spatula, from two orthogonal steps, according to a back and forth movement.

2.3 Mortar Air Content

The amount of incorporated air it is calculated from the value found according to Eq. (1), according NBR 13278.

$$A = 100 \times \left(1 - \frac{d}{d_t}\right) \quad (1)$$

where: d is the value for the density of mortar determined from the test (g/cm^3); d_t is the value for the theoretical density of mortar (g/cm^3), and A is the value of percentage of incorporated air (%).

2.4 Workability

The consistency index was measured according to the NBR 13276 standard. The procedure consisted of cleaning and drying the table top for measuring consistency index and also the wall of the trunk-conical mold with an appropriate tissue.

Freshly prepared mortar was filled into the mold by keeping it into the center of the table until it was completely filled. The mold was held firmly while it was filled with three successive layers of approximately equal heights. Then each layer was hit 15, 10, and 5 times by using a metal socket in order to make it uniform until there was a small excess volume of mortar compared to the mold. Then, the excess mortar flattened off by using a metal ruler by following the edge of the mold, from back and forth movements all over the surface.

Simultaneously, any particles around the mold were removed by using a clean and dry tissue. Then, the trunk-conical mold was vertically removed, and the process was initiated by moving the lever from the table to measure consistency index. The table underwent 30 strokes in approximately 30 s. After the last stroke, a metal ruler was employed to measure mortar spreading, by considering three diameters from pairs of points uniformly distributed over the perimeter. Results from consistency index were the average from three measurements (mm) and rounded to the nearest integer.

2.5 Bending Tensile Strength and Compressive Strength

The test method for determining flexural strength was carried out according to NBR 13279. Samples were prepared by using metal prismatic molds, which have open frames with removable walls, capable of forming three compartments when mounted, one for each of the three specimens measuring (4 × 4 × 16) cm. The age of the specimen was 7 and 28 days. The test consisted of positioning the specimen into the sample holder of the universal testing machine in such a way its flat surface would not touch the support or load devices from the universal testing machine. Then a load of 2 mm/min was applied until failure. Their bending strength was calculated from Eq. (2).

$$R_t = \frac{1.5 \times F_f \times L}{40^3} \tag{2}$$

where: R_t is the bending strength (MPa); F_f is the maximus vertically applied load at the center of the prism (N); L is the distance between the supports (mm).

2.6 Axial Compressive Strength

Axial compressive strength tests were carried out after those for bending strength measurements, since the remaining halves from the fractured specimens were used herein. The fractured specimens were placed into the testing equipment in a way that flat surface would not touch neither the sample holder nor the load device. Then a load of 2 mm/min was applied until failure. Their axial compressive strength was calculated from Eq. (3).

$$R_c = \frac{F_c}{A} \tag{3}$$

where: R_c is the compressive strength (MPa); F_c is the maximum applied load (N); A is the cross-sectional area of the specimen resisting the load (mm^2).

2.7 Statistical Analysis of the Results

Data from density measurements from fresh and hardened mortar, bending strength, axial compressive strength and workability were analyzed. Many effects were analyzed, such as aging time (7–28 days) and content of proteins. The data were submitted to the analysis of variance (ANOVA) to evaluate the effect of parameters and their interaction. The comparison of averages was carried out by means of the Tukey test, at a significance level of 5%.

3 Results and Discussion

3.1 Preliminary Tests

During the development of this project, the first drawback was trace determination, which could meet the requirements for protein addition, since protein incorporation would interfere into the empty spaces from the mortar composition. In order to overcome this difficulty, many mortar compositions were prepared with percentages of 2–10% of protein, at ratii of W/C of 0.6, 0.7 and 0.8 aiming at unveiling their behavior. Therefore, only a small amount of cement (25 g) was added to a porcelain tile, with different amounts of sand, by using 3 parts of sand (#30, #50, and #100 sieve) at different percentages to compare the behavior of the resulting fresh mortar.

It was noticed that sand #100 sieve required a slightly higher percentage compared to the other sand samples, which allowed a better behavior compared to the mortar compositions from the other sand samples. Then it was concluded that percentages

Table 1 Density of fresh mortar analyzed

Mortar	Density (g/cm^3)
Control	2.11 ± 0.01
WP 2%	1.56 ± 0.01
WP 10%	1.65 ± 0.06
ALB 2%	1.65 ± 0.06
ALB 10%	1.72 ± 0.09

Table 2 Mortar air content analyzed

Mortar	Air content (%)
Control	1.6
WP 2%	27.3
WP 10%	23.1
ALB 2%	23.1
ALB 10%	19.8

of 20, 20, and 60% of sand #30, #50, and #100 sieve, respectively, would be worth testing. In terms of W/C ratio, it was found that 0.7 would be appropriate for the mortar samples in this project, except for the addition of proteins.

3.2 *Fresh Mortar*

Three tests were performed for calculating the density of each mortar composition. The results shown in Table 1 are the mean values for 3 measurements for each mortar composition.

The statistical analysis by means of ANOVA evidenced the rejection of the null hypothesis for the tests performed in this work, which allowed the acceptance of the hypothesis that at least one sample is different from the others. Moreover, the analysis from the Tukey test evidenced that the control mortar sample had a different behavior from the others, since its parameters were considerably higher, as shown in Table 1. The WP 2% mortar sample evidenced an analogous behavior to the other mortar compositions containing proteins, with the exception of ALB 10%, which provided a considerably higher result compared to WP 2%.

The results from the other samples in terms of mortar air content are summarized in Table 2.

Results reported elsewhere, despite the different traces, such as that from other article, provided similar values to those found herein for the mortar samples, which evidences this is an acceptable result. However, data from other article are significantly different from those achieved in this work for mortar air content. Regarding the additive-modified mortar compositions, despite their similar results, a high amount of

Table 3 Density from the hardened state of mortar analyzed

Mortar	Density (g/cm^3)
Branca	1.92 ± 0.06
WP 2%	1.36 ± 0.03
WP 10%	1.27 ± 0.03
ALB 2%	1.63 ± 0.03
ALB 10%	–

Table 4 Results from bending strength and compressive strength for mortar compositions aged for 7 days

Mortar	Bending strength (MPa)	Compressive strength (MPa)
Control	2.83 ± 0.3	10.58 ± 1.1
WP 2%	1.27 ± 0.1	3.21 ± 0.3
WP 10%	–	–
ALB 2%	0.63 ± 0.1	2.99 ± 0.7
ALB 10%	–	–

incorporated air when compared to the control mortar sample was found, evidencing poor results [10].

3.3 *Hardened Mortar*

In terms of density, the results are summarized in Table 3.

The hyphen from Table 3 means the analysis could not be performed, since the resulting mortar broke when removed from the mold and was significantly soft, which did not allow appropriate measurements.

The results for the density from hardened mortar by the statistical analysis by means of ANOVA evidenced that the hypothesis that at least one of the samples has a different behavior was true. Therefore, the Tukey test proved that the control mortar is significantly better, as shown in Table 3. WP 2% and WP 10% mortar samples work homologously, with a different behavior and poorer result compared to ALB 2%.

The results for the bending strength for mortar samples aged for 7 days are shown in Table 4 and the results for the bending strength for mortar samples aged for 28 days are shown in Table 5.

Comparing the results achieved herein with those from other article, the mortar compositions from those authors provided a significantly lower result compared to the control mortar sample studied herein, which shows that the chosen trace in this work enables a better resistance performance [10]. In terms of mortar compositions containing additives, it was noticed that mortar compositions with 2% additives had a different behavior when aged for 7 days, even though their results were slightly

Table 5 Results from bending strength and compressive strength for mortar compositions aged for 28 days

Mortar	Bending strength (MPa)	Compressive strength (MPa)
Control	3.88 ± 0.7	16.94 ± 2.5
WP 2%	1.56 ± 0.2	4.83 ± 0.6
WP 10%	0.77 ± 0.1	1.06 ± 0.2
ALB 2%	1.46 ± 0.2	4.82 ± 1.1
ALB 10%	–	–

similar when aged for 28 days. Mortar compositions from 10% of additives are not recommended for improving strength, since they have a fragile behavior when compared to the other compositions studied in this work. Therefore, the presence of additives did not improve the strength performance of the mortar compositions when compared to the control mortar samples.

The statistical analysis from samples aged for 7 days evidenced the hypothesis that at least one of the samples did not behave in a homologous manner was true. Therefore, the Tukey test showed that the results for the bending strength for all samples were significantly different from each other. The control mortar had a higher value, followed by the WP 2% sample. The results for axial compressive strength evidenced that the control sample had a different behavior from those containing proteins, since it provided a higher value, as shown in Table 4. WP 2% and ALB 2% presented a similar behavior.

The statistical analysis for the mortar samples aged for 28 days evidenced the hypothesis that at least one of the samples has a different behavior. Therefore, the Tukey test showed that only the control mortar samples had a different behavior from the other protein-modified mortar samples according to bending strength data (Table 5), since a higher bending strength was measured for the former. Nevertheless, the control samples had a different behavior from the other samples according to axial compressive strength data, since a higher compressive strength was found for those samples, as shown in Table 5. Among the protein-modified mortar samples, WP 2% and ALB 2% samples had a homologous behavior, since they provided higher and very different results compared to WP 10% samples.

3.4 *Workability*

The results for the consistency index from the mortar samples are shown in Table 6.

Hyphens in Table 6 mean those measurements could not be performed, since the corresponding mortar samples did not spread as expected. Instead, these samples remained practically the same way after the test begun, i.e., the ALB-modified mortar samples provided satisfactory workability, even though WP 2% and WP 10% samples had good and poor workabilities, respectively.

Table 6 Results for the consistency index from mortar samples analyzed

Mortar	Spreading (mm)
Control	173 ± 3
WP 2%	219 ± 2
WP 10%	160 ± 3
ALB 2%	–
ALB 10%	–

The statistical analysis took into account that at least one element would behave differently. Therefore, the Tukey test evidenced that all mortar samples evaluated herein worked heterogeneously, whereas WP 2% mortar samples had significantly higher results compared to the others, as well as the control mortar samples, as shown in Table 6.

4 Final Remarks

Protein-modified mortar samples in this work, particularly those from whey and albumin, had a poor performance, since their mechanical resistance was significantly lower than that for control mortars prepared for comparison purposes. The conclusion was that it is not recommended the addition of those proteins for modifying mortar.

References

1. Da Silva, N. G.: Argamassa de revestimento de cimento, cal e areia britada de rocha calcária, pp. 9–10 (2006)
2. Monteiro, D.M.P.: Avarias em argamassas – causas, prevenção e reparação, pp. 7–28 (2008)
3. Silva, R.D.P.: Argamassas com adição de fibras de polipropileno – Estudo do comportamento reológico e mecanico, pp. 20–35 (2006)
4. Nunes, L.R., Rodolfo Jr, A., Ormanji, W.: Tecnologia do PVC (2002)
5. Vilar, W.: Química e Tecnologia dos Poliuretano (2002)
6. Lima, J.: O contributo das argamassas de barro para a qualidade do ambiente interior dos edifícios: o caso das argilas do sotavento algarvio. In: 2nd Congresso Internacional da Habitação no Espaço Lusófono, pp. 1–11, Lisboa (2013)
7. Nunes, E.C.D., Lopes, F.R.S.: Polímeros: Conceitos, Estrutura Molecular, Classificação e Propriedades, 1st edn. Saraiva, Érica, São Paulo (2014)
8. Bentur, A., Mindess, S.: Fibre reinforced cementicious composites, p. 603 (2006)
9. Mohamad, G., Lourenço, P.B., Roman, H.R.: Propriedades mecânicas das argamassas sob compressão triaxial – Análise e Previsão, In: 32th Jornadas Sulamericanas de Engenharia Estrutural, pp. 2954–2963, Campinas (2006)
10. Silva, N.G., Buest, G., Campiteli, V.C.: Argamassas com areia britada, influência dos finos e da forma das partículas, pp. 12–21 (2005)

Retrieval of Effective Electromagnetic Properties in Terahertz: From Organics to Metamaterials

Paloma E. S. Pellegrini, Lucas H. Gabrielli, Livia Martins dos Santos and Rafael Luis Ribessi

Abstract In this work, the terahertz frequency range is exploited. From organic materials to metamaterials, different algorithms able to extract their effective electromagnetic properties were developed. In order to characterize different concentrations of lactose, such samples were submitted to a terahertz time-domain spectroscopy (TTDS) to experimentally acquire their transmission spectrum. With only these input data, the retrieval method employed parametrization of the refractive index by using the Lorentz model. Extending the studies to metamaterials, single and hybrid geometries based on split ring resonator (SRR) were simulated and characterized utilizing the Kramer-Kronig relations in an algorithm whose response was the effective permittivity and permeability of this class of materials.

Keywords Electromagnetic properties · Terahertz · Organic materials · Metamaterials

1 Introduction

Since early applications in astronomy, [1, 2], the terahertz range—usually defined between 0.3 and 30 THz, [3]—has proved suitable to several fields in science. By providing much greater bandwidths than the millimeter waves, this part of the spectrum promises higher transmission speed and capacity storage in communication systems, specially wireless transmission, which is considered fundamental in satisfying today's increasing demand [4, 5].

Another appealing characteristic of such waves is their deep penetration ability in organic materials. As they have low ionization energy, terahertz waves are suitable for non hazardous and non destructive techniques used to characterize such organic

P. E. S. Pellegrini (✉) · L. H. Gabrielli
School of Electrical and Computer Engineering,
University of Campinas, Campinas, Brazil
e-mail: palomapellegrini@gmail.com

L. M. dos Santos · R. L. Ribessi
Institute of Chemistry, University of Campinas, Campinas, Brazil

Y. Iano et al. (eds.), *Proceedings of the 4th Brazilian Technology Symposium (BTSym'18)*, Smart Innovation, Systems and Technologies 140,
https://doi.org/10.1007/978-3-030-16053-1_3

samples [6, 7]. One of these methods consists on a TTDS, where physical parameters such as the complex refractive index can be obtained from the electromagnetic signal over time with no prior knowledge of the sample to be analyzed, as long as a resonance is found in the considered frequency range.

This advantage of being able to characterize unfamiliar samples opens up the path to characterizing and understanding another range of materials: the metamaterials. Being artificial structures formed by arrays of unity cells, the dimension of metamaterials are quite smaller than the propagating wavelength [8] (in the order of millimeters for terahertz frequencies). This scale condition restricts the probability of diffraction and refraction. The incident radiation in a metamaterial effectively interacts with a homogeneous medium with electromagnetic properties taylored by the structure of the unit cell.

They are commonly constituted by metallic patterns due to the elevated concentration of free electrons and their effective medium influences the propagation of the electromagnetic fields according to the geometry of the material and therefore alters the effective permittivity and permeability. It is even possible to achieve values not found in nature such as magnetic response in non magnetic materials and negative refractive index, as first studied by Veselago [9] and further developed by Pendry [10]. New effective parameters widens the possibilities for the development of new photonic devices such as superlenses and resonators [11, 12]. Extending the development of metamaterials to terahertz frequencies opens up the possibility of improvement of technology in this frequency range, still in its early stages. Moreover, the fabrication process of terahertz metamaterials is facilitated in comparison to their optical counterparts due to the larger sizes of the structures that form each unit cell.

As the geometry of the structure plays a significant role in tuning the effective properties, hybrid unit cells composed of slightly different metal patterns can exhibit interesting results due to the superposition of the spectra [13, 14], being specially well-suited to the development of tunable broadband filters. The design of filters, lenses, phase-plates and myriad other components employing metamaterials depends on the correct characterization of the effective medium they constitute. Several different methodologies to retrieve the effective parameters of metamaterials can be found in the literature [15, 16]. In particular for passive media one that fully explores the Kramer-Kronig relations (KKR) is desirable to help enforce causality in the homogenized medium.

This paper proposes two different methods which are able to retrieve the effective electromagnetic properties one for organic samples and the other for metamaterials. Concerning the approach considering organic and non-magnetic materials, the Fresnel transmission coefficients based on Lorentz model were used in a developed algorithm whose only input data is the transmission spectrum obtained through TTDS. Lactose samples in polytetra uoroethylene (PTFE) were analyzed experimentally and both their effective permittivity and thickness were successfully retrieved.

As for metamaterials, two different geometries are presented in this paper: a SRR and a hybrid metamaterial, composed of SRR with different gap sizes and strong coupling that can be tunned into a filter.

Using a Kramer-Kronig based technique, we retrieve the effective material permittivity and permeability of the considered materials from transmittance and reflectance data of the single and hybrid structures, showing strong coupling between neighboring cells.

2 Extraction of Effective Parameters

The models for extraction of material parameters for both organic and metamaterials have the same basis. The Lorentz model is used to parameterize the effective complex refractive index in the S-parameters, obtained from the scattering equations; In all cases, normal incident plane waves are assumed. The throughout developments of the methods are shown in Sects. 2.1 and 2.2.

2.1 *Organic Samples*

The approach for organic samples considers the materials to be non-magnetic and it takes into account the experimental data that will be available for analysis, which is only the transmittance spectrum obtained through TTDS. For this reason, the algorithm developed is iterative.

$$\epsilon_r(f) = \epsilon_\infty + \sum_{p=1}^{P} \frac{\Delta\epsilon_p f_p^2}{f_p^2 - f^2 - \frac{i\gamma_p f}{2\pi}} \tag{1}$$

The dielectric function, in the Lorentz model, describes the temporal dispersion of the medium and it is found in Eq. 1 in which there are P resonant modes at frequencies f_p, their strengths and damping coefficients are, respectively, $\Delta\epsilon_p$ and γ_p. Each mode is comprehended as an absorption peak in the transmittance spectrum.

With the information provided by Eq. 1, the refractive indexes $\tilde{n}_a$ and $\tilde{n}_b$ of the surrounding media, air and nitrogen in our experiment, are modelled. As for the transmittance, a theoretical function based on Fresnel equations (Eqs. 2 and 3) is used for a numerical fitting on the experimental data acquired, T_{exp}. This function, T_{model}, is explicit in Eq. 4. The fitting minimizes the transmission error, given by $\left| |T_{\text{model}}| - |T_{\text{exp}}| \right|^2$, over the whole spectrum and results into the complex refractive index of the sample as well as its thickness, d.

$$t_{ab} = \frac{2\tilde{n}_b}{\tilde{n}_a + \tilde{n}_b} \tag{2}$$

$$r_{ab} = \frac{\tilde{n}_b - \tilde{n}_a}{\tilde{n}_b + \tilde{n}_a}, \quad \{a, b\} \subset \{0, 1\} \tag{3}$$

$$T_{\text{model}} = \frac{t_{01}t_{10}e^{-ik_1d}}{1 + r_{01}r_{10}e^{-2ik_1d}} \tag{4}$$

2.2 Metamaterials

The methodology to extract effective permittivity and permeability from metamaterials begins by obtaining their reflectance and transmittance coefficients, S_{11} and S_{21}. This data is easily acquired with the aid of a time domain solver and is dependent on the effective impedance, Z_{eff}, and complex refractive index, $\tilde{n}_{\text{eff}}$, as well as the thickness d of the simulated slab and the wavenumber in free-space k_0, which is considered to be the medium surrounding the metamaterial. The relations between these quantities can be seen in the following, where the term R_{01} carries the effective impedance dependency:

$$S_{11} = \frac{R_{01}(1 - e^{i\tilde{n}_{\text{eff}}k_0d})}{1 - R_{01}^2 e^{i2\tilde{n}_{\text{eff}}k_0d}} \tag{5}$$

$$S_{21} = \frac{(1 - R_{01}^2)e^{i\tilde{n}_{\text{eff}}k_0d}}{1 - R_{01}^2 e^{i\tilde{n}_{\text{eff}}k_0d}} \tag{6}$$

$$R_{01} = \frac{Z_{\text{eff}} - 1}{Z_{\text{eff}} + 1} \tag{7}$$

Inversion of (5) and (6) results in closed-form expressions for the complex impedance and refractive index of the effective medium [16]. Note, however, that it is still necessary to choose the correct sign for the impedance, and the real part of the refractive index is not uniquely defined: it is necessary to explicitly choose the appropriate branch of the complex logarithm defined by the integer value m.

$$Z_{\text{eff}} = \pm\sqrt{\frac{(1 + S_{11})^2 - S_{21}^2}{(1 - S_{11})^2 - S_{21}^2}} \tag{8}$$

$$\tilde{n}_{\text{eff}} = n_{\text{eff}} - ik_{\text{eff}} \tag{9}$$

$$n_{\text{eff}} = \frac{1}{k_0d}\left[\Im\left\{\ln\left(\frac{S_{21}}{1 - S_{11}R_{01}}\right)\right\} + 2m\right] \tag{10}$$

$$k_{\text{eff}} = \frac{1}{k_0d}\Re\left\{\ln\left(\frac{S_{21}}{1 - S_{11}R_{01}}\right)\right\} \tag{11}$$

As the metamaterial in question is a passive medium, the issues mentioned earlier do not raise concerns to the retrieval of parameters. The passivity condition states that the imaginary part of the permittivity and permeability are positive [15], thus the impedance sign is chosen according to:

$$\Re\{Z_{\text{eff}}\} \geq 0 \tag{12}$$

$$\Im\{\tilde{n}_{\text{eff}}\} \geq 0 \tag{13}$$

Passivity also implies that the permittivity and permeability satisfy the KKR, which allows us to obtain the real part of the function once the imaginary one is known over all frequencies. Because we are only able to simulate and measure k_{eff} over a limited frequency window, the straight use of KKR would produce inaccurate results for n_{eff}. Nonetheless, we can use that value as an approximation to correctly find the branch number m in (10) [16]:

$$n_{\text{KK}}(\omega) = 1 + \frac{2}{\pi} \mathrm{P} \int_0^\infty \frac{\omega' k_{\text{eff}}(\omega')}{\omega'^2 - \omega^2} \, \mathrm{d}\omega' \tag{14}$$

$$m = \arg\min |n_{\text{eff}} - n_{\text{KK}}| \tag{15}$$

We note that, although not explicitly indicated, the value of m is not constant for all frequencies.

3 Characterization of Organic Samples and Metamaterials

Using the methodology previously described, samples of PTFE homogeneously mixed with 20, 30 and 40% of lactose were experimentally measured with TTDS and characterized. Also, SRR metamaterial geometries were simulated: single and hybrid unit cell designs, the first is a simple structure whose gap size is 5 μm and the last one consists on a composition of 4 single structures with different gap sizes as depicted in. Their effective electromagnetic properties were retrieved with the algorithm proposed. The result for each characterization can be found in the following subsections.

3.1 Lactose Samples

The organic samples to be characterized were made through a powder mixture of PTFE and lactose pellets, they were made into a cylindric shape whose thickness is about 5 and 10 mm in diameter. As already mentioned before, the experiment consists on a TTDS analysis, the system utilized is the High Speed Asynchronous Sampling

Terahertz Transmission Spectroscopy System (HASSP) [17]: Each sample was put in the focal point of the terahertz beam and it is worth noting that the beam sweeps the whole surface of the sample in order to guarantee an homogeneous measure. The terahertz beam was generated a photoconductive antenna fed by 30 fs probe pulses at a 1 GHz repetition rate which are stimulated by Ti:sapphire femtosecond oscillators. Finally, electro-optic crystals collect the signal and the time domain transmittance is acquired.

The spectrum for each lactose concentration (20, 30 and 40 %) and their respective comparison between experimental and theoretical transmittance, modelled by the method proposed in 2.1, can be seen in Fig. 1. Note that there is a strong resonance at 1.37 THz and a less pronounced one at 2.54 THz. Also, the experimental data includes a considerable level of noise, reaching peaks up to 20% of the signal, nevertheless, the algorithm is able to fit the model and correctly identify the strongest resonances.

The complex refractive index for each sample was extracted and their profiles are shown in Fig. 2a, b, real and imaginary, respectively. In literature [18], some studies of lactose in terahertz were performed and the results are compatible. As for the thickness of samples, the values obtained with the proposed algorithm are organized in Table 1. Considering the techniques used to fabricated the samples, the existent deviation in such results are acceptable and thus they are considered in agreement with the expected one, which is 5 mm.

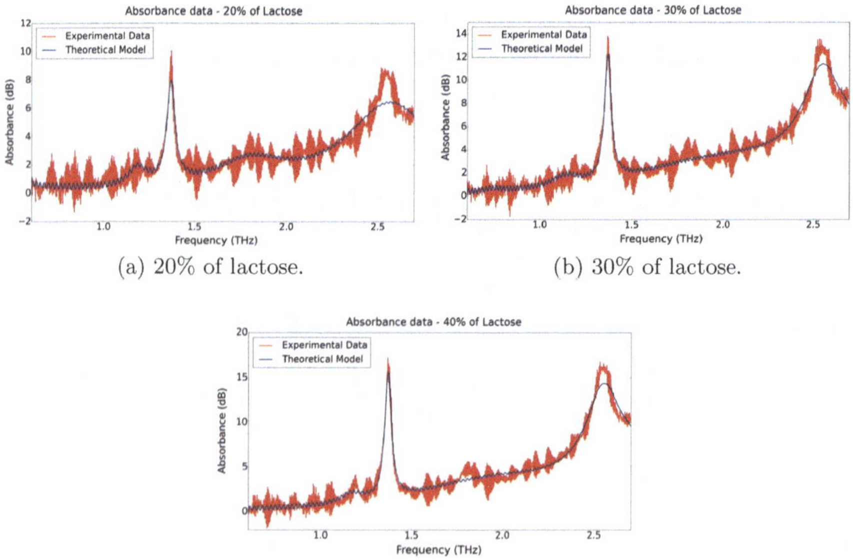

(a) 20% of lactose.

(b) 30% of lactose.

(c) 40% of lactose.

Fig. 1 Measured and modeled absorbance spectra for samples with different concentrations of lactose

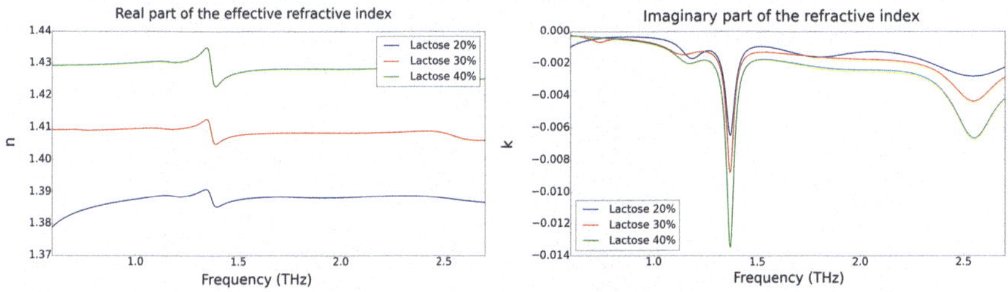

(a) Real part of the effective refractive index.

(b) Imaginary part of the effective refractive index(absorptioncoefficient).

Fig. 2 Profiles of the effective refractive indexes for different concentrations of lactose

Table 1 Thickness of the lactose samples retrieved from the model fit

Concentration of lactose (%)	Thickness (mm)
20	4.8
30	5.5
40	4.6

3.2 Metamaterials

As already known in literature [19, 20], the SRR geometry presents a resonant profile when the incident electric field is parallel to its gap; due to the excitation of a circular current, a magnetic dipole normal to the surface of the structure is generated. The understanding of this kind of structure can be analogous to an LC-circuit, [21] The approach to metamaterials began with simulations of structures with such unit cell designs, on the commercial software CST utilizing time domain electromagnetic solver. To validate the resonances, initially, a single unit cell based on the SRR was designed (Fig. 3a) and its dimensions can be found on Table 2, the metal used was copper and its thickness is represented by t. Later, a different unit cell was considered, the hybrid one (Fig. 3b). This is a composition of single SRR geometries with gap spaces varying from 5 to 7 μm, as also shown on Table 2.

The main objective of simulating both metamaterials was to compare their transmittance spectra and check if the hybrid geometry presented a superposition of the single ones, resulting in a larger and tunable bandwidth. The transmittance for each individual unit cell can be seen in Fig. 4a. Note that by increasing gap size (from 5

Table 2 Unit cell dimensions for SRR geometries

Parameter	(μm)	Gap	(μm)
a	25.5	g_1	5
l	24	g_2	6
w	2.25	g_3	6.5
g	5	g_4	7

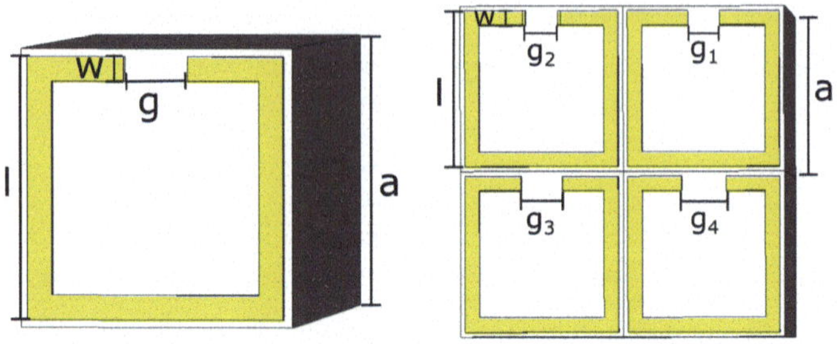

(a) Single unit cell. (b) Hybrid unit cell.

Fig. 3 Unit cell designs for metamaterials in terahertz

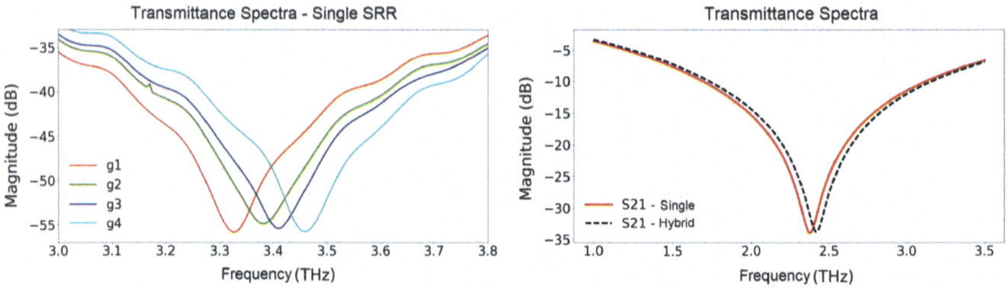

(a) Transmittance spectra for single SRR structures with varying gaps. (b) Transmittance spectra for single geometry with g_1 (solidline) e for the hybrid one(dashedline).

Fig. 4 Comparison of transmittance spectra for single unit cell with varying gaps in **a** and for hybrid geometries in **b**

to 7 μm), the resonant peak assumes higher frequency values, therefore, a blue shift. Such results highlights the tunability aspects of this geometry: by slightly changing the gap size, it is possible to predict the resonant region.

Now, by comparing the transmission of the single 5 μm unit cell with the hybrid metamaterial in Fig. 4b, it was noted a 11 % increase in bandwidth of the composed geometry. Although it was only a small increase, it is still a promising technique to finely tune the metamaterial's resonant bandwidth; It is worth emphasizing that the values for the gaps composing the hybrid unit cell must exhibit close values, otherwise the superposition will not be smooth enough so the result is a unique larger peak.

Once the simulations of the designs were analyzed, their effective electromagnetic properties were obtained using the algorithms developed and presented in Sect. 2.2. As the transmittance spectra for the single and hybrid metamaterials are alike, only distant in 11 % in their bandwidth, it is already expected that these absolute values will also be similar. For this reason, only the results for the hybrid structure are presented in this work.

Thus, the complex effective permittivity and permeability were retrieved and these profiles are shown in Fig. 5a, b. Both real and imaginary parts of the parameters exhibit negative behavior for certain frequency ranges which broaden even more their application in unconventional transmission and reflection filters.

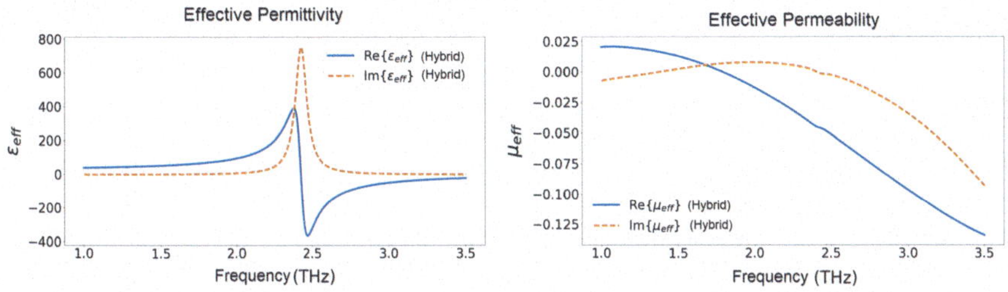

(a) Effective permittivity. (b) Effective permeability

Fig. 5 Effective electromagnetic parameters for the hybrid structure SRR

4 Conclusion

The development of the terahertz frequency range is profitable for several fields in science. In this work, the resonances in organic materials were explored. The algorithm developed was able to successfully extract the complex refractive index of samples with different lactose concentrations and, therefore, characterize them. Moreover, the proposed method requires only the transmittance data of the samples, acquired experimentally through TTDS. The method can be applied to different organic materials which present resonances in the proper frequency range and it is a non-destructive technique.

Another great advantage of terahertz waves lies on scaling limits. Due to the micro dimensions, these waves are suitable to receive these new technologies and and significant advances are expected in this range of frequencies once the manipulation of waves is getting more and more known. For example, the fabrication of metamaterials could open up a new class of resonators with distinct properties to be explored.

In this work, two different geometries were studied: the SRR and the hybrid SRR. By studying their transmittance spectra and proposing a methodology to effectively extract their complex permittivity and permeability, interesting results were found: it is feasible to tune the resonant frequency and the bandwidth of a metamaterial by adjusting their unit cell geometry and combining single geometries. By making a wider bandwidth, bandpass filters can be engineered and applied in several fields, the terahertz frequency range makes them specially attractive to imaging and wireless communications.

Besides, metamaterials exhibit unique electromagnetic properties which can be tuned and retrieved with the method presented in this work. It already provides progress in the development of devices, but by exploring different combinations in their unit cells, it is possible to spread even more this field of science and study the response of their superposition.

References

1. Kulesa, C.: Terahertz spectroscopy for astronomy: from comets to cosmology. IEEE Trans. Terahertz Sci. Technol. **1**(1), 232–240 (2011)
2. Wiltse, J.: History of millimeter and submillimeter waves. IEEE Trans. Microw. Theory Tech. **32**(9), 1118–1127 (1984)
3. Sirtori, C.: Applied physics: bridge for the terahertz gap. Nature **417**, 132–133 (2002). May
4. Koenig, S., Lopez-Diaz, D., Antes, J., Boes, F., Henneberger, R., Leuther, A., Tessmann, A., Schmogrow, R., Hillerkuss, D., Palmer, R., Zwick, T., Koos, C., Freude, W., Ambacher, O., Leuthold, J., Kallfass, I.: Wireless sub-THz communication system with high data rate. Nat. Photonics **7**, 977–981 (2013)
5. Lin, C., Li, G.Y.L.: Terahertz communications: an array-of-subarrays solution. IEEE Commun. Mag. **54**, 124–131 (2016)
6. Lee, S.-H.L., Kang, B.J., Yoo, B.-W., Lee, S.-C., Lee, S.-J., Jazbinsek, M., Yun, H., Rotermund, F., Kwon, O.-P.: Terahertz phonon mode engineering of highly efficient organic terahertz generators. Adv. Funct. Mater. **27**(14), 1605583 (2017)
7. Parrott, E.P.: Terahertz time-domain and low-frequency Raman spectroscopy of organic materials. Appl. Spectrosc. **69**(1), 1–26 (2015)
8. Min, L., Huang, L.: Perspective on resonances of metamaterials. Opt. Express **23**, 19022–19033 (2015)
9. Veselago, V.G.: The electrodynamics of substances with simultaneously negative values of ϵ and μ. Sov. Phys. Usp. **10**, 509–514 (1968)
10. Pendry, J.B., Holden, A.J., Robbins, D.J., Stewart, W.J.: Magnetism from conductors and enhanced nonlinear phenomena. IEEE Trans. Microw. Theory and Tech. **47**, 2075–2084 (1999)
11. Zhang, X., Liu, Z.: Superlenses to overcome the diffraction limit. Nat. Mater. **7**, 435–441 (2008)
12. Zhou, Q., Shi, Y., ad Lei Li, A.W. , Zhao, D., Liu, J., Sun, H., Zhang, C.: Ultrafast optical modulation of terahertz metamaterials. J. Opt., **13**, 125102–125109 (2011)
13. Pan, Z.Y., Chen, P.Z.Z.C., Vienne, G., Hong, M.H.: Hybrid srrs design and fabrication for broadband terahertz metamaterials. IEEE Photonics J. **4**, 1267–1272 (2012)
14. Lim, C.S., Hong, M.H., Chen, Z.C., Han, N.R., Lukyanchuk, B., Chong, T.C.: Hybrid metamaterial design and fabrication for terahertz resonance response enhancement. Opt. Express **18**, 12421–12429 (2010)
15. Chen, X., Grzegorczyk, T.M., Wu, B.-I., Joe Pacheco, J., Kon, J.A.: Robust method to retrieve the constitutive effective parameters of metamaterials. Phys. Rev. E, **70**, 0166081–016607 (2004)
16. Szabo, Z., Park, G.-H., Hedge, R., Li, E.-P.: A unique extraction of metamaterial parameters based on kramer-kronig relationship. IEEE Trans. Microw. Theory Techn. **58**, 2646–2653 (2010)
17. Giga Optics Femtosecond Technologies: High-speed asynchronous sampling THz transmission spectroscopy system (2012)
18. Vazquez-Caboa, J., Chamorro-Posada, P., Fraile-Pelaez, F.J. Rubinos-Lopez, J. M. Lopez-Santos, Ó., Martin-Ramos, P.: Windowing of THz time-domain spectroscopy signals: a study based on lactose. Opt. Commun. **366**, 386–396 (2016)
19. Rockstuhl, C., Lederer, F., Etrich, C., Zentgraf, T., Kuhl, J., Giessen, H.: On the reinterpretation of resonances in split-ring-resonators at normal incidence. Opt. Express **14**, 8827–8836 (2006)
20. Li, C., Zhoul, Q., Shi, Y., Yang, Z., Shi, L., Zhang, C.: Manipulating the resonant behaviors in the asymmetric terahertz metamaterials. Opt. Commun. **391**, 77–81 (2017)
21. Linden, S., Enkrich, C., Dolling, G., Klein, M.W., Zhou, J., Koschny, T., Soukoulis, C.M., Burger, S., Schmidt, F., Wegener, M.: Photonic metamaterials magnetism at optical frequencies. IEEE J. Sel. Top. Quantum Eletron. **12**, 1097–1105 (2006)

SpaceWire Receiver: Synchronization for Different Clock Domains

Alexsander Deucher, Angela Alves dos Santos, Antonio Carlos da Costa Telles and Saulo Finco

Abstract The SpaceWire standard, used in aerospace embedded applications, was developed to connect devices that operate at high data rates, having as goals the traffic control and payload transmission. This standard enables flexibility on transmission rates, which brings some challenges related to clock domains. In an effort to fulfill the essential system requirements, such as reliability and versatility, synchronization techniques are introduced to avoid possible problems related to the clock domain crossing. The objective of this work is to present a technique applied to SpaceWire Receiver that enables its operation at many signaling rates.

Keywords Receiver · SpaceWire · Synchronism · Clock domain

1 Introduction

It was always a challenge obtaining a standard that enables the connection of devices from different manufacturers, such as data processing, mass memory and remote command, since each manufacturer had its own communication interface. Being each link different, there was the need for interfaces to establish the interoperability of the devices. This fact increased the design time and the development costs. It was clearly necessary the creation of a link to be adopted by organizations and developers of onboard products worldwide.

A. Deucher (✉) · A. A. dos Santos · A. C. da C. Telles · S. Finco
Centro de Tecnologia da Informação Renato Archer, Campinas, Brazil
e-mail: adeucher@cti.gov.br

A. A. dos Santos
e-mail: angela.santos@cti.gov.br

A. C. da C. Telles
e-mail: antonio.telles@cti.gov.br

S. Finco
e-mail: saulo.finco@cti.gov.br

Y. Iano et al. (eds.), *Proceedings of the 4th Brazilian Technology Symposium (BTSym'18)*, Smart Innovation, Systems and Technologies 140,
https://doi.org/10.1007/978-3-030-16053-1_4

This way, the European Cooperation for Space Standardization created, based on IEEE-1355-1995 standard, the ECSS-E-ST-50-12, which defines SpaceWire as the pattern for data communication targeted to aerospace and aircraft applications. The main space agencies involved in the elaboration of this standard were ESA (European Space Agency), NASA (National Aeronautics and Space Administration) and JAXA (Japan Aerospace eXploration Agency).

The directives adopted by SpaceWire search for the simplification of the development of interfaces between high data rate devices, diminishing the development costs and the interoperability and compatibility of devices from distinct space missions. Among its features, it can be highlighted: 2–200 Mbps transmission rates, fullduplex serial communication, system of point-to point link and low voltage differential signals (LVDS) [1]. Additionally, there are some punctual characteristics that determine the design architecture. In this article it was discussed the implementation of the coded signal recovery by the SpaceWire reception module.

There are two possible architecture concepts of digital design to make the coded signal recovery using the SpaceWire standard. One of them is implementing the recovery by using a clock that is twice the highest reception rate of the system, described by the sampling theorem [2]. This solution has as disadvantage the high-power dissipation, even in low reception rates, due to the clock switching. As advantage, it is easier to be implemented because it has only a clock domain.

The other possibility is using the Data/Strobe codification to generate the receiver clock. This implementation has as disadvantage a higher architecture complexity with many clock domains, e as advantage the reduction of the power consumption due to clock switching at lower rates. The tradeoff of both solutions is between complexity and consumption, and can be attributed as best option the Data/Strobe coding from the point of view of the final application and this work study.

2 SpaceWire Standard Architecture

The generic architecture of the SpaceWire standard is composed by a CODEC, which codes and decodes a serial bit-stream through a link interface. The CODEC is constituted by a Transmitter, a Receiver and a State Machine, as shown in Fig. 1.

2.1 Transmitter Module

The Transmitter module is responsible by coding and transmitting the data through SpaceWire link using the Data/Strobe coding technique.

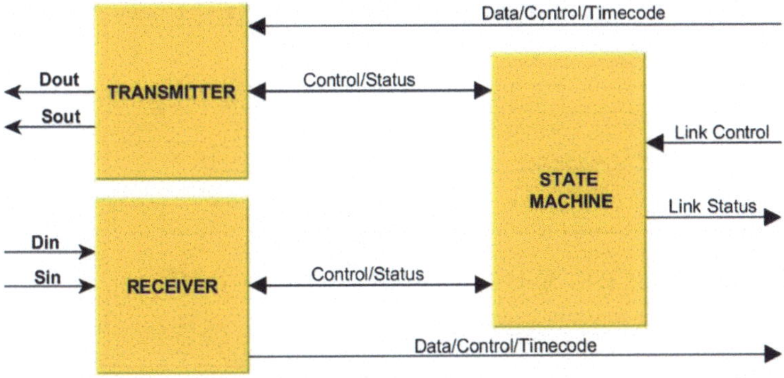

Fig. 1 Block diagram of a SpaceWire generic interface

2.2 *State Machine Module*

The State Machine module controls all the operations of the interface link. This module is responsible to initializing, disabling or leaving the link in auto-start.

2.3 *Receiver Module*

The receiver module is responsible for decoding the Data/Strobe signals and producing a sequence of characters for a Host-Interface, in addition to the detection of the link disconnection.

2.4 *Data/Strobe Coding*

The SpaceWire uses the Data/Strobe coding scheme for serial transmission and reception, as illustrated in Fig. 2. This coding uses a signal called Data for coding the data serially and other called Strobe. Both signals are used to determine the reception clock.

To recover the clock a XOR operation should be done, as illustrated in Fig. 3, between Data-Strobe, and the receiver should decode one data bit from the line in the positive edge and the negative edge of the recovered clock.

Since the data should be recovered at the positive and negative edges of the recovered clock, the reception domain should have flip-flops sensitive to positive edge and others sensitive to negative edge. Another possibility would be using flip-flops sensitive to positive edge and an inverted Reception clock. This alternative was chosen for the system implementation and Fig. 4 shows the data recovery process.

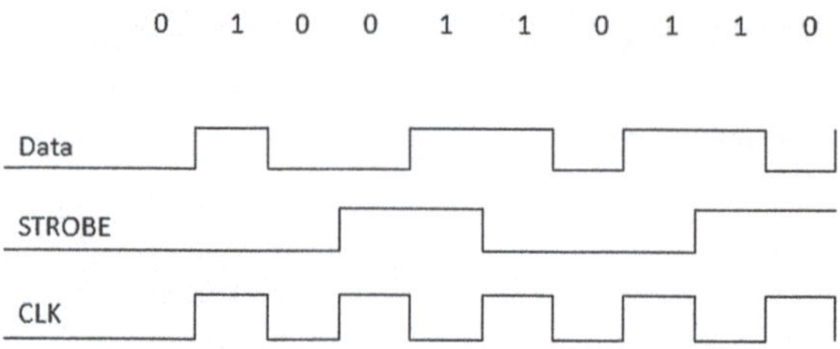

Fig. 2 DS coding (ECSS-E-ST-50-12C)

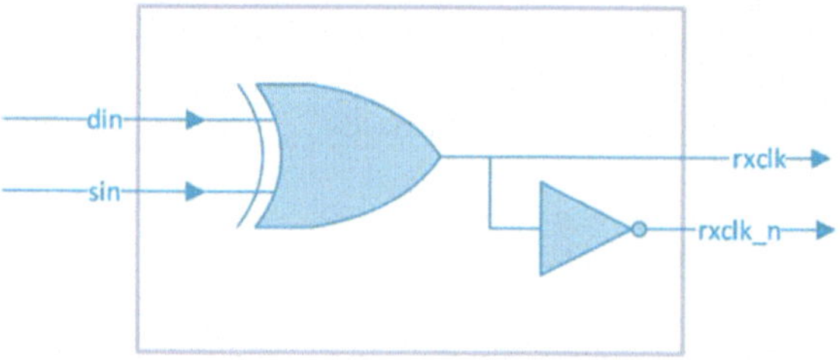

Fig. 3 XOR of D/S input signals

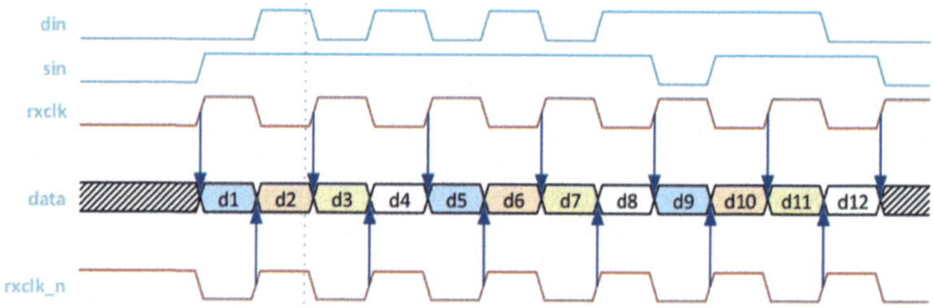

Fig. 4 Data recovery

It can be noted that d1 is recovered at the positive edge of rxclk while d2 is recovered at positive edge of the inversion of rxclk, called rxclk_n.

From the transmitter point of view, the signals Data/Strobe should not change simultaneously, i.e., while data change in a clock cycle, the strobe signal remains with the previous value and so on. However, in certain situations this simultaneous transition is possible (e.g., transmission system reset, physical disconnection…) and the receiver should be robust to this condition.

2.5 Data Signaling Rate

The minimal data signaling rate at which SpaceWire operates is 2 Mbps. The maximum will consider phenomena like skew and jitter, but usually reaches 400 Mbps.

3 Receiver Clock Domains

The use of many clock domains in a circuit has been a challenge to verification, simulation and to the use of other tools and methodologies like Static Timing Analysis.

New methods have been explored for dealing with this problem [3].

The clock domains used in this work were:

Reception domain, which works in the frequency determined by the reception rate, in the range of 2–200 MHz.

System domain, which contains the memories where the received data will be storaged and the external interface.

The clock domain crossing occurs when a datum is transferred from a flip-flop of one domain to a flip-flop from other domain [4]. Depending on the relation between the clocks of the domains, different problems can occur at the datum transference. Some of these problems and their solving approaches are described below.

3.1 Metastability

When a time violation happens, or an input datum changes improperly before or after a clock rising, metastability appears [5]. This metastability or metastable signal is the propagation of an undesired value from a flip-flop output, as shown in Fig. 5.

If the transition of [AB out] occurs much closer to the positive edge of [CLK CD], a violation of the setup/hold of the flip-flops can result, turning [CD Out] unstable or producing a glitch.

If this datum is transmitted to several elements, there might be a current rising which at the worst case will lead to the component loss at short or long term.

Other potential consequence is that the logic elements can interpret the metastable signal as a "0" or a "1", which can cause failures in the operation of the circuit.

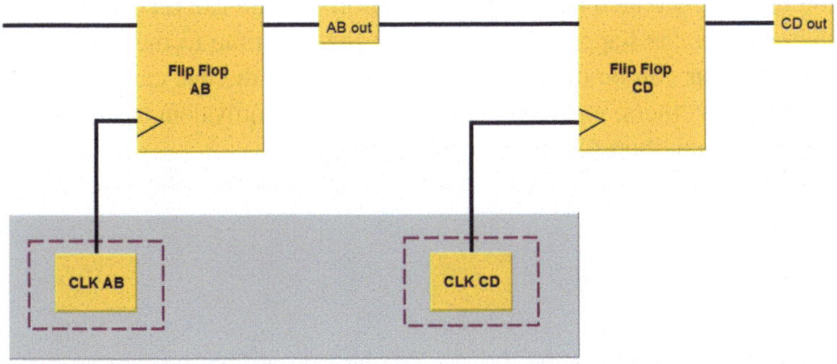

Fig. 5 Metastability

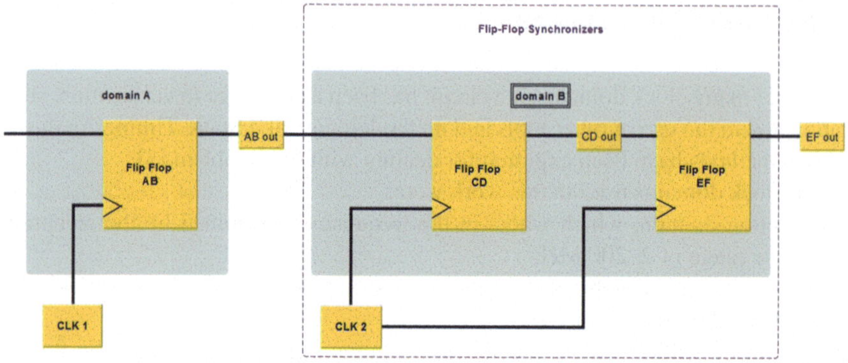

Fig. 6 Synchronizers inserted in the circuit

A third consequence is that the transition of the metastable signal can cause timing problems, violating the setup/hold times of other flip-flops that are in the circuit sequence, thus propagating the metastability.

The problem can be avoided adding structures called synchronizers at the beginning of the destination clock [6]. This kind of structure is usually composed by a sequence of interconnected flip-flops, as illustrated in Fig. 6.

This solution can be applied in control signals of one or multiple bits, or data signals from one single bit. Other techniques should be used for the data buses treatment.

3.2 Failures

Even when metastability is treated, data can be lost or present incoherencies. Data loss can occur when the origin domain is faster than the destination one, this way the datum can be generated before the destination domain can recover it [7].

Incoherent data can happen due to the fact that the time to the synchronization of each bit in a bus can be different, thus some bits of the bus can last more to be synchronized than others, making the domain data not equivalent to those generated in the origin domain, therefore turning them invalid.

3.3 Failure Mitigation

To prevent data loss, at the origin domain the data should be kept for several cycles until its reception be assured at the destination domain. This technique is called pulse stretching.

On the other hand, the handshaking technique can be used to prevent the data incoherence [8]. The origin domain generates a validation signal that should be received by the destination domain. This one should wait one or more cycles for the signal stabilization, store it and generate an answer signal to the origin domain warning that the signal was received.

4 Methodology

The applied methodology took in consideration the synchronization techniques already discussed, in such a way that the data can be received in different rates, varying from 2 Mbps up to the implemented limit, considering as the clock of the Reception domain that one recovered from the line (Data/Strobe).

It is assumed that the exact moment in that the datum is concluded is unpredictable to the system domain. Thus the edge detection was chosen to inform the system domain about the received datum [9].

There are three formats of the SpaceWire character: Control, with four bits; Data, with 10 bits and Timecodes with 14 bits. In each one of them one bit represents the correct parity.

The system has a fifo of 36 bits by word, which represent four data characters of SpaceWire. These four characters should be written simultaneously at the fifo. This demands that the receiver be able to store at least four data characters at the System domain. Intending to fulfill the fifo with data and considering that these data can arrive at high transmission rates, the System and Reception domains have a group of distinct buffers. Three buffers for the System domain and four buffers for the Reception domain were created with this intention. These buffers enable the Receiver to operate without overwriting the data not yet recorded in the fifo.

The data transference between Reception and System domains is done through a technique similar to pulse stretching. The difference is that, unlike the utilization of pulses to inform the System domain that the signal is ready, the edges of the control signal related to each buffer are used (d0, d1 and d2), as shown in Fig. 7.

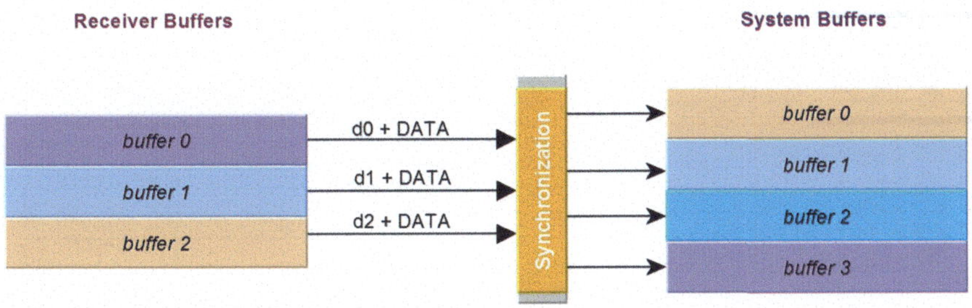

Fig. 7 Communication between buffers

Each buffer is synchronized and detected at the System domain, and the information is then recorded in the 36 bits buffer, further transferred to the fifo. Using this implementation, it is possible, considering only the Receiver, to reach recovery rates that enable the recovered clock be four times faster than the System clock. Example: for a System clock of 50 MHz it is possible to receive data at a rate of 400 Mbps, which represents a recovered clock of 200 MHz, i.e., recovery of bits at the rising and falling edges of the Reception clock.

5 Results

The following simulations show the data transference between the buffers from the Reception domain to System domain. The tool applied for simulation was the Incisive Enterprise Simulator 13.10-s030 from Cadence Design System, Inc.

The signals data_0[8:0], data_0[8:0], data_2[8:0] represent the data signals from the buffers at the Reception domain. The edges of signals d0, d1 and d2 mean the signaling at the Reception domain that they are ready. The signals sys_d0, sys_d1, sys_d2 mean the signaling recognition at the System domain. Finally sys_buffer_d0, sys_buffer_d1, sys_buffer_d2 and sys_buffer_d3 represent the received data, illustrated in the Fig. 8.

Figure 9 illustrates the moment the data are recorded in the System fifo.

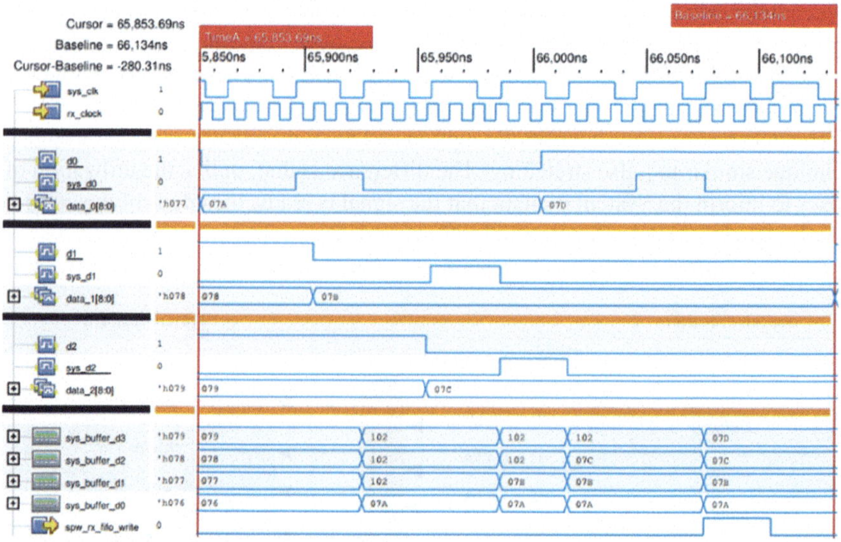

Fig. 8 Simulation of data recovery

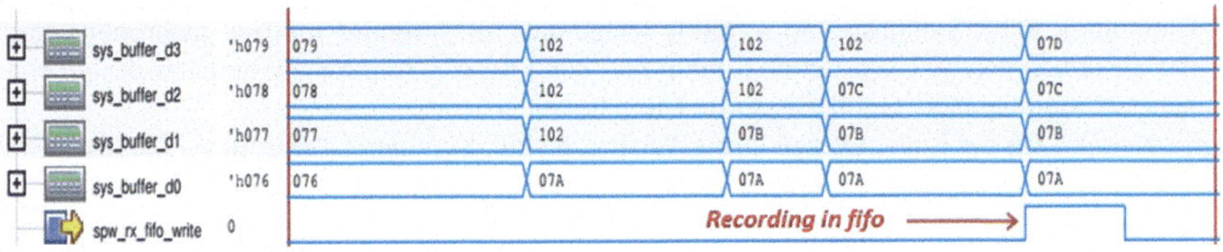

Fig. 9 Fifo storage

The clock at the system domain has a frequency of 33 MHz, while the clock at Reception domain is at the frequency of 100 MHz, which represents a Reception rate of 200 Mbps.

6 Conclusion

It was shown that the present technique is functional in the aspects of reliability and versatility, besides following the demands of the SpaceWire protocol in the aspect of data recovery, this way enabling flexibility when working with many domains and clock crossing.

Seeking for robustness, the Receiver was designed in order to be able to assure the data transference between the clock domains without incoherencies, even at high transmission rates.

References

1. Parkes, S.M. et al.: SpaceWire: Links, Nodes, Routers and Networks. European Cooperation for Space Standardization, Standard No. ECSSE50-12A, Issue 1, January 2003
2. Fey, A.F., Gauer, R.R.: Fundamentos de Telecomunicações e Comunicação de Dados. 1ª edição 2016. p 395. Caxias do Sul. ISBN 978-85-919835-3-7
3. Chen, S., Huang, W., Shen, C.: Lessons Learned from Multi-clock Design of SpaceWire Interface. PKP Open Conference Systems
4. Cummings, C.: Clock Domain Crossing (CDC) Design & Verification Techniques Using SystemVerilog. In: Proceedings Synopsys User Group Meeting (SNUG) (2008). http://wwwsunburst-design.com/papers/CummingsSNUG2008Boston_CDC.pdf. Oct 2016
5. Arora, M.: The Art of Hardware Architecture—Design Methods and Techniques for Digital Circuits, pp. 1–2. Springer Science & Business Media, Octiber (2012)
6. Ginosar, R.: Metastability and Synchronizers: A Tutorial. Technion Israel Institute of Technology. Copublished by the IEEE CS and the IEEE CASS (2011)
7. Fuchs, G., Fugger, M., Steininger, A.: On the threat of metastability in an asynchronous fault-tolerant clock generation scheme. In: 15th IEEE Symposium Asynchronous Circuits and Systems, ASYNC'09 (2009)

8. Cummings, C.E.: Synthesis and scripting techniques for designing multi-asynchronous clock designs. In: Synopsys Users Group Conference, San Jose, CA. http://www.sunburst-design.com/papers/CummingsSNUG2001SJ_AsyncClk.pdf, Oct 2016
9. Taraate, V.: Digital Logic Design Using Verilog, p. 450. Springer, United States (2016)

A Solution for Dubins Path Problem with Uncertainties Using World Cup Optimization and Chebyshev Polynomials

Navid Razmjooy, Mehdi Ramezani and Vânia Vieira Estrela

Abstract In this paper, an efficient numerical approach is developed for solving the baseline problem with interval uncertainties. Interval arithmetic is also utilized for developing the proposed method in the presence of uncertainties with only lower and upper bounds of parameters. In the proposed method, the equation of motion, performance index, and boundary conditions are first changed into some algebraic equations. This process converts the problem into an optimization problem. The presented technique is based on an interval extension of Chebyshev polynomials where its coefficients are achieved by world cup optimization algorithm, as a new optimization algorithm. The proposed method approximates the control and state variables as a function of time. The proposed solution is based on state parameterization, such that the state variable is approximated by the proposed interval Chebyshev polynomials with unknown interval coefficients. Finally, by solving the baseline problem in the presence of interval uncertainties, the reliability and effectiveness of the proposed method are demonstrated.

Keywords Chebyshev polynomials · Optimal control · Dubins path problem · Interval analysis · World cup optimization algorithm · Uncertainty

1 Introduction

Optimal control theory (OCP) is about dealing to find a control law for a considered system such that a specific optimality criterion is achieved. Recently, OCPs have become to mathematically challenging discipline [1].

N. Razmjooy (✉) · M. Ramezani
Department of Electrical and Control Engineering, Tafresh University, 39518 79611 Tafresh, Iran
e-mail: navid.razmjooy@hotmail.com

V. Vieira Estrela
Departamento de Engenharia de Telecomunicações, Universidade Federal Fluminense, Rio de Janeiro, Brazil

Y. Iano et al. (eds.), *Proceedings of the 4th Brazilian Technology Symposium (BTSym'18)*, Smart Innovation, Systems and Technologies 140,
https://doi.org/10.1007/978-3-030-16053-1_5

Direct methods are a group of OCP solution techniques which achieve the control law by direct minimization of the performance index, subject to the constraints. Pontryagin's maximum principle and dynamic programming methods are introduced as the most popular techniques for OCP solutions [2, 3].

Due to being inaccessible of the analytical solutions for some OCPs, solutions approximations have been introduced.

Numerical methods have been known as the popular methods for the researchers for solving the complicated OCPs [4].

In this field, numerous techniques have been proposed which employ the orthogonal functions for solving the time-varying LQR OCPs [5, 6].

The introduced methods are founded based on either converting the dynamic OCP into a quadratic programming problem or converting the TPBVP into a set of algebraic equations.

El-Gindy et al. introduced a deterministic method for solving OCPs based on orthogonal methods [7]; in that research, the control variable is first approximated by the Chebyshev polynomials with unknown coefficients and then, the problem is transformed into an OCP. This made the OCP as a parameter optimization problem of the performance index. By this idea, the performance index can be also solved by transforming it into an unconstrained optimization problem and solving it by gradient-based methods or meta-heuristics [8, 9].

Basically, popular OCP solutions are based on model constructions. However, there is always one problem which makes a lot of issues; constructing an accurate model for the real state systems is very complicated and there is always some assumptions and some unconsidered phenomena in the model which have a significant effect on the system model (uncertainties) [10].

Utilizing the traditional deterministic methods for solving these OCPs can result invalid and even non-convincing solution which makes them into unprofitable methods in the presence of uncertainties.

Different methods have been developed to guarantee the OCPs robustness in the presence of uncertainties [11]. In the meantime, fuzzy programming [12], stochastic methods, and interval arithmetic methods [13–15] are the most popular methods.

Stochastic methods need information about probabilistic distribution and fuzzy methods need to an expert with enough information about system changes. Generally, there is often no information about distribution and memberships. Fortunately, interval methods need only lower and upper bounds of uncertainties [16, 17].

This paper presents a baseline OCP method based on Chebyshev polynomials and a new meta-heuristic algorithm, WCO, for solving the baseline OCPs to be an effective choice for accommodating the interval uncertainties present in flight control applications and for ensuring satisfactory reference tracking in the presence of uncertainties.

The rest of the paper is organized as follows: in Sect. 2, the concept of the Chebyshev polynomials and their technique to approximate a function is introduced. Section 3 presents a brief theory of interval analysis. In Sect. 4, the optimal control problems in the presence of Interval uncertainties are explained. In Sect. 5, world cup optimization as an optimization algorithm is illustrated. In Sect. 6, Interval Dubins

Path Problem is solved and analyzed to validate the proposed method. The paper is finally concluded in Sect. 7.

2 Chebyshev Polynomial Approximation

A Chebyshev polynomial (C_{i}) in the interval $t \in [-1, 1]$ of degree i can be defined as follows [18]:

$$C_i(t) = \cos i\theta, \theta = \arccos(t) \in [0, \pi] \tag{1}$$

where i describes the nonnegative integer.

Truncated Chebyshev polynomial is a better way to approximate the system practically; truncated Chebyshev approximating of the function f(x) with an accuracy of degree k is as follows:

$$f(t) \approx p(t) = \frac{1}{2} f_0 + \sum_{i=1}^{k} f_i C_i(t) = \frac{1}{2} f_0 + \sum_{i=1}^{k} f_i \cos(i\theta) \tag{2}$$

where, f_i is the ith constant coefficient. By evaluating the formula using the numerical Mehler (Gaussian-Chebyshev) integration formula [19], the coefficient of Chebyshev polynomial can be achieved by the following formula:

$$f_i = \frac{2}{p} \sum_{j=1}^{p} f\left(\cos\theta_j\right) \cos i\theta_j \tag{3}$$

where, p is the order of truncation. Truncated Chebyshev polynomial can be achieved by substituting the equation above in Eq. (2).

3 Interval Analysis

Interval arithmetic is a way to define the interval variables. Consider the interval integers $[x]$ and $[y]$ with lower and upper bounds where $[x] = \left\{x | x \in \mathbb{R} \cup \{-\infty, \infty\}, \underline{x} \leq x \leq \overline{x}\right\}$ and $[y] = \left\{y | y \in \mathbb{R} \cup \{-\infty, \infty\}, \underline{y} \leq y \leq \overline{y}\right\}$. In this case, the basic arithmetic operations can be defined as follows:

$$[x] + [y] = [\underline{x} + \underline{y}, \overline{x} + \overline{y}], \tag{4}$$

$$[x] - [y] := [\underline{x} - \overline{y}, \overline{x} - \underline{y}]. \tag{5}$$

$$[x]\times[y]=[\min\{\underline{xy},\bar{x}\underline{y},\underline{x}\bar{y},\overline{xy}\},\ \max\{\underline{xy},\bar{x}\underline{y},\underline{x}\bar{y},\overline{xy}\}],\tag{6}$$

$$[x]/[y]=[x]\times\frac{1}{[y]},$$
$$\frac{1}{[y]}=[\frac{1}{\bar{y}},\frac{1}{\underline{y}}],0\notin[\underline{y},\bar{y}]\tag{7}$$

More detailed about interval arithmetic can be found in [20–22].

4 OCP with Interval Uncertainties

Consider the following OCP with uncertainties:

$$\underset{u(t)}{\text{Min}}\ J(X(t),U(t),\Delta)=\int_{t_0}^{t_f}G(t,X(t),U(t),\Delta)\,dt\tag{9}$$

where, $J(X(t),U(t),\Delta)=\left[\underline{j}(X(t),U(t),\Delta),\bar{j}(X(t),U(t),\Delta)\right]$ is the interval performance index of the system. The system dynamic and the initial and the final conditions are as follows:

$$\dot{X}(t)=f(t,X(t),U(t),\Delta),t\in(t_0,t_f),\tag{9}$$

$$X(t_0)=X_0,\ X(t_f)=X_f.\tag{10}$$

where, U(t) and X(t) are the control and the state variables, respectively. $\Delta=[\delta_1,\delta_2,\ldots,\delta_n]$ defines the system uncertainties, and X_0 and X_f are the initial and final states, respectively.

For the simplicity of the work without losing the generalities, assume that the time interval is located in the interval $t_0\neq-1$ and $t_f\neq 1$. i.e.

$$J(X)=\frac{t_1-t_0}{2}\int_{-1}^{1}G\left(\frac{t_1-t_0}{2}t+\frac{t_1+t_0}{2},X(t),\dot{X}(t),\Delta\right)dt\tag{11}$$

where,

$$U(t)=F\left(\frac{t_1-t_0}{2}t+\frac{t_1+t_0}{2},\ X(t),\dot{X}(t),\Delta\right)$$

And the initial conditions are:

$$X(t_0) = X_{-1}, X(t_f) = X_1 \tag{13}$$

By changing the time interval to the [−1,1] and by considering unknown interval coefficients ($\alpha \equiv (a_0, a_1, \ldots, a_n)$), the state variable can be approximated by the following Chebyshev polynomial:

$$[X_m]([t]) \approx \frac{1}{2}[a_0] + \sum_{j=1}^{m} [f]([a_i]T_i), i = 0, 1, \ldots, m \tag{14}$$

where, $[X_m] = [\underline{x}_m, \overline{x}_m]$. Finally, based on the relationship between the state and control variables, the system's control law will be achieved as follows:

$$a_j = \frac{2}{K} \sum_{j=1}^{m} f(\cos(\theta_j))\cos(\theta_j), j = 0, 1, \ldots, m \tag{15}$$

$$\theta_j = \frac{2i - 1}{2K}\pi, i = 0, 1, \ldots, K \tag{16}$$

where, $K>m$.

In the direct solving method for traditional OCPs, these coefficients can be obtained as follows:

$$[X_m]([t]) \approx \frac{1}{2}[a_0] + \sum_{j=1}^{m} f([a_j], t), j = 0, 1, \ldots, m \tag{17}$$

When the system has interval uncertainty, two performance indexes have been made. So in this case, if we want to use the classical method to solve this system, we will have to transform these two performance indexes into a single function (Lagrange coefficients) which the results are very inefficient. The other method is to use multi-objective optimization methods instead of using the classical method.

In the single-objective optimization problems, the main purpose is to improve a single performance index, whose minimum or maximum value fully reflect the quality of the response. But in some cases, one cannot just rely on an index to rely on a hypothetical answer to the optimization problem. In this type of problem, we have to define several performance indexes and, simultaneously, optimize all of them. Multi-objective optimization is one of the most effective and highly applied research techniques in optimization. In this research, the WCO method has been used to solve the multi-objective optimization problem.

5 World Cup Optimization Algorithm

In the last decades, intelligent algorithms which are inspired from natural and social processes have been developed for solving the optimization problems. Some of the popular intelligent algorithms like Genetic algorithm, particle swarm optimization, and quantum invasive weed optimization algorithm have been utilized in most of engineering and science fields to solve the complicated problems and achieve the optimized solution [23–25].

Recently, a new algorithm has been inspired from the FIFA world cup competitions which have good results for different works. This algorithm is called World Cup Optimization (WCO) algorithm. WCO is a competitive algorithm; the main target in this algorithm is to challenge the teams with each other until one team could reach the best score and achieves the championship cup. In the WCO algorithm, 'rank' has a direct impact on the team's success.

The reason is that the team's strength defines the seeding. From the rank scores, n first strong teams have been categorized as the first seed, the second seed comprises the teams weaker than the first seed and the others have been categorized as the second team hierarchically.

At first, strong teams rise to the next level with no competitions. The competition starts by competing the teams separately in their seeds to win the competition and achieve more scores and upgrade their rank for the next games and cups. After the preliminary competitions in seeds, the best two teams from each group arise to the next level and the rest has been eliminated.

The third place of each competition in the seeds has another chance to rise to the next level by winning the other same score teams from the other seeds which are called Play-Off. The final competition is held between two teams with the most scores to define the champion of the competitions. The flowchart of the multi-objective WCO algorithm is shown in Fig. 1.

6 Solving Interval Dubins Path Problem

The main purpose of the Dubins Path Problem is to determine the control law so that it can take a vehicle in the shortest time from a geographical position to another geographical location. This is an important issue for goals such as steering UAVs and preventing accidents.

By limiting the transverse motion problem for a vehicle, the kinematic system based on the Dubbin model is as follows:

$$\dot{x}_1 = \delta_1 V(t)\cos(\theta(t))$$
$$\dot{x}_2 = \delta_1 V(t)\sin(\theta(t))$$
$$\dot{V} = u_1(t)$$

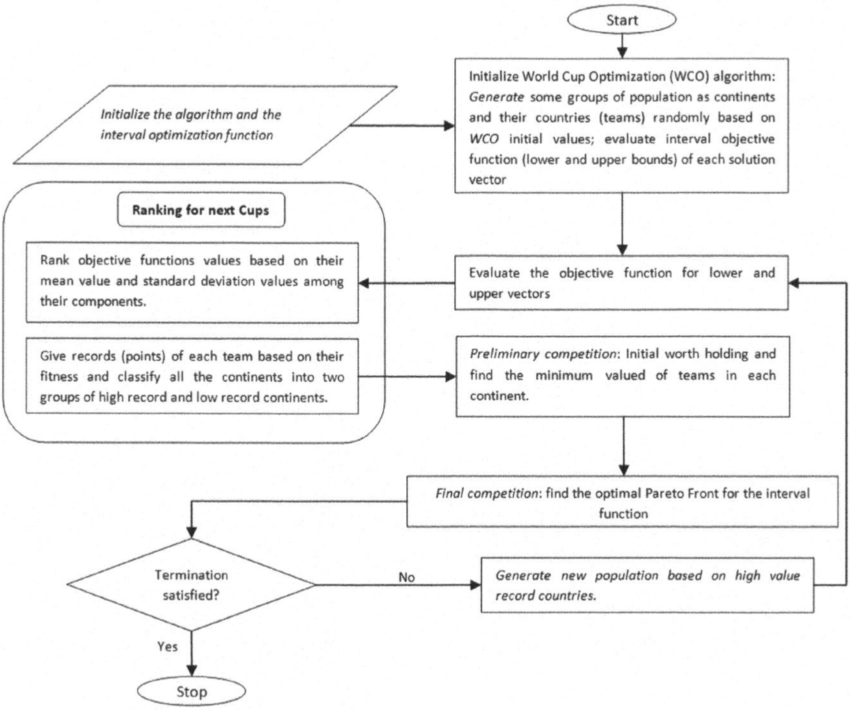

Fig. 1 Flowchart of world cup optimization algorithm

$$\dot{\theta} = \delta_2 u_2(t) \tag{18}$$

where, $p = (x, y)$ is the vehicle position in two directions, V is the initial velocity, $\theta(t)$ describes the motion angle and $u = (u_1, u_2)$ shows the control variables. Therefore, $V = \sqrt{\dot{x}^2 + \dot{y}^2}$

OCP for this system is formed as follows:

$$p_{\min} = \min_{u(t),\, t\in[0,t_f]} \int_0^{t_f} \delta \times J(p)dt \tag{19}$$

where, $z = [x(t), y(t), \theta(t)] \in \mathbb{R}^2$ and $u(t) = [\theta(t)] \in \mathbb{R}^1$. The velocity of the vehicle is considered 470 m/s and uncertainties $\delta = \delta_1 = \delta_2 = [0.7, 1.4]$, and t_f can be free, indefinite or interval.

$$p_0 = (0, 0),\ V_0 = 0.5,\ \psi_0 = \frac{\pi}{4},\ p_f = (1, 1),\ V_f = 0.5,\ \psi_f = -\frac{\pi}{4}$$

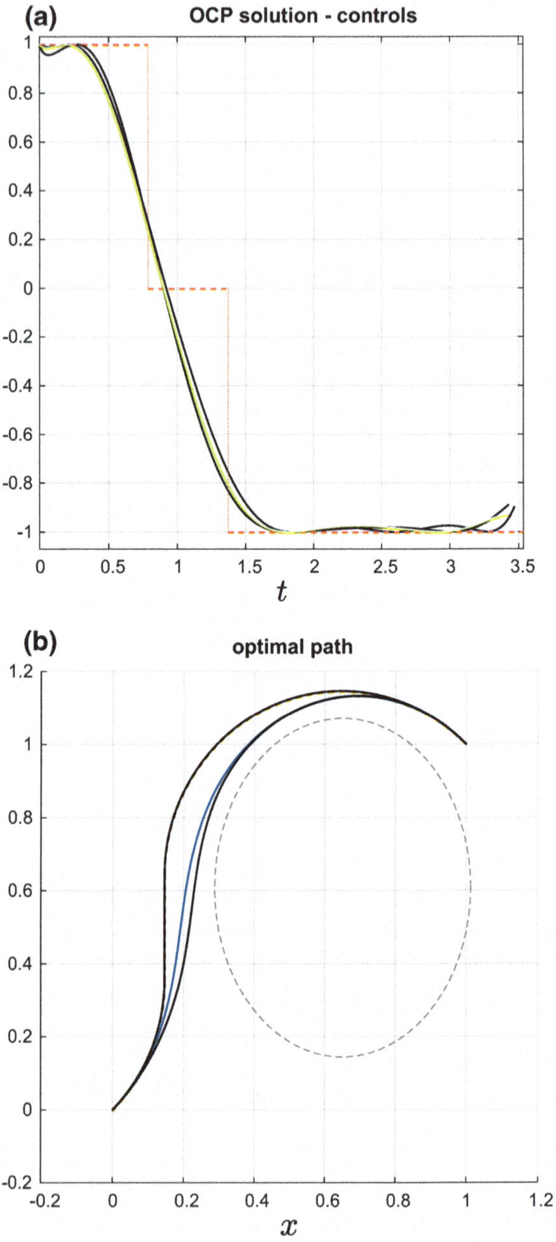

Fig. 2 **a** Tracking problem control rate and Dubins confident path and **b** the confident path response for the Dubins confident path

These boundaries are used to restrict the vehicle's reachable operating area. Therefore, with the help of this constraint, the prohibited areas can be induced according to the desired area.

If $v(t)$ is constant and $u_1(t) = 0$, then the solution method for the problem is the dubins path. This problem can be solved in a straightforward manner by parameterizing the control variables ($u(t)$) using the Chebyshev functions with open-field partition coefficients. The time for diving in the dubins path is considered 3.727 s. Figure 2 shows the control of the system and the optimal path.

From the figure, it is clear that by applying a random input to the system, the confidence interval is taken on the guaranteed path. In general, this example provides a more specific application for optimal control problems. Therefore, by determining the boundaries of the interval in the optimal control system, we only need to maneuver on the methods in which the amount falls within this interval. Otherwise, the selection method is wrong.

7 Conclusions

A new topology has been proposed for finding the confidence interval of the optimal control problems with uncertain-but-bounded parameters. The method is based on the Chebyshev polynomials and world cup optimization algorithm.

In this study, the Chebyshev polynomial is extended to an interval polynomial based on interval analysis and approximating the control variable of the OCP, the unknown parameters have been evaluated with WCO multi-objective optimization algorithm. The numerical study is applied to the Dubin's path problem with interval uncertainties to show the proposed method's efficiency.

References

1. Razmjooy, N., Ramezani, M.: Analytical solution for optimal control by the second kind Chebyshev polynomials expansion. Iranian J. Sci. Tech. **41**, 1017–1026 (2017)
2. Kopp, R.E.: Pontryagin maximum principle. Math. Sci. Eng. **5**, 255–279 (1962)
3. Foster, I., Kesselman, C.: The Grid: Blueprint for a New Computing Infrastructure. Morgan Kaufmann, San Francisco (1999)
4. Chen, E., Cao, H., Wang, K., Jafar, S., He, Q.: Technological updating decision-making model for EcoFactory through dynamic programming. Adv. Green Energy Sys. Smart Grid **21**, 129–138 (2018)
5. Bryson, A.E.: Applied Optimal Control: Optimization, Estimation, and Control. Routledge (2018)
6. Ashpazzadeh, E., Lakestani, M., Razzaghi, M.: Nonlinear constrained optimal control problems and cardinal hermite interpolant multiscaling functions. Asian J. Control. **20**, 558–567 (2018)
7. Razmjooy, N., Ramezani, M.: Robust optimal control of two-wheeled self-balancing robot using Chebyshev inclusion method. Majlesi J. Elec. Eng. **12**, 13–21 (2018)
8. El-Gindy, T., El-Hawary, H., Salim, M., El-Kady, M.: A Chebyshev approximation for solving optimal control problems. CAMWA **29**, 35–45 (1995)

9. Razmjooy, N., Ramezani, M.: An improved quantum evolutionary algorithm based on invasive weed optimization. Indian J. Sci. Res. **4**, 413–422 (2014)
10. Razmjooy, N., Khalilpour, M., Ramezani M.: A new meta-heuristic optimization algorithm inspired by FIFA World Cup competitions: theory and its application in PID designing for AVR system. J. Control, Autom, Electr. Sys. 1–22 (2016)
11. Razmjooy, N., Ramezani, M., Nazari, E.: Using LQG/LTR optimal control method for car suspension system. SCRO Res. Annu. Rep. **3**, 1–8 (2015)
12. Hosseini, H., Tousi, B., Razmjooy, N.: Application of fuzzy subtractive clustering for optimal transient performance of automatic generation control in restructured power system. J. Intel. Fuzzy Sys. **26**(3), 1155–1166 (2014)
13. Yang, J., Zhang, C., Li, S., Niu, X.: Semi-global exquisite disturbance attenuation control for perturbed uncertain nonlinear systems. Asian J. Control. **19**(4), 1608–1619 (2017)
14. Chaouech, L., Soltani, M., Dhahri, S., Chaari, A.: Design of an optimal fuzzy sliding mode control using scalar sign function. Asian J. Control. **19**, 1019–1033 (2017)
15. Hayes, N.T.: Introduction to modal intervals. Available at grouper (1788). Ieee.org/groups/1788/Material/HayesModalInterval.pdf
16. Yuta, O., Kashima, K., Ohta, Y.: Iterative path integral approach to nonlinear stochastic optimal control under compound poisson noise. Asian j. Control. **19**, 781–786 (2017)
17. Alipouri, Y., Poshtan, J.: Robust minimum variance lower bound estimation by uncertainty modeling using interval Type2 fuzzy set. Asian J. Control. **19**, 47–56 (2017)
18. Olegovna, O.V.: Interval optimal control problem in a Hilbert space. Comp. Math. Math. Phys. **53**, 389–395 (2013)
19. Leal, U.A.S., Silva, G.N., Lodwick. W. A.: Necessary condition for optimal control problem with the interval-valued objective function. Proc. Ser. Braz. Soc. Comput. Appl. Math. **3** (2015)
20. Rivlin T.J.: An Introduction to the Approximation of functions: Courier Corporation (2003)
21. Salem, N.B., Trimeche, K.: Mehler integral transforms associated with Jacobi functions with respect to the dual variable. J. Math. Anal. Apps. **214**, 691–720 (1997)
22. Jaulin, L.: Applied Interval Analysis: With Examples in Parameter and state estimation. Robust Cont. and Robotics. 1: Springer Science Business Media (2001)
23. Lai, C.-K., Lone, M., Thomas, P.: Whidborne J., Cooke A.: On-board trajectory generation for collision avoidance in unmanned aerial vehicles. In: Aerospace Conference, 2011, pp. 1–14. IEEE (2011)
24. Madadi, A., Razmjooy, N., Ramezani, M.: Robust control of power system stabilizer using world cup optimization algorithm. Int. J. Inf. Secur. Syst. Manag. **5**(1), 519–526 (2016)
25. Jesus de, M.A., Estrela, V.V., Saotome, O., Stutz, D.: Super-resolution via particle swarm optimization variants. In: Biologically Rationalized Computing Techniques for Image Processing Applications, pp. 317–337 (2018)

Efficacy Comparison of Satellite Attitude Control Systems Using a Spherical Air Bearing

Rômulo Fernandes da Costa, Osamu Saotome and Elvira Rafikova

Abstract This work presents practical and simulated experiments designed to compare the efficacy of two satellite attitude control methods, using the satellite simulation platform developed by ITA, the MuSat, as a testbed. The first algorithm is based on feedback linearization, and is designed to suppress precession effects caused by the momentum wheels. The second algorithm is a controller based on the State Dependent Riccati Equation (SDRE) technique, which aims to optimize convergence iteratively. An attitude maneuver in the three axes of rotation was executed using the MuSat platform, using the two controller algorithms. A numerical simulation of these experiments using a computational model of the MuSat is also presented, providing a baseline for the experiment. Results indicate that while both algorithms perform similarly well under idealized conditions, the SDRE algorithm outperforms the feedback linearization controller, due to its higher precision and robustness characteristics.

Keywords SDRE control · Feedback linearization · Satellite attitude control system · Practical comparison · Spherical air bearing

1 Introduction

The Attitude Control System (ACS) of a satellite is responsible for adjusting its orientation towards a given reference. It must point the satellite with sufficient accuracy to fulfill its mission, such as image acquisition, communications, and guid-

R. F. da Costa (✉) · O. Saotome
Instituto Tecnológico de Aeronáutica (ITA), São José dos Campos/SP, Brazil
e-mail: Elromulo2006@yahoo.com.br

O. Saotome
e-mail: osaotome@gmail.com

E. Rafikova
Universidade Federal do ABC (UFABC), Santo André/SP, Brazil
e-mail: elvira.ufabc@gmail.com

Y. Iano et al. (eds.), *Proceedings of the 4th Brazilian Technology Symposium (BTSym'18)*, Smart Innovation, Systems and Technologies 140,
https://doi.org/10.1007/978-3-030-16053-1_6

ance. The ACS must also be capable of correcting any disturbances caused by the movement of mobile parts, such as actuators or flexible parts.

Given its importance, errors in ACS design may result in several other problems, such as communication issues due to antennas not being accurately pointed, drifting away from a referential trajectory, and preventing the batteries from recharging properly, due to solar panels not being correctly aligned with the sun. These problems can prevent the satellite from completing its missions or even cause the loss of the equipment [1]. This justifies the need for evaluating the attitude control algorithms through simulations in laboratory, either through computational or physical simulators [2].

Several institutions [3, 4] have built their own air-cushion based simulators, in order to evaluate the performance of ACS while running under embedded hardware.

This paper presents a practical and simulated experiment using the satellite simulation framework developed by ITA, the MuSat, comparing the performance of two different satellite attitude control methods. The first algorithm is based on feedback linearization method, and the second algorithm is a controller based on the State Dependent Riccati Equation (SDRE) technique. The maneuver was also numerically simulated using a computational model of the MuSat, providing a baseline for the experiment results.

2 Related Work

Accordingly to the historical review provided by Schwartz [5], the use of air bearings as satellites simulators can be traced back as early as 1960, built by governmental institutions with the purpose of validating technology embedded on real satellites. Air bearing testbeds built by universities tend to be smaller, as those are built for the development of new control algorithms rather than hardware implementation.

Examples include devices built by UNAM, as reported by Prado [3], and at University of Sydney, as reported by Kwan [4]. Both works present disk-shaped air bearings measuring 76 and 30 cm of diameter respectively, built specifically for educational purposes.

This paper continues the works of Silva [1] and by Costa [6] on the MuSat simulator. M.A.C Silva presents the mathematical model for the simulator used in this work for use in a HIL simulator. In 2016, the work presented by R.F Costa validates the controller's design by demonstrating its performance in the MuSat platform in a physical experiment.

3 Musat Framework

3.1 *MuSat Testbed*

The MuSat testbed is a device designed for performing satellite attitude control experiments on laboratory. It consists on an air bearing that maintains an aluminum sphere floating over a very thin layer of air. This creates a nearly frictionless environment in which the sphere is free to rotate in any axis, and is only limited by the edges of the bearing [1, 6].

Inside the sphere there are three orthogonally orientated momentum wheels, which act as actuators for attitude control. Torque over the sphere can be applied by varying the wheels speed, which causes the sphere to rotate over the bearing. Figure 1 is a photograph taken of the sphere over the air bearing.

A single Arduino Mega Board is used to manage the three actuators and the IMU (Inertial measurement unit) contained atop the lid of the sphere. It also receives commands from an external computer, which hosts the control algorithm.

The algorithm executed in external computer is implemented in MATLAB and Simulink, and is comprised of three sequential modules, being the first responsible for data reception, the second module contains the control laws, and the last module manages data transmission. This segmented architecture makes it easier for testing new controller schemes [6].

3.2 *Computational Model*

The sphere dynamics can be described by Eq. 1:

$$\dot{\omega} = \boldsymbol{I}^{-1}\left[-(\omega \times \boldsymbol{I}\omega) - \boldsymbol{J}\dot{\Omega} - (\omega \times \boldsymbol{J}\Omega)\right] \tag{1}$$

Fig. 1 The MuSat testbed

where $\boldsymbol{\omega} = [\omega x\ \omega y\ \omega z]^T$ is the angular velocity vector, $\boldsymbol{I}$ is the matrix containing moments of inertia of the platform, J is the moment of inertia of each momentum wheel and $\boldsymbol{\Omega} = [\Omega x\ \Omega y\ \Omega z]^T$ is a vector containing the spin velocity of each wheel. This equation describes a summation of all torques acting on the sphere, not accounting torques caused by air friction or gravity's influence.

In order to represent attitude using Euler angles as a function of angular rates, the following conversion matrix must be applied.

$$\begin{bmatrix} \dot{\phi} \\ \dot{\theta} \\ \dot{\psi} \end{bmatrix} = \begin{bmatrix} 1 & sin\,\phi\, tan\,\theta & cos\,\phi\, tan\,\theta \\ 0 & cos\,\phi & -sin\,\phi \\ 0 & \frac{sin\,\phi}{cos\theta} & \frac{cos\,\phi}{cos\,\theta} \end{bmatrix} \begin{bmatrix} \omega_x \\ \omega_y \\ \omega_z \end{bmatrix} \tag{2}$$

In which $\phi\ \theta\ \psi$ are the Euler angles: roll, pitch and yaw respectively.

From Eqs. 1 and 2, it was possible to develop a computational model of the sphere using Simulink, which will allow for testing the controller before its implementation on the physical simulator.

4 Control Algorithms

4.1 Feedback Linearization Algorithm

Feedback Linearization is a popular design concept in nonlinear control. The objective of a controller employing this design is to convert a nonlinear system into a linear system by generating an appropriate input, and then stabilizing the dynamics with tools developed for linear systems.

While it is a simple and often used design, it has some drawbacks, as it may require large inputs from actuators and can be strongly influenced by uncertainties and unmodeled dynamics [7].

A controller based on feedback linearization for the MuSat was designed in [1] as defined in Eq. 3. It consists in a non linear controller that calculates angular acceleration for each momentum wheel based on the sum of two components, $\boldsymbol{a}$ (defined in Eq. 4) and $\boldsymbol{b}$ (defined in Eq. 5).

$$\dot{\boldsymbol{\Omega}}_{\mathbf{ref}} = \mathbf{a} + \mathbf{b} \tag{3}$$

$$\mathbf{a} = \frac{1}{\mathrm{J}}\left[\boldsymbol{K}_r\boldsymbol{\omega} + \boldsymbol{K}_p(\boldsymbol{\Theta} - \boldsymbol{\Theta}_{\mathbf{ref}})\right] \tag{4}$$

$$\mathbf{b} = \frac{1}{\mathrm{J}}[-(\boldsymbol{\omega} \times \boldsymbol{I}\boldsymbol{\omega}) - (\boldsymbol{\omega} \times \boldsymbol{J}\boldsymbol{\Omega})] \tag{5}$$

Component $\boldsymbol{b}$ is non linear and it is designed to suppress the nonlinear effects caused by coupling and precession seen on Eq. 1. Component $\boldsymbol{a}$ is linear, being designed to impose a damped response on the system. The constants K_r and K_p define the torque applied to the system proportionally to angular rate and angular error respectively.

4.2 State Dependent Riccati Equation Algorithm

A new control algorithm was designed for the MuSat testbed, based on the State Dependent Riccati Equation (SDRE) technique. This control technique is considered to be a counterpart of the Linear Quadratic Regulator (LQR) technique for non-linear systems, as it consists in finding an input $\boldsymbol{u}$ that minimizes the cost function $\boldsymbol{J}$, defined on Eq. 6, for any given iteration of the algorithm.

$$J(u) = \frac{1}{2}\int_{0}^{\infty}\left[\boldsymbol{x}^T \boldsymbol{Q}(\boldsymbol{x})\boldsymbol{x} + \boldsymbol{u}^T \boldsymbol{R}(\boldsymbol{x})\boldsymbol{u}\right]dt \tag{6}$$

where $\boldsymbol{x}$ is a state vector of a nonlinear system given by:

$$\dot{\boldsymbol{x}} = \boldsymbol{A}(\boldsymbol{x})\boldsymbol{x} + \boldsymbol{B}(\boldsymbol{x})\boldsymbol{u} \tag{7}$$

In the SDRE algorithm, all elements of matrices $\boldsymbol{A}(\boldsymbol{x})$ and $\boldsymbol{B}(\boldsymbol{x})$ in Eq. 7 are numerically calculated, and the system is then treated as linear. Then, the optimal input $\boldsymbol{u}$, which minimizes Eq. 6, has the form of Eq. 8:

$$\boldsymbol{u} = -\boldsymbol{R}^{-1}(\boldsymbol{x})\boldsymbol{B}^T(\boldsymbol{x})\boldsymbol{P}(\boldsymbol{x})\boldsymbol{x} \tag{8}$$

For obtaining $\boldsymbol{u}$, it is necessary to solve the Algebraic State Dependent Riccati Equation (ARE), shown in Eq. 9 [8].

$$\boldsymbol{P}(\boldsymbol{x})\boldsymbol{A}(\boldsymbol{x}) + \boldsymbol{A}^T(\boldsymbol{x})\boldsymbol{P}(\boldsymbol{x}) - \boldsymbol{P}(\boldsymbol{x})\boldsymbol{B}(\boldsymbol{x})\boldsymbol{R}^{-1}(\boldsymbol{x})\boldsymbol{B}^T(\boldsymbol{x})\boldsymbol{P}(\boldsymbol{x}) + \boldsymbol{Q}(\boldsymbol{x}) = 0 \tag{9}$$

By solving Eq. 9 numerically, one finds P(x).

Replacing P(x) in Eq. 8 yields the optimal input $\boldsymbol{u}$ for that single iteration of the algorithm, and by repeating the process at a high enough rate, the system will converge towards its origin in a suboptimal trajectory.

Given that this method requires continuously recalculating an optimized controller at a high rate, it requires a significantly larger amount of computational power than other control designs. It is estimated in [9] that at least a 10 MFLOPS processor is required to be able to execute a 5 state SDRE controller at a rate of 1 kHz.

For implementing the SDRE algorithm the MuSat, a vector $\boldsymbol{x}_{\boldsymbol{sphere}} = \left[\phi\,\theta\,\psi\,\omega_x\omega_y\omega_z\right]^T$ containing all the system variables was considered. The derivatives of the each one of the variables can be obtained directly from Eqs. 1 and 2.

By defining a second system as $\boldsymbol{x_{ref}} = \left[\phi_{ref}\theta_{ref}\psi_{ref}\omega_{xref}\omega_{yref}\omega_{zref}\right]^T$, acting as referential state and modeled in the same way as $\boldsymbol{x_{sphere}}$, one can define the error between the system and reference as $e = x_{sphere} - x_{ref}$, and use e as the state vector for the SDRE algorithm.

The derivative of the e vector can then be parameterized in space-state representation, as seen in Eq. 10.

$$\dot{e} = \dot{x_{sphere}} - \dot{x_{ref}} = \boldsymbol{A}(\boldsymbol{e})\begin{bmatrix} e_\phi \\ e_\theta \\ e_\psi \\ e_{\omega_x} \\ e_{\omega_y} \\ e_{\omega_z} \end{bmatrix} + \boldsymbol{B}(\boldsymbol{e})\begin{bmatrix} u_x \\ u_y \\ u_z \end{bmatrix} \tag{10}$$

where matrixes $\boldsymbol{A}$ and $\boldsymbol{B}$ are defined as:

$$\boldsymbol{A}(\boldsymbol{e}) = \begin{bmatrix} 0\,0\,0 & 1 & \mathrm{k}_{1e_{\omega y}} & \mathrm{k}_{1e_{\omega z}} \\ 0\,0\,0 & 0 & \mathrm{k}_{2e_{\omega y}} & \mathrm{k}_{2e_{\omega z}} \\ 0\,0\,0 & 0 & \mathrm{k}_{3e_{\omega y}} & \mathrm{k}_{3e_{\omega z}} \\ 0\,0\,0 & 0 & \mathrm{k}_{4e_{\omega y}} & \mathrm{k}_{4e_{\omega z}} \\ 0\,0\,0 & \mathrm{k}_{5e_{\omega x}} & 0 & \mathrm{k}_{5e_{\omega z}} \\ 0\,0\,0 & \mathrm{k}_{6e_{\omega x}} & \mathrm{k}_{6e_{\omega y}} & 0 \end{bmatrix} \tag{11}$$

$$\boldsymbol{B}(\boldsymbol{e}) = \begin{bmatrix} & 0\,0\,0 & \\ & 0\,0\,0 & \\ & 0\,0\,0 & \\ \mathrm{k}_{4u_{\Omega x}} & 0 & 0 \\ 0 & \mathrm{k}_{5u_{\Omega y}} & 0 \\ 0 & 0 & \mathrm{k}_{6u_{\Omega z}} \end{bmatrix} \tag{12}$$

Each coefficient in **A** and **B** is dependent on the current state of the system, obtained from rearranging the error equations derived from Eq. 10.

The matrices Q and R were defined as:

$$\boldsymbol{Q} = diag\left(k_{Q_{att}}, k_{Q_{att}}, k_{Q_{att}}, k_{Q_w}, k_{Q_w}, k_{Q_w}\right); \tag{13}$$

$$\boldsymbol{R} = diag\left(k_{R_w}, k_{R_w}, k_{R\_w}\right) \tag{14}$$

where k_{Q_att}, k_{Q_w} and k_{R_w} are constants that define the penalties associated of the attitude error, angular velocity error and control effort in Eq. 8.

From Eqs. 10–12, it was possible to create a SDRE controller in Simulink. Elements from the ***A(e)*** matrix are recalculated at each iteration of the control algorithm. Signal $\boldsymbol{u}$ can be calculated directly from the MATLAB function "lqr", which solves the ARE and calculates the product $\boldsymbol{R}^{-1}(\boldsymbol{x})\boldsymbol{B}^{T}(\boldsymbol{x})\boldsymbol{P}(\boldsymbol{x})$ seen in Eq. 8, after all elements in $\boldsymbol{A}$ and $\boldsymbol{B}$ have been calculated.

5 Results

5.1 Experiment Description

A test experiment was performed for both algorithms, using the MuSat platform and its computational model. These experiments consist in rotating the sphere from a leveled attitude to a new attitude vector defined by $\boldsymbol{\Theta}_{\mathbf{ref}} = \left[\phi_{ref}\ \theta_{ref}\ \psi_{ref}\right] = [5°, 5°, 5°]$, while under control of one of the algorithms. All momentum wheels had their initial velocity set to 2100 RPM. The algorithms were activated for a period of 60 s.

The values of the parameters used on the simulations and controllers are listed in Table 1.

5.2 Simulation Results

Figure 2 shows the response obtained by simulation for system attitude to the left and momentum wheel speed to the right.

Table 1 System parameters

Parameter	Value	Units
$\boldsymbol{I}$	$\begin{bmatrix} 0.1358 & -0.0000 & -0.0002 \\ -0.0000 & 0.1359 & 0.0002 \\ -0.0002 & 0.0002 & 0.1227 \end{bmatrix}$	Kg m^2
J	6.2757e−04	Kg m^2
$\boldsymbol{K}_r$	$\left[0.2174\ 0.2175\ 0.1964\right]$	
$\boldsymbol{K}_p$	$\left[0.0149\ 0.0150\ 0.0135\right]$	
k_{Q_att}	14.4	
k_{Q_w}	600	
k_{R_w}	0.0206	

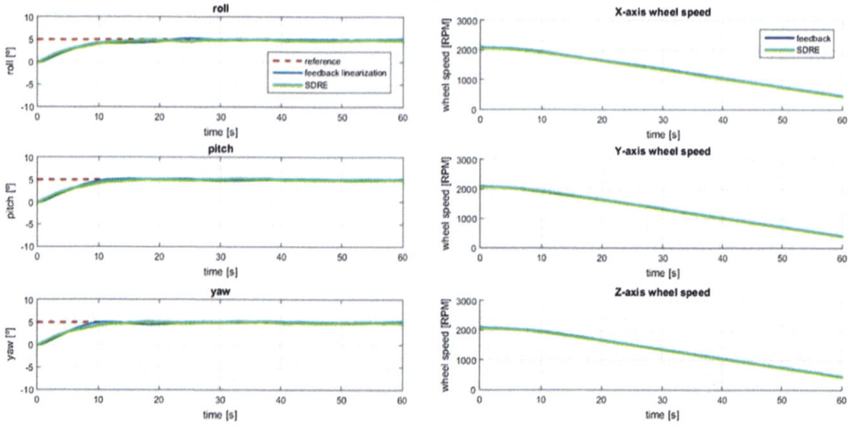

Fig. 2 Attitude and wheel response obtained under simulation

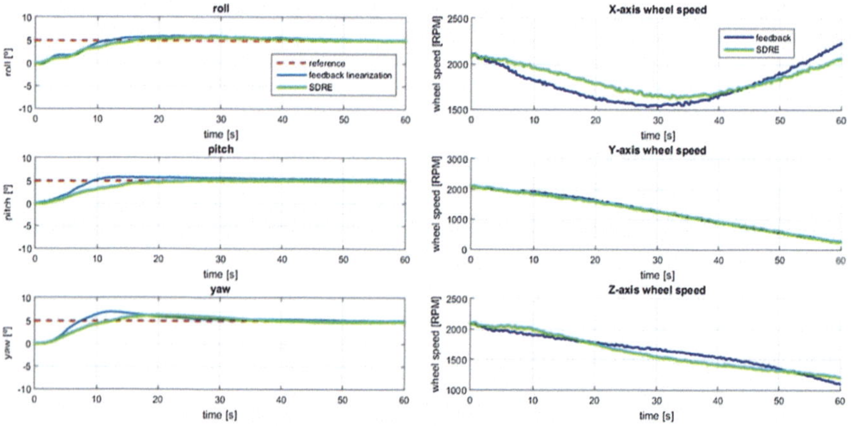

Fig. 3 Attitude and wheel response obtained using the MuSat testbed

5.3 *Experimental Results*

Figure 3 shows the attitude response of the sphere in the experiment to the left, and momentum wheel speed to the right.

5.4 *Results Discussion*

While both algorithms had similar performances in simulation, it can be better seen on the experimental results that the SDRE algorithm is more precise than the feedback linearization algorithm, as in both cases the SDRE presented less oscillation and overshoot. This is due to disturbances in the testbed that cannot be completely removed from the environment, such as gravity induced torque and air friction. Feedback linearization controllers tend to be quite susceptible to disturbances, as opposed to SDRE-based controllers, which are significantly more robust due to its design.

6 Conclusion

The SDRE controller presented in this work outperformed the feedback linearization controller in both cases, as its response is slightly more precise, with lower overshoot and oscillation frequency, which demonstrates this control technique's efficacy.

However, the constant recalculation of the ARE in the SDRE algorithm requires a larger amount of computational operations, which makes it harder to implement on simpler embedded systems. As the feedback linearization method required fewer computational operations, it can be embedded more easily.

Future work can further improve SDRE controller design by using offline trained neural-optimal controllers, in order to reduce the computational cost associated with these controllers. Furthermore, matrices $\boldsymbol{Q}$ and $\boldsymbol{R}$ can be designed as state dependent as well, in order to force the controller to meet certain requirements and constraints, such as maximum control effort and settling time.

Acknowledgements We would like to acknowledge the financial support provided by CAPES during the development of the new control algorithm, and the support from CNPq and Equatorial Sistemas during the development of the MuSat platform.

References

1. Silva, M.dA.Ce., de Figueiredo, H.V., Boglietti, B.G.N., Saotome, O., Villani, E., Kienitz, K.H.: A framework for development of satellite attitude control algorithms. J. Control, Autom. Electr. Syst. **25**(6), 657–667 (2014)
2. Yu, Y., Yang, Z.: A semi-physical simulation platform of attitude determination and control system for satellite. Adv. Mech. Eng. **8**(5), 1–11 (2016)
3. Prado, J., Bisiacchi, G., Reyes, L., Vicente, E., Contreras, F., Mesinas, M.: Three-axis air-bearing based platform for small satellite attitude determination and control simulation. Revistas Unam **3**(3), 222–237 (2005)
4. Kwan, T.H., Lee, K.M.B., Yan, J., Wu, X.: An air bearing table for satellite attitude control simulation. In: 2015 IEEE 10th Conference on Industrial Electronics and Applications (ICIEA), pp. 1420–1425 (2015)
5. Schwartz, J.L., Peck, M.A., Hall, C.D.: Historical review of air-bearing spacecraft simulators. J. Guid. Control Dyn. **26**(4), 513–522 (2003)
6. Costa, R.F., Saotome, O.: Satellite attitude control system validation in an air beared sphere. In: 2016 Brazilian Technology Symposium, pp. 1–3, Campinas, Brazil (2016)

7. Erdem, E.: Analysis and Real-Time Implementation of State-Dependent Riccati Equation Controlled Systems. University of Illinois at Urbana-Champaign, Urbana, IL (2001)
8. Cloutier J.: State-dependent Riccati equation techniques: an overview. In: Proceedings of the 1997 American Control Conference (Cat. No. 97CH36041), vol. 2, pp 932–936 (1997)
9. Menon, P.K., Lam, T., Crawford, L.S., Cheng, V.H.: Real-time computational methods for SDRE nonlinear control of missiles. Proc. Am. Control Conf. **1**, 232–237 (2002)

A Comparative Study Between Methodologies Based on the Hough Transform and Watershed Transform on the Blood Cell Count

Ana Carolina Borges Monteiro, Yuzo Iano, Reinaldo Padilha França, Rangel Arthur and Vânia Vieira Estrela

Abstract It is increasingly common to use engineering techniques in the areas of health, in order to solve simple problems or even create new diagnostic methods. In the last decade, the Hough Transform has been widely used as a tool for segmentation of blood smear images for the purpose of counting blood cells. However, it is noted that the Watershed transform has been applied to perform the same function. Based on this, a methodology based on the Hough Transform was created, aiming to perform the detection and counting of erythrocytes and leukocytes and verify the applicability of the methodology when compared to others. The study was conducted based on the determination of accuracy and simulations performed on different hardware platforms and subsequent comparison with the WT-MO methodology. The results demonstrated that both methodologies are able to perform the task of detection and counting of blood cells in digital images of blood smear. However, the methodology based on the Watershed Transform best meets the criteria of speed and reliability (counts), which are indispensable to medical laboratory routine.

Keywords Image segmentation · Comparative study · Watershed transform · Hough transform · Blood cells

A. C. B. Monteiro (✉) · Y. Iano · R. P. França · R. Arthur
School of Electrical and Computer Engineering (FEEC), University of Campinas, UNICAMP, Av. Albert Einstein- 400, Barão Geraldo, Campinas, SP, Brazil
e-mail: monteiro@decom.fee.unicamp.com

Y. Iano
e-mail: yuzo@decom.fee.unicamp.com

R. P. França
e-mail: padilha@decom.fee.unicamp.com

R. Arthur
e-mail: rangel@ft.unicamp.br

V. Vieira Estrela
Department of Telecommunications, Fluminense Federal University (UFF), Rio de Janeiro, Brazil
e-mail: vania.estrela.phd@ieee.org

Y. Iano et al. (eds.), *Proceedings of the 4th Brazilian Technology Symposium (BTSym'18)*, Smart Innovation, Systems and Technologies 140,
https://doi.org/10.1007/978-3-030-16053-1_7

1 Introduction

The human blood consists of approximately 45% blood cells. These cells are responsible for homeostasis of the organism, through transport gases, defense of the organism against distinct etiological agents, tissue repair and blood coagulation [1, 2].

Through observations in optical microscopy, it was possible to visualize biconvex disc-shaped structures without cell nucleus. These cells were called erythrocytes. Its main function is the transport of gases by the organism [3]. Other structures can also be observed, however, they have a larger size, and the presence of segmented nuclei in lobules, and may or may not be covered by cytoplasmic granules. Such structures are called leukocytes and are responsible for the body's defense against external agents.

Over time, the study of blood cells that was restricted to just the observation of cell morphology began to use specific methods. This new methodology responsible for the analysis of blood cells was called hemogram [4].

The hemogram is a laboratory exam for the quantification and qualification of blood cells. In this exam, the erythrocytes level should be between 4.0 and 6.0 $\times 10^6/mm^3$ of blood. Values lower or higher than this parameter is indicative of genetic anemia or iron deficiency anemia, leukemia or polycythemia [1, 4, 5].

The automation of the hemogram results in greater agility in performing the exams and in the release of reports. However, it is a more expensive methodology because the automation is based on impedance and flow cytometry methods for the differentiation and counting of blood cells [4, 6].

However, before acquiring hematological equipment it is necessary to take into consideration the following parameters: automation equipment versus the type of patient attended; demand for daily exams, cost of each hemogram, interfacing, and training of employees [4–6].

The high cost of the methodologies used for the confection of hematological equipment implies a high cost of the final product [4, 6]. This high-cost results in the exclusion of the right to health of many people from underdeveloped and developing countries [7].

Often new methodologies are created in the area of engineering, being able to solve problems in other areas of knowledge [7]. The image segmentation is used as a step in general object recognition, being useful in many applications for identifying regions of interest. This process consists in the act of fractionating an image into groups of spatially connected pixels homogeneously. In the last decades can be noticed a growing increase in the use of techniques of segmentation of images for the resolution of problems or improvement of techniques already existing in the medical areas.

2 Literature Review

In the year 2000, Soltanzade carried out a study related to the segmentation of cells and cell nuclei using membrane markers. Segmentation of individual cell nuclei from microscope images usually involves labeling of nuclei with a marker of DNA. However, this methodology often presents flaws when applied to clusters of nuclei, since the distinction of the nuclear borders becomes difficult to identify in these conditions. Thus, this work presented a methodology that solved this limitation and allowed the segmentation of whole cells [8].

This segmentation was accomplished through the identification of each nucleus and/or cell, and the boundaries were expanded until reaching the cell as a whole. This process relied on the use of staining, using curvature gradients and flow techniques. The algorithm developed was tested on computer-generated objects to evaluate its robustness against noise and subsequent application to cells in culture. In all cases presented, the algorithm obtained considerable precision results [8].

In the year of 2002, Walsh and Raftery, developed a work related to precise and efficient curve detection in images emphasizing the importance of the Hough transform. In this work, we investigated the use of grouping techniques aiming at the simultaneous identification of multiple curves in the image. We also used probabilistic arguments to develop the stopping conditions for the algorithm. A blood smear image containing 26 red cells was used, and 23 cells were counted by the algorithm in a Processing Time greater than 4 min [9].

In 2003, Ji Y. Xie introduced the Hough randomized transform to improve the accuracy and robustness detection curve as well as computational efficiency. Robustness and accuracy improvements were achieved by the analytical propagation of errors with image pixels to estimate curve parameters. The errors with the curve parameters were then used to determine the contribution of pixels in the accumulated arrangement. Computational efficiency was achieved by mapping a set of points near certain selected sites to the space frontier [10].

A year later, Jiaqiang Song, concluded that the Hough transform is recognized as a powerful tool for extracting graphics elements for images due to its global vision and robustness in noise or degradation of the environment. However, the application of the Hough transform has been limited to small size images, since the peak detection and the scan line consumed much more time for large images [11].

In 2005, Baggett and colleagues carried out a study aimed at the segmentation of cells into solid tissues, in order to understand the cellular and molecular structures, which are the basis of tissue development and function. For this, a software was developed that through segmentation of two-dimensional microscopic images found the cell membrane in better state of conservation. The samples were labeled with a fluorescent cell surface marker so that the membranes became brighter than the other cell structures. The developed algorithm requires the user to mark two dots per cell, one in the center and one at the edge, ensuring a segmentation practically 100% correct [12].

In 2006 Wu and colleagues developed their study using Otsu based on the circular histogram for the leukocyte segmentation. Thus, for the detection of blood cells, the detection of the border by this method is weak, because not all the limits are well determined, being difficult to obtain all the border information and to locate the cells with precision [13].

In the year of 2007, Kharma and collaborators realized a work based on the automatic segmentation of cells coming from microscopic images using the ellipse detection. For this, they developed a method of automatic extraction of the cells of microscopic images, being the process divided in 2 steps. The first step used a certain threshold to identify and mark foreground objects, and presented an accuracy >97%. The second step of the method used an ellipse detection algorithm to identify cells quickly and reliably [14].

A year later, Marcin Smereka, modified the Hough transform, proposing an improvement in the detection of low contrast circular objects. The original Hough transform and its numerous modifications were discussed and compared, concluding that both can be improved in the efficiency criteria and computational complexity of the algorithm [15].

In 2009, Siyu Guo stated that the Hough transform has been a method often used to detect lines in images. However, when applying the Hough transform and algorithms derived from real-world images, the method often undergoes considerable degradation in performance. Especially in detection rate because of the large amount of dots provide a complex background or texture [16].

In 2010, Roy A. Diamayuga and colleagues, used a histogram to distinguish the nuclei of the leukocyte cytoplasm from the remaining cells of the images [17]. In turn, Soltanzadeh, through three experiments with blood cells, proposed a technique based on the extraction of morphological characteristics. Thus, through the central mass of each cell found the distance of each pixel at a central point [18].

In the following year, Kareem and colleagues described in their work a new methodology to identify the total number of erythrocytes as well as their location on Giemsa dye stained slides. The method uses the basic knowledge related to the structure and brightness of the components, since the Giemsa dye is able to stain the sample and assist in detecting the location of the cells in the image [19]. In turn, SaVkare and Narote, carried out a study whose focus was the detection of red blood cells infected by the intracellular parasite Plasmodium falciparum, which is the etiologic agent of malaria [20].

Arivu et al. In 2012, performed the counting of red blood cells and leukocytes by differentiating the distinct morphological characteristics that these cells present with each other. The red cell count was performed through the following steps: conversion to the grayscale; segmentation; noise removal; removal of edge elements and counting by algorithm. In turn, the leukocyte count came from other steps such as image conversion into binary image; noise reduction; space filling and removal of cells with poorly defined edges. The result shown demonstrated 75–80% accuracy in cell counts [21].

Mohammed in 2013 carried out a study of 140 images with blood cells related to Acute Lymphocytic Leukemia. For this, an algorithm was used to convert the image into RBG for grayscale. Then, the cell nuclei were segmented using the Otsu's approach, which is a method of grayscale delimitation. In this way, the nuclei were detected by means of their darker coloration when compared to the surrounding components. The results showed 99.85% accuracy in the detection of a specific type of leukocyte called lymphocytes [22].

Mogra et al. [23] and Kaur et al. [24], in the year of 2014, performed comparative studies between K-Means method and Hough transform in the extraction and counting of red blood cells and leukocytes. The K-Means algorithm is a simple technique that can be used for cluster observations within clusters of observations reported without any knowledge of their relationships. The image processing through K-means methodology consists of: image input with agglomerates; histogram equalization; image segmentation; blood cell extraction and counting.

One year later, Nasreen and colleagues carried out a study using a digital microscope interfaced to the computer, where the images of blood cells were obtained in the form of digital images, and the image input in the RBG format. This image was later converted to the grayscale format. Then the medium filtration was used to remove noises from the time of image capture [25].

In addition, image input has been improved for better segmentation. In turn, the image segmentation was performed through gray outlines, through the "graytresh" function in MATLAB. This function uses the Otsu's method, which chooses a limit to minimize the classes entering variance of black and white pixels. The detection of blood cells was performed by the Hough transform, which has the function of finding circles in images. As a result, this methodology was able to count 334 red blood cells [25].

In 2016, Bhagavathi et al. developed a work on an automatic system for the detection and counting of red cells and leukocytes using the Fuzzy Logic methodology. In this study, the RGB image obtained was converted to the grayscale and points were detected using the Fuzzy logic rules. The counting was performed through the centers of the circles, where the radii were determined for each cell. Another point used for the differentiation of blood cells was based on the presence or absence of cell nuclei [26].

One year later, Ghane et al. developed a study of leukocyte segmentation from microscopy images. For this purpose, a new combination of thresholding, modified k-means, and Watershed clustering algorithms was presented. The study has three stages: (1) segmentation of leukocytes from a microscopic image, (2) extraction of nuclei from the cellular image and (3) separation of superimposed cells and nuclei. The results of the evaluation of the proposed method showed that the measures of similarity, precision and sensitivity, respectively, were 92.07, 96.07 and 94.30% for the nucleus segmentation and 92.93, 97.41 and 93.78% to the cell segmentation [27].

In the year 2018, Shahin et al. developed a novel targeting algorithm for leukocytes in blood smear images. For this, it was proposed to measure the similarity between the different color components of the blood smear image. Two segmentation methodologies were proposed: one for the nucleus segmentation and the other for the cytoplasm. The results were obtained through public data sets with different resolutions, evaluating the performance of the system based on qualitative and quantitative measurements. The quantitative results indicate high accuracy rates of the segmentation performance measure. The average results of segmentation performance for different types of white blood cells reach 97.6% [28].

2.1 Transform Hough

The Hough Transform (HT) was introduced by P.V.C. Hough in 1962 in the form of a patent. Its application was based on particle physics, aiming to detect lines and arcs in the photographs obtained in cloud chambers. The Hough transform is classified in the middle range of the image processing hierarchy. This methodology is applied to images previously treated and free of irrelevant details. Thus, this method is dependent on the processes of filtering, boundary and edge detection. The method assigns a logical label to an object that until then existed only as a collection of pixels. Therefore, it can be classified as a segmentation procedure [29, 30].

The idea behind the method is simple: parametric shapes in an image are detected through points accumulated in the parameter space. In this way, if a particular shape is present in the image, the mapping of all its points in the parameter space must be grouped around the parameter values that correspond to that shape. This approach maps distributed and disjoint elements of the image to a localized accumulation point. Such a feature provides benefits and drawbacks to the method. Partially occluded forms are still detected, in evidence of their visible parts, that is, all segments of the same circle contribute to the detection of this circle, regardless of the gaps between them. On the other hand, local information inherent to shape points, such as adjacency, is lost—the endpoints of circular arcs and line segments must be determined in a subsequent step [29, 30].

The computational load of the method increases rapidly with the number of parameters that define the detected form. The number of parameters is distinct according to the form: rows have two parameters, circles have three, and ellipses have five. The Hough method has been applied to all of them, but the ellipse is probably at its maximum practicality limit. Its greatest strength lies in specialized insight, such as manufacturing quality control, aerial photo analysis and data analysis in particle physics [29].

Algorithms developed in Matlab software often have wide applicability as the Hough Transform, which is considered in academic scientific milieu an established methodology. In this way, the present work aims to evaluate the best method of counting and detecting blood cells present in blood smear images.

3 Proposal

For the detection and counting of blood cells, 30 blood smear images containing red blood cells and leukocytes were used. These images were acquired through a camera

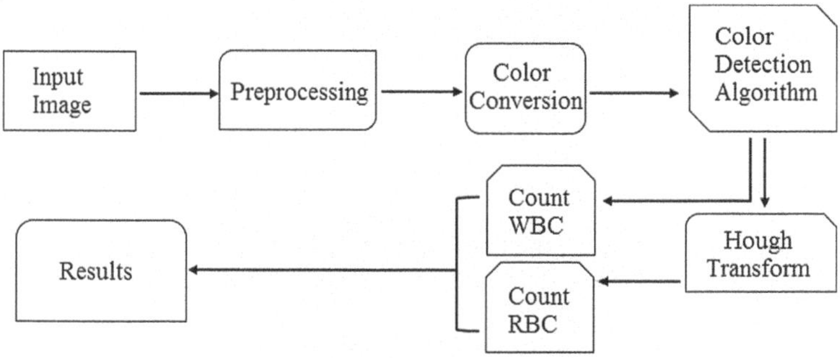

Fig. 1 The logic of the HT-DC methodology

attached to a conventional optical microscope. These images were then used on a physical machine with MATLAB software version (2014a), previously installed. In this simulation environment, a medical algorithm was developed based on the Hough transform and on the detection of objects through coloration. Thus, the developed methodology was named HT-DC (Hough transform and Detection Color). Figure 1 shows the logic used to create the HT-DC methodology.

As it is known that images from fields of optical microscopy are subject to deficiencies in brightness, contrast and even sharpness, the images underwent a preprocessing, aiming at correcting and/or improving such possible failures.

Considering the existence of 3 color planes (RGB) responsible for containing the image color information, the green color component is extracted from the image because it contains the maximum value needed for this type of segmentation. Due to the morphological characteristic of the red cells, the Hough Transform is applied for the detection of circular objects in the images.

The Hough Transform detects edge points in each circle and draws a circle with that point as its origin and radius. It also uses a three-dimensional matrix, the first two dimensions being responsible for representing the coordinates of the matrix, which increases each time the circle is drawn around the rays over each edge point. An accumulator maintains the proper count [31].

This methodology can be applied in a way isolated, or be accompanied by filtering and image enhancement processes before counting circular objects. Thus, the present study aims only to evaluate the Hough Transform without the addition of other components, evaluating only its ability to detect and count red blood cells, marking them with a surrounding circle, as shown in Fig. 2 [31].

Subsequently, the image was submitted to the color detection algorithm. This tool aims at detecting and counting leukocytes, which are detected and counted through their azurophilic staining. Finally, the results of erythrocyte and leukocyte counts are released separately.

Detection by staining was used to avoid erroneous counts of blood cells, since erythrocytes and leukocytes are circular in shape and can easily be counted without distinction by the Hough Transform. The blue staining of the leukocyte nuclei can be

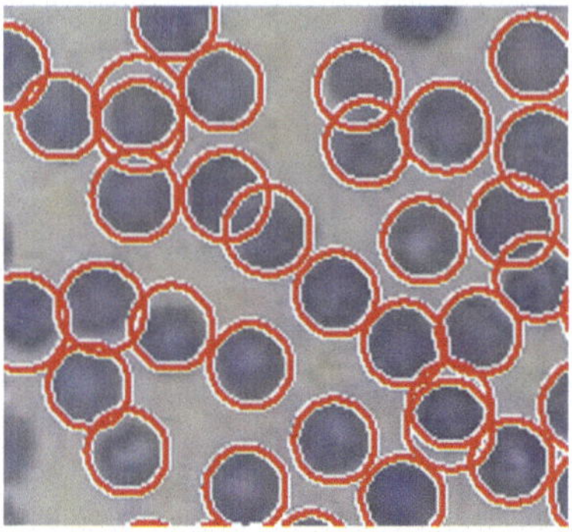

Fig. 2 Detection and counting of erythrocytes through the HT-DC methodology

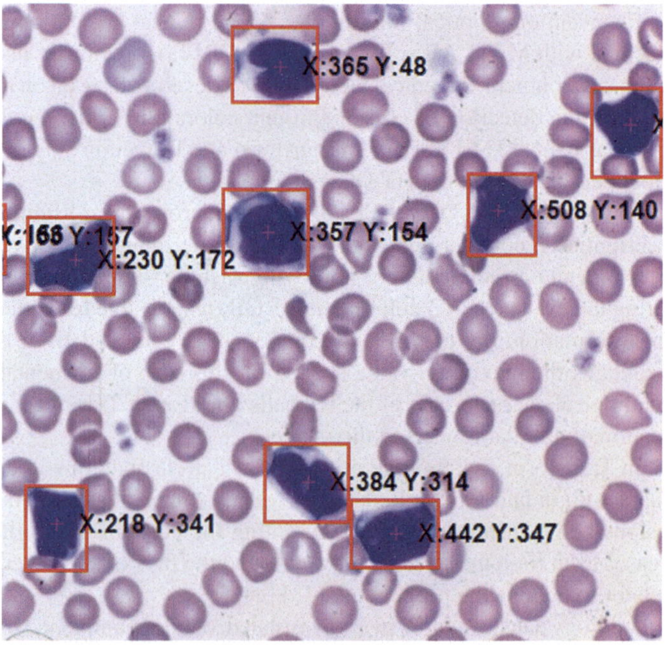

Fig. 3 Detection and counting of leukocytes by the HT-DC methodology

easily separated from the image by means of the RGB staining separation process, preventing leukocytes from being counted as erythrocytes.

Considering that a digital image is defined by means of the *x*, *y* coordinates, the detection algorithm by coloring marks the leukocytes, which indicates the center of each detected form, as shown in Fig. 3. The detection of objects by staining can be performed through the MATLAB software's Image Processing Toolbox.

4 Results and Discussion

The selected 30 images of blood smear contain cells in healthy state. These images were submitted to manual counts by a health professional and were later sent to the MATLAB software, where through the HT-DC methodology the number of erythrocytes and leukocytes were counted simultaneously and separately. The values counted by the HT-DC methodology and by the health professional were compared to each other, aiming to determine the accuracy of the methodology developed. These results can be observed in Figs. 4 and 5.

The comparison of the traditional methodology of cell counting with the developed methodology demonstrated an accuracy of 73% in the detection of erythrocytes and 60% in the detection of leucocytes. However, when the HT-DC methodology is compared with the WT-MO, presented in the studies [32, 33], these results do not indicate that the applicability of the proposal can be used to the routine of the clinical analysis laboratories without losing the quality and accuracy of the exams.

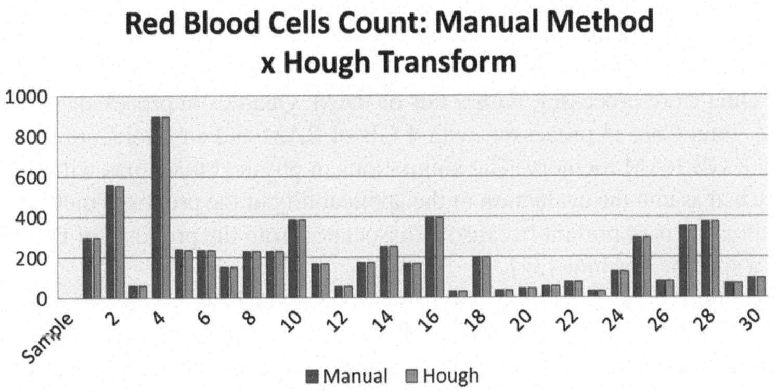

Fig. 4 Comparison of the traditional methodology with image segmentation methodology for the erythrocyte count

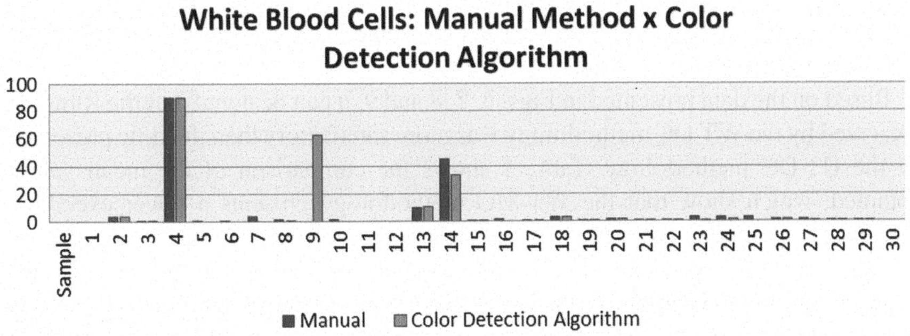

Fig. 5 Comparison of the traditional methodology with image segmentation methodology for the leukocyte count

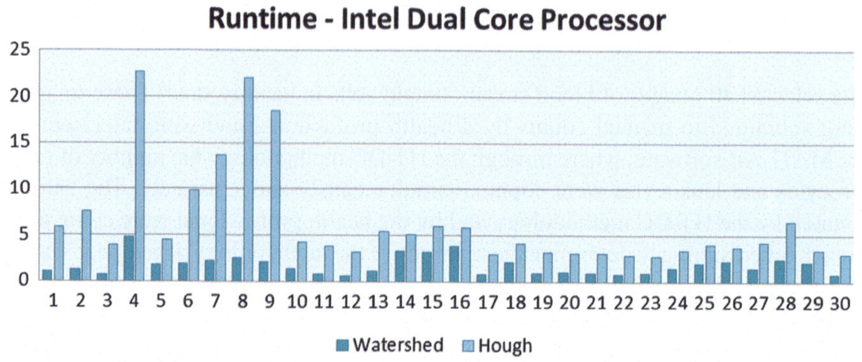

Fig. 6 Runtime (in seconds) of the detection and erythrocyte counting methodology in Intel Dual Core processor

This is because the WT-MO methodology has an accuracy greater than 90% in the determination and counting of erythrocytes and leukocytes.

The developed methodology was also submitted to evaluations of runtime. For this, were made executions in physical machines with different hardware configurations, a Dual Core processor, with 2 GB of RAM, Quad Core processor with 6 GB of RAM, Intel Core i3 processor, with 4 GB of RAM and an Intel Core i5 processor with 8 GB RAM memory. The simulations in physical machines with different software had as aim the evaluation of the applicability of the proposed methodology. This evaluation is important because it dissociates from the proposal of the need to acquire a specific machine [34].

The runtime is an important parameter for the analysis of simulations, being responsible to prove the applicability of the proposed methodology. For this, the "tic toc" command was used via the command line of the MATLAB prompt. This function is responsible to measure the time spent by each of the methodologies during their execution (in seconds) [34].

The results obtained in each of the simulations were compared with previous studies [32, 33], which address a hybrid methodology for the detection and counting of erythrocytes and leukocytes present in blood smear images. The comparisons are presented in the Figs. 6, 7, 8 and 9.

Based on the data presented in Figs. 6, 7, 8 and 9, it can be noted that the Runtime presented by the WT-DC methodology was more satisfactory than the time presented by the HT-DC methodology. Table 1 shows the comparison of the mean values obtained, which show that the WT-MO methodology presents a lower execution time for each sample analyzed.

In environments hospital, time is an extremely important variable because the faster the test result is obtained, the faster the patient can be referred for treatment or complementary tests. In cases of patients with leukocytosis (number of leukocytes above the reference values), even if the hemogram was performed on a hematological device, it is recommended that these reports be confirmed by performing a

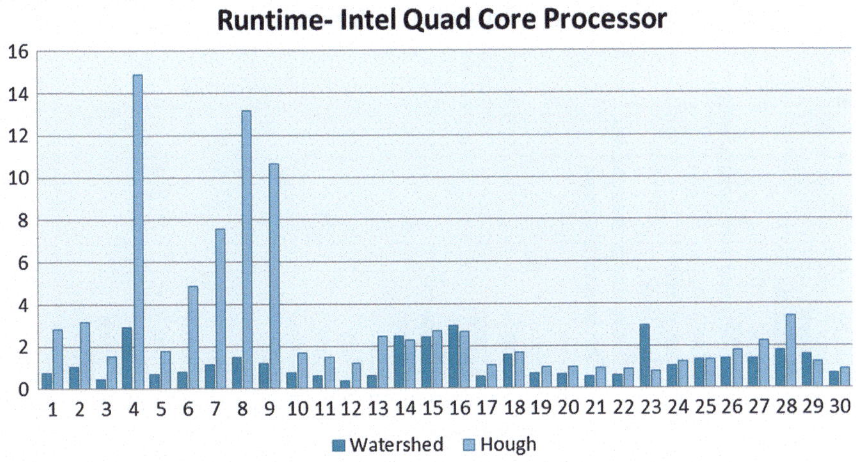

Fig. 7 Runtime (in seconds) of the detection and erythrocyte counting methodology in Intel Quad Core processor

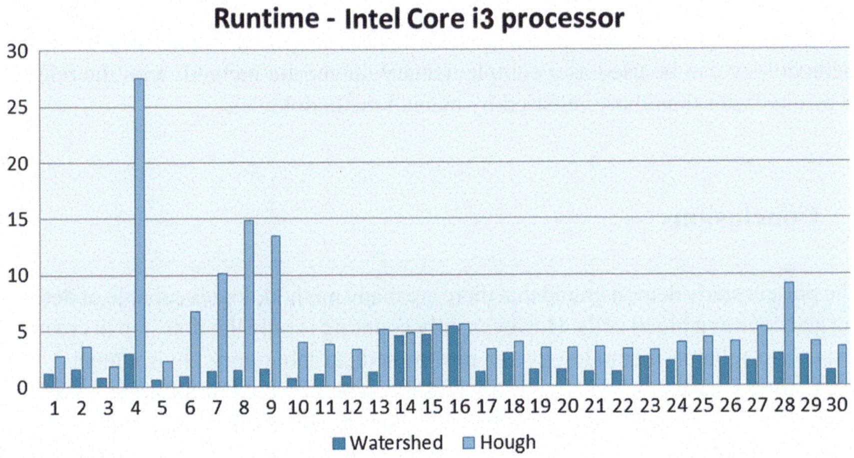

Fig. 8 Runtime (in seconds) of the detection and erythrocyte counting methodology in Intel Core i3 processor

Table 1 The average execution time of the methodologies in physical machines with different processors

Average runtime		
Processor	Watershed	Hough
Dual Core—2 GB of RAM	1.75″	6.41″
Quad Core—6 GB of RAM	1.25″	3.15″
Intel Core—i3–4 GB of RAM	1.44″	3.57″
Intel Core—i5–8 GB of RAM	1.27″	2.81″

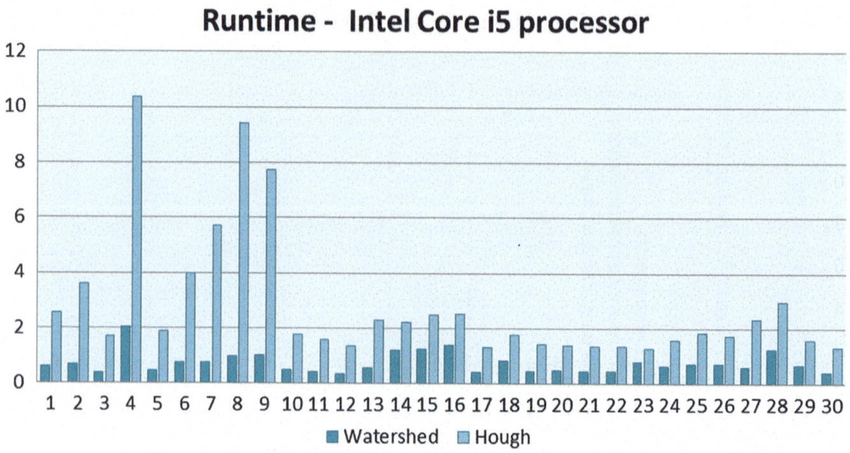

Fig. 9 Runtime (in seconds) of the detection and erythrocyte counting methodology in Intel Core i5 processor

blood smear and the manual counting of the cells. The WT-MO image segmentation methodology can be used as a complementary diagnostic method, with the release of results faster than those released by manual methodologies.

5 Conclusions

The present study demonstrated that there are many methodologies capable of detecting and counting blood cells. However, only counting is only the first step in creating a safe and reliable methodology for use in medical laboratory environments. The methodology based on the Hough transform can be executed in different hardware platforms, however, the time spent for the counting process of images with a larger number of cells becomes great when compared as time spent by the methodology WT-MO.

Thus, more research should be carried out in the area of image processing aiming to create an increasingly accurate methods in the count of detection of leukocytes, so that in future the hemogram is no longer dependent on a specific equipment and has its final cost reduced, as well assisting medical professionals who want to set up their own laboratory and patients who may have a quality examination and low cost.

References

1. Abbas, A.K., Lichtman, A.H., Pillai, S.: Cellular and Molecular Immunology, 7th edn. Elsevier, Amsterdam, The Netherlands (2012)
2. Hoofman, R., Benz, E.J.., Silberstein, L.E., Heslop, H., Weitz, J., Anastasi, J.: Hematology: Basic Principles and Practice, 6th edn. Elsevier, Canada (2013)
3. Abbot Laboratorie de México S.A.: Atlas com Interpretacion Histogramas y Escatergramas, Buenos Aires, Argentina, E.G (2002)
4. Failace, R.: Hemograma. Manual de Interpretação, 5th edn. Artemed, Porto Alegre, Brazil (2009)
5. Ciesla, B.: Hematology in Practice. 2nd edn. Davis Company (2012)
6. Turgeon, M.L.: Clinical Hematology Theory, and Procedures, 4th edn. Lippincott Williams and Wilkins, Philadelphia (2004)
7. Mohammed E.A. et al.: Chronic lymphocytic leukemia cell segmentation from microscopic blood images using a watershed algorithm and optimal thresholding. In: Proceeding 26th IEEE Canadian Conf. of Electrical and Computer Engineering (CCECE) (2013)
8. Soltanzade, R.: Classification of Three Types of Red Blood Cells in Peripheral Blood Smear Based on Morphology. In: Proceedings of ICSP (2010)
9. Walsh, D., Raftery, A.E.: Accurate and e cient curve detection in images: the importance sampling Hough transform. Pergamon J. Pattern Recogn. Soc. (2001)
10. Li, Q., Xie, Y.: Randomised hough transform with error propagation for line and circle detection. In Springer-Verlag London Limited accepted at 21/06/02 (2003)
11. Song, J., Lyu, M.R.: A Hough transform based line recognition method utilizing both parameter space and image space in Department of Computer Science & Engineering, The Chinese University of Hong Kong, Shatin, N.T., Hong Kong SAR, P.R. China (2004)
12. Baggett, D., Nakaya, M., McAuliffe, M., Yamaguchi, T.P., Lockett, S.: Whole cell segmentation in solid tissue sections. Int. Soc. Anal. Cytol. Cytometry. **67A**: 137–143 (2005)
13. Wu, J., Zeng, P., Zhou, Y., Oliver, C.: A novel color image segmentation method and its application to white blood cell image analysis. In: 8th International Conference on Signal Processing (2006)
14. Kharma, N.N., Moghnieh, H., Rouleau, G.A., Cheriet, M.: Automatic segmentation of cells from microscopic imagery using ellipse detection. In: IEE Proceeding Vision, Image & Signal Processing (2007)
15. Smereka, Marcin, Duleba, I.: Circular object detection using a modified hough transform. Int. J. Appl. Math. Comput. Sci. **18**(1), 85–91 (2008). https://doi.org/10.2478/v10006-008-0008-9
16. Guo, S., Pridmore, T., Kong, Y., Zhang, X.: An improved Hough transform voting scheme utilizing surround suppression. Pattern Recogn. Lett. **30**, 1241–1252 (2009)
17. Dimayuga, R.A., Ong, G.T., Perez, R.C.S., Siy, G.O., Langroudi, S.C.S., Miguel O. Gutierrez. "Leukemia Detection Using Digital Image Processing in Matlab". ECE Student Forum, De La Salle Univ, Manila. March 26, 2010
18. Soltanzadeh, R.: Classification of three types of red blood cells in peripheral blood smearbased on morphology. In: Proceedings of ICSP (2010)
19. Kareem, S., Morling, R.C.S., Kale, I.: A novel method to count the red blood cells in thin blood films. In: 978 -1-4244- 9474-3/11©2011 IEEE (2011)
20. Savkare, S.S., Narote, S.P.: Automatic classification of normal and infected blood cells for parasitemia detection. IJCSNS Int. J. Comput. Sci. Netw. Secur. **11**(2) (2011)
21. Arivu, S.K., Sathiya, M.: Analyzing blood cell images to differentiate WBC and counting of linear & non-linear overlapping RBC based on morphological features. Elixir Comp. Sci. Engg. **48**, 9410–9413 (2012)
22. Mohammed, E.A et al.: Chronic lymphocytic leukemia cell segmentation from microscopic blood images using watershed algorithm and optimal thresholding. In: 26th IEEE Canadian Conference Of Electrical And Computer Engineering (CCECE) (2013)

23. Mogra, M., Srivastava, V.: A comprehensive review of analysis of counting blood cells using different image processing algorithms. Int. J. Eng. Sci. Invention ISSN (Online): 2319–6734, ISSN (Print): 2319–6726 (2014)
24. Kaur, R., Kaur, H.: Comparative analysis of white blood cell by different segmentation methods using knowledge based learning. Int. J. Adv. Res. Elect. Electron. Instrum. Eng. (An ISO 3297: 2007 Certified Organization. **3**(9) (2014)
25. Nasreen, N., Kumar, C., Nabeel, A.P.: Counting of RBC using circular hough transform with median filtering. ISBN 978-93-85477-33-1/2015 Bonfring
26. Bhagavathi, S.L., Thomas, N.S.: An automatic system for detecting and counting RBC and WBC using fuzzy logic. J. Eng. Appl. Sci. 2006–2016 Asian Research Publishing Network (ARPN). **11**(11) (2016)
27. Ghane, N., Vard, A., Talebi, A., Nematollahy, P.: Segmentation of white blood cells from microscopic images using a novel combination of K-means clustering and modified watershed algorithm. J. Med. Signals Sensors. **7**(2), 92–101 (2017)
28. Shahin, A.I., Guo, Y, Amin, K.M.: A novel white blood cells segmentation based on adaptive neutrosophic similarity socre. Health Inf. Sci. Syst. **6**(1), 1 (2018)
29. Antolovic, D.: Review of the hough transform method, with an implementation of the fast hough variant for line detection. Sch. Inf. Comput. Eng. Indiana Univ. Bloomington (2008)
30. Alejandro, N.F.: The intersective Hough transform for geophysical application. Geofís. Int. **53**(3), 321–332 (2014)
31. Sahastrabuddhe, A.P, Ajij, S.D (2016) Blood group detection and RBC, WBC counting: an image processing approach. Int. J. Eng. Comput. Sci. **5**(10): 18635–18639, ISSN: 2319-7242
32. Monteiro, A.C.B., Iano, Y., Franca, R.P.: Detecting and counting of blood cells using watershed transform: an improved methodology. In: Iano Y., Arthur R., Saotome O., Vieira Estrela V., Loschi H. (Org.). Proceedings of the 3rd Brazilian Technology Symposium Emerging Trends and Challenges in Technology, 1st edn, vol. 1, pp. 301–310. Springer (2018)
33. Monteiro, A.C.B., Iano, Y., Franca, R.P.: An improved and fast methodology for automatic detecting and counting of red and white blood cells using watershed transform. In: VIII Simpósio de Instrumentação e Imagens Médicas (SIIM)/ VII Simpósio de Processamento de Sinais da UNICAMP, 2017, São Bernardo do Campo. Anais do VIII Simpósio de Instrumentação e Imagens Médicas (SIIM)/VII Simpósio de Processamento de Sinais da UNICAMP. vol. 1 (2017)
34. Gonzalez, R.C., Woods, R.E., Eddins, S.L.: Digital image processing using MATLAB. 2nd edn. Gatesmarki Publishing (2009)

Methodology of High Accuracy, Sensitivity and Specificity in the Counts of Erythrocytes and Leukocytes in Blood Smear Images

Ana Carolina Borges Monteiro, Yuzo Iano, Reinaldo Padilha França and Rangel Arthur

Abstract The evaluation of human blood is an important diagnostic method for the detection of diseases. The analysis of the erythrocytes contained in the blood contributes to the detection of anemia and leukemia, whereas the leukocyte analysis allows the diagnosis of inflammation and/or infections. The blood is analyzed through of the complete blood count test (CBC), which is dependent on automated and/or manual methodologies. The dependence of medical areas on new technologies leads the present study to the goal of developing an image segmentation algorithm that meets the criteria of efficiency and reliability for detection and counting of blood cells. The algorithm was developed through the Matlab software, being the image processing methodology based on the union of the Watershed transform and Morphological Operations, originating the WT-MO methodology. For the simulations, 30 blood smear images containing erythrocytes and leukocytes were used in a non-pathological state. The results showed that the WT-MO methodology presents high sensitivity (99%), specificity (96%) and accuracy (98,3%) when compared with the manual methodology. Therefore, the WT-MO methodology is an accurate, reliable and low-cost technique and can be applied as a third more accessible methodology to perform of the complete blood count test (CBC) in populations of underdeveloped and developing countries.

Keywords Blood cells · Watershed transform · Morphological operations · Blood smear · Image processing · Algorithm · ROC curve

A. C. B. Monteiro (✉) · Y. Iano · R. P. França · R. Arthur
School of Electrical and Computer Engineering (FEEC), University of Campinas – UNICAMP, Av. Albert Einstein – 400, Barão Geraldo, Campinas, SP, Brazil
e-mail: monteiro@decom.fee.unicamp.com

Y. Iano
e-mail: yuzo@decom.fee.unicamp.com

R. P. França
e-mail: padilha@decom.fee.unicamp.com

R. Arthur
e-mail: rangel@ft.unicamp.br

Y. Iano et al. (eds.), *Proceedings of the 4th Brazilian Technology Symposium (BTSym'18)*, Smart Innovation, Systems and Technologies 140,
https://doi.org/10.1007/978-3-030-16053-1_8

1 Introduction

During the embryonic stage, the yolk sac is responsible for the formation of the first blood cells, later during fetal development, this function is transferred to the liver and spleen. During adulthood, the interior of the bone marrow is composed of a hematopoietic tissue, being present in the long bones and in the axial skeleton. Bone marrow presents a suitable microenvironment for the development of hematopoietic cells and proliferation of primitive cells and progenitor cells [1, 2]. Hematopoiesis is the process responsible for the formation of blood cells inside the bone marrow, from a cell mother called stem cell. This pluripotent cell originates from the different blood cell lines: erythrocytes, leukocytes and platelets. This phenomenon is dependent on a series of actions and responses that lead to the processes of duplication, differentiation and maturation, resulting in the production and release of mature cells into the bloodstream [3, 4].

As a result, human blood consists of approximately 55% plasma and 45% blood cells, with their total volume represented by approximately 7% of the individual's body weight. Its function consists in the conduction of gases and nutrients to the tissues, aid in the excretion of metabolites, hormonal distribution, passage of chemical messages between distant organs, regulation and maintenance of body temperature, basic acid balance and osmotic equilibrium [5]. Plasma is an aqueous solution formed by molecules of different molecular weights, being responsible for the transport of substances such as water, plasma proteins, inorganic salts, amino acids, hormones, glucose, albumin, immunoglobulins (IG), components of the blood coagulation cascade, among others. For its part, the blood cells are responsible for the homeostasis of the organism, transport of gases, defense of the organism against distinct etiological agents, tissue repair and blood coagulation. However, such observations could only occur through the invention of instruments capable of visualizing the cells [6].

The erythrocytes are also called red blood cells. They are anucleated, biconcave disc-shaped cells, formed by a tetram (two α chains and two β chains) responsible for the transport of oxygen through four iron molecules [7]. When erythrocytes reach the lungs, oxygen molecules are attached to the iron molecule, generating oxyhemoglobin. This binding is disrupted only when the erythrocytes reach the tissues, where the oxygen pressure is lower. In tissues, erythrocytes deposit oxygen and withdraw carbon dioxide through carbo-hemoglobin binding. Thus, carbon dioxide is either taken directly to the lungs or is dissolved in plasma [3].

Leukocytes are the defense cells of the body, constituting the innate immune response and the adaptive immune response. They are also called white cells, are classified into 2 distinct classes: leukocytes granulocytes and leukocytes agranulocytes. Granulocytic leukocytes are the defense cells that have granules dispersed in the cytoplasm, which have anti-inflammatory and antimicrobial action. In addition to the presence of cytoplasmic granules, these cells present variable numbers of lobes, being called polymorphonuclear cells (neutrophils, eosinophils and basophils). Agranulocyte leukocytes are those that do not have granules visible in optical microscopy, and have only one lobe, being denominated monomorphonuclear. Agranulocyte leukocytes are monocytes and lymphocytes [3, 8].

During hematopoiesis, cells destined to form the granulocytic series (neutrophils, eosinophils and basophils) synthesize proteins and cytoplasmic granules. The primary granules have azurophilic staining and are responsible for the conversion of precursor cells called myeloblasts into promyelocytes. Subsequently, there is the appearance of specific granules, responsible for the progression to myelocytes of the type: neutrophils, eosinophils and basophils. After determining the granulocytic lineage, the following cells are indivisible by mitosis, being characterized by the presence of segmented nucleus, ability of motility, phagocytosis and microbial destruction. Depending on the lineage, mature leukocytes are able to adhere and traverse the wall of the venules, for the purpose of performing tissue defense and reconstruction [3, 9].

Given the importance of blood cells, the analysis morphologically, quantitatively qualitatively, is an important task, since they can indicate anemia, leukemias, virus infections, bacteria or parasites, thromboses, allergies, among other diseases. This analysis is performed through an examination called complete blood count test (CBC), which is a highly requested medical examination in the medical routine, as it provides the diagnosis directly or is indicative of several diseases. This test consists of the erythrogram, leukogram, and platelet, which evaluate the quantity and morphologies of red blood cells, leukocytes and platelets, respectively [10].

2 Problematic

Currently, the complete blood count test (CBC) can be performed through two methodologies: the manual and the automated. The manual methodology is totally dependent on the human performance combined with the use of non-automated equipment. It is a cheaper but more time consuming and less reliable exam because it depends on the counts and calculations performed by health professionals. It is considered a good alternative for small laboratories, where the demand for exams is small and the cost with the acquisition with hematological equipment and reagents does not match the cost-benefit of the process [10].

For this, it is necessary to make a blood smear, by sliding a few microliters of blood on a glass slide and subsequent use of dyes, which allow the visualization of cellular structures. These dyes have the function of staining the nuclear and cytoplasmic structures of blood cells. Only the final portion of the slide is used to perform the counts, since the anterior portions have clustered and/or overlapping cells, preventing a reliable count. This blood smear is used to analyze the morphology and staining of red blood cells, leucocytes and platelets [10, 11].

The automation of the hemogram implies a greater agility in the accomplishment of the exams and in the release of the reports, however, they are a more expensive methodology when compared to the manual methodology. In the 1950s, Coulter Electronic, Inc. introduced the impedance principle for cell counts. The principle of impedance is based on the fact that the electrically conductive cells are diluted in a conducting solution of electricity. This cell suspension is weighed through an orifice with a diameter of about 100 μm, where there is an electric current passing through. This electric current originates from two electrodes: one located on the inner side of the hole and positively charged, and another located on the outer side of the hole, negatively charged. In this way, each time the cell passes through the hole it interrupts the electric current and there is a change in the conductance, consequently, each interruption is counted as a particle [10, 11].

The impedance principle, over the years, was enabled with counters capable of measuring cell volume. Such evolution was the result of the correlation of the proportionality of the magnitude of the interruption of the electric current (pulses) as the cellular volume. Thus, it was observed that small pulses correspond to small volumes, whereas large pulses result from larger volumes. From this correlation between the magnitude of the electric current and the cellular volume, a new concept was created called the threshold concept. The threshold concept is responsible for classifying cells according to their volume, thus allowing the detection of globular volume. The globular volume corresponds to the hematocrit performed in the manual blood count, however, it receives this name because it is performed without the need for microcentrifugation. Both the impedance principle and the threshold concept are responsible for the introduction of multi-parameter devices on the market. These devices are able to perform simultaneous cell counts using separate channels for the counts [10, 11].

In the 1970s, laser light scatters and hydrodynamic fluid techniques were introduced. Both techniques preserve nuclei and granulation of leukocytes, retracting only the cytoplasmic membrane. The techniques are based on the principles of diffraction, refraction and reflection of the light emitted. However, in these techniques, the erythrocytes are undetectable, as a solution the erythrocytes are counted by means of flow cytometry and hydrodynamic focus, where the erythrocytes are counted one by one through an extremely fine capillary. These cells are subjected to a laser beam, where the light scattering is analyzed at different angles of deviation, where at zero degree is indicated the cell size, the ten-degree indication of the internal structure and the 90-degree indication of leukocytes and their characteristics of looseness and granulation content [10, 11].

Over the years, interest in digital imaging methods has increased due to its two main areas of application: improvement of pictorial information for human interpretation and processing of image data for storage, transmission and representation for autonomous machine perception. This way, new technology has been developed in the engineering branch, being of great utility in the medical areas, such as x-ray and tomography. These examinations are based on the capture of images of the patient's body, thus facilitating the diagnostic imaging of various diseases [12]. The aim of the present study was to develop a blood smear segmentation algorithm capable of detecting and counting erythrocytes and leukocytes accurately and with high precision, sensitivity and specificity.

3 Proposal

The experiments were conducted through digital images acquired in hematology databases. The blood smear is made by depositing a few microliters of blood on a glass slide. The fields of the blood smear contained erythrocytes in different sizes, colors, and quantities [13]. It is important to emphasize that only the final portion of the slide was subjected to analysis and image capture, because in this place the cells do not show overlap and/or hyper coloring by the use of dyes. The images were obtained in jpeg, jpg e png format and were transferred to Matlab® software simulation environment, version 8.3 of 64 bits (2014a). In this environment, image segmentation algorithm was developed for counting and detection of erythrocytes. This process was performed through the union of the Watershed Transform image processing techniques and morphological operations, which originated the WT-MO methodology for the detection and counting of erythrocytes and leukocytes. The logic of the algorithm is shown in Fig. 1.

In the image preprocessing and color conversion step, image quality is being improved. This occurs because most of the images of this type can find illumination problems. Thus, it passes through processes of filtering, image enhancement, color conversion and the segmentation process itself for this digital quality of these images [14, 15]. In the process of image segmentation, the morphological technique is the most used, because the mathematical morphology offers a powerful tool for the segmentation of images, being used to describe the format of the region, such as limits, skeleton and texture.

The segmentation process consists in the act of fractionating an image into groups of spatially connected pixels homogeneously [14]. Based on these characteristics the Watershed transform was chosen as an image segmentation tool, responsible for the detection and counts of blood cells present in the digital images of blood smears.

The segmentation of images by the Watershed transform is employed as a process of pixel labeling, where all pixels belonging to the same homogenous region are marked with the same label. The definition of homogeneity of a region of the image

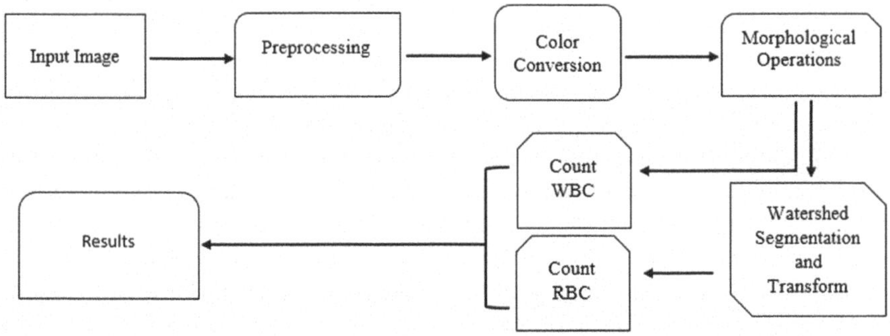

Fig. 1 Diagram demonstrating the logic used in the development of the WT-MO methodology

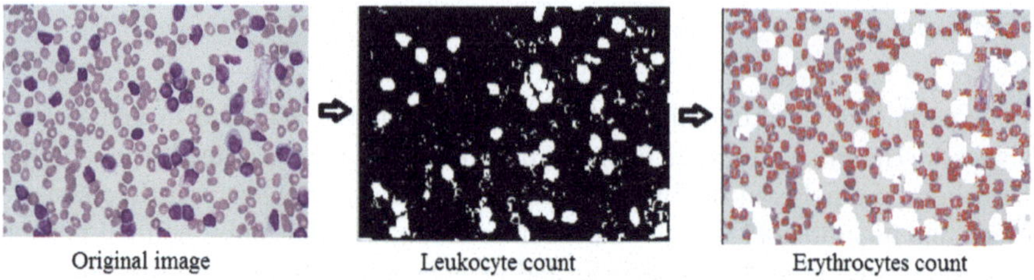

Fig. 2 Detection and counting of erythrocytes and leukocytes simultaneously by the WT-MO methodology

presents a particular concept of the segmentation process, since each image presents its own pattern of pixels [16, 17]. Considering that erythrocytes in a healthy state have color and homogeneous sizes, the Watershed transform presents a good performance in the labeling of sets of pixels representing these blood cells.

For the labeling of the red cells the pixel is assumed as a variable directly related to the morphological characteristics of the cells under analysis. In that case, the labeling should take into account the size and color of the erythrocytes. As the pixels are grouped according to their characteristics, the algorithm counts and labels each cell in the image. The labeling consists of inserting a number on top of each cell and according to the counting order established by the WT-MO methodology, as shown in Fig. 2.

Morphological operations constitute a broad set of image processing operations based on the shape of binary images [18]. Removal of leukocytes from the image is an important process for the detection and counting of blood cells. Considering the morphological similarity of leukocytes and erythrocytes, referring to the rounded form, it is possible that the algorithm performed an erroneous count. Another factor that could cause a failure to count would be the size of the blood cells, since an erythrocyte measures about 8 μm of diameter, while a newly released lymphocyte (leukocyte type) by the bone marrow measures about 10 μm.

Besides the shape and diameter, the amount of cells produced and released by the bone marrow also justifies the choice of Morphological Operations as an image segmentation tool. The erythrocytes are expressed on the order of 10^6, while the leukocytes are expressed in 10^3 [4]. The lower amount of leukocytes facilitates the action of morphological operations in their removal of the digital image. Consequently, this action interferes positively with the results of precision.

Morphological operations are applied to the structural element at the input of an image creating an output of the same size [16]. In this way, the value of the pixels in the image output is based on the comparison of pixels corresponding to the image pixels and adjacent pixels. The number of pixels added or removed from objects in an image depends on the size and format of the structural element used to render the image. The morphological operations were used to segment erythrocytes. This process is characterized in the concepts of size, shape, structure, and connectivity of objects in the image, involving erosion, dilation, opening, closing and reconstruction.

The dilation consists of adding pixels to the edges of objects in an image, aiming to repair breaks in the image. The erosion consists of removing the pixels from the edges of the image, being used to divide objects. The opening is applied to smooth out contours. The closing is used to merge intervals and fill spaces. The reconstruction is responsible for extracting relevant information from the image [18].

Therefore, this entire process involves 2 images and a structural element. The two images: one acts as a starting point for transformation and another act as a mask that restricts the transformation of the image. Finally, morphological reconstruction aims at restoring the original forms of the object that remains after the erosion process [19, 20]. This process is applied in the detection and counting of leukocytes by the WT-MO methodology, as shown in Fig. 2.

4 Results and Discuss

Were selected 30 images of microscopy fields, containing 6453 red cells and images of microscopy fields containing 187 leukocytes. Both fields were submitted to manual counts, with subsequent counting by the algorithm of detection and counting of blood cells. The values are then compared with each other to determine the accuracy of the proposed methodology. To obtain the results of this study, more than 300 simulations were performed.

The digital images were first submitted to a traditional manual counting methodology. Afterward, these images were sent to the Matlab software, where the red cells were detected and counted by the Watershed transform and the leucocytes were quantified through the morphological operations technique. The results obtained were compared with the manual methodologies, as shown in Figs. 3 and 4.

The development of systems, methods or tests involving the detection, diagnosis or prediction of results, presents the need to validate their results in order to quantify their discriminative power and to identify a procedure or method as appropriate or

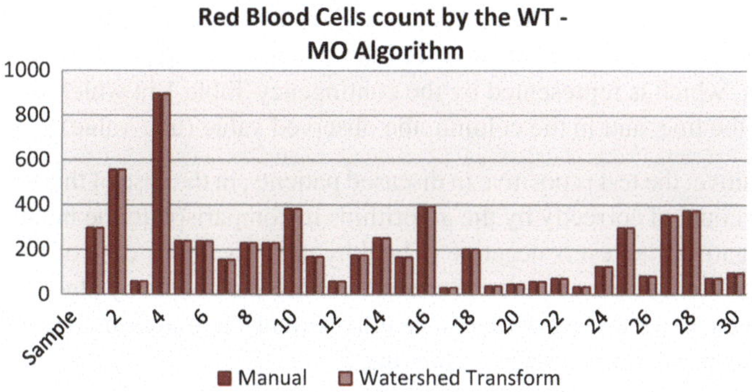

Fig. 3 Comparison of manual counting of erythrocytes x counting by the WT-MO methodology

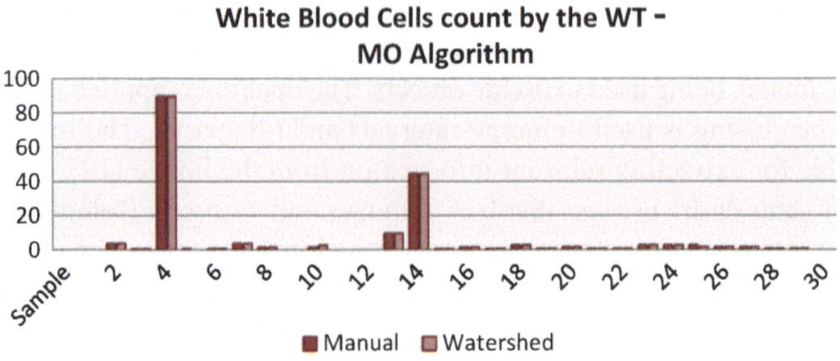

Fig. 4 Comparison of the manual count of leukocytes x count by the WT-MO methodology

Table 1 Matrix of confusion

True value (observed value)		
Predictive value	Y = 1	Y = 0
	TP (True positive)	**FP** (False positive)
	FN (False negative)	**TN** (True negative)

inappropriate for a particular type of analysis. However, the simple quantification of hits in a test group does not adequately reflect the efficiency of a system, since the quantification is dependent on the quality and distribution of the data in this test group [21].

Thus, when the test presents binary response variables (1 if the individual is an event and 0 otherwise), it is necessary to choose a prediction rule that determines where the individual should be framed (cutoff point). In the case of this research, the classification is based on the values obtained during the manual counts of red blood cells and leukocytes. A widely used way to determine the cutoff point is through the Receiver Operating Characteristic Curve (ROC) curve. The ROC curve plots sensitivity versus specificity for all possible cut-off points between 0 and 1 [22, 23].

After determining the cut-off point, it is necessary to evaluate the discriminating power of the model, that is, to discriminate between the events of the non-events. For that, the metrics were created: Accuracy, Sensitivity, Specificity, True Predictive Positive and True Predictive Negative. All these metrics are dependent on the confusion matrix, which is represented by the contingency Table 1 in which the expected value is in the line, and in the column, the observed value (true value) [21–23].

- True Positive: the test is positive in diseased patients, in the case of this research the cells are counted correctly by the algorithms in comparison to the manual counts;
- True Negative: the test is negative in healthy patients, in the case of this research are leukocyte-free images in both manual counts and counts by algorithms;
- False Positive: the test is positive in healthy patients, in this case, they are counted more cells in relation to manual counting;

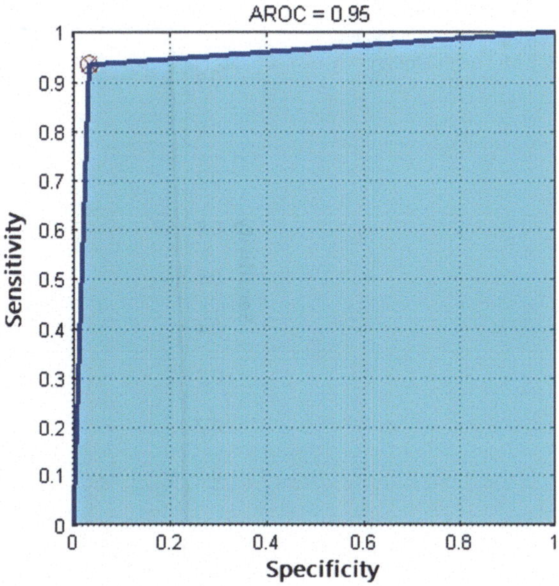

Fig. 5 Roc curve referring to red cell count by the WT-MO methodology

- False Negative: the test is negative in diseased patients, in this case, they are leukocytes counted by the algorithms, and in manual counting these leukocytes are non-existent.

Thus, 2 binary matrices were created: (1) matrix for erythrocytes quantified by the Watershed transform; and (2) matrix for leukocytes quantified by morphological operations. Later, these matrices were transferred to Matlab software, where the values of: Sensitivity, Specificity, AROC (area of the ROC curve), VPP—True Predictive Positive (the probability of an individual evaluated and with a result positive) and VPN—True Negative Predictive—(probability that an evaluated individual with negative result is actually normal). The ROC curve of the Watershed Transform is shown in Figs. 5 and 6, with your parameters showed in the Tables 2 and 3. All values were given by function used in software Matlab.

Thus, it is possible to note that the WT MO methodology can be seen as a tool with high accuracy in erythrocyte detection (95%), high sensitivity (93%) and high specificity (96%). Regarding leukocyte detection, the algorithm also presented highly satisfactory values of accuracy (98,3%), sensitivity (99%) and specificity (96%). The highest accuracy, sensitivity and specificity in white blood cell counts is due to human physiology, as the white blood cells are synthesized by the bone marrow in a smaller quantity when compared to the red blood cells.

When compared these results with the works [24–27], the WT-MO methodology also presents a high accuracy. In this way, the WT-MO methodology developed in this research is a viable alternative for the counting of erythrocytes and leukocytes in laboratories of clinical analysis, since the precision and reliability of the tests are indispensable criteria.

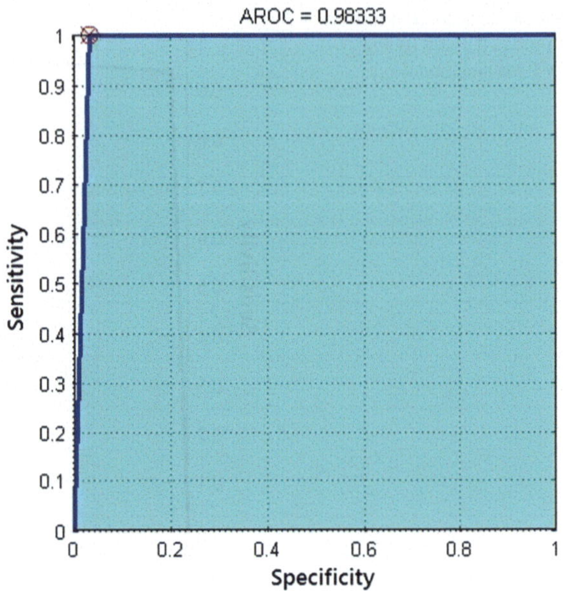

Fig. 6 Roc curve referring to white cell count by the WT-MO methodology

Table 2 Parameter analysis of the ROC curve for erythrocyte counting

Parameters of the ROC curve	
RBC—*Watershed Transform*	
Distance	0.074
Threshold	0.5
Sensitivity	0.93
Specificity	0.96
AROC	0.95
Accuracy	95.00%
PPV	96.60%
NPV	93.54%

Table 3 Parameter analysis of the ROC curve for erythrocyte counting

Parameters of the ROC curve	
WBC—*Morphological Operations*	
Distance	0.033
Threshold	0.5
Sensitivity	0.99
Specificity	0.96
AROC	0.98
Accuracy	98.33%
PPV	96.77%
NPV	100%

5 Conclusions

The confirmation of the diagnosis or even the accomplishment of laboratory tests through algorithms, provides greater reliability of the results to both health professionals and patients, because the algorithms reduce the chances of human failures. Due to its high sensitivity, accuracy and specificity in the simultaneous counting of erythrocytes and leukocytes, the WT-MO methodology can be seen as a future tool of medical laboratory routine, and can be used to confirm altered reports in laboratories with higher purchasing power, or It can be used with a method of substitution of the manual methodology in smaller laboratories, where the demand for examinations per day does not entail the acquisition of a high-cost hematological equipment.

It is important to emphasize that techniques result in the reduction of the cost during the production/supply of products/services, it directly impacts the final value. High values of products and services are important factors for the exclusion or removal of populations from access to health. Thus, the use of blood cell detection algorithms can make the blood count a more accessible test for less favored populations around the world.

References

1. Molinaro, E., Caputo, L., Amendoeira, R.: Métodos Para a Formação de Profissionais em Laboratórios de Saude. Fio Cruz, Rio de Janeiro (2009)
2. Verrastro, T., Lorenzi, T.F., Neto, S.W.: Hematologia e Hemoterapia: Fundamentos de Morfologia, Patologia e Clínica. Atheneu, Rio de Janeiro (2005)
3. Abbot laboratorie de México S.A. Atlas com interpretacion Histogramas y Escatergramas. Buenos Aires: E.G (2002)
4. Fischbach, F., Duynning, M.B.A.: Manual of Laboratory an Diagnostic Tests. Lippincott Williams e Wilkins, Pennsylvania (2009)
5. Junqueira, L.C., Carneiro, J.: Biologia Celular e Molecular. Guanabara Koogan, São Paulo (2012)
6. Junqueira, L.C., Carneiro, J.: Histologia Básica. Guanabara, São Paulo (2004)
7. Monteiro, A.C.B et al.: Sickle Cell Anemia, a genetic disorder characterized by the change in shape of red blood cells. Saúde em Foco, Edição nº: 07/Ano (2015)
8. Abbas, A.K., Lichtman, A.H., Pillai, S.: Cellular and Molecular Immunology. 7th edn. Elsevier (2012)
9. Lofller, H., Rastetter, J., Hafferlach, T.: Atlas of clinical hematology.6th edn. Springer, Berlin (2005)
10. Failace, R et al. Hemograma, manual de interpretação. 5ª. Artemed, Porto Alegre (2009)
11. Bernadette, F., Rodak, G.A., Fristma, K.D.: Hematology—Clinical principles and aplications. Elsevier, New York (2015)
12. Gonzalez, R.C., Woods, R.E., Eddins, S.L.: Digital Image Using Matlab. 2nd. Gatesmark (2009)
13. Hoofman, R., Benz, E.J., Silberstein, L.E., Heslop, H., Weitz, J., Anastasi, J.: Hematology: Basic Principles and Practice, Canada, 6th edn. Elsevier (2013)
14. Comaniciu, D., Meer, P.: Robust analysis of feature spaces: color image segmentation. In: Proceedings of IEEE Conference on Computer Vision and Pattern, pp. 750–755 (1997)
15. Deng, Y., Kenney, C., Moore, M.S., Manjunath, B.S.: Peer group filtering and perceptual color image quantization. Proceedings of IEEE International Symposium on Circuits and Systems, vol. 4, pp. 21–24 (1999)

16. Mogra, M., Srivastava, V.: A Comprehensive review of analysis of counting blood cells using different image processing algorithms. Int. J. Eng. Sci. Invention. **3**(6), 29–31 (2014). www.ijesi.org. ISSN (Online): 2319–6734, ISSN (Print): 2319–6726
17. Mogra, M., Srivastava, V.: A comprehensive review of analysis of counting blood cells using different image processing algorithms. Int. J. Eng. Sci. Invention. ISSN (Online): 2319–6734, ISSN (Print): 2319–6726 (2014)
18. Soltanzadeh, R.: Classification of three types of red blood cells in peripheral blood smearbased on morphology. In: Proceedings of ICSP (2010)
19. Chourasiya, S., Rani, G.U.: Automatic red blood cell counting using watershed segmentation. (IJCSIT) Int. J. Comput. Sci. Inf. Technol. **5**(4), 4834–4838 (2014)
20. Hemant, T., Sexena, S., Vadak, N.: Segmentation using morfological watershed transformation counting blood cells. Int. J. Comput. Appl. Inf. Technol. **2**(3) (2013)
21. Kawamura, T.: Interpretação de um teste sob a visão epidemiológica: eficiência de um teste. Arq. Bras. Cardiol., São Paulo, **79**(4), 437–441 (2002)
22. Sabattini, R.M.E.: Um Programa para o Cálculo da Acurácia, Especificidade e Sensibilidade de Testes Médicos. Revista Informédica **2**(12), 19–21 (1995)
23. Chow, S.C.: Statistical Design and Analysis of Stability Studies. CRC Press, Boca Raton (2007)
24. Arivu, S.K., Sathiya, M.: Analyzing blood cell images to differentiate WBC and counting of linear & non-linear overlapping RBC based on morphological features. Elixir Comp. Sci. Engg. **48**, 9410–9413 (2012)
25. Berge, H., Taylor, D., Krishnan, S., Douglas, T.S.: Improved red blood cell counting in thin blood smears. In: Proceedings of ISBI, pp. 204–207 (2011)
26. How, KB., Bin, A.S.K, Bin, A.S.K, Siong, N.T., Soo, K.K.: Red blood cell segmentation utilizing various image segmentation techniques. In: Proceeding of International Conference on Man-Machine Systems, September 15–16, 2006, Langkawi, Malaysia (2006)
27. Zack, G.W., Rogers, W.E., Latt, S.A.: Automatic-measurement of sister chromatid exchangebfrequency. J. Histochem. Cytochem. **25**, 741–753 (1977)

Toxoplasmosis Gondii: From Discovery to Advancens in Image Processing

Ana Carolina Borges Monteiro, Yuzo Iano, Reinaldo Padilha França and Rangel Arthur

Abstract Toxoplasmosis is one of the most frequent parasitic diseases in the world, and the Toxoplasma gondii causes it, an obligate intracellular parasite that infects its hosts through the consumption of raw or under-cooked meat, vertical transmission, milk consumption without pasteurization, blood transfusions and contact contaminated environments. Given the wide geographical distribution of this parasite, many people may be infected in the latent phase. Because of the above, this study aims to carry out a bibliographical review of the literature demonstrating the parasitic cycle and especially the ways of diagnosis of the disease. Immunological diagnoses are of great importance, however, and the recent creation of new magnetic resonance imaging tools has been promising and may be used as a detection method for *Toxoplasma gondii* cysts in the future. Only the creation of new technology can lead to increasingly accurate, reliable diagnoses, being able to lead patients to the right pathways faster and even avoiding more serious consequences of the disease, caused by the delay in diagnosis.

Keywords *Toxoplasma gondii* · Toxoplasmosis general aspects · Toxoplasmosis diagnostic · Toxoplasmosis treatment · Image processing · Parasitology

---

A. C. B. Monteiro (✉) · Y. Iano · R. P. França · R. Arthur
School of Electrical and Computer Engineering (FEEC), University of Campinas, UNICAMP, Av. Albert Einstein—400, Barão Geraldo, Campinas, SP, Brazil
e-mail: monteiro@decom.fee.unicamp.com

Y. Iano
e-mail: yuzo@decom.fee.unicamp.com

R. P. França
e-mail: padilha@decom.fee.unicamp.com

R. Arthur
e-mail: rangel@ft.unicamp.br

Y. Iano et al. (eds.), *Proceedings of the 4th Brazilian Technology Symposium (BTSym'18)*, Smart Innovation, Systems and Technologies 140,
https://doi.org/10.1007/978-3-030-16053-1_9

1 Introduction

The discovery of the etiological agent of toxoplasmosis (aka toxo) was carried out in 1908. However, its nomenclature was given only a year later. In 1939, the medical significance of *T. gondii* remained unknown, coming to light with the identification of congenitally infected child tissue. Its veterinary importance was only discovered in 1957 when it was associated with abortions in sheep.

It was just in 1948, through Sabin-Feldman's dye test, that *T. gondii* was recognized as a parasite [1]. Currently, toxoplasmosis is classified as the most common parasitic zoonosis in the world. It is caused by an intracellular parasite called *T. gondii,* which is a poloxenic parasite that is optionally heteroxenic, belonging to the *Protista* kingdom, *Apicomplexa* phylum, *Eucoccidiidae* order, and *Sarcocystidae* family. It is an obligate intracellular coccidia, which has developed several transmission routes between hosts of several species. It attacks warm-blooded animals, infects humans, wildlife, domestic animals, and birds. Felids are definitive hosts, while humans, other mammals, and birds are intermediate hosts. Transmission may occur through the transplacental route, organ transplantation, blood transfusion, ingestion of raw or undercooked meat, food or water intake with sporulated oocysts [2–5].

For many years, the life cycle of *T. gondii* has remained unclear. In 1970, it was discovered that felines are the definitive host and an environmentally resistant stage (oocyst) is excreted in the feces of infected cats. Nowadays, it has been discovered that this infection can affect certain marine animals, such as otters. This finding indicates that even today, the seas may be contaminated by *T. gondii* oocysts washed from the ground [1].

The biological cycle of *T. gondii* consists of the development of morphologies: tachyzoites or trophozoites, bradyzoites, sporozoites, and oocysts. The tachyzoites or trophozoites are rapidly proliferating organisms found in the acute phase. They have one of the edges sharper and the other more rounded and a large core. They can be found throughout the body of the host [5, 6].

Bradyzoites are slow-growing organisms, in which cysts may be latent and develop during chronic infection. They are mainly found in cardiac muscle, skeletal muscle, and nervous system. When found in the brain they present a spherical shape, whereas in the heart muscle and the skeletal muscle they are more elongated [5, 6].

Sporozoites and oocysts are found in the feces of definitive hosts: domestic or wild felids. They affect the intestinal epithelium of the felids, where asexual reproduction occurs, with subsequent elimination in the feces. They may contain around 10 million oocysts at peak elimination. Depending on the environmental conditions, such as humidity and temperature, these oocysts become infective after a period of one to five days [5, 6].

The oocysts are spherical, have a double wall, and inside they have 8 infecting sporozoites. The oocyst ruptures in the gut after gastric processing, releasing sporozoites that divide rapidly into intestinal cells and lymph nodes [7].

When *T. gondii* is contracted during the gestation period, the vertical transmission may occur, which is caused by the passage of tachyzoites through the placenta.

Another form of contagion of the pathology is through horizontal transmission, caused in several ways, such as ingestion of infectious oocysts presents from the environment or by ingestion of tissue cysts or tachyzoites present in meat and animal viscera. In addition, transmission can still occur through blood transfusions, organ transplantation consumption of milk without pasteurization, where the tachyzoites are responsible for the invasion of the human body [8].

Organisms with viable conditions can remain in the state of cysts within the host, indefinitely, and can extend for a lifetime. These hosts can serve as reservoirs of *T. gondii* in these cases. Even though it is of great medical importance, the mechanisms used by the parasite to subvert host cell processes to their advantage remain obscure. In this context, *T. gondii* infects all nucleated cells and remains within them until the dividing parasites are capable of breaking host cells. What intrigues many researchers, is that these stresses caused in host cells could trigger the process of cellular apoptosis. Recent research has shown that cells infected with *T. gondii* are resistant to various apoptosis inducers. This mechanism requires that the intracellular parasite be alive and the presence of protein synthesis [9].

The prevalence of infection in humans is relatively high, with estimates of chronic infection in adult individuals at 90%, depending on the geographic region and the food habit of the population [7, 10, 11]. In the last decades, several global outbreaks of human toxoplasmosis have been related to small groups of individuals or families, the involvement of large population groups is infrequent. Studies show that the sources of infection and transmission routes vary in the different populations studied. Infection is related to the dependence of the eating habits of the population and the degree of soil contamination by cat feces. Oocysts present on soils seem to be the factor responsible for the higher prevalence of infection in children. On the other hand, the habit of eating raw or undercooked meat is related to infection in adults [7].

In ancient times, the habit of consuming meat from pigs and sheep in a raw or improperly cooked state was pointed out as one of the main forms of infection by the disease. However, studies show that in the last 20 years the prevalence of *T. gondii* infection has been decreasing, since today the breeding of cutting animals has stricter laws related to animal health. An example of this is currently seen in countries of the European Union where the prevalence of the etiological agent is less than 1% in animals intended for human consumption on a commercial scale. However, this does not apply to developed or underdeveloped countries. In the Americas, cases of disease outbreaks are still common, associated with environmental contaminations by the oocyst [8].

The most at-risk population, in particular, are patients with HIV among whom the prevalence of toxoplasmosis is about 10%. In approximately 5% of AIDS (Acquired Immune Deficiency Syndrome) cases, toxoplasmosis is the first indication of the human immunodeficiency virus. Toxoplasmosis in immunocompromised individuals can be an aggressive and often fulminating disease, causing visceral infection, infection of the lymph nodes and infection of the central nervous system. Toxoplasmosis has been shown to be an important cause of infectious retinitis, with increasing numbers of retinal lesions associated with AIDS. In the central nervous system, it is the most frequent cause of focal brain injury [12].

In immunocompetent patients, toxoplasmosis has a rapid humoral and cellular development, effectively restricting the pathogenic action of the parasite, with the formation of aggregates in microcysts that characterize the form of latency of the process in its chronic, permanent form throughout life. Infection of the acute phase is common and remains at subclinical levels or only mononucleosis-like symptoms: a headache, fever, lymphadenopathy, malaise, and apathy [13].

The toxoplasmosis test is of great importance because of the congenital toxoplasmosis results from intrauterine infection, ranging from asymptomatic to lethal, depending on fetal age and factors not yet known. In placental transmission, the fetus is usually infected by tachyzoites that cross the placenta from the maternal circulation during the primary infection. However, inactive tissue cysts of past infection may also restart the cycle when pregnant women are immunocompromised. Congenital toxoplasmosis results in prematurity, low weight, post-maturity chorioretinitis, strabismus, hepatomegaly, and jaundice. If the infection occurs in the last trimester of pregnancy, the newborn may develop pneumonia, hepatitis with jaundice or myocarditis, thrombocytopenia, and anemia. If it occurs in the second trimester, the baby may be born prematurely, showing signs of encephalitis with seizures, cerebrospinal fluid, and cerebral calcifications, and may present with Sabin tetrad: microcephaly with hydrocephalus, chorioretinitis, mental retardation and intracranial calcifications [6].

Blood transfusions also present a risk of *T. gondii* infection, as the parasite can survive in refrigerated blood. The first case of infection by blood transfusion was reported in 1971, and the leukocyte concentrate was detected as the focus of the infection. In 1989, another toxoplasmosis infection was described, but this was due to platelet transfusion. After 51 days of the procedure, the patient presented chorioretinitis due to *T. gondii*. However, in both the detection of the transmission form occurred by the exclusion of facts [14].

Toxoplasma gondii develops through two distinct phases: the sexed phase and the asexual phase. The asexual phase is that which occurs in intermediate hosts, where there is no development of the cycle in the intestinal epithelium. Thus, when the infection occurs in cows or goats, there may be the release of tachyzoites through the milk. Ingestion of the milk takes the tachyzoites into the stomach, where the acidity of the gastric juice destroys the morphologies. However, the tachyzoites that can penetrate through the oral mucosa can continue the cycle [15].

The felines are the major Toxoplasmosis vector, since they are responsible for the development of the sexual phase of *T. gondii* that occurs in the intestinal epithelium of cats, with the release of oocysts into the environment [2, 15]. However, contact with them is not a prime factor for the development of the disease, since the parasite infects cats that deposit their feces in the sand. If T. gondii contaminates these, the parasite will be in an environment resistant morphology, presenting a double wall [7, 15].

2 Proposal

This survey carries out a bibliographic review of the main forms of transmission, diagnosis, and treatment of toxoplasmosis as well as evidence of cases of disease outbreaks over time to provide an objective and concise pathology view to students and health professionals.

3 Results and Discussion

In Brazil, the parasite has a wide geographical distribution ranging from 50 to 80% [7, 11]. Handling of carcasses and viscera of contaminated animals poses a risk of Toxoplasmosis infection. In Paraná, 150 blood samples from people working in the refrigerator were analyzed, and 70% were positive for Toxoplasmosis [13].

Numerous outbreaks of toxoplasmosis have been reported in Brazil, resulting from the consumption of meat contaminated by cysts of the parasite, or from the consumption of food and water contaminated with *T. gondii* oocysts. Among the contaminations, stands out, the outbreak occurred in Santa Isabel do Ivaí-PR, considered the largest outbreak in the world [7].

The Erechim/RS region is an area in which ocular toxoplasmosis has become endemic. On September 13, 1993, in the city of Bandeirantes/PR, 17 people became contaminated by the consumption of raw quibe (fried meatballs) of mutton during an Arab party. The affected individuals had a clinical picture and serological profile suggestive of acute toxoplasmosis. The people affected were in the age range of 6–57 years [2].

Most toxoplasma infections are subclinical, and the diagnosis is usually based on the immunological criteria. The studies are based on the detection of specific IgG, IgM, and IgA immunoglobulins, associating them with the IgG avidity test [16].

The Sabin-Feldman dye test is considered the greatest advance in the field of toxoplasmosis [17]. This test is classified as highly sensitive and specific, being able to identify *T. gondii* infections based on a serological test. These tests are based on the specific neutralization of the live parasite in the presence of IGg antibodies and complements, being very useful in the detection of the acute or chronic phase. The specificity of this test related to non-crossing with results of other pathologies. However, this diagnostic method is not frequently used in the laboratory routine, since it is necessary to manipulate the parasite in its infecting form. This test primarily detects IgG antibodies and is currently being replaced by others, especially by Indirect Immunofluorescence, which is safe, economical, while also detecting IgG and IgM antibodies [18–20].

The passive hemagglutination test is widely used to indicate the prevalence of toxoplasmosis, but it is not indicated for the diagnosis of neonatal or acute infection in pregnant women since it presents a great possibility of a false positive result. It is based on the detection of IGg antibodies later [18–20].

The immunofluorescence test has advantages over the Sabin-Feldman dye test because it is performed with preserved parasites, fixed on a microscope slide, which makes it more practical. Another advantage is to allow the identification of antibodies according to immunoglobulin classes, IgG or IgM by the use of specific conjugates. For the standardization of these tests, the World Health Organization distributes anti-toxoplasma reference serum, the titles of which are expressed in International Units (UI/ml) thus allowing the uniformity of the results obtained in different laboratories [18–20].

In turn, the indirect immunofluorescence test is more suitable for the detection of toxoplasmosis in the acute phase and particularly in the congenital form of the disease. Its positivity is 25%, and false positivity can occur due to "placental escape," collagenosis and infections, false negative, ocular form and due to the saturation of antigenic receptors by IgG. Serum should be given special treatment to avoid erroneous results. The IgG indirect immunofluorescence test is a 95% sensitivity method that can be a false positive for Anti-Nuclear Factor and a false negative for low IgG titers.

The Enzyme-Linked Immunosorbent Assay (ELISA) method, in turn, is an enzyme-linked immunosorbent assay with 80% positivity, which translates into an early infection. It eliminates the interference of IgG and rheumatoid factor, present in indirect immunofluorescence. It is especially important in the diagnosis of congenital infection [18].

When toxoplasmosis reaches the individual in the fetal stage, it is the result of acute parasitic infection in uninfected mothers. Consequences of infection are most severe if infection occurs during the first trimester of gestation. Effective prenatal diagnosis allows the initiation of fetal treatment in the uterus. The current diagnosis of the disease, still in the uterine phase, is dependent on a combination of specific immunoglobulin M detection, amniotic fluid culture, liquid and fetal blood analysis, and other nonspecific infection measurements. The only definitive diagnosis is per culture, which takes up to 3 weeks to the issue of the report definitive [21].

Studies in the last decades have used molecular biology as an ally in the diagnosis of toxoplasmosis. Animal studies demonstrate that the DNA fragment named 529 bp is repeated 200–300 times over the *T. gondii* genome. The discovery of this genetic sequence was employed for the development of Polymerase Chain Reaction (PCR) with high sensitivity in the detection of cysts present in the brains of mice with chronic infection by the parasite. Through the development of this technique, in the future, the number of bradyzoites present in the human host can be estimated [22].

Another study indicates that the B1 gene is conserved and present in six strains of *T. gondii*, including in HIV-infected patients. Thus, the PCR technique is based on the detection of the parasite through the identification of this gene, which is repeated 35 times along the genome of the etiological agent [23].

Currently, the diagnosis of toxo in patients with HIV positive serology is not reliable when performed by conventional immunological methods. In this context, medical reports become dependent on computed tomography or magnetic resonance imaging. However, such tests are expensive and may not be accessible to all. It is also possible to perform biopsies of brain tissue to identify the parasite in the cyst stage. However, such a procedure presents a considerable risk to the patient [21].

Magnetic resonance imaging (MRI) is a noninvasive medical imaging technique characterized by the rapidity and use of non-ionizing radiation to acquire high quality and high contrast images of any anatomical structure present in the human organism. In this process, structures formed by soft tissues, such as lungs, heart, liver, and encephalon, are presented in lighter colorations when compared to denser tissues. When it comes to brain analysis, MRI can detect cysts, tumors, swellings, bleeding, inflammation, vascular problems, anatomical abnormalities, and lesions. Thus, MRI has been a great ally in the diagnosis of various pathologies [24].

Given its wide applicability, MRI has been the subject of several improvements investigations. Considering the large amount of data present in a single image, the use of manual methods of image analysis and interpretation becomes a tedious, time-consuming process and subject to errors. Given the above, investing in tools capable of extracting relevant information from magnetic resonance imaging in a fast, efficient and reliable way is an important step in the medical areas, and may even aid in the diagnosis of toxoplasmosis [24].

Recent researches deal with the creation of an automatic and high accuracy methodology capable of classifying MRI in a pathological and non-pathological state of the human brain. This technique used the Ripplet transform Type-I (RT) and a multiscale geometric analysis (MGA) for digital images, which was used to represent the main characteristics of MRI of the human brain. In turn, the classification of the images was performed through the creation of a computational support vector. All datasets submitted to the new methodology showed high accuracy (>99%), and the results obtained were compared with the latest technologies. Therefore, this methodology demonstrated that the automation of brain magnetic resonance diagnostics would bring significant benefits to the medical community [25].

Considering the stages of development of *T. goondi* is an important task to take the patient to the most appropriate treatment. Knowing that cysts are fundamental morphology for the survival of the parasite, allowing these to escape the action of the host's immune system, a study was developed to allow the automatic detection of cysts. For this, a medical algorithm was developed based on digital image processing techniques of native Colombian strains isolated from meat samples for human consumption. This methodology was developed in Matlab 2013 software, specifically through the Image Processing Toolbox [26].

In the last decade, the adaptive techniques were combined with optical coherence tomography. This union has undergone several improvements, resulting in benefits such as increased resolution and sensitivity. These modifications gave the ability to optical coherence tomography to obtain and analyze cellular images at the level of ultra high-resolution 3D and gave a higher speed to the process. The union of these techniques has been used in capturing structures that make up the retina. At other times, such analyzes were only possible through histological material [27].

Super-resolution, which is a method that increases image resolution using low resolution (LR) images [28], was used to reconstruct high-resolution eye fundus video from multiple LR video frames of the retinal fundus. It is important to consider that natural eye movements during a test are used as a suggestion for super-resolution [29]. To compensate for the heterogeneous illumination in the fundus of the eye, the researchers integrated the retrospective illumination correction for the photometric record to the underlying image model. The method used the quality self-assessment to provide objective quality scores for reconstructed images. Parameters were adjusted automatically, with improvements in resolution and sharpness in 74%, as well as improvements in the segmentation of ocular blood vessels in 13% [29]. Based on this, ultra-resolution techniques seem as a great ally to the diagnosis and control of retinal lesions in patients affected by toxoplasmosis. Some studies appoint that the retinochoroiditis and strabismus were outstanding as important sequelae of congenital toxoplasmosis [30].

3.1 Treatment

The treatment against *T. gondii* infection is basically performed through the administration of antibiotics and antiparasitics. The first reports concerning the treatment of this pathology occurred in 1942 when Sabin and Warren described the efficacy of sulfonamides against murine toxoplasmosis. In 1995, Eyles and Colema reported the synergistic effect of the combination of sulfonamides and pyrimethamine, which is considered the standard therapy for humans. Four years after these findings, they pointed to spiramycin as antitoxoplasmic activity in rats.

In 1954, spiramycin had the characteristic of non-toxicity and the ability to cross the transplacental barrier. Given these characteristics, it has been used as a prophylactic measure in women during pregnancy, aiming to reduce vertical transmission.

In 1974, the discovery was made on clindamycin, which showed antitotoxoplasmatic activity and was chosen as a good alternative for the treatment of patients with toxo. This drug is primarily indicated for individuals allergic to sulfonamides.

4 Conclusions

Even with more than a century since the first description of *T. gondii*, millions of people around the world are still affected by toxoplasmosis, many of which are still latent. In order to avoid severe harm to both adult and fetal-aged individuals, it is important to invest in effective and accurate diagnostic means. Only early diagnosis can minimize chronic and severe damage, which becomes costly both to the public health organs and to each of the affected individuals. Many advances have been made in the diagnostic aspect, even involving molecular biology techniques. Yet, the detection of *T. gondii* is an important ally in the arrest and clinical management of the disease. Therefore, investing in automated detection methods of cysts by MRI can be a major step for the diagnosis and early treatment of the pathology with strong cloud-based processing and well-designed databases [31–34].

Image processing is a promising tool for the direct diagnosis of toxo as well as lesions due to secondary damage caused by the etiological agent. Thus, it is vital to invest in interdisciplinary research that develops effective solutions and high reliability for medical diagnostics.

The development of high precision technologies such as super-resolution are of great importance not only for parasitological diseases but also for any disease that generates small lesions in organs of difficult access. Fluorescent microscopy together with the late advances in computational intelligence and automated diagnosis can overcome the challenges related to the parasite study and the disease diagnosis/treatment [33, 35–38]. Only the early diagnosis of any pathology can increase the quality of life and the life expectancy of individuals.

While this step is not yet given in the academic field, it is imperative to invest in campaigns that raise the awareness of the less favored populations regarding the prophylaxis measures and the importance of prenatal screening for all pregnant women, thus avoiding that more people go through emotional and physical damage caused by toxo.

References

1. Jitender, P.D.: The history of *Toxoplasma gondii*—the first 100 years. J. Eukaryot. Microbiol. **55**(6), 467–475 (2008). https://doi.org/10.1111/j.1550-7408.2008.00345.x
2. Dias, R.A.F., Freire, R.L.: Surtos de toxoplasmose em seres humanos e animais. Sem. C Agr. **26**(2), 239–248 (2005). https://doi.org/10.5433/1679-0359.2005v26n2p239
3. Fialho, C.G., Teixeira, M.C., Araujo, F.A.P.: Toxoplasmose animal no Brasil. Rev. Acta Sci. Veterinariae **37**(1), 1–23 (2009)
4. Negri, B.: Surto de Toxoplasmose no município de Santa Isabel do Ivaí – Paraná Funasa –Bol Eletr Epidemiológico **2**(3) (2002)
5. Lopes, C.C.H., Berto, P.: Aspectos associados à toxoplasmose: uma referência aos principais surtos no Brasil. Saúde & Amb. Rev. D de Caxias **7**(2), 01–07 (2012)
6. Souza, S.T.: Toxoplasmose congênita: uma revisão bibliográfica. Universidade Federal do Rio Grande do Sul, Escola de Enfermagem (2010)
7. Ekman, C.C.J.: Influência da forma infectante do *Toxoplasma gondii* na doença aguda humana: revisão sistemática de surtos epidêmicos. M.Sc. Thesis Instituto de Medicina Tropical de São Paulo, USP (2012)
8. Astrid, M., Tenteraanja, R., HeckerothaLouis, W.M.: *Toxoplasma gondii*: from animals to humans. Int'l J. for Parasit **30**(12–130), 1217–1258 (2000)
9. Paul, B.N., Matthew, B.P., Leon, R.P., Clarke, P., Duke, R.C., Curiel, T.J.: *Toxoplasma gondii*-infected cells are resistant to multiple inducers of apoptosis. J. Immunol. **160**, 1824–1830 (2018)
10. Oréfice, F., Filho, R.C., Barboza, A.L., Oréfice, J.L., Calucci, D.: Toxoplasmose ocular adquirida, Toxoplasmose ocular pós-natal. Rev. Bras. Oft **69**(3), 184–207 (2010)
11. Carmo, E.L., Póvoa, M.M., Monteiro, N.S., Nascimento, J.M., Freitas, N.A., Bichara, C.N.C.: Surto de toxoplasmose humana no Distrito de Monte Dourado, Município de Almeirim, Pará, Brasil. Rev. Pan-Amaz Saude **1**, 61–66 (2010). https://doi.org/10.5123/S2176-62232010000100009
12. Zajdenweber, M., Muccioli, C., Belfort, R.J.R.: Acometimento ocular em pacientes com AIDS e toxoplasmose do sistema nervosa central – antes e depois do HAART. A Bras. Oft **68**(6), 773–775 (2005). https://doi.org/10.1590/S0004-27492005000600012

13. Prado, A.A.F., De Almeida, G.F., Gontijo, L.S., Torres, M.L.M.: Toxoplasmose: o que o profissional da saúde deve saber. Enciclopedia Biosfera **7**(12), 1 (2011)
14. Amorim, L.: Toxoplasmose e transfusão de sangue. Rev. Bras. Hematol. Hemoter **30**(4), 264–265 (2008). https://doi.org/10.1590/S1516-84842008000400006
15. Neves, D.P.: Parasitologia humana. 11th. Atheneu, São Paulo (2005)
16. Tsukuda, L.R.: Imunidade humoral na toxoplasmose ocular. D.Sc. Thesis, Instituto de Ciências Biomédicas, USP (2007)
17. Sabin, A.B., Feldman, H.A.: Dyes as microchemical indicators of a new immunity phenomenon affecting a protozoon parasite (Toxoplasma). Sci. New Series **108**(2815), 660–663 (1948)
18. Jobim, E.M., Silva, J.D.P.: Toxoplasmosis, a congenital disease. Saude **30**(1–20), 50–56 (2004)
19. Male, D.: Imunologia: um resumo ilustrado, 3rd. Manole, São Paulo (1988)
20. Abbas, A.K., Lichtman, A.H., Pillai, S.: Cellular and molecular immunology, 7th edn. Elsevier (2012)
21. Burg, J.L., Grover, C.M., Pouletty, P., Boothroyd, J.C.: Direct and sensitive detection of a pathogenic protozoan, *Toxoplasma gondii*, by polymerase chain reaction direct and sensitive detection of a pathogenic protozoan, Toxoplasma *gondii,* by polymerase chain reaction. J. Clin. Microbiol. **27**(8), 1787–1792 (1989)
22. Homan, L., Vercammen, W., De Braekeleer, M., Verschueren, H.: Identification of a 200- to 300-fold repetitive 529 bp DNA fragment in *Toxoplasma gondii,* and its use for diagnostic and quantitative PCR. Int'l J. Parasitol. **30**, 69–75 (2000). https://doi.org/10.1016/S0020-7519(99)00170-8
23. Mesquita, R.T., Vidal, J., Pereira-Chioccola, V.L.: Molecular diagnosis of cerebral toxoplasmosis: comparing markers that determine *Toxoplasma gondii* by PCR in peripheral blood from HIV-infected patients. Braz. J. Infect. Dis. **14**(4), 346–350 (2010)
24. Srinivasan, K., Kanakaraj, J.: A review of magnetic resonance imaging techniques. Smart Comp. Rev., 3358–366 (2013). https://doi.org/10.6029/smartcr.2013.05.006
25. Sudeb Das, C.M., Kundu, M.K.: Brain MR image classification using multi-scale geometric analysis of Ripplet. Prog. Electromagn. Res. **137**, 1(17) (2013)
26. Quinones, A., Graciela, J.: Identification of *Toxoplasma gondii* cysts in samples from Colombia using digital image processing. In: 2014 IEEE ANDESCON, pp. 1–1 (2014). https://doi.org/10.1109/andescon.2014.7098536
27. Miller, D.T., Kocaoglu, O.P., Wang, Q., Lee, S.: Adaptive optics and the eye (super resolution OCT). Eye (London, England) **25**(3), 321–330 (2011)
28. de Jesus, M.A., Estrela, V.V., Saotome, O., Stutz, D.: Super-resolution via particle swarm optimization variants. In: Hemanth, J., Balas, V. (eds.) Biologically Rationalized Computing Techniques for Image Processing Applications. LNCVB, vol 25. Springer, Cham (2018)
29. Köhler, T., et al.: Multi-frame super-resolution with quality self-assessment for retinal fundus videos. In: Golland, P., Hata, N., Barillot, C., Hornegger, J., Howe, R. (eds.) In Medical Image Comp and Computer-Assisted Intervention—MICCAI 2014. LNCS, vol. 8673. Springer, Cham (2014)
30. Soares, J.A.S., Násser, L.S., Carvalho, S.F.G., Caldeira, A.P.: Ocular findings in children with congenital toxoplasmosis. Arq. Bras. Oftalmol. **74**(4), 255–257 (2011)
31. Estrela, V.V., Monteiro, A.C.B., França, R.P., Iano, Y., Khelassi, A., Razmjooy, N.: Health 4.0: applications, management, technologies and review. Med. Technol. J. **2**(4), 262–276 (2019). https://doi.org/10.26415/2572-004x-vol2iss1p262-276
32. Coelho, A.M., Estrela, V.V.: A study on the effect of regularization matrices in motion estimation. Int'l J. Comp. Appl. **51**(19), 17–24 (2012). https://doi.org/10.5120/8151-1886
33. Herrmann, A.E., Estrela, V.V.: Content-based image retrieval in remote clinical diagnosis and healthcare. In: Cruz-Cunha, M., Miranda, I., Martinho, R., Rijo, R. (eds.) Encyclopedia of e-health and Telemedicine. IGI Global, pp. 495–520 (2016). https://doi.org/10.4018/978-1-4666-9978-6.ch039
34. de Jesus, M.A.D., Estrela, V.V.: An introduction to data mining applied to health-oriented databases. OJCST **9**(3). https://doi.org/10.13005/ojcst/09.03.03

35. Razmjooy, N., Mousavi, B.S., Khalilpour, M., Hosseini, H.: Automatic selection and fusion of color spaces for image thresholding. Sig. I Vid. Proc. **8**(4), 603–614 (2014)
36. Razmjooy, N., Mousavi, B.S., Fazlollah, S., Hosseini Khotbesara, M.: A computer-aided diagnosis system for malignant melanomas. N Comp. Appl. **23**(7–8), 2059–2071 (2013)
37. Fernandes, S., Estrela, V.V., Magalhaes, H.A., Saotome, O.: On improving sub-pixel accuracy by means of B-Spline. In: Proceedings of the 2014 IEEE International Conference on Imaging System and Techniques (IST 2014). Santorini, Greece (2014). https://doi.org/10.1109/ist.2014.6958448
38. Hemanth, D.J., Estrela, V.V.: Deep learning for image processing applications. In: Advances in Parallel Comp Series, vol 31, IOS Press, ISBN 978-1-61499-821-1 (print), ISBN 978-1-61499-822-8 (online) (2017)

Development of a Kernel: A Deeper Look at the Architecture of an Operating System

Reinaldo Padilha França, Manoel Peluso, Ana Carolina Borges Monteiro, Yuzo Iano, Rangel Arthur and Vânia Vieira Estrela

Abstract The operating systems (OSs) created the possibility for people to interact with computer hardware, these same are made with an enormous number of lines of code for this kernel, usually developed with the languages of C and Assembly language programming. Thus, the objectives of this study are in the area of knowledge, development, and learning of construction, approach and design in the development of complex codes for the creation of an OS. In this present study, the build process from C code compulsorily uses a compiler. An assembler generates the machine code. Assembly language is crucial for real-time operation. Nevertheless, most of OS can be implemented with an additional language. After Assembly, the leanest Kernel code can be obtained with a low-level language like C/C++. The use of a proper boot manager or an existing one, such as Grand Unified Bootloader (GRUB), this itself once programmed, adds knowledge about the hardware developer. This project's Kernel has been loaded into two virtual machines (one with Linux and one with Windows) and on a physical machine. These results demonstrated that the

R. P. França (✉) · A. C. B. Monteiro · Y. Iano · R. Arthur
School of Electrical and Computer Engineering (FEEC),
University of Campinas – UNICAMP, Av. Albert Einstein - 400,
Barão Geraldo, Campinas, SP, Brazil
e-mail: padilha@decom.fee.unicamp.com

A. C. B. Monteiro
e-mail: monteiro@decom.fee.unicamp.com

Y. Iano
e-mail: yuzo@decom.fee.unicamp.com

R. Arthur
e-mail: rangel@ft.unicamp.br

M. Peluso
Regional University Center of
Espirito Santo do Pinhal (UNIPINHAL), Espírito Santo do Pinhal, Brazil

V. Vieira Estrela
Department of Telecommunications, Fluminense Federal University (UFF),
Rio de Janeiro, Brazil
e-mail: vania.estrela.phd@ieee.org

Y. Iano et al. (eds.), *Proceedings of the 4th Brazilian Technology Symposium (BTSym'18)*, Smart Innovation, Systems and Technologies 140,
https://doi.org/10.1007/978-3-030-16053-1_10

developed software is viable, relevance and academic potential, with high learning power, being able to serve as a starting point for several pieces of academic research.

Keywords Operating systems · Kernel · Middleware development · C Language · Assembly · Computer architectures · Embedded systems

1 Introduction

To develop a source code base as large as the kernel, it is undoubtedly a substantial educational, technical, and academic gain. The construction of an Operating System (OS) arouses interest in the matter of actually knowing how a computer system behind the graphical screen and even behind the black screen of the well-known command terminal [1–6].

A Virtual Machine (VM) consists of an emulated OS running as an application on a computer. It is a computer environment software in which an OS or program can be installed and run, in a very simplified way; it functions as a "computer inside the computer" [10, 11]. VMs are extremely useful, allowing to run other OSs, having access to all the necessary software for the desired task, as in the case of this project, OSs still in the development stage [10, 11]. A VM behaves like a complete physical machine that can run its OS, similar to a traditional OS that is in control of the machine.

This project is geared towards the development of kernels for particular machines architectures (e.g., an FPGA-based reconfigurable processor) and to develop courseware on OS [1–4, 7–9].

Using a VM to test the development, besides being recommended is much more advantageous than restarting the computer with every change or change, or having to transfer the files from the development (host) machine to the testing (target) machine [10, 11].

For the development of the Kernel was used a VM with the Linux Ubuntu OS. After the Kernel development, then it was loaded on a Linux VM, another Windows VM, and on a physical machine with no loaded OS.

The present paper is organized as follows: Sect. 2 discusses the kernel. Section 3 presents and describes the materials and methods. Section 4 presents the methodology of the experiment. Section 5 the results and, finally, in Sect. 6, the conclusions are presented as the potential of the research.

2 Kernel

The union of the kernel and the other software is what makes the computer usable (drivers, communication protocols, among others), forming the OS itself, can be

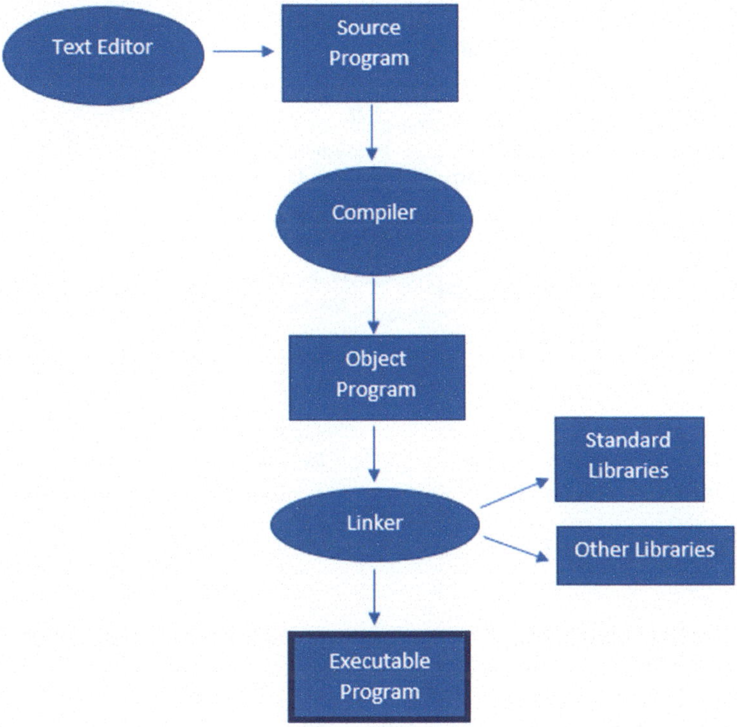

Fig. 1 Steps for generating a code [18]

understood in simplicity as a series of files and instructions written in C language and Assembly language [6–9].

In principle, at least for those who run a program written in C/C++, Java, Fortran or even Python, that is regardless if the language is compiled or interpreted. It makes no difference if the hardware of the machine executes the code directly because a translation is made for the code that is finally executed (the machine language) [18].

The Kernel is a component of the OS, its core, considered the main item of the OSs, being so hidden among the many lines of code, not due to its insignificance but to its importance, being so essential for the operation of a computer it is best to keep it safe [6–8].

The Kernel links the hardware and the software (logical part). It is responsible for the interaction among these layers, managing the system resources, allowing the programs to use them while making the connection between data processing and programs [5–9], as shown in the Figs. 1 and 2 of this section.

The bootloader can be interpreted as the disk access function of the computer and load the kernel into memory to take control of the device [5–9].

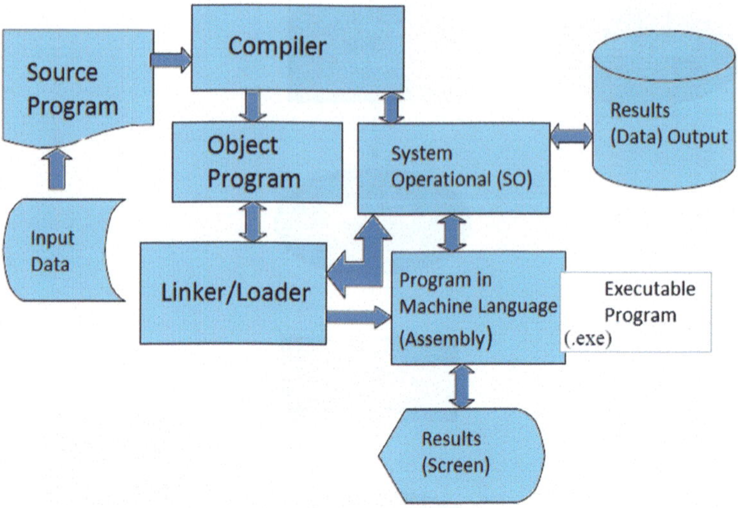

Fig. 2 Stages of the compilation and assembly generating processes [18]

The GRUB is a program that can load any executable file with a multi-boot header in its first 8 Kb, with a sequence of flags bits and initialization bits as well as bits that reference the file image executable.

The Kernel starts working as soon as the computer is turned on, starting to detect all the hardware that is essential to the machine operation (monitor, video, audio, keyboard, mouse, among all other components). Soon after, the OS is loaded, and from then on, the Kernel starts to administer the main functions within the OS, among the main ones: memory management, processes, files and all peripheral devices [5–8].

It is responsible for ensuring that all programs have access to the resources they need simultaneously (among them the RAM, for example), that is, a concurrent sharing, having the responsibility to ensure that the RAM is used in the best way possible, not offering any risk to the computer [1–9].

3 Materials and Methods (Linux Environment)

This project required the creation of a development environment with the necessary tools and packages for Kernel construction containing all the necessary items to develop and deploy the system, that is, the infrastructure for creation.

Wine is software that eliminates the heterogeneity between Windows and Linux being an implementation of the Windows Application Programming Interface (API) on the Unix platform [12].

The Software Construction (Scons) tool, an open source construction software, is an Open Source tool for building software being a utility with integrated functionality for an easier, more reliable and faster way to build software [13]. The GNU Compiler Collection (GCC) is a union of compilers of programming languages elaborated for construction of an OS similar to Unix [14]. The Netwide Assembler (NASM) is an assembler designed for portability and modularity supporting a variety of object file formats, among these binary disk image files, used to compile OSs [15]. UltraISO is a software developed for Windows platform that creates, edits, converts and/or records image files. The Quick EMUlator (QEMU) is an emulator software of processor that uses the dynamic translation technique [16].

4 Experiment Methodology

```
kernel.c ×
void dmain (void* mbd, unsigned int magic)
{
        if(magic != 0x2BADB002)
        {
                print("Error: Invalid magic number!", 0x04);
        }
        else
        {
                clearscreen();
                print("Padilha!", 0x07);
        }
}
void print(char* message, int color)
{
        char*mem = (char*)(0xb8000);
        while(*message != 0)
        {
                *mem = *message;
                mem++;
                message++;
                *mem = (char*)color;
                mem++;
        }
}

void clearscreen(void)
{
        char*mem = (char*)(0xb80000);
        while(*mem != 0)
        {
                *mem = 0;
                mem++;
        }
}
C ▾    Largura das tabulações: 8 ▾    Lin 34, Col 2    INS
```

Fig. 3 Kernel.c

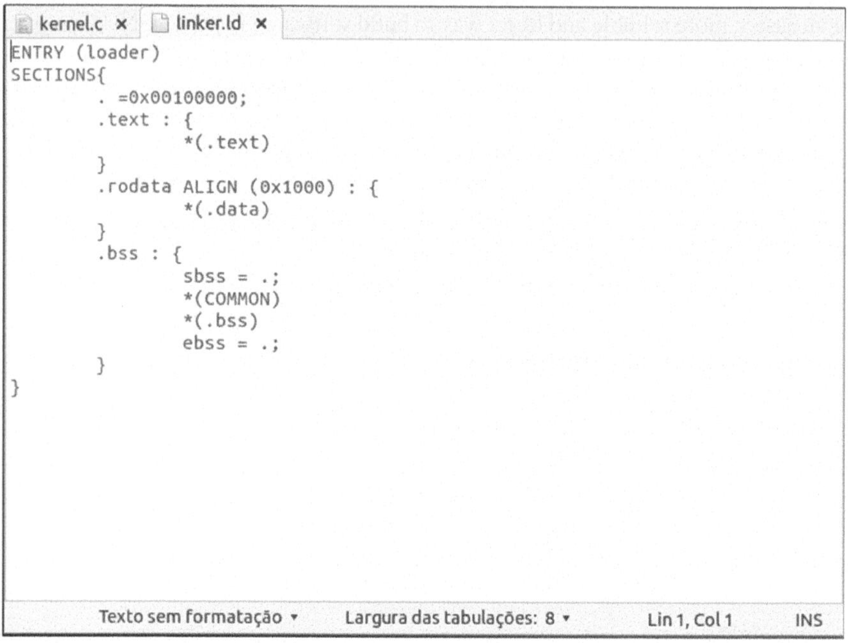

Fig. 4 Linker.ld

This experiment has as primary objective the development of a small OS to start equipment loading a kernel in memory. The development was performed using Linux systems Ubuntu and Windows. Virtualized in a VM, using the Free Software VirtualBox [17].

In this platform, all codes and commandos were searched through literature, through academic records, the Internet itself, and developed and implemented to create the gcc, nasm, and qemu development environment, essential for the compilation of the project, as shown in the figures of this Section.

It is necessary to have installed the tools gcc, nasm and qemu in the Ubuntu platform native notepad for typing the basic instructions of the kernel, written in C language and saved with file extension .c, in Fig. 3, this code prints the first kernel instructions on the computer screen.

The next step was to create a file extension .ld and extension .asm, which contains the loading instructions and have the purpose of joining the compiled kernel objects files, to get the binary file. The code of the example is shown in Figs. 4 and 5.

Object codes for the previously program presented were generated through of the tools gcc and nasm in the Linux environment. These object codes have .o file extension. Then, they were merged into a single binary file containing all the instructions.

For memory management, using Linux and Windows, it is necessary to implement the python development tool, being a compiler of the Python language. For the generation of the files and compilation process using the Scons tool.

```
global loader
extern dmain
MODULEALIGN equ 1<<0
MEMINFO equ 1<<1
FLAGS equ MODULEALIGN | MEMINFO
MAGIC equ 0x1BADB002
CHECKSUM equ -(MAGIC + FLAGS)

section .text
align 4
MultiBootHeader:
dd MAGIC
dd FLAGS
dd CHECKSUM

STACKSIZE equ 0x4000

loader:
mov esp, stack+STACKSIZE
push eax
push ebx

call dmain

cli

hang:
hlt
jmp hang

section .bss
align 4
stack:
resb STACKSIZE
```

Fig. 5 Loader.asm

The Scons compiles all the codes in language C generating the respective object files doing the process of binding consequently as to the kernel. The result this compilation was the file named as “tst.”

A file with extension .lst containing instructions that reference the kernel was created, with a base on greater flexibility in the development and generation of the kernel in the image, as shown in Fig. 6.

With the software UltraISO, it was possible to construct an image file, where the file in question containing in your innermost nucleus the instructions implemented in this kernel basic, being added together with the structure of the image file, as shown in Fig. 7.

Through literature, through academic records, the Internet itself, were also searched functions in C language, being developed and implemented, which making communication with the hardware (keyboard) and memory management.

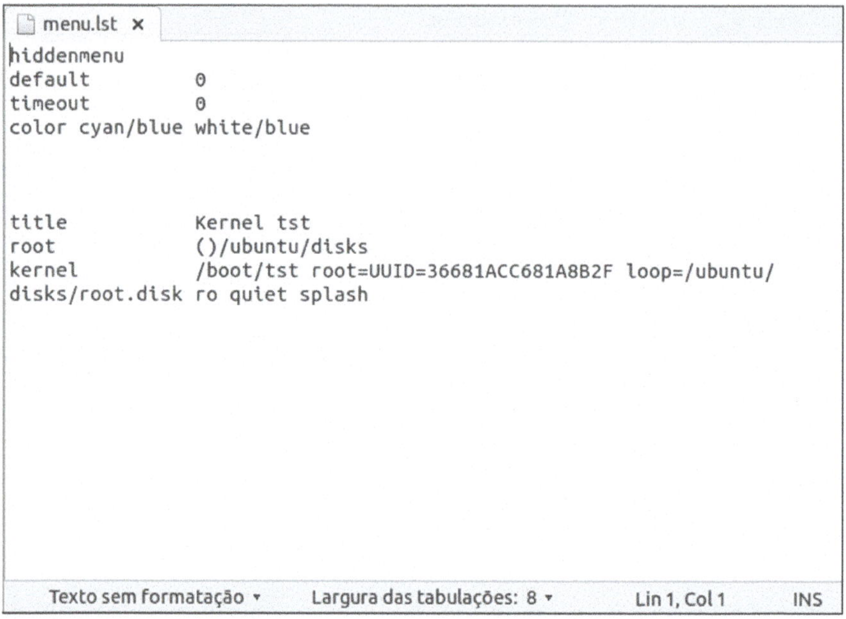

Fig. 6 Menu.lst in Linux Ubuntu

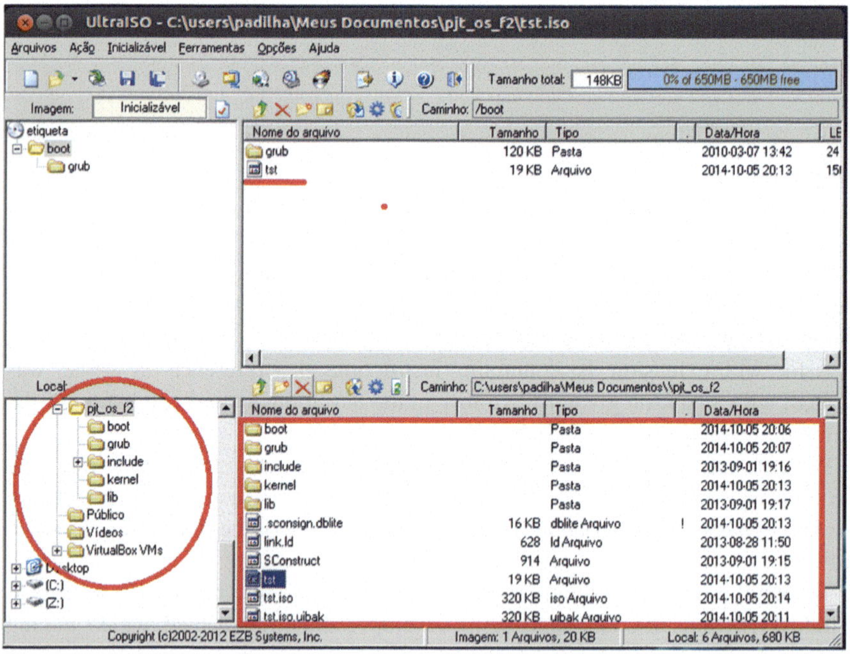

Fig. 7 UltraISO

5 Results

From the kernel generated in Ubuntu, the project was initialized and virtualized in Linux and Windows environments where the OS creates an environment where users can make programs and run them without worrying about hardware details [18]. An OS is an intermediary between the programmer and the hardware, which offers libraries with high-level functions for reading and writing data on these devices. Examples of OSs are Linux, Windows, and MS-DOS itself [18].

VirtualBox software was used to virtualize the developed project, the creation of the VM follows the characteristics of a DOS system, still used today, and its configurations follow the software standard for this type of system.

Figure 8 shows the screen with the file manipulation options, where the Kernel is loaded, and it controls and manages the VM loaded from the codes contained in the Kernel.

Following the same logic and methodology in the Linux environment, the VM had the same settings and the same characteristics displayed in the previous Section, Fig. 9 shows that the developed kernel had the same behavior.

The developed kernel can also be written to a common CD-R or memory stick and boot into a real machine with no OS loaded on the machine, and having the same behavior as expected, as shown in Fig. 10.

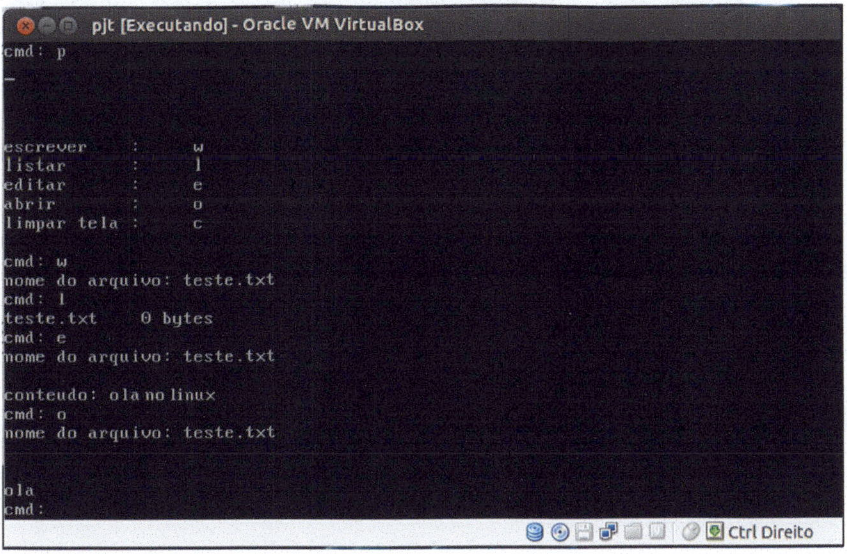

Fig. 8 Experiment Functions in Linux

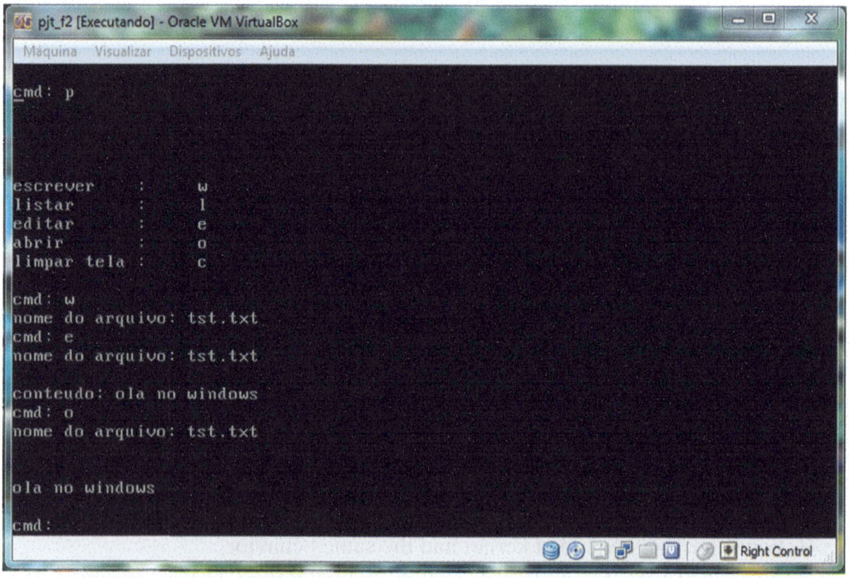

Fig. 9 Experiment Functions in Windows

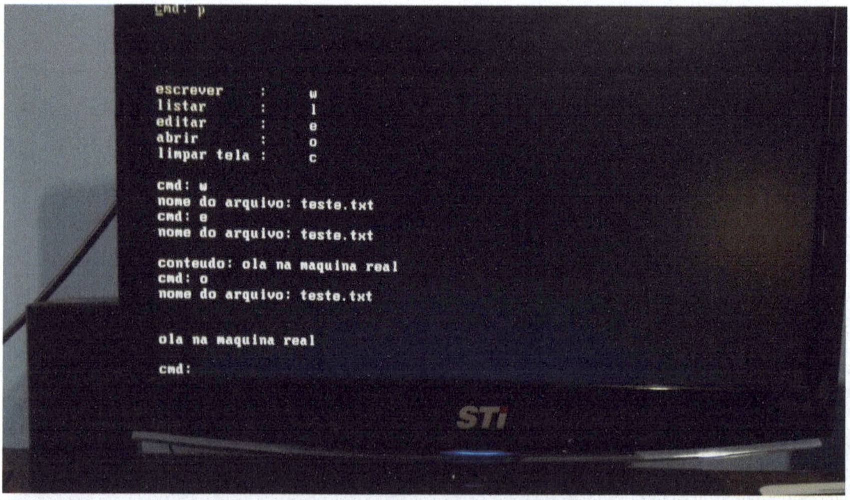

Fig. 10 Experiment options in a physical machine

6 Conclusions

The development of a kernel on Windows and Linux is a long process of code writing, pure programming, as well as debugging various system components and much research, and from that, it acquires the deepest and most real knowledge of the architecture of computers. As well as a deeper understanding of a development environment with the tools needed to write and develop it.

Since the developed software has an application, relevance and academic potential, with high learning power, it is aimed at future work the implementation of new functionality in the software.

Despite the recent advances regarding the field of quantum computing is still veiled in relative mystery and myth, referring to the use of phenomena like superposition, entanglement, among others, even within the field of data science and technology. Moreover, even those within the field of quantum computing and quantum machine learning are still learning the potential for progress and the limitations of current systems from this area. However, quantum computing has arrived in its infancy and has grown to the point where many major companies are pouring money into related R&D efforts, and other systems have been opened for research purposes with quantum machine learning. However, quantum machines set many surprises for the future.

References

1. Tanenbaum, A.S., Bos, H.: Modern Operating Systems, 4th edn. Prentice Hall, Upper Saddle River (2004)
2. Silberschatz, A.: Operating System Concepts, 9th edn. Wiley, Hoboken (2012)
3. Deitel, H.M., Deitel, P.J., Choffnes, D.R.: Operating Systems, 3rd edn. Prentice Hall, Upper Saddle River (2003)
4. Stallings, W.: Operating Systems: Internals and Design Principles, 8th edn. Pearson (2004)
5. Negus, C.: Linux Bible, 9th edn. Wiley, Hoboken (2015)
6. Schimmel, C.: UNIX Systems for Modern Architectures: Symmetric Multiprocessing and Caching for Kernel Programmers, 1st edn. Addison-Wesley Professional (1994)
7. Bovet, D., Cesati, M.: Understanding the Linux Kernel, 3rd edn. O'Reilly Media (2008)
8. Lixiang, Y.: The Art of Linux Kernel Design: Illustrating the Operating System Design Principle and Implementation, 1st edn. Auerbach Publications (2014)
9. Love, R.: Linux Kernel Development, 3rd edn. Addison-Wesley Professional (2010)
10. Portnoy, M.: Virtualization Essentials, 1st edn. Sybex (2012)
11. Golden, B.: Virtualization for Dummies, 1st edn. For Dummies (2007)
12. WINEHQ. Wine Developer's Guide. Wine Project Official Page. http://www.winehq.org/site/docs/winedev-guide/index
13. SCONS. SCons User Guide 2.3.4. http://www.scons.org/doc/HTML/scons-user.html (2004)
14. GCC. The GNU Compiler Collection. http://gcc.gnu.org/
15. NASM. The Netwide Assembler. http://www.nasm.us/xdoc/2.10.09/html/nasmdoc1.html#section-1.1 (2012)
16. QEMU Documentation. https://wiki.qemu.org/Documentation/
17. The VirtualBox PC virtualizer. http://virtualbox.org/

18. Estrela, V.V., Herrmann, A.E.: Conceitos Básicos Sobre Programação Programação Orientada a Objetos (POO). https://www.researchgate.net/publication/268510321_CONCEITOS_BASICOS_SOBRE_PROGRAMACAO_Programacao_Orientada_a_Objetos_POO (2010)

Betterment Proposal to Multipath Fading Channels Potential to MIMO Systems

Reinaldo Padilha, Yuzo Iano, Ana Carolina Borges Monteiro, Rangel Arthur and Vânia Vieira Estrela

Abstract The behavior of a system is studied through a simulation model. Once developed and validated, the simulation model can be used to investigate a wide variety of issues, as well as propose improvements. This research objective is to propose a novel simulation paradigm to improve the transmission of content in wireless telecommunication systems. The simulation environment employs a pre-coding process of bits based on the application of a technique relying on discrete events and signals before the modulation process. The signal transmission on the channel occurs in the discrete domain with the implementation of discrete entities in the process of bit generation applied at a low level of abstraction in a wireless telecommunication system. The simulation considered the advanced differential binary phase shift keying (DBPSK) as the modulation format for signal transmission in an AWGN channel, using different hardware platforms and in the studies. The results show improvements ranging from 9 to 35% in memory utilization, related to information compression, in the context of the research.

Keywords Multipath fading channels · Discrete events · Simulation · Precoding · DBPSK modulation · IEEE 802.11

R. Padilha (✉) · Y. Iano · A. C. B. Monteiro · R. Arthur
School of Electrical and Computer Engineering (FEEC),
University of Campinas – UNICAMP, Av. Albert Einstein – 400,
Barão Geraldo, Campinas, SP, Brazil
e-mail: padilha@decom.fee.unicamp.com

Y. Iano
e-mail: yuzo@decom.fee.unicamp.com

A. C. B. Monteiro
e-mail: monteiro@decom.fee.unicamp.com

R. Arthur
e-mail: rangel@ft.unicamp.br

V. Vieira Estrela
Department of Telecommunications, Fluminense Federal University (UFF),
Rio de Janeiro, Brazil
e-mail: vania.estrela.phd@ieee.org

Y. Iano et al. (eds.), *Proceedings of the 4th Brazilian Technology Symposium (BTSym'18)*, Smart Innovation, Systems and Technologies 140,
https://doi.org/10.1007/978-3-030-16053-1_11

1 Introduction

A simulation allows to build knowledge practically and offers a systemic view of a given problem, with the necessary design capabilities, allowing several parameter changes in a wireless system, evaluating it without physical experimental setups. It also stimulates the critical data analysis. The formulation of questions, problem-solving, and the implementation of different system architectures to analyze different layers, such as physical, transport, transmission, and higher layers, improves and validates the system for different applications [1–6].

Simulation permits to study the performance of a wireless system, reproducing as accurately as possible its characteristics. Manipulation and analysis of the model help to trace several factors impacting the system, e.g., multipath fading, shadowing, and noise among other interferences to facilitate the development of new methodologies addressing caveats [2, 6–8].

Fading degrades the communication system performance due to a loss of signal power without decreasing the noise power over some or all of the signal bandwidth. Fading can change over time, which demands communication systems capable of accommodating to such impairments. However, the adaptations cannot be able of following fading. The probability of experiencing fading with the concomitant bit errors as the Signal-to-Noise Ratio (SNR) drops on the channel limits the link performance.

The fading effects can be mitigated by inserting diversity in signal transmission over multiple channels that undergo independent fading as well as at the receiver, coherently joining them. The probability of suffering a fade in this type of channel is proportional to the probability that all the component channels simultaneously experience a fade, which is a much more unlikely event.

The adequate simulation of multipath fading channels (MFCs) is a fundamental issue in the development and evaluation of wireless systems. Since the received signal is contingent to several mutable factors, statistical models typically help to simulate fading. Many fading models have been developed relying on many statistical distributions such as Rayleigh, and Rician, with different degrees of complexity.

Multipath fading (MF) affects most forms of radio communications links in one way or another. MF occurs in an environment where there is multipath propagation, and the paths change for some reason, resulting of propagating multiple versions of signals transmitted across different paths before they reach the receiver. This changes not only their relative strengths but also their phases, as the path lengths vary, also distorting the radio signal. These attenuations suffered can be represented by statistical distributions such as Rayleigh and Rician [6–8].

The term discrete event mainly denotes the model representing the system as a sequence of operations performed on entities (transactions) of certain types such as data packets, bits, among others. These entities are discrete in a discrete event simulation [1, 2].

This technique is usually used to model concepts having a high level of abstraction, such as clients in a queue, emails on a server, flow of vehicles, transmission of data packets, and so forth [2–11].

This paper proposes a wireless system model simulation built upon an AWGN channel with the advanced modulation format called differential binary phase shift keying (DBPSK) to improve the transmission capacity of information content through the channel and to compensate to the additional complexity posed by multipath techniques.

A bit treatment with discrete events methodology inserted in the bit generation step applied in a low abstraction level is the differential and main contribution of this research. The results show better computational performance regarding memory utilization of the simulation model.

The present paper is organized as follows: Sect. 2 discusses the traditional method of modeling an AWGN transmission channel. Section 3 presents and describes the proposal. Section 4 presents the results and, finally, in Sect. 5, the conclusions are presented as also the potential of the research.

2 Traditional Method

A digital communication system performance corresponds typically to the probability of bit detection errors (BER) in the presence of thermal noise. The so-called additive white Gaussian noise (AWGN) channel model (Fig. 1) is widely used due to its mathematical modeling simplicity, and application to a large set of physical [12–14].

The basic AWGN model emulates the effect of many naturally occurring random processes. This noise is additive because it is added to any signal (the received signal equals to the transmitted signal plus noise). The term white refers to the fact the noise is assumed intrinsic to the system. In the frequency domain, white noise interferes with all frequencies, which leads to a uniform power spectrum across for all frequencies. Finally, the word Gaussian refers to the fact the probability distribution of the noise samples in the time domain follows a normal distribution [12–14].

In the milieu of wireless communications, the primary source of noise is the thermal activity arising from to the vibration of atoms in the receiver [12–14]. However, the mobile wireless channel is susceptible to several interferences including multipath, fading, shadowing, and other types of noise due to general deficiencies that cause a considerable degradation in the performance of the system [15–20].

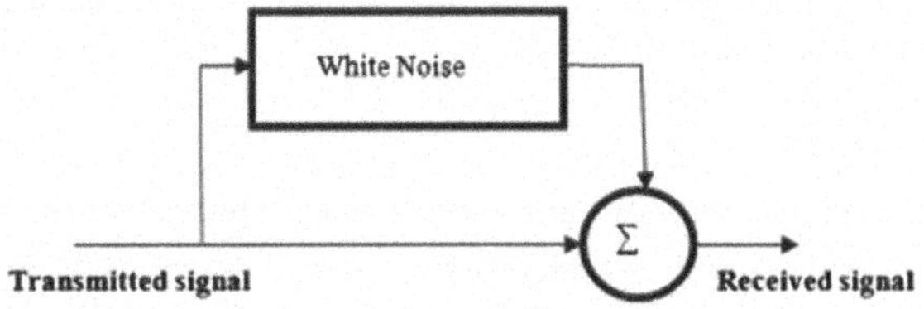

Fig. 1 Representation of AWGN channel

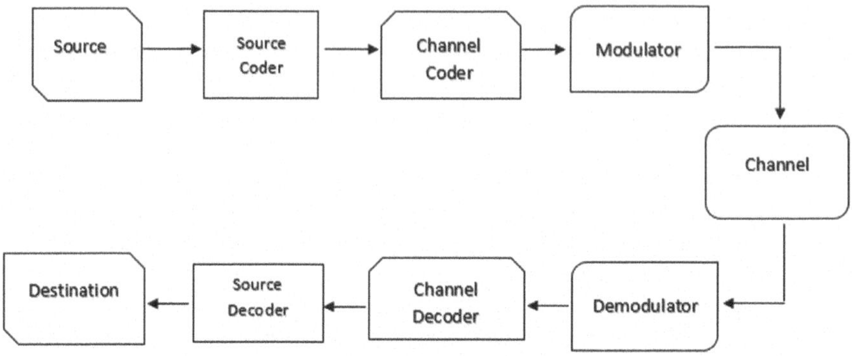

Fig. 2 Demonstration of traditional model

The Rayleigh fading is an ideal model for heavily built urban environments, in a way where there is no dominant propagation along a line of sight between the transmitter and the receiver [18–20].

Rician fading is also a useful model of real-world phenomena in wireless communications, being a stochastic model for the radio signal propagation anomaly, caused by the partial cancellation of a radio signal by itself. The signal reaches the receiver exhibiting multipath interference, and at least one of the paths is changing, lengthening or shortening. This factor occurs when one of the paths, typically a line-of-sight signal, is much stronger than the others [18–20].

The DBPSK modulation eliminates phase ambiguity and the need for phase acquisition and tracking and consequently reduces the energy cost. Using a non-coherent approach to circumvent the need for coherent reference signal at the receiver, the phase of the signal changes between two angles separated by 180° (identical BPSK) with one phase represented by the binary *1* and the other phase by *0* in its constellation. Since it does not require phase synchronization, it is very attractive for digital communications systems and widely used by wireless LANs compliant with the IEEE 802.11 standard [12–14, 17, 18].

The research presented in this section show an AWGN transmission channel with DBPSK modulation. Being used the Simulink simulation environment of the MATLAB® software in its version 8.3 of 64 bits (2014a). Figure 2 presents a model where the signals corresponding to bits *0* and *1* are generated and then modulated in DBPSK, passing through a multipath Rayleigh fading channel with Jakes model with Doppler shift defined at *0.01* Hz, as also inserted a block incorporated which has a math function 1/u.

Such a function is required to track the channel time-variability where the receiver implementation ordinarily incorporates an automatic gain control (AGC). Posteriorly, an AWGN channel follows according to the parameters specified at a sample time of *1* s, with an input signal power of 1 W, initial seed in the generator of the 37th and the 67th channels, with Eb/No ranging from *0* to *14* dB. Then the signal is demodulated to perform the bit error rate (BER) of the channel. The values obtained referring the

BER are sent to the MATLAB® workspace, for further processing and generating of the signal BER graph.

3 Proposed Method

The discrete event model follows in the footsteps of the previous section. Differentiating that now was added the discrete events process of pre-coding (low-level abstraction), being the treatment performed on the signal relative to bit 0, converted into discrete entities, and forwarded for a First-In-First-Out (FIFO) queue with infinite capacity. Furthermore, there is no limit of capacity and retention, with the ordering the bits following the order of arrival, and thus driving data to a server, which has the configuration of service time equal to the simulation time.

This research tackles the bit generation stage. After the signal passes by the server, is converted back to its original format respecting the original format and data type specified, just like maintaining the sampling period. In this way, the signal is modulated in DBPSK and inserted into the AWGN channel, and then demodulated to calculate the BER of the signal.

The model from Fig. 3 incorporates the traditional method with the proposed innovation, and it also highlights the part modeled using discrete events in blue.

The proposed method was implemented in a wireless system with a mobile wireless channel susceptible to several impediments like multipath fading, shadowing, and noise among other interferences [6–8] as seen in Fig. 4. The signals corresponding to the bits *0* and *1* are generated and then modulated in DBPSK. Posteriorly, it passes through a multipath Rayleigh fading channel and other with multipath Rician fading, both containing Jakes model with a Doppler shift defined at *0.01* Hz, as also inserted a block incorporated which has a math function *1/u* [42].

Such a function is required to track the channel time-variability, where the receiver implementation ordinarily incorporates an automatic gain control (AGC). Next, an AWGN channel follows according to the parameters specified at a sample time of

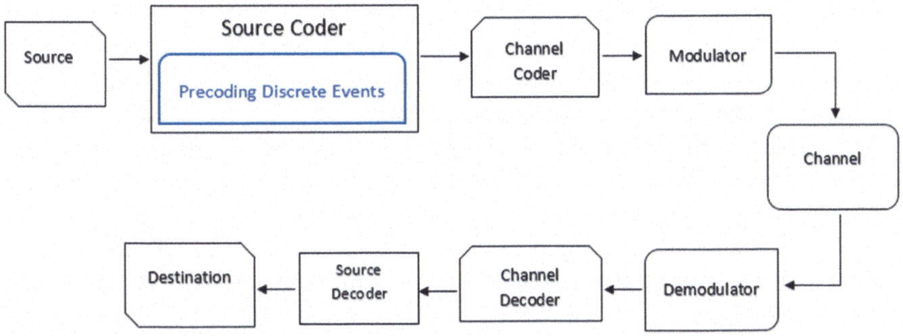

Fig. 3 Model with the proposal of this study

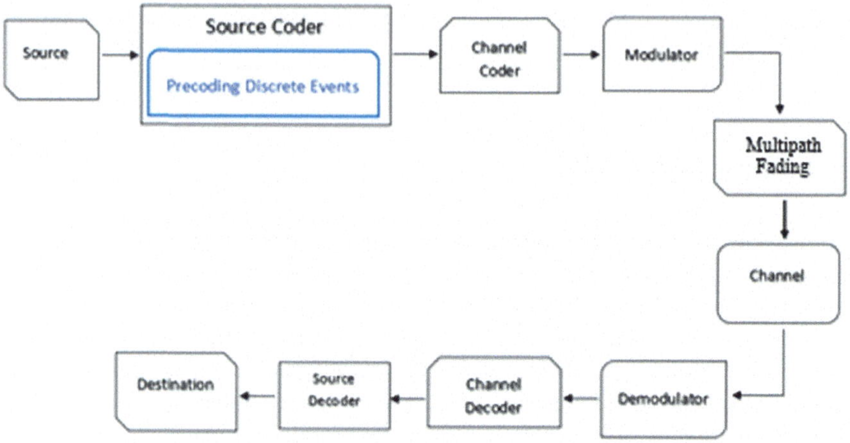

Fig. 4 Wireless model with the proposal of this study

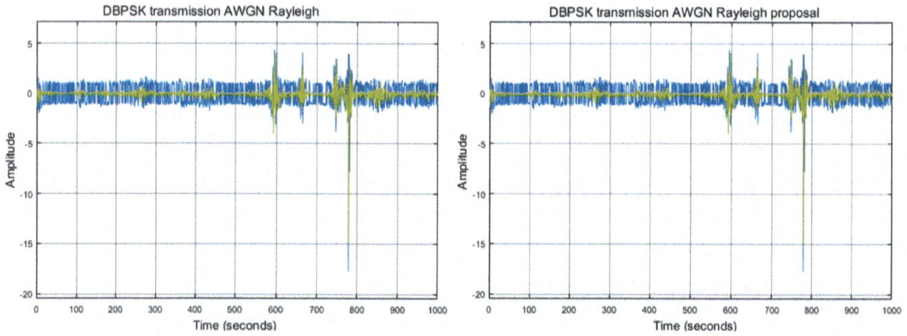

Fig. 5 Transmission flow for DBPSK with Rayleigh distribution

1 s, with an input signal power of 1 W, initial seed in the generator of the 37th and the 67th channels, with Eb/No ranging from *0* to *14* dB. Then, the signal is demodulated to obtain the channel BER.

Figures 5 and 6 use *10,000* s of simulation time. They also show the transmission flow for the DBPSK signal in the proposed (right) and the traditional methods (left) for comparison using multipath fading with Rayleigh and Rician distributions. Note that both methodologies generated the same result.

Also used was the scatterplot command, being a scatter diagram or constellation diagram, viewing the constellation of the modulated digital signal, useful for comparing the performance of one system with the other.

Figures 7, 8 and 9 display the constellations for *5, 10*, e *15* dB for the proposed (below) and the traditional methods (top) in multipath fading with Rayleigh and Rician distributions.

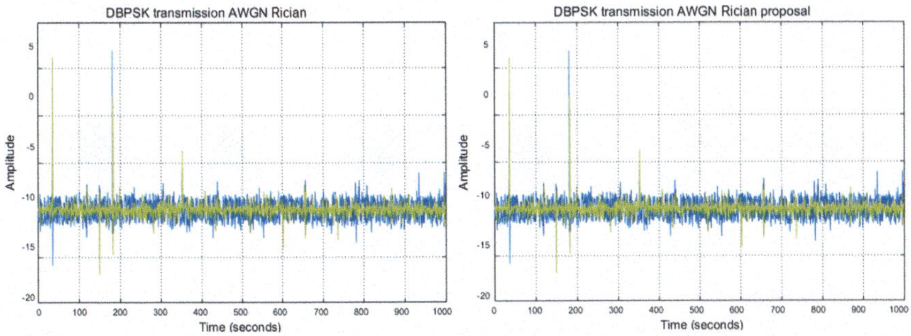

Fig. 6 Transmission flow for DBPSK with Rician distribution

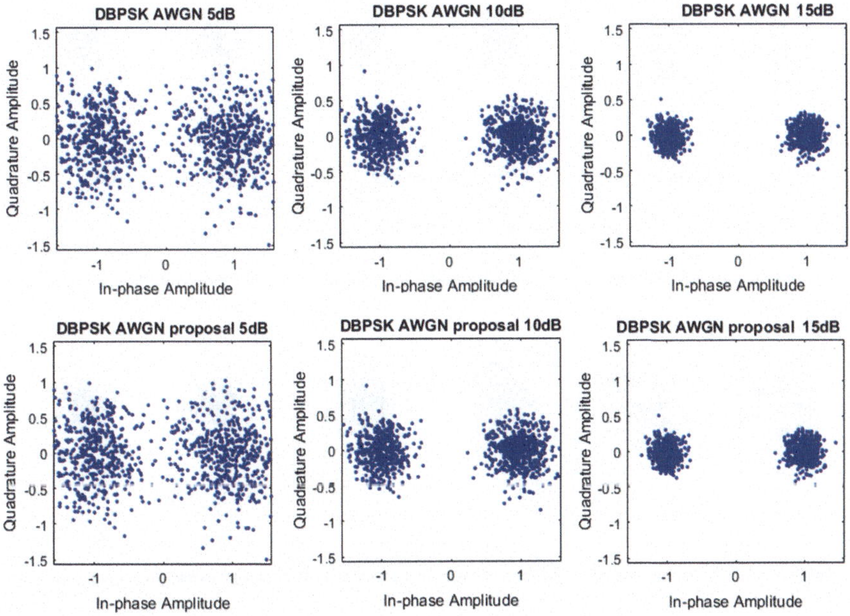

Fig. 7 Simulated DBPSK constellations

4 Results

The results correspond to 5 sequential simulations performed with the models presented previously, on physical machines with different hardware configuration, for an Intel Core i5 processor and 8 GB RAM, and another for an Intel Core i3 processor and 4 GB RAM.

`sldiagnostics` function allows displaying the diagnostic information about the modeling system in Simulink. The `ProcessMemUsage` parameter obtains the sum all of the memory consumption for all model processes in the entire simulation.

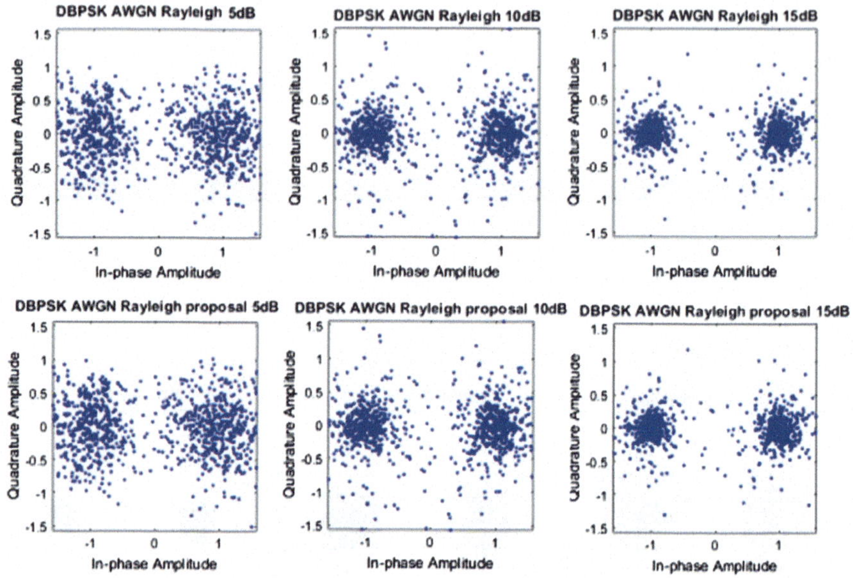

Fig. 8 Simulated DBPSK constellations for the Rayleigh distribution

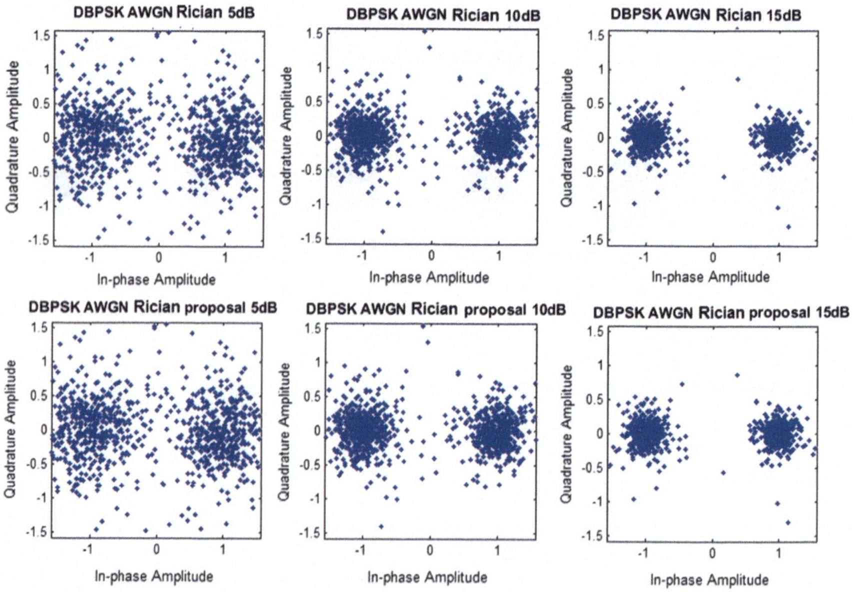

Fig. 9 Simulated DBPSK constellations for the Rician distribution

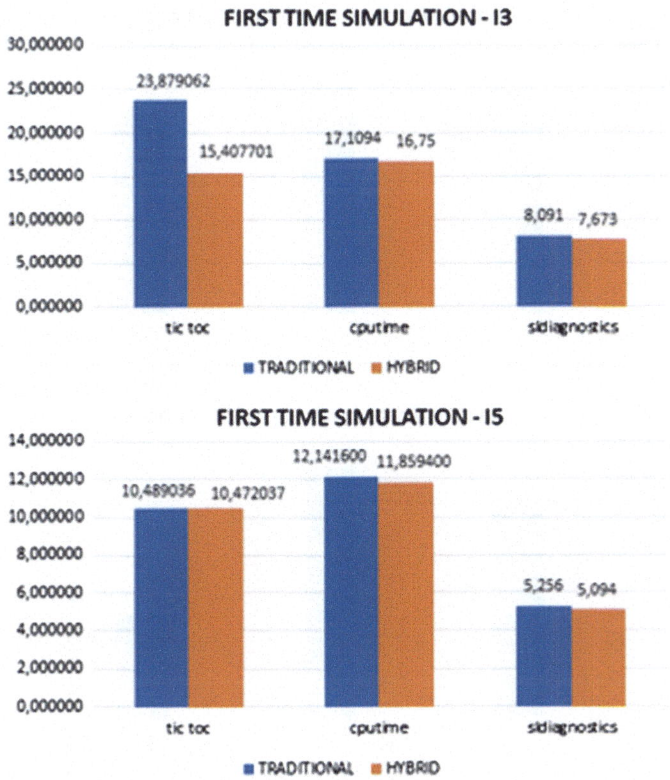

Fig. 10 Time simulation for the DBPSK

`ProcessMemUsage` counts and returns the total amount of memory utilized in each phase of the model in MB. It takes into account that the first simulation of both models, the variables are allocated, and the memory is reserved for the execution of the model according to the Figs. 10, 11, 12, 13, 14 and 15.

So, can be understood that if in a transmission channel containing the proposal and in another the traditional method, they passed the same information content (quantity of bits), without any loss (signal and constellation) and with the same quality (BER). Table 1 and the Figs. 13, 14 and 15 display the results related to the memory consumption of the proposed technique related to the transmitted information compression.

The proposed framework reduced practically all the memory consumption resulting from Rayleigh and Rician MF in models used in simulations with equalization of its resource consumption to a channel without the MF techniques while maintaining all the existing benefits and characteristics, as shown in the Figs. 16 and 17.

To analyze the relationship between the simulation methodology (proposed × traditional method), and the impact on the physical layer of the channel, scripts were made in the MATLAB® for processing of the graph BER.

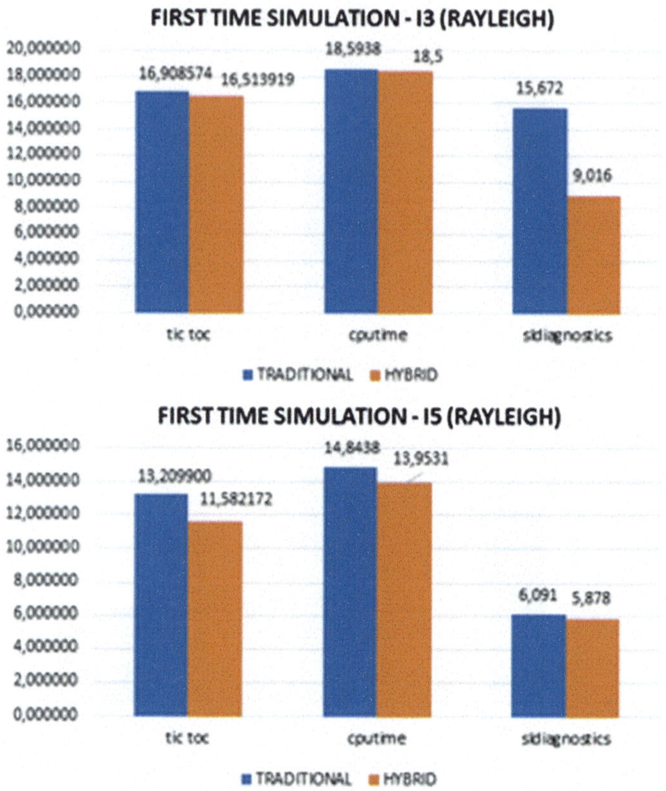

Fig. 11 Time simulation for the DBPSK with Rayleigh distribution

Table 1 Computational improvement

Memory consumption		
Machines	i3	i5
DBPSK Model (%)	22.01	9.22
DBPSK Rayleigh Model (%)	29.72	9.36
DBPSK Rician Model (%)	34.35	11.48

Figures 18, 19, and 20 display the performance of the models during transmission with noise ranging from 0 to *60* dB.

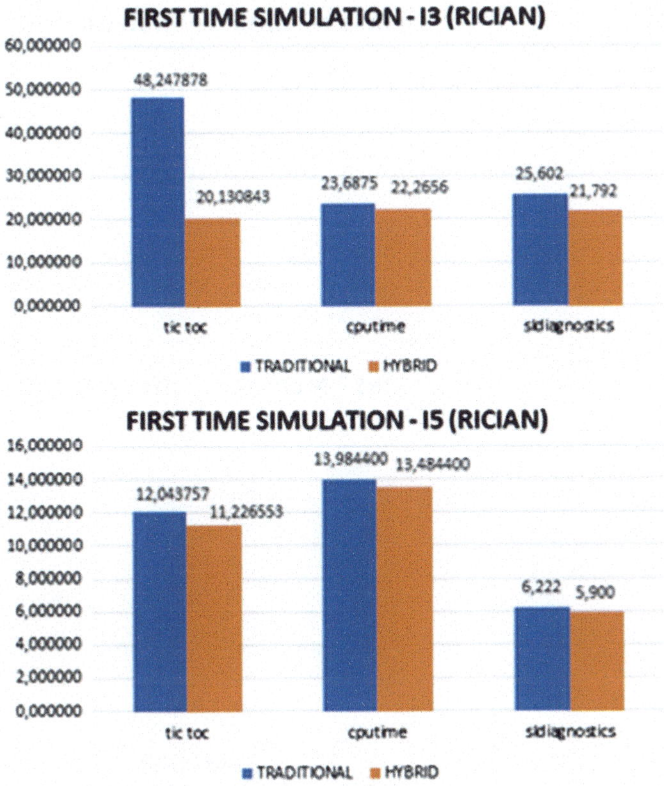

Fig. 12 Time simulation for the DBPSK with Rician distribution

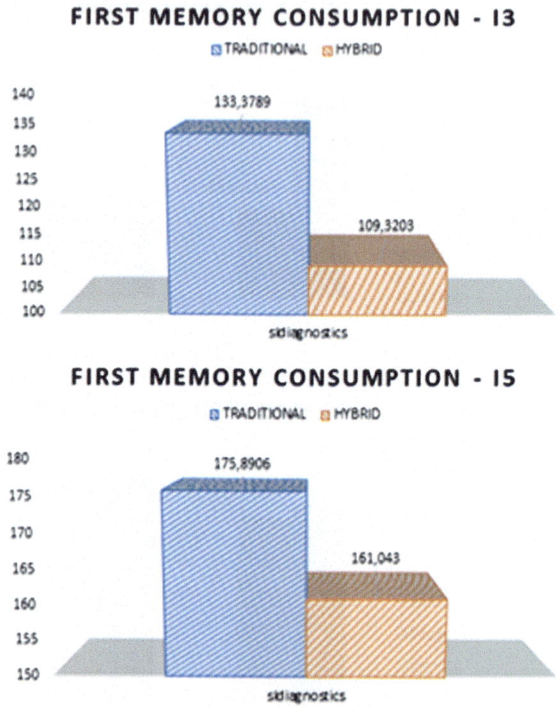

Fig. 13 Memory consumption for the DBPSK

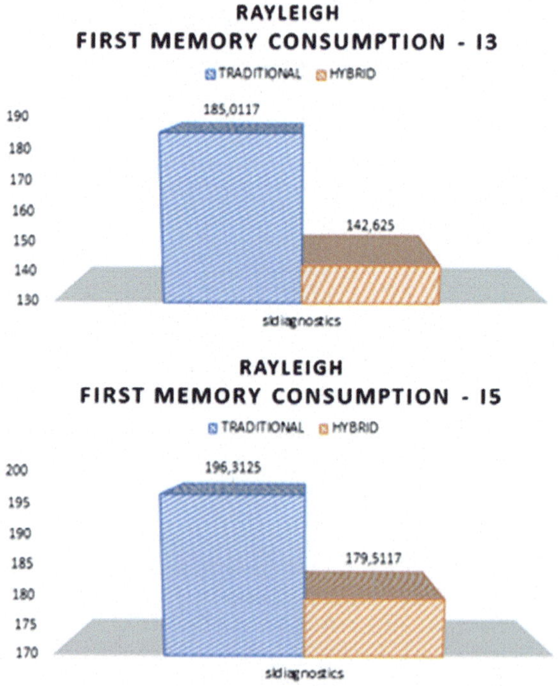

Fig. 14 Memory consumption for the DBPSK with Rayleigh distribution

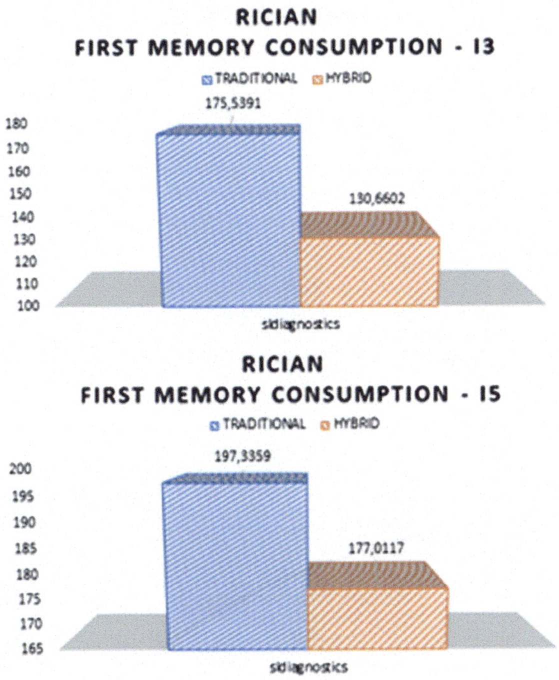

Fig. 15 Memory consumption for the DBPSK with Rician distribution

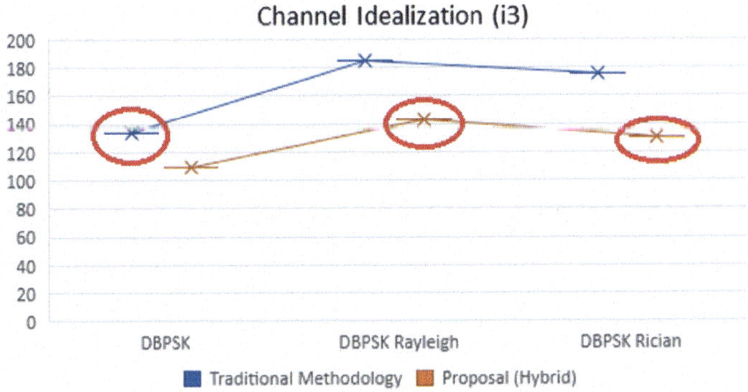

Fig. 16 Idealization for the DBPSK using Intel Core i3 processor

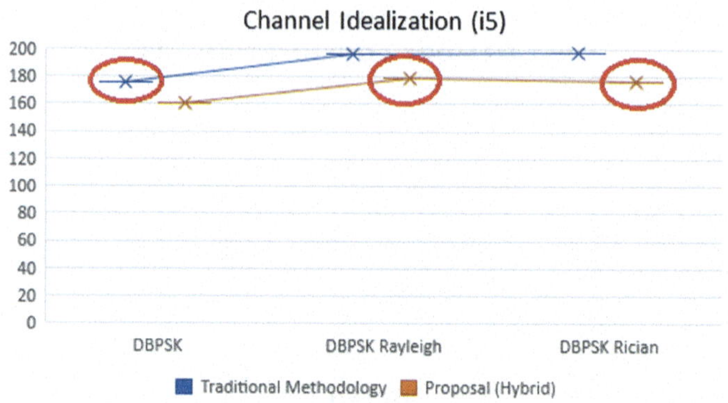

Fig. 17 Idealization for the DBPSK using Intel Core i5 processor

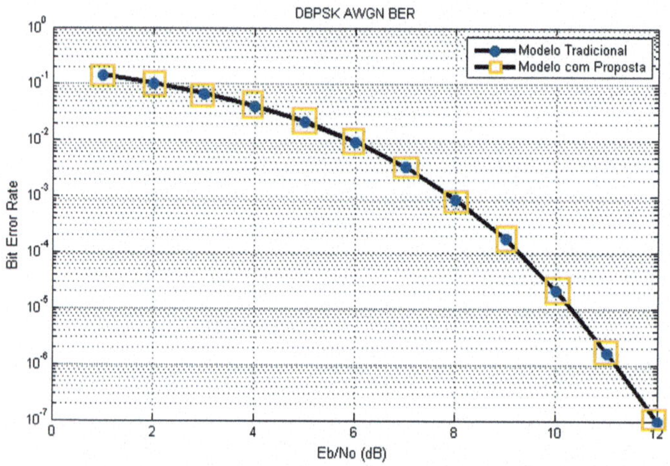

Fig. 18 BER between the models

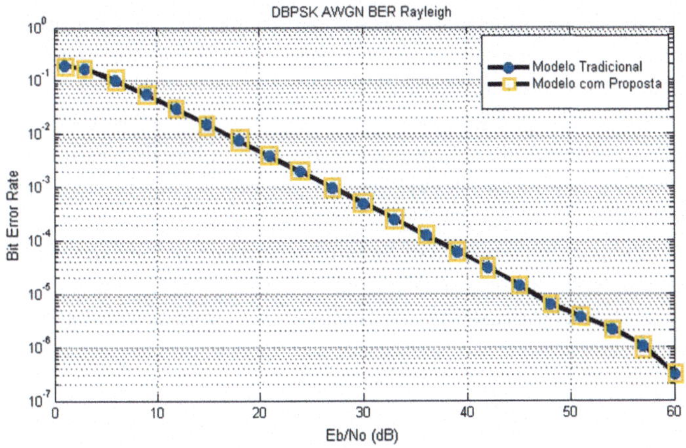

Fig. 19 BER between the models for Rayleigh fading

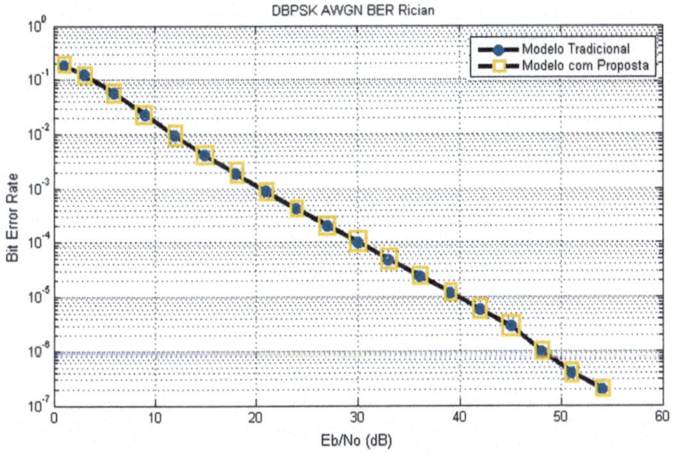

Fig. 20 BER between the models for Rician fading

5 Conclusions

The intent of this research was the use of discrete events applied in the lowest level of abstraction possible in a telecommunication system, such as bit generation. Being this a treatment of the bits before of the modulation process, functioning as a pre-coding process differentiated.

Another critical factor that the proposal achieves is the compensation obtained concerning to the complexity that the multipath fading, Rayleigh, and Rician in this research, related to channel idealization.

The information compression is a byproduct since that the proposal acts on the bits, having a substantial impact on the compression methods performed in higher layers (e.g., HEVC, MPEG-4 AVC/H.264, among others) in a broadcasting system, for example.

Therefore, the presented results of the proposed methodology show an enormous potential for the Non-Orthogonal Multiple Access (NOMA) contexts, credited as the future 5G, thus compensating the additional complexity brought by the MF techniques.

References

1. Digital modulation in Communications Systems. An Introduction. Agilent Technologies (2001)
2. Padilha, R., Martins, B.I., Moschim, E.: Discrete event simulation and dynamical systems: a study of art. BTSym'16, Campinas, SP–Brasil, December (2016)
3. Pereira, F.T., Takano, A.M., Leal, F., Pinho, F.A.: Aplicação da Simulação a Eventos Discretos em um Ambiente Hospitalar Visando a Melhoria no Processo de Atendimento. XLVSBPO, Natal, RN–Brasil (2013)
4. Sharda, B., Bury, J.S.A.: Discrete event simulation model for reliability modeling of a chemical plant. In: Winter Simulation Conference (2008)
5. Hu, W., Sarjoughian, H.S.: Discrete-event simulation of network systems using distributed object computing. In: SPECTS'05 (2005)
6. Sasaki, N.K., Moschim, E.: Simulação de Sistemas de Comunicação Óptica Baseada em Simulação a Eventos Discretos. Universidade Estadual de Campinas. Campinas, SP–Brasil, July (2007)
7. Pissinelli, J.G., Risso, L.A., Picanco, S.R.A., Ignacio, A.S.P., Silva, L.A.: Modelo de Simulação de Eventos Discretos para Análise de Fluxo de Veículos. ENEGEP, Fortaleza, CE–Brasil (2015)
8. Rangel, J.J.A., Costa, J.V.S., Laurindo, Q.M.G., Peixoto, T.A., Matias, I.O.: Análise do fluxo de operações em um servidor de e-mail através de simulação a eventos discretos com o software livre Ururau. Produto & Produção **17**(1), 1–12 (2016)
9. Gomes, E.N., Fernandes, M.S.R., Campos, C.A.V., Viana, A.C.: Um Mecanismo de Remoção de Mensagens Obsoletas para as Redes Tolerantes a Atrasos e Interrupções. CSBC (2012)
10. Godoy, E.P., Lopes; W.C., Sousa, R.V., Porto, A.J.V.: Modelagem e Simulação de Redes de Comunicação Baseadas no Protocolo CAN - Controller Area Network. Revista SBA: Controle & Automação, vol. 21, no. 4 (2010)
11. Forrester, J.W.: Industrial dynamics—after the first decade. Manage. Sci. **14**(7), 398–415(1968)
12. Freeman, R.L.: Fundamentals of Telecommunications. Wiley, Hoboken (1999)
13. Freeman, R.L.: Telecommunication System Engineering, 4th edn. Wiley, Hoboken (2004)
14. Proakis, J.G.: Digital Communications, 5th edn. McGraw-Hill (2008)
15. Tozer, E.P.: Broadcast Engineer's Reference Book. FOCAL PRESS (2012)
16. Whitaker, C.J.: Standard Handbook of Broadcast Engineering. McGraw-Hill (2005)
17. Couch II, L.W.: Digital and Analog Communication Systems, 8th. Prentice Hall (2013)
18. Freeman, R.L.: Fundamentals of Telecommunications. Wiley, Hoboken (1999)
19. Freeman, R.L.: Telecommunication System Engineering, 4th. Wiley, Hoboken (2004)
20. Proakis, J.G.: Digital Communications, 5th edn. McGraw-Hill (2008)

Development of a Hybrid Heating System Based on Geothermal–Photovoltaic Energy to Reduce the Impact of Frosts on Inhabitants of Rural Areas in the Ring of Fire, Southern Peru

Dana Chavarria, Rubi Ramos and Carlos Raymundo

Abstract The recent climate change has forced people to live in extreme conditions, either excessive heat or cold, implying that they must adapt to survive in these situations. However, there are people who, because of their geographical condition and lack of resources, lack the means and tools to combat these climate changes. The context of this study is provided in a rural town located in the Arequipa region (Peru), whose inhabitants have to fight against frosts of up to −20 °C in an area without electricity. A viable solution to this problem is found through the design and implementation of a heating system using geothermal and photovoltaic energy, which are resources found in the area, according to a report of the Ministry of Energy and Mines. This study analyzes and researches the geographical and meteorological conditions of the region, for validating, through theory and simulations, whether the proposed system can supply the thermal energy required to maintain the indoor temperature at a minimum of 15 °C under extreme conditions. The system is designed after analyzing the best technological options and techniques currently available in the context studied for its ultimate financing and establishing guidelines and indicators for monitoring results.

Keywords Renewable energy · Geothermal · Photovoltaic · Frost · Heat exchanger · Outdoor · Efficiency

D. Chavarria (✉) · R. Ramos
Escuela de Ingeniería Industrial, Universidad Peruana de Ciencas Aplicadas (UPC), Lima, Peru
e-mail: u201319437@upc.edu.pe

R. Ramos
e-mail: u201319425@upc.edu.pe

C. Raymundo
Dirección de Investigación, Universidad Peruana de Ciencas Aplicadas (UPC), Lima, Peru
e-mail: Carlos.raymundo@upc.edu.pe

Y. Iano et al. (eds.), *Proceedings of the 4th Brazilian Technology Symposium (BTSym'18)*, Smart Innovation, Systems and Technologies 140,
https://doi.org/10.1007/978-3-030-16053-1_12

1 Introduction

Every year, Peru goes through a serious climatic event known as frosts. Frosts are natural weather phenomena consisting of an abrupt drop in temperature to below 0 °C, which is below the freezing point of water, measured at heights between 1.25 and 2 m above ground level [1]. This phenomenon is similar to cold spells or "friajes," which instead mostly affect high Andean areas. The frost season depends on the climate conditions of each country. For Peru, according to Senamhi, the frost season occurs between May and September, where the lowest temperatures of the year (reaching up to −20 °C) are recorded, especially in southern areas of the country [2]. In this area, the most affected departments included Moquegua, Puno, Tacna, and Arequipa. Frosts impact the inhabitants of these regions in terms of both health and their economic activities (agriculture, cattle farming, etc.) [3]. In fact, in this region, more economic losses are reported due to crop freezing and camelid-related losses, at 10%, than for any other type of meteorological risk [4]. Likewise, cases of chronic respiratory diseases (CRDs) and acute respiratory infections (ARIs) are reported in 20 and 60% of the population, respectively [5]. These same issues have been reported in the highland areas of other countries.

Although the economic losses are important, other secondary effects, such as low income due to a decrease in exports, low cash flow and reserves, limited education, basic services, and transaction difficulties, are also caused in the communities.

This contribution paper is divided into the following four sections: Introduction, which introduces the context and the main problem; State of the Art or Literature, where relevant work and technical studies on the proposed technology, previously compiled and classified, are reported; Methodology and Proposal, which explains the development and calculations for the proposed geothermal–photovoltaic heating system, which we consider to be the option available currently; and System Validation, using simulators such as Homer Energy and Trnsis, which display the system result scenarios. This last part also includes an analysis of the project at the economic level and its social and environmental impacts.

2 State of the Art

Recently, research figures worldwide with respect to renewable energy have increased considerably. The first prototype of the geothermal heating system was generated in France in the XIV century [6] and was composed of only a single part. By 1969, only air pumps were available. Through the years, other pumps were developed for different functionalities. Simultaneously, their performance started to be measured through their Coefficient of Performance (COP).

Merging geothermal energy with another renewable resource was first proposed in 1962 and performed in 1970 by Penrod and Prasanna scientists, in the same country of origin: France. Their purpose was to improve the performance of the application in this country. Solar energy was aggregated to geothermal energy through collectors that store energy in the soil [7]. Since that date, several studies have been conducted focusing on raising the COP and the system efficiency through integrating the energies and simulating their viability.

In Iran, the USA, and China, the evaluation and optimization of a thermal energy system using photovoltaic and geothermal panels have been performed. They performed a simple design to heat an entire building story using vertical systems [8], without issues caused by the external climate. On the other hand, in the Mediterranean, performance was evaluated using only the electrical power supplied by the photovoltaic modules during March, when solar radiation is very unstable, in the range of 7 to 13 UV, which affected the available electric power produced by PFV. Later, another study in Switzerland integrated hydrogen into the system. As a result, during partially cloudy days, the electrolyzer operated intermittently with several failures in hydrogen production. Therefore, the exposure of the climate to the system is a critical factor for its effectiveness and technical success. For these reasons, the system must be completely studied and assessed according to the area where it will be implemented. Depending on the results of these studies and their corresponding validation, the addition of an additional component can be assessed only if the system cannot fully supply itself.

This fact provides enough resources for the implementation of projects. However, studies of geothermal application in Latin American soils are limited. They are mostly projects and/or prototypes in Europe, places that do not have the same geographical conditions or the context of American highlands. Asia is the only developed country with similar conditions. The foregoing validates the importance of this study, which attempts to assess the technical, financial, and environmental feasibility of implementing a carefully selected system in the community of Arequipa in the southern area of Peru [9].

3 Contribution

The proposal consists of three core parts and an additional section. The core parts were selected based on the following matrix, and the additional section is a research appendix dealing with the economic and financial analyses of the project Fig. 1.

The following chart displays the sequence of data or studies that must be performed in advance to make an optimal choice of the types and components of the hybrid system Fig. 2.

Diagnosis

Concept: Analysis performed to determine any situation. This determination is made based on data and facts systematically gathered and organized, which allow for better insights on the current situation.

Importance: An assessment of the study location is very significant, since it provides insights on the geographical and meteorological conditions. This information is extremely important for the feasibility study of installing a hybrid heating system.

Interpretation: In this study, this item is very important since, without it, the assessment of energy demands of the system cannot be initiated, nor can the minimum power required or produced be determined.

Energy Demands

Concept: Demand means the quantity of goods or services acquired by consumers in a specific unit of time (a day, a month, a year, etc.), since without a time parameter, it is not possible to determine whether the demand grows or decreases. On the other hand, Energy refers to the science that deals with the production, use, and transformation of energy.

Importance: Energy demand calculation, both the minimum and maximum energies required, provides a notion of how much energy the system must produce to comply with the heating conditions required by the place of study, against different climate changes, variations, and conditions.

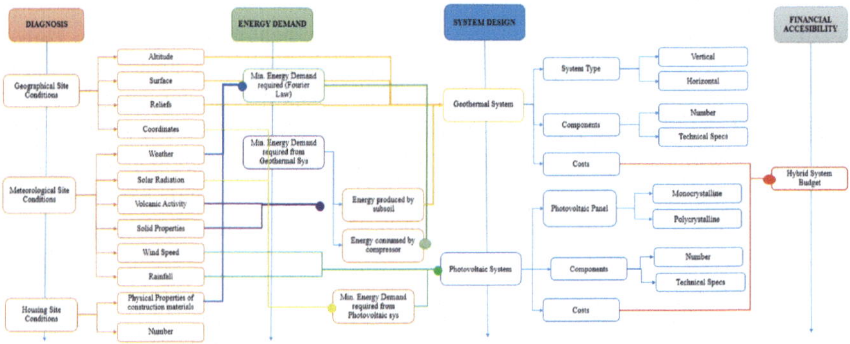

Fig. 1 Abstract flowchart for the proposal

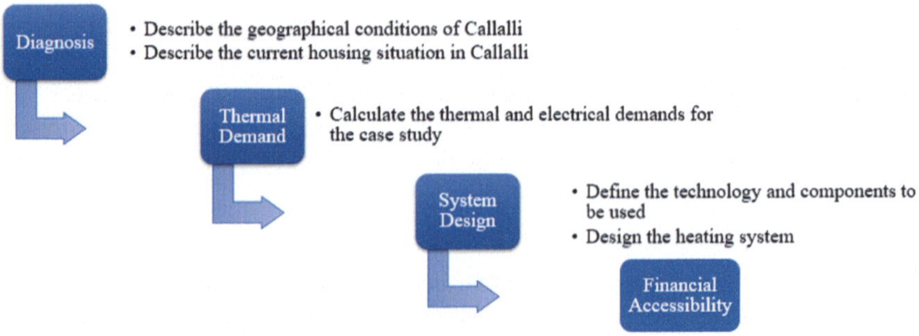

Fig. 2 Data and study sequence matrix for the proposal

Interpretation: In the case study, the information on energy demand is essential to determine if the proposed hybrid system works, that is, if it can supply the minimum energy required to maintain the Callalli homes at a minimum room temperature of 15 °C, regardless of the outdoor temperature.

System Design
Concept: Design indicates the final result of a process, which seeks to find a solution to a problem that aims to project useful and aesthetic objects. On the other hand, System refers to an ordered set of procedures, standards, etc., that regulate the operation of a group of processes.

Importance: Defining our system is essential, because with this analysis, we can identify the number of components needed and their technical specifications, and therefore, can determine system costs, such as maintenance costs.

Interpretation: In this case study, the geothermal hybrid system comprises polycrystalline solar panels and a vertical geothermal system.

Once the proposal was defined, we proceeded with the case study. As already mentioned, the location is the Callalli district in Peru.

4 Implementation

To validate the performance calculation of the hybrid system, we used a TRNSYS simulator. For these purposes, we first designed the model house in Google Sketch Up (Fig. 3), and then, exported it to TRNBuild and added information (type 56)

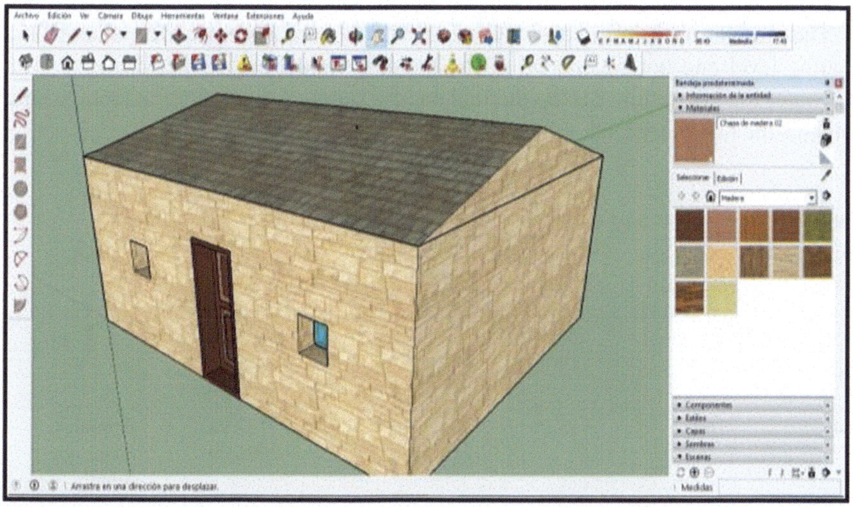

Fig. 3 Design of the model house in Google sketch-up

regarding the thermal behavior of the construction model, and finally, simulated this behavior in TRNSYS through the 3D model.

The model simulated in TRNSYS is of a geothermal/photovoltaic hybrid system, which consists of the following:

Photovoltaic Energy:

First, we obtained the weather file of Callalli (Peru) using Type 109-TMY2. These data were recorded in °C, so a K degrees converter was placed since the polycrystalline panel (type 94a) works with these units. displays the simulation of the photovoltaic system, since we need to verify whether the calculated theoretical energy supply is close to that simulated. That is why, it is programmed in the output printer (type 65c), for which the Power vs. Efficiency chart is shown below.

The maximum power generated in W by the photovoltaic panel of 24 is 169, which only satisfies the hybrid system (compressor) when the outdoor temperature is less than −10 °C. Therefore, it would generate thermal shortage to the compressor, since this component requires electrical energy to work if the outdoor temperatures fall within the range of −10 to −20 °C in Callali. However, we are interested in determining whether the hybrid system (photovoltaic/geothermal) will generate enough energy to maintain the room temperature at 15 °C (Table 1).

If the outdoor temperature is −10 °C, the polycrystalline photovoltaic panel requires 118% of energy supply, and at −20 °C, it requires 84% of the supply.

Geothermal Energy:

Once the energy supply has been simulated and validated, the design of the hybrid system is finalized. This hybrid system comprises a closed-loop earth heat exchanger (GHE) (Type 557), serving as a connection between the solar collector fields, and the GHE evaporator (Type 668); four fan coil tubes, two tubes for the heating circuit and another two for cooling (type 987); a water heat pump (type 927), (type 560); heat storage with variable inputs and uniform losses (type 64); and a pump (type 3d).

The simulation determines whether the heat source system corresponds to the heating level and hot water load of the building model and its performance. The hot water load determines if the thermal energy supplied from the heat storage tank satisfies the hot water load.

Table 1 Energy supply analysis

Panel (m^2)	Simulated electrical power (W)	Simulated electrical power provided (Wh/day)	Outdoor temperature (°C)	Electrical power consumed by the compressor (Wh/day)	Condition
24	169	2535	−10 −20	2132.92 2986.09	Supplies Does not supply

Note This table compares the energy supply between the energy power consumed by the compressor and the simulated power of the photovoltaic panel. Prepared by the Authors, 2018

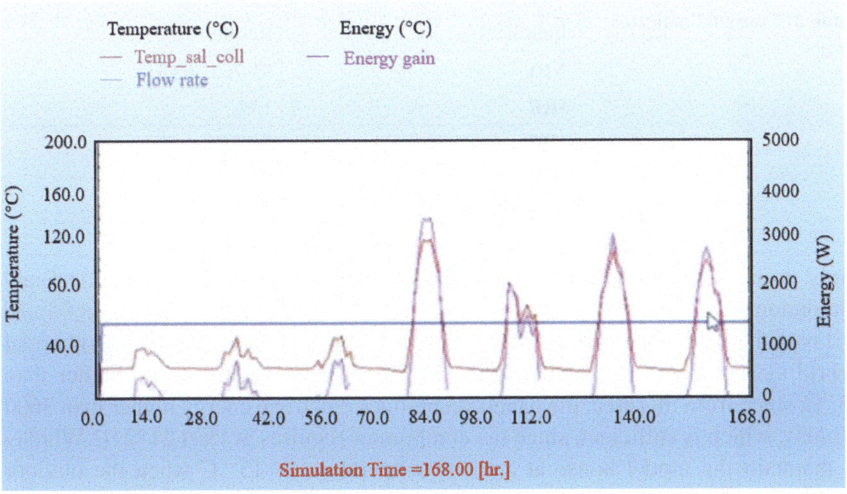

Fig. 4 Temperature versus energy simulation chart

In addition, the calculator is used to describe the quantity of hot water supplied by the heat storage tank per day. Hot water is delivered at 45 °C by the storage tank, where the water temperature is 15 °C because the storage tank is buried in the ground, and therefore, fixed at subsoil temperature. Based on the demand of hot water per hour, the storage tank is filled again to maintain the tank level with water at 15 °C and replace the hot water used Fig. 4.

On the other hand, the minimum electrical power generated by the system is 0.97 KWh and the maximum is 3997 KWh. Theoretically, 1324 KWh is required to maintain the house at 15 °C at an outdoor temperature of −20 °C, so the system in Callalli does generate enough energy.

Even though the hybrid system supplies the thermal energy required to maintain the house at 15 °C at a temperature of −20 °C, the simulation of photovoltaic energy evidences that the module 24 does not produce the energy required by the compressor.

5 Results

Technical Level

The simulation aimed to assess the results obtained from the simulators with the theoretical data, in order to validate the calculation of thermal supply for the hybrid geothermal/photovoltaic system in Callalli. According to the report provided by the Ministry of Energy and Mines and the Volcano Observatory of the INGEMMET (OVI) [34], this town is considered a viable land for the extraction of geothermal energy, because in this area, four active volcanoes are located. However, technical

Table 2 Financial indicator results

Indicators	Results
NPV	S/. 766,762.31
IRR	13%
CBR	S/. 1.09
PRD	9.43 years

study projects are hard to find, so it was necessary to conduct a theoretical and simulation study.

From the studies performed, it was found that the photovoltaic and geothermal hybrid system is efficient if the solar panel area is at a temperature higher than 33 °C, since from this area, the panel can produce electrical energy higher than 3600 Wh/day, which is sufficient, since the compressor requires at least 2132.92 Wh/day to maintain the model house at a room temperature of 15 °C when the outdoor temperature is −10 °C and 2986.09 Wh/day if the temperature falls below −20 °C. In conclusion, the scenarios assessed the evidence that with a photovoltaic panel of 33, the supply of electric power between the photovoltaic panel and the compressor is more efficient, for outdoor temperatures of −10 and −20 °C with 169% and 121% efficiency, respectively. That is, the electrical energy produced by the panel will be greater than that consumed by the compressor.

Economic Level

Financial indicators were calculated and assessed through the construction of the cash flow based on the income and expenses already determined, throughout the 10-year lifespan assigned. The following results were obtained (Table 2):

The NPV is positive, meaning that the government would not only stop investing, but also save much more. The IRR is 13%, which is higher than the discount rate (9%). That is, the Internal Rate of Return would cover the minimum rate of return required for the investment. The Cost–Benefit Relation is greater than 1, which means that for every sol that is inverted, S/1.09 is returned. PRD is set at 9.3 years.

6 Conclusions

Using the simulation, we could verify that the hybrid system satisfies the necessary thermal energy, so that the house is at a minimum temperature of 15 °C at any outdoor temperature. The system produces a minimum energy of 0.97 KWh and maximum of 3997 KWh, and according to the theoretical calculations, 1.324 KWh is needed to maintain the house at 15 °C, considering that the outdoor temperature is the most critical (−20 °C). On the other hand, with the implementation of the hybrid system, when supplying thermal energy to the 288 families, 502,396.95 kg of CO_2 that would be generated if another heating method were used, which would go directly to the ozone layer, would be prevented.

References

1. Organización de las naciones unidas por la Alimentación (FAO).: El daño producido por las heladas. Roma: FAO. Recuperado de http://www.fao.org/docrep/012/y7223s/y7223s02.pdf [Consulta: 16 de Junio del 2018] (2010)
2. Servicio Nacional de Meteorología e Hidrología del Perú (SENAMHI).: Documentación de distritos afectados por la ola de friajes. Lima: SENAMHI. Recuperado de https://www.senamhi.gob.pe/?p=distrito-helada-friajes [Consulta: 16 de Noviembre del 2017] (2013)
3. Instituto Nacional de Estadística e Informática (INEI).: Mapa de pobreza provincial y distrital 2013. Lima: INEI. Recuperado de https://www.inei.gob.pe/media/MenuRecursivo/publicaciones_digitales/Est/Lib1261/Libro.pdf [Consulta: 16 de Noviembre del 2017] (2014)
4. Ministerio de Agricultura y Riego (MINAGRI).: Monitoreo de efectos de las heladas y friajes en el Perú. Lima: MINAGRI. Recuperado de http://www.senasa.gob.pe/senasacontigo/minagri-monitorea-efectos-de-heladas-y-nevadas-en-el-sur-del-pais/ [Consulta: 16 de Junio del 2018] (2018)
5. Ministerio de Salud (MINSA).: Registros de casos de enfermos y descensos. Lima: MINSA. Recuperado de http://www.minsa.gob.pe/ [Consulta: 16 de Mayo del 2018] (2018)
6. Kim, M.B.: Tax evasion, income inequality and opportunity costs of compliance. In: Proceedings, 3. (2003) https://doi.org/10.2307/41954396
7. Rodríguez, V.M.: Ver para leer: diseño y análisis técnico-económico de un sistema de climatización urbana con aprovechamiento de geotermia de baja entalpía en un proyecto de viviendas de integración social en chile. (Tesis de licenciatura, Universidad de Chile, Facultad de ciencias físicas y matemáticas, Santiago de Chile, Chile) (2015)
8. Androulakis, D., Armen, N.G., Bozis, K.A., Papakostas, K.: Simulation of the thermal performance of a hybrid solar-assisted ground-source heat pump system in a school building. Int. J. Sustain. Energ. **37**, 1–14 (2016). https://doi.org/10.1080/14786451.2016.126186
9. Arévalo, J., Quispe, G., Raymundo, C.: Sustainable energy modal for the pro-duction of biomass briquettes basad on rice huskin low-income agricultura! areas in Peru. Energy Procedia **141**, 138–145 (2017)

Reuse Method for Deposits of Polymetallic Tailings in a State of Abandonment Through the Application of Mineral Flotation

Anthony Anchiraico, José Bazo, Vidal Aramburú and Carlos Raymundo

Abstract In the mining industry, due to the extraction and processing of ore, a significant amount of tailings are produced which are discharged into deposits. In some cases, these are in contact with effluents and generate acid waters that pollute the environment and affect the health of the inhabitants. Additionally, these tailings within their composition contain valuable metals that can be reused through a process of mineral flotation. As a part of this research, a study based exclusively on laboratory tests was performed on the tailings deposit located in the Recuay–Peru district, where the presence of concentrations of Pb, Zn, and Ag at 48.36% was obtained, 23% and 250 g/TM. Thus, this study aims to take advantage of polymetallic tailings deposits that are in an abandonment state by extracting valuable ore through the flotation process based on their chemical composition and mineralogical characterization.

Keywords Mining tailings · Tailing deposits · Reutilization · Mineral flotation

1 Introduction

The global mining industry has exhibited deficiencies in its ability to dispose 5–7 trillion tons of tailings that are produced annually [1]. The aforementioned statistics contextualize a real and current situation that affects Peru, a mining country and one

A. Anchiraico (✉) · J. Bazo · V. Aramburú
Escuela de Ingeniería de Gestión Minera, Universidad Peruana de Ciencas Aplicadas (UPC), Lima, Peru
e-mail: u201416512@upc.edu.pe

J. Bazo
e-mail: u201314111@upc.edu.pe

V. Aramburú
e-mail: pcgmvara@upc.edu.pe

C. Raymundo
Dirección de Investigación, Universidad Peruana de Ciencas Aplicadas (UPC), Lima, Peru
e-mail: Carlos.raymundo@upc.edu.pe

Y. Iano et al. (eds.), *Proceedings of the 4th Brazilian Technology Symposium (BTSym'18)*, Smart Innovation, Systems and Technologies 140,
https://doi.org/10.1007/978-3-030-16053-1_13

of the largest ore producers in the world. This situation requires the analysis of the possible impacts generated by the large amounts of tailings that are not managed properly, such as the filtration of pollutants, collapse of dams by over accumulation, and occupation of large tracts of land. This requires the assessment of the tailings deposited in past decades because there are great possibilities for finding important metallic concentrations for future extraction.

In Peru, 8616 abandoned properties are currently registered of which 7531 currently lack an owner, thus leaving the State as the sole responsible party [2]. By the beginning of the 20th century, several mining companies started operating in the Peruvian territory without any environmental regulations. This background, added to the technological limitations of the time, accounts for the poor waste disposal, which is known as Mining Environmental Liabilities (PAM). Mining tailings are a type of PAM that, according to the current Peruvian regulations, must be identified and assessed to establish the best strategy for their treatment. The main characteristic of tailings in the state of abandonment in Peru is their incorrect disposition since they have been present for more than 60 years and possess a great risk of contamination to the environment and human health.

This study aims to overcome these environmental issues, since, in Peru, there are ~450 polymetallic mines in a state of mining liabilities, which represents an alteration to the physical environment, such as soil, water, and air [3]. For this reason, this study aims to propose a reuse method to deposit polymetallic tailings in a state of abandonment by applying mineral flotation. This research study is divided into 5 Sections: Section 2 contains the state of the art regarding the research topic. Section 3 covers the development of the method proposed. Results and validation are described in Sect. 4. Finally, conclusions are listed in Sect. 5.

2 State of the Art

The initial analysis of an inactive tailings deposit is directly linked to its geochemical and physical characteristics to determine the type of disposal it requires, which may be technical or final closure, application of control and isolation techniques, reuse as a hydraulic landfill for underground works, and reuse for brickworks or management [4].

Just the way the geochemical study is relevant to the present study, in a similar manner, the physical stability analysis of a tailings deposit. This requires an analysis of the local geology and the structural behavior generated in the deposit slope caused by the tailings it contains since they generate a pressure and there is a linear deformation that can modify its angle of inclination [5].

In China and other countries, the methodology used to reuse tailings starts with a chemical analysis and mineralogical characterization, which are performed through laboratory studies and ray diffraction (XRD), respectively. This stage selects the appropriate method to recover valuable ore [6]. Likewise, different metallurgical processes have been observed to provide economic value to an ore found as waste, such as the bioleaching process, which is a technology that uses bacteria to extract valuable ore [7].

The mineral flotation technique may be applied to polymetallic tailings in a state of abandonment or produced in past decades, depending on their mineralogy and tonnage. One example is a tailings deposit located in Chile, which is still active but has tailings stages discharged since 1930. These have been assessed based on the information, i.e., no previous tailings studies have been performed. Instead, existing data have been used. In addition, the authors designed a classic mineral flotation circuit at the laboratory level, which consists of Rougher and Scavenger phases [8]. According to this precedent, the tailings deposit reuse feasibility analysis does not necessarily have to be in-the-field and experimental but can also be extracted from previous information and theoretically.

The supporting previous studies show a pattern in the methodology to demonstrate that a deposit can be reused, which is the chemical verification of the tailings from mineralogical characterizations, chemical analysis, and metallurgical tests. These procedures are performed using field collected data or data from historical records. However, the feasibility of the tailing treatment requires both a chemical and a physical analysis, since, if waste may be transferred for the second time to a processing plant, it must be extracted from the area where it is disposed. Therefore, studying its stability becomes critical.

3 Contribution

The technique implemented herein is the recovery of valuable ore from different abandoned tailings deposits through the mineral flotation, i.e., to perform laboratory tests with an objective of generating business opportunities and economic benefits for mining companies (Table 1).

3.1 Proposed Methodology

The study focuses on the reuse of abandoned polymetallic tailings so that they can be sustainably extracted and treated in a metallurgical plant, i.e., generate business opportunities from Mining Environmental Liabilities (PAM), which currently generate maintenance costs to the State and are prone to contaminate their areas of influence. For these purposes, a work methodology is proposed, as described in Fig. 1.

Table 1 Analysis of current techniques and this proposal

Mining environmental liabilities (currently)	
Solution	Disadvantage
Use for progressive closure of mining operations	Pollution
	Costs due to selection processes
	Much of the tailings do not comply with the appropriate granulometry
Remediation of affected area	High maintenance costs assumed by the Peruvian State
Mining environmental liability (applied technique)	
Proposed solution	Advantages
Recover valuable ore	Generate business opportunities
	Economic, social and environmental benefits

4 Validity

The tailings deposit in the state of abandonment studied is located in the district of Ticapampa, province of Recuay and Aija, department of Ancash on the banks of

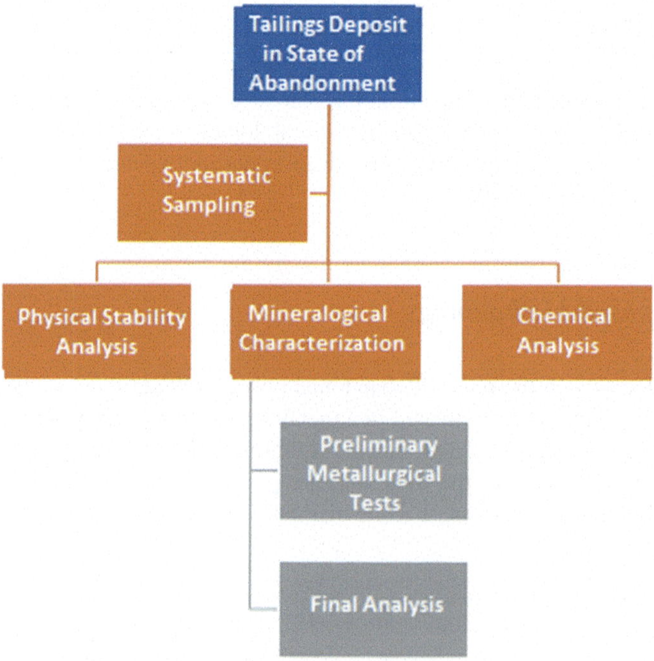

Fig. 1 Proposed methodology

the Santa River, whose receiving body is in direct contact with tailing drains, thus limiting its use for human consumption.

As for its sizing, it has an extension of 750 × 200 m and a maximum height of 19 m, which would represent approximately 4 MT [2]. In addition, its geochemical composition presents heavy metals in important concentrations, which are attributed to 2 aspects.

First, there is a presence of As, Cd, and Pb (Fig. 2) in proportions above the maximum permissible limits established by the National Standards for Environmental Water Quality for human consumption, which generates a concern regarding water leaks from the tailings to the Santa River.

The second aspect refers to the opportunity of extracting certain ore due to their high concentrations, which opens the option to remedy current mining liabilities in a profitable way.

4.1 Systematic Sampling

The amounts required are approximately 40–50 kg of tailings. For the current study, not only representative samples were needed but also an adequate amount was randomly extracted from each Ticapampa tailings deposit zone. This consisted of a selection of fragments from the tailings.

4.2 Physical Stability Analysis

In the case of tailings, the Ticapampa deposit has a cohesion value of 18.5 kPa, a weight of 1.8 T/m^3, which is equivalent to 17.65 kN/m^3 and an angle of friction at 22°. From these data obtained from empirical studies, the depth of the stress crack was calculated as follows:

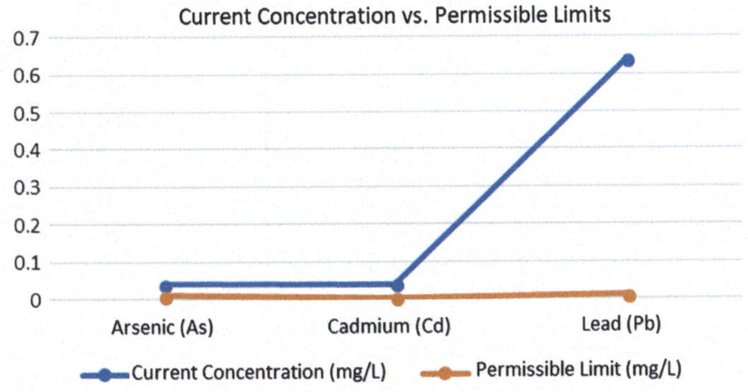

Fig. 2 Comparative heavy metal concentration chart

$$Zc = \frac{2\left(18.5\frac{KN}{m^2}\right)}{17.65\frac{KN}{m^3}} \tan^2\left(45 + \frac{1}{2}\left(22^{\circ}\right)\right) = 4.6072\,\mathrm{m}^{-1}$$

$$Zc = 0.2171\ \mathrm{m} = 21.71\ \mathrm{cm}$$

In this way, it can be deduced that due to the horizontal forces acting on the slope, cracks with a maximum length of 21.71 cm are generated, and their behavior may be affected by the filtration of materials.

The type of soil present in the tailings deposit of Ticapampa is clayey silt, which has a consolidation coefficient of 5 m^2/min, an average drainage time of 15 min and a drainage length of 4.5 m. From the data described, it can be calculated that the resulting dimensionless factor is 3.70, which indicates that the tailings deposit under study is in drained condition. Then, the analysis method must consider the effects of water on the physical stability of the structure. This procedure is the one applied by Bishop. Considering a stem base length of 0.5 m, a height of 2 m, and a vertical angle of 26.5°, these values are replaced in the formula, and we obtain FS = 1.14.

4.3 Mineralogical Characterization

In Fig. 3, the metal with greater participation within the composition is lead, which is present as galena or lead glance, anglesite, and litharge.

Fig. 3 Mineralogical tailings composition

MINERALS	FORMULA
Quartz	SiO_2
Pyrite	FeS_2
Chlorite	$(Mg,Fe^{2+})_5Al(Si_3Al)O_{10}(OH)_8$
Calcite	$(CaCO_3)$
Sodalite	$Na_4(Si_3Al_3)O_{12}C$
Diaspore	AlO(OH)
Chloromagnesite	$MgCO_3$
Pyrophillite	Al2Si4O10(OH)2
Spharelite	ZnS
Nantokite	CuCl
Galena	PbS
Anglesite	$PbSO_4$
Anorthite	$CaAl_2Si_2O_8$
Litharge	PbO

Table 2 Chemical tailings composition

Concentration of metals of interest (%)			
Pb (%)	Zn (%)	Cu (%)	Ag (g/TM)
4.97	0.23	0.11	26.80
Other metals (ppm)			
Al (%)	As	Ba	Be
5.99	10,000	200	0.5
Bi	Ca (%)	Cd	Co
11	0.76	0.5	3
Cr	Fe (%)	Ga	K (%)
13	6.53	10	2
La	Mg (%)	Mn	Mo
10	0.51	564	1
Na (%)	Ni	P	S (%)
0.15	1	680	1.67
Sb	Sc	Sr	Th
221	7	69	20
Ti (%)	Au	V	U
0.24	0.514	77	10

4.4 *Chemical Analysis*

As mentioned above, the chemical analysis of the tailings under study was performed in a laboratory from the representative sample obtained. In Table 2, the metal with the largest share in terms of concentration is lead, with a grade of 4.97%.

4.5 *Preliminary Metallurgical Tests*

Based on the results above, the mineral flotation metallurgical process was selected, since the metal of interest is lead and is mostly contained in galena (PbS), which is a sulfide and may be treated efficiently with this technique. The proposed metallurgical process consists of 2 stages, with the objective of obtaining lead and Zinc concentrates, respectively, with the following arrangement of the reagents:

Pb flotation-first stage	
• Cell pH	8.5
• Lime	3.0 g
• $ZnSO_4$ (5%) (depressor)	7.0 ml
• NaCN (1%) (depressant)	3 ml
• AP-3418 (collector)	1 drop
• MIBC (foaming)	1 drop
• Conditioning time	5 min
• Flotation time	8 min

Zn flotation-second stage	
• $CuSO_4$ (5%) (reactivator)	5 ml
• Lime	4.0 g
• pH	11
• Z11 (1%) (collector)	3 ml
• MIBC (foaming)	1 drop
• Conditioning time	5 min
• Flotation time	6 min

5 Conclusions

Herein, the proposal assessed is associated to the Ticapampa tailings deposit, which is abandoned and a potential risk for the surrounding inhabitants. As shown in Figs. 2 and 3, the Ticapampa deposit can be reused using the mineral flotation technique, since this deposit contains ore of economic value today. In addition, the operating costs of a tailings deposit are much lower than a deposit.

According to the calculated head from the metallurgical balance, the lead grade is 5%, zinc is 0.87%, copper is 0.3%, and silver is 29.53 g/MT. This means that the lead grade is attractive for recovery, while zinc and copper are relatively low. In the case of the copper grade, which is 0.11% in the laboratory analysis, its metallurgical recovery by mineral flotations is not feasible, since other tailings studies present copper grades close to 0.30% using the same technique.

Based on the metallurgical balance, the quality of the lead concentrate obtained is 48.36%, which is close to the marketing conditions. The silver grade is 250 g/T, which gives it a very important economic value. Lead and silver concentrate prices are held in different international markets at an average of 2200 USD/MT and 17 USD/Oz, respectively. Regarding the quality of lead concentrate, there is no minimum standard for its recovery because the said activity is performed based on the percentage obtained.

The application of the mineral flotation technique to recover valuable ore in tailings deposits that are in a state of abandonment is quite attractive since business opportunities are generated by working a mining liability that generates a significant expense close to 20 million soles for the Peruvian State. Furthermore, the money obtained from the recoveries of valuable ore may generate development for the surrounding community, as well as environmental remediation of the affected area.

References

1. Liu, S., Zhao, Y., Wang, W., Wen, S.: Beneficiation of a low-grade, hematitemagnetite ore in China. Miner. Metall. Process. **31**(2), 136–142 (2014)
2. Gestión, D.: Cartera de proyectos mineros en Perú asciende a US$ 45,596 millones, ¿en qué situación está?. [En línea] Disponible en: https://gestion.pe/economia/cartera-proyectos-mineros-peru-asciende-us-45596-millones-que-situacion-esta-2170852 (2016)
3. Romero, A.A., Flores, S., Pacheco, W.W.: Estudio de la calidad de agua de la cuenca del río Santa. Revista del Instituto de Investigación de la Facultad de Ingeniería Geológica, Minera, Metalurgica y Geográfica **13**(25), 61–69 (2010)
4. Espín, D., Jarrín, J., Escobar, O.: Manejo, gestión, tratamiento y disposición final de relaves mineros generados en el proyecto Río Blanco. Revista de Ciencias de Seguridad y Defensa. 1–13 (2017)
5. Aguirre, E.: Análisis de la Estabilidad Física del Depósito de Relaves N°5 de la Concesión de Beneficio de Minera Titán del Perú S.R.L., título de pregrado, Universidad Nacional San Agustín de Arequipa. Arequipa, Perú (2017)
6. Zheng, Q., Bian, X., Wu, W.Y.: Iron recovery and rare earths enrichment from Bayan Obo tailings using Coal-Ca (OH) 2-NaOH roasting followed by magnetic separation. J. Iron. Steel Res. Int. **24**(2), 147–155 (2017)
7. Falagán, C., Grail, B.M., Johnson, D.B.: New approaches for extracting and recovering metals from mine tailings. Miner. Eng. **106**, 71–78 (2017)
8. Alcalde, J., Kelm, U., Vergara, D.: Historical assessment of metal recovery potential from old mine tailings: A study case for porphyry copper tailings. Chile. Miner. Eng. 1–12 (2018)

Techniques of Conversion Gain Improvement for CMOS Active Mixers

Phillipe Oliveira Menezes, Eduardo Rodrigues Lima and Julián Alberto Herrera

Abstract This paper covers the design of double balanced active mixers based on the Gilbert cell, exploring methods of conversion gain enhancement such as capacitive cross-coupling, PMOS current bleeding and a hybrid topology using both methods.

Keywords Active mixer · Gilbert cell · Cmos technology

1 Introduction

Mixers are devices responsible for frequency translation in RF communication systems. Gilbert cell mixers aim to ease signal processing complexity where RF signals are converted to a base-band signal while providing a high spur suppression due to its double-balanced operation [1]. Mixer performance can be measured by several parameters, such as its base-band signal gain when compared to the RF original amplitude (i.e., conversion gain), and noise figure, which indicates noise contribution of a mixing stage to a given system.

With continuous advancement of telecommunication and widespread interest in smart utility networks and Internet of Things (IoT), improving mixing devices might help to loosen base-band circuitry constraints, such as gain dynamic range requirements that might get as high as 75 dB [2], requiring several amplifying stages. Higher conversion gain might also heighten system sensitivity [3], and therefore, coverage of telecommunication networks.

P. Oliveira Menezes (✉) · E. Rodrigues Lima · J. Alberto Herrera
Eldorado Research Institute, Campinas, Brazil
e-mail: phillipe.menezes@eldorado.org.br

E. Rodrigues Lima
e-mail: eduardo.lima@eldorado.org.br

J. Alberto Herrera
e-mail: julian.herrera@eldorado.org.br

Y. Iano et al. (eds.), *Proceedings of the 4th Brazilian Technology Symposium (BTSym'18)*, Smart Innovation, Systems and Technologies 140,
https://doi.org/10.1007/978-3-030-16053-1_14

Section 2 introduces the concept behind the operation of a double-balanced active mixer. Section 3 discusses circuit improvement through transconductance stage refinement, and Sect. 4 discusses simulation and results using TSMC's 65 nm MS/RF CMOS process design kit on Virtuoso® Analog Design Environment.

2 Gilbert Cell Active Mixers

The operation of a down-conversion active mixer based on the Gilbert cell consists of three functions: a G_m stage, that converts a RF signal V_{rf} to current I_{rf}; a switching stage, which commutes this current by the Local Oscillator signal V_{lo} resulting in a new frequency signal I_{if}; and a load stage, which converts I_{if} back to a voltage level V_{if}. Figure 1 illustrates these three stages [1].

A typical CMOS Gilbert Cell, as shown in Fig. 2, can be implemented using six transistors and two resistors in a rather simple arrangement, therefore, due to practicality, is a widely used topology.

The transconductance stage (G_m) is composed of two common-source transistors M1 and M2, where RF differential input signal $Vrf+$ and $Vrf-$ is applied. The current resulting from this amplifying stage flows through a switching quad transistor stage (M3-M6), which is driven by complementary square waves toggling between zero and VDD ($Vlo+$ and $Vlo-$) and a differential resistive load (R) converts the current back to a readable voltage level.

The resulting first order frequency components are given by $\omega_{RF} \pm \omega_{LO}$. In down-conversion systems, the sum of these two frequencies ($\omega_{RF} + \omega_{LO}$) is filtered out, leaving only the intermediary frequency component V_{IF} (Eq. 1).

$$V_{IF}(t) = \frac{2}{\pi} g_{m1} R V_{RF} \cos(\omega_{RF} - \omega_{LO})t, \tag{1}$$

where g_{m1} is the transconductance of the common-source transistors M1 and M2 in Fig. 2, and R the load resistance. This yields a conversion gain [4, 5] of:

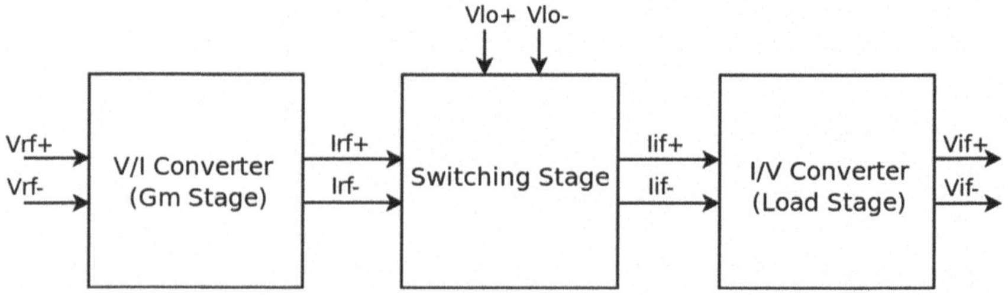

Fig. 1 Three stages of a Gilbert cell

Fig. 2 The Gilbert cell

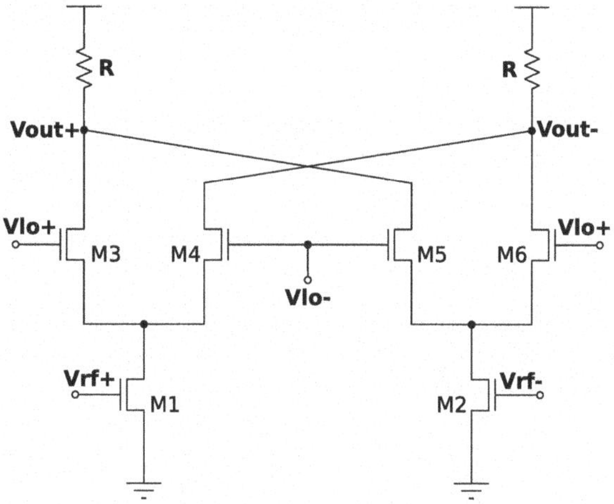

$$\frac{V_{IF}}{V_{RF}} = \frac{2}{\pi} g_{m1} R. \quad (2)$$

Noise figure in a Gilbert Cell is also a function of transconductance g_{m1} and is given by:

$$NF_{(SSB)} = \frac{\alpha}{c^2} + \frac{(\gamma_1 + r_{g1} g_{m1}) g_{m1} \alpha + 2\gamma_3 \overline{G} + r_{g3} \overline{G}^2 + \frac{1}{R}}{c^2 g_{m1}^2 R_s}. \quad (3)$$

Equation 3 shows that changing increasing g_{m1} of the input stage also contributes positively to noise reduction, since g_{m1} and NF are inversely proportional. The constant α is equal to one for a square LO wave, and c represents the conversion gain of the switching quad alone, which is usually suppressed due to its negligible contribution to the overall conversion gain. The parameters r_{g1} and r_{g3} are the poly-silicon gate resistance of the G_m stage and switching stage transistors respectively. Factors γ_1 and γ_3 are technology dependent constants, usually 2/3 for long channel transistors. $\overline{G}$ is the time average small-signal transconductance of the switching quad transistors, R is the load resistance seen in Fig. 2, and R_s is the input source resistance which drives the RF port [6].

3 Gain Enhancement Methods

3.1 Capacitive Cross-Coupling (CCC)

By taking advantage of the differential structure of the Gilbert Cell (Fig. 2), a capacitive cross-coupling (CCC) configuration can be implemented by using shunting gate-source capacitors, as shown in Fig. 3. By cross-coupling the differential inputs of each side of the structure, a typical common-source stage is superposed by a common-gate stage.

By ignoring both gate-bulk and gate-drain capacitance, the effective transconductance of the input stage can be written [7] as:

$$G_{m,eff} = \frac{2Cb1}{Cgs + Cb1} g_{m1}, \tag{4}$$

where Cgs is gate-source capacitance, and $Cb1$ is the cross-coupling capacitance. Choosing a cross-coupling capacitor value sufficiently large ($Cb1 \gg Cgs$), it is possible to achieve:

$$G_{m,eff} = \frac{2Cb1}{Cb1} g_{m1} = 2g_{m1}, \tag{5}$$

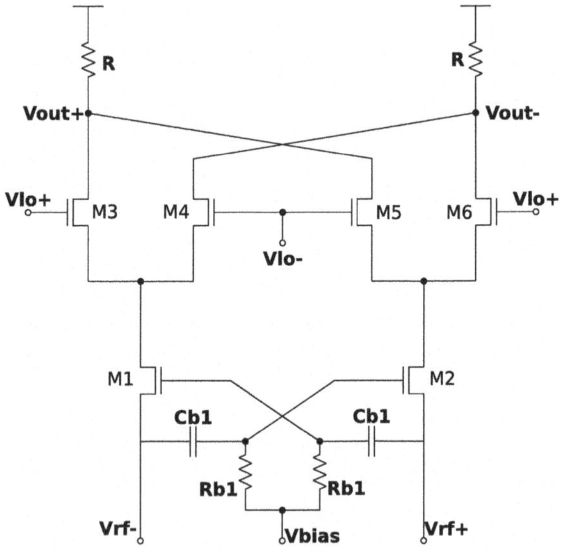

Fig. 3 Gilbert cell with CS-CG cross-coupling capacitance structure

Thus, conversion gain [7] is:

$$\frac{V_{IF}}{V_{RF}} = \frac{4}{\pi} g_{m1} R. \tag{6}$$

Even though this method doubles circuit transconductance, it has some constraints regarding cases of use. In this topology, current is drained by the input terminals ($Vrf+$ and $Vrf-$). If the previous block in the receiving chain does not provide a current path to ground, this topology cannot be used. An example where this enhancement method can be employed is when RF signal is fed to the circuit through an external Balun.

3.2 *PMOS Current Bleeding*

Looking at Eq. 2, it seems intuitive to use larger load resistor (R) values, therefore increasing conversion gain. But a problem emerges from this method. The load resistors R has to drive all current drained by the switching stage. As R increases, the voltage drop across it also increases, resulting in a substantial change in common mode voltage levels. In extreme cases, it might even compromise voltage headroom, therefore, having a negative impact on linearity. This DC dependency can be avoided by introducing current bleeding devices as shown in Fig. 4.

A current bleeding mixer [4, 5] has devices (M7 and M8) connected between VDD and drain nodes of M1 and M2 to provide an alternative current path (I_3). In a typical Gilbert cell, $I_1 = I_2$, but current bleeding introduces a new current path, making $I_1 = I_2 + I_3$. It allows M1 and M2 to operate as in the conventional Gilbert cell while mitigating the negative effects caused by large current in both switching and load stages, such as its noise contribution.

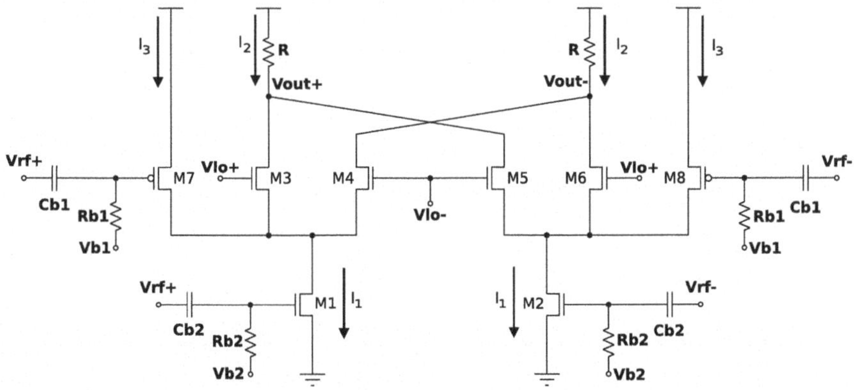

Fig. 4 Gilbert cell with CMOS current bleeding

The transistors M7 and M8 not only creates a current path to avoid DC dependency between the G_m stage and other stages, but the RF input signal can also drive its gate terminals to increase the total transconductance of the circuit.

Assuming a square wave drives LO, conversion gain can be approximated by [4, 5]:

$$\frac{V_{IF}}{V_{RF}} = \frac{2}{\pi}(g_{m1} + g_{mp})R, \tag{7}$$

where g_{mp} is the PMOS stage transconductance. This equation is similar to Eq. 2, although it accounts for the PMOS contribution to conversion gain. Decoupling capacitors $Cb1$ and $Cb2$ allows independent biasing between NMOS (M1, M2) and PMOS (M7, M8) stages, and $Vb2$ can be adjusted to achieve a higher gm/ID ratio for the PMOS stage.

3.3 Current Bleeding and Capacitive Cross-Coupling Hybrid Circuit

Current bleeding and capacitive cross-coupling can be implemented together, resulting in the a hybrid circuit shown in Fig. 5. If a perfect square wave drives LO, its conversion gain can be approximated by Eq. 8.

$$\frac{V_{IF}}{V_{RF}} = \frac{4}{\pi}\left(g_{m1} + \frac{g_{mp}}{2}\right)R. \tag{8}$$

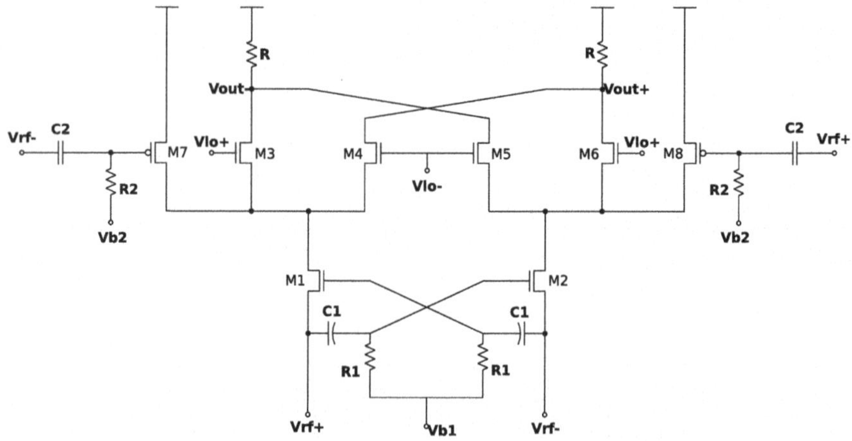

Fig. 5 Hybrid capacitive cross-coupling + PMOS current bleeding mixer

This circuit gets rid of the DC dependency between input and output stages, as the conventional current bleeding circuit does, but also inherits the constraints of the capacitive cross-coupling topology, having specific cases of use.

4 Simulation and Results

In order the compare the performance of a Gilbert cell typical topology and its different methods of conversion gain enhancement, a simulation test-bench was designed to measure: conversion gain, noise figure, 1dB compression point, input referred third harmonic intersection point (IIP3) and power consumption for a down-conversion mixing stage. Design and simulation were carried out on Virtuoso® using TSMC's 65 nm MS/RF CMO technology with 1.2 V supply.

To guarantee fairness during the comparison, the dimension of the transistors in the core cell shown in Fig. 2 where fixed to the values in Table 1. Transconductance stage dimensions (M1 and M2) were set to achieve higher g_m/I_D ratio, and load resistances is fixed to $R = 325\Omega$.

In the simulations, RF input is a sine wave having $f_{RF} = 2.4$ GHz, LO input is a square wave toggling between VDD and GDD with $f_{LO} = 2.38$ GHz to generate an output which translate to $f_{IF} = f_{RF} - f_{LO} = 20$ MHz.

4.1 Conversion Gain

Conversion gain at 20 MHz over an RF input ranging from −60 to −10 dBm is shown in Fig. 6, comparing the three enhancement techniques discussed in Sect. 3 and the typical Gilbert cell.

The Gilbert cell shows conversion gain peak value is 5.5 dB for lower amplitude input, since for higher input power, voltage headroom limitations degrades linearity. Current Bleeding and CCC techniques have conversion gain peaking at 12.5 and 12.9 dB consecutively. The hybrid circuit achieves conversion gain as high as 17.5 dB, outperforming other circuits. Besides its good conversion gain for low power input, PMOS current bleeding gain deteriorates faster when RF input increases, which indicates poor linearity. In contrast, CCC shows a better performance for higher amplitude RF input.

Table 1 Gilbert cell (Fig. 2) transistor dimensions

Transistor	W (width)	L (length)
M1, M2	17.6 μm	0.1 μm
M3, M4, M5, M6	15.4 μm	0.08 μm

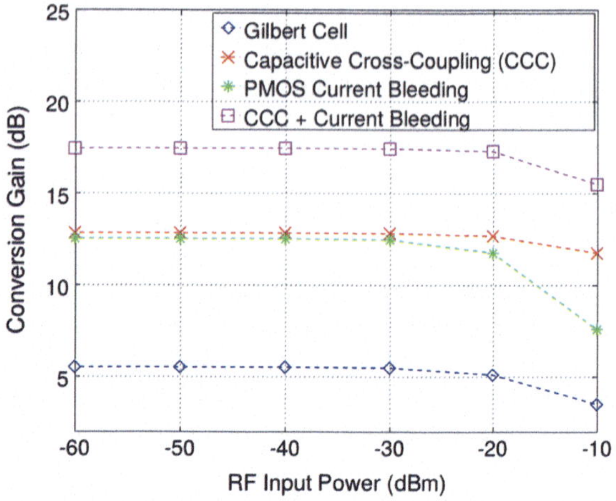

Fig. 6 Conversion gain (at 20 MHz) curves for different enhancement techniques

4.2 *Noise Figure*

Figure 7 shows noise figure up to 100 MHz. At 20 MHz (IF), typical Gilbert cell reaches 9.8 dB. Capacitive cross-coupling shows a 0.7 dB improvement, reaching 9.1 dB while PMOS current bleeding shows the best noise figure performance with 7.5 dB. The Hybrid circuit, however, shows the worst noise figure performance with 10.4 dB at 20 MHz in contrast with its higher conversion gain as observed in Fig. 6.

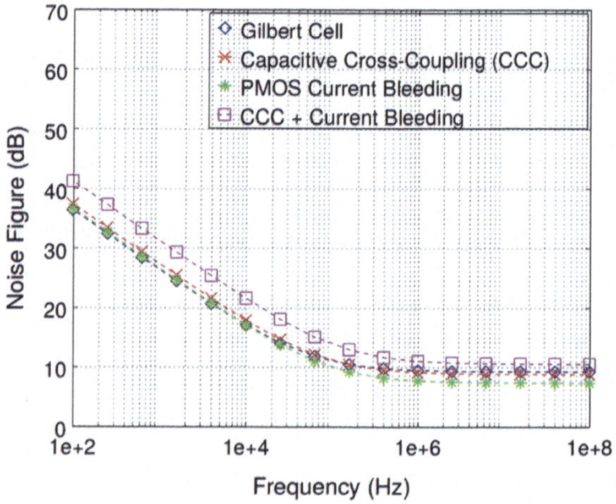

Fig. 7 Noise figure curves for different enhancement techniques

Table 2 Performance comparison table

Topology	G [dB]	NF [dB]	P [mW]	1dB Comp. [dBm]	IIP3 [dBm]
Gilbert cell	5.5	9.8	8.84	−16.2	−2.6
CCC	12.9	9.1	9.06	−8.6	5.3
Current bleeding	12.5	7.5	10.66	−19.5	−3.7
CCC + Current bleeding	17.5	10.4	11.05	−13.7	4.4

4.3 Comparison Table

Table 2 lists conversion gain (G) at 20 MHz, noise figure (NF) at 20 MHz and power consumption (P). Linearity indicators such as the 1 dB compression point (1dB Comp.) and input referred third-order intercept point (IIP3) also are listed for comparison.

5 Conclusions

The goal of this paper is to propose and compare methods of conversion gain enhancement. Equations 6, 7 and 8 numerically expresses this gain improvement, and to validate the theoretical results, conversion gain and other performance indicators such as noise figure were estimated in a professional simulation environment.

Results show that, as expected, all proposed circuits outperforms the typical (or not enhanced) Gilbert cell regarding conversion gain, with the hybrid topology (CCC + current bleeding) showing the highest performance. However, noise figure did not follow the same pattern since the hybrid circuit, on the contrary of the other circuits, had a slight noise figure degradation when compared to the typical Gilbert cell. Among the results, PMOS current bleeding topology exhibits the lowest noise figure.

Concerning power consumption, current bleeding (both, regular and hybrid) are worst power cases, which is expected considering the addition of an alternative current path. As for linearity, the PMOS current bleeding option has the poorest linearity, while all other methods linearity improves when compared to the Gilbert cell.

The methods proposed clearly shows trade-offs that have to be taken into account depending on the needs of a given system, if higher conversion gain is imperative, regardless noise figure or power consumption, the hybrid circuit is a good choice. If noise is critical, PMOS current bleeding might be a good option, and if linearity is a concern, capacitive cross-coupling is an option to be considered.

References

1. Razavi, B.: RF Microelectronics, 2nd edn. Prentice Hall, Upper Saddle River, NJ, USA (2012)
2. Kim, B. et al.: A 9dBm IIP3 direct-conversion satellite broadband tuner-demodulator SOC. In: 2003 IEEE International Solid-State Circuits Conference, 2003. Digest of Technical Papers. ISSCC, pp. 446–507. IEEE, San Francisco (2013)
3. Friis, H.T.: Noise figures of radio receivers. Proc. IRE **32**(7), 419–422 (1944)
4. Lai, D., Chen, Y., Wang, X., Chen, X.: A CMOS single-differential LNA and current bleeding CMOS mixer for GPS receivers. In: 12th IEEE International Conference on Communication Technology (ICCT), pp. 677–680. IEEE, Nanjing (2010)
5. Douss, S., Touati, F., Loulou, M.: Design optimization methodology of CMOS active mixers for multi-standard receivers. Int. J. Electron. Circuits Syst. **1**(1), 1–9 (2007)
6. Terrovitis, M.T., Meyer, R.G.: Noise in current-commutating CMOS mixers. IEEE J. Solid-State Circuits **34**(6), 772–783 (1999)
7. Fan, X., Zhang, H., Snchez-Sinencio, E.: A noise reduction and linearity improvement technique for a differential cascode LNA. IEEE J. Solid-State Circuits **43**(3), 588–599 (2008)

Cross-Platform Enterprise Application Development Framework for Large Screen Surfaces

Miguel Cuadros, Alonso De la Fuente, Rosario Villalta and Alfredo Barrientos

Abstract In recent years there has been an increasing interest in the development of applications on large surfaces that allow interaction with the user through gestures and multitouch events, this is how born the idea of Large Screen Surfaces Devices. Due to the current task of developing applications that rely on privative frameworks and having to rewrite the code to port the project to multiple platforms, we have seen as an opportunity the implementation of a cross-platform framework for the development of enterprise applications designed to support this technology. This paper presents the procedures for the creation of the framework and the implementation of two applications that validate the worked functionalities. Additionally, this Framework presents a low cost implementation compared to actual devices in the market.

Keywords Large screen devices · Framework · Cross-platform · Enterprises

1 Introduction

In the last years, a technology called Large Screen Computers or Large Screen Surfaces (called LSS) has appeared and it was defined by Gartner as screens with big sizes that stand the direct interaction with touch and gestures. These screens can be placed in horizontal or vertical position and include multitouch interaction elements that can be found in handheld devices as smartphones or tablets but unlike these, they

M. Cuadros (✉) · A. De la Fuente · R. Villalta · A. Barrientos
Escuela de Ingeniería de Sistemas Y Computación, Universidad Peruana de Ciencias Aplicadas (UPC), Lima, Peru
e-mail: U811176@upc.edu.pe

A. De la Fuente
e-mail: U201013634@upc.edu.pe

R. Villalta
e-mail: rosario.villalta@upc.pe

A. Barrientos
e-mail: pcsiabar@upc.edu.pe

Y. Iano et al. (eds.), *Proceedings of the 4th Brazilian Technology Symposium (BTSym'18)*, Smart Innovation, Systems and Technologies 140,
https://doi.org/10.1007/978-3-030-16053-1_15

recognize multiple users allowing to do collaborative works. Usually, larger screens stand gestures where the user no needs to touch the screen physically [1].

Currently, it is increasing the quantity of applications and LSS devices without a standard. Each development framework is focus exclusively in the device of each company and that limits the deploy of the applications and their scope. There are efforts to solve this problem like Tactive, a Framework Cross Platform for development in Tabletop [2] that uses web technologies as HTML5, CSS3 and JavaScript, but its only valid for 42″ devices or more and to interact with the LSS in a horizontal position. In this article, we present a framework called Large Screen Framework (LSF), designed and build for the development of LSS applications without restrictions to its orientation or size. LSF is a framework cross-platform that allows to facilitate the development of business applications so that, they can be deployed in different platforms, saving resources in the development of an application and giving the option to increase its scope. For this purpose, we use web technologies such HTML5, CSS3 y JavaScript, so we can reduce the difficulty of the frontend cover and simplify the logic of the connection with the backend.

The article has 6 sections. Section 2 describes works related to frameworks for LSS. Section 3 presents LSF and details the use of design guidelines, the construction of the LSS device, important for testing and for its design. Section 4 explains LSF validation. Section 5 presents the conclusions. Finally, the 'Acknowledgments' section presents the gratitude to all the persons that support us with our solution.

2 State of the Art

2.1 LSS Devices

Related in LSS Devices, Martinez-Maldonado [3] presents a conceptual work called Tabletop Supported Collaborative Learning (TSCL), that provides the basis of system development based in tabletops that help the collaborative work.

Nowadays exists devices developed by enterprises like Microsoft [4] that shows devices like the Microsoft Surface Hub, with an initial cost of $8999 (S/30,596 approximately) (see Table 1).

2.2 Frameworks for LSS

On the other hand, Ho Lee [5] developed a software that allows a clear display in Large Screen Surfaces device and works using a content visualization method called Spread View.

As far as he is concerned, Gaggi [2] proposes a Tactive, framework cross platform, that allows, under web technologies, encapsulate applications in widgets prepared

Table 1 Microsoft surface Hub specifications and costs

Characteristic	Value
RAM	8 GB
Storage	128 GB SSD
Display	55 in.
CPU	4th generation core i5
Graphic card	Intel graphics 4600
Total cost USD dollar	8999.00
Total cost PEN Peruvian Nuevo sol	30,596.00

for multitouch interfaces, permitting to abstract the hardware and software complexity giving a JavaScript API oriented to the gesture control. It can manage up till five gestures. The features are: gesture recognition, screen orientation and windows management.

2.3 LSS User Experience

As can be seen in [6], Scott presents eight guidelines to develop multiuser interaction based in tabletops. These guidelines suggest that technology must support: (1) natural interpersonal interaction, (2) transitions between activities, (3) transitions between personal and group work, (4) transitions between tabletop collaboration and external work, (5) the use of physical objects, (6) accessing shared physical and digital objects, (7) flexible user arrangements, and (8) simultaneous user interactions.

3 Proposal

In this chapter will show the hardware, software and the guidelines that the LSF framework require in the development stage.

- *Screen Orientation*: The screen orientation depends of the type of program required. The vertical mode suggests apps that involves showing the content to multiple people. An example of these types of applications are the audience or class presentations apps. On the other hand, the horizontal mode it's oriented to collaborative environments that many users interact with the device.
- *Multiuser*: One of the main features of these devices is to work in groups simultaneously in contrast of desktop or mobile apps that the screen size and the multitouch capacity can't allow make collaborative apps.
- *Gestures*: The gestures are the language that permits the user interact with the interface only with hand or finger movements without touch specific buttons. These gestures reflect daily actions and behaviors.

- *Private Workspaces*: LSS Private Workspaces are defined regions that users can have and manipulate their own elements. These environments can identify their own child elements and have exclusive control of them.

3.1 Logic Framework Implementation

The objective of the logic framework is proposing a set of generic functionalities that are needed to develop apps for this kind of devices. They are divided in four modules that will be explained below:

- *Connectivity*: This module brings the capacity of connect with WebRTC or Web sockets with any server with the vanilla standard. With this module, the developer can connect in real-time several instances of the same app allowing interact among them.
- *Gestures*: This module can manage the native gestures of the browser and overrides them. With this module, the developer has capacity to manage multiple gestures in the same time and has total control of them.
- *Speech Recognition*: It provides the capacity of use the voice as command to boost the productivity and do tasks more easily. With this module the developer can activate sections of the device that the user can't reach or hide specific sections of the workspace.
- *Workspaces*: This module brings the capacity of manage multiple spaces, one for each user that control the LSS, this module provides a virtual keyboard for each space. This feature allows the users to interact with the device simultaneously.

This logic implementation that showed in the picture (see Fig. 1) allow it to see in which part the Framework connect with the web application. This layer be in the top level, over the implementation of RAW WebRTC and WebSocket API's, and over the libraries of jQuery and Annyang.

3.2 Implementation Process

This section presents the necessary software and hardware requirements to build and deploy the LSF Framework. The first step is the construction of a 60″ multitouch device with an infrared touch frame, supported by a wood frame with a tempered glass with a projection film to enable see the images by the projector placed in the rear side (see Fig. 2).

1. The hardware requirements are:
2. The Software requirements are:

   - Google Chrome 47+
   - Chrome para Android 49+

  - Firefox 44+
  - NodeJS 8+ (for Web socket Server).
3. Download the framework from https://bitbucket.org/mikicegal/large-screen-framework/downloads repository.
4. Unzip the LSF CSS and JS files in your project.
5. Check the lsf\doc\index documentation.

The total costs of the implementation are shown in the Table 2.

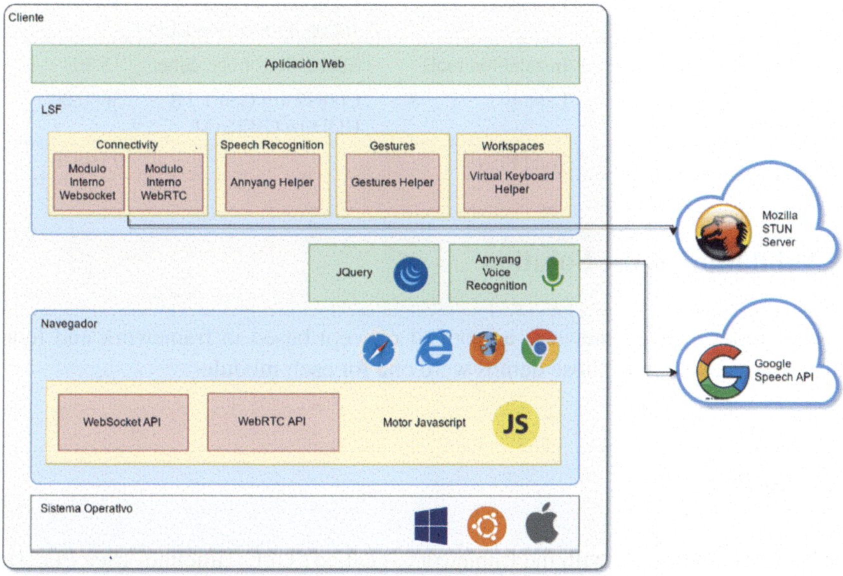

Fig. 1 Logic architecture of LSS

Fig. 2 Explode view of LSS device

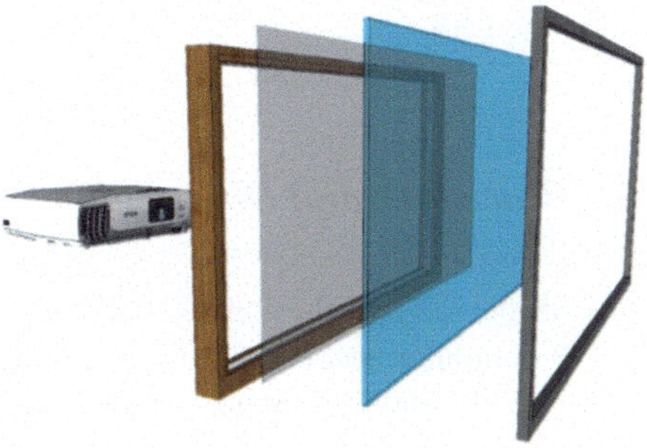

Table 2 LSF hardware costs table

Product	Description	Price S/.
Infrared touch frame	Defytech, @1920 × 1080, 10 touch points, 60″	2709.80
Projection film	2.5 m * 1.5 m	724.20
Tempered glass	6 mm one fase tempered glass	195.00
Wood frame	Made of Tornillo wood, 6 cm depth	180.00
Projector	ExcelVan, 1920 × 1080p, Proyecci6n LED	306.00
Installation tools	Stick, Microfiber, cutter	15.30
Laptop	Core i7 4th Gen 1 TB HDD 16 GB RAM	3600.00
Total PEN		7550.30

4 Validation of the Solution

To validate the LSF framework, a proof of concept based in framework and four demos were developed. Those demos were one for each module.

4.1 Study Case

The application develop with the framework is called "UPC Timeline" (See Fig. 3). It's is based in LSF framework and its goal is to show an interactive timeline where allows the user to navigate among many timelines and give the ability to move the information contained in the timeline to the cellphone since the last event that has seen in the LSS.

4.2 Procedure and Observation

The LSS device and the logical framework is installed in a core i7, 8 GB RAM, 500 GB computer according to Sect. 3.6 *Implementation process*.

The UPC Timeline application and the demos of the framework specific components are running to verify their operation that will show that the capabilities of web technologies allow the development of applications for LSS that can be used in any operative system with the support of the mentioned in the implementation process. Each component will fulfill a specific role in the development process and will show how each module works in an independent mode by a specific example of functionality where the use that is given in the example will be explained.

4.3 Results

First, the device was constructed the system was built, that means, to place the LSS at 1.5 m from the projector connected to the computer (see Fig. 4).

Once the physical connections are validated, the "Timeline UPC" application was running. It uses the Connectivity component.

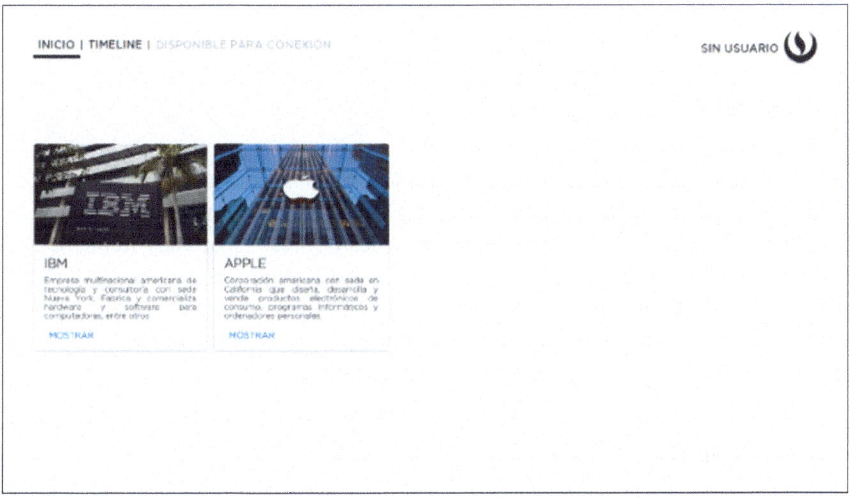

Fig. 3 UPC timeline home view

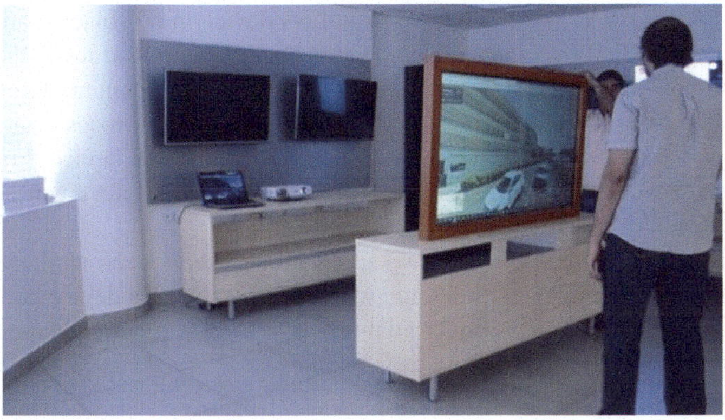

Fig. 4 Model of mount of the large screen surface

The Connectivity module allows to send and receive information in a JSON format and can find the local IP computer and with this IP find other servers to connect. It also can send texts, files, images, error plots and custom plots.

Additionally, the gesture and workspaces module run. The first module permits to implement the basic gestures in a native way (scroll, zoom, simple y multiple selection, rotation and elimination) independently.

The second module of the framework used in this Project allows to activate the workspaces, which each one permits each user manage their own objects. Additionally, the framework provides a virtual keyboard for each user. As seen in the example, there are 2 workspaces, that have their own keyboard and they can write only in the input of their own space.[1]

It also exists a fourth module called Speech Recognition that using the Annyang library allows the voice recognition commands and execute an action.

4.4 Benefits

The LSF framework can save coding and implementation time in the development of LSS applications. This framework gives a standardization in app development due to reduce the complexity of the basic functionalities because its already implemented in this software layer. In other words, this framework brings a set of common functionalities to all LSS solutions and allows make the code more scalable and readable. The API has been built to make future versions and reduce this limitation.

5 Conclusions

In this article we present LSF, a framework that allows to offer functionalities grouped in 4 modules, connectivity in real time, recognition of voice commands, multiuser capacity in parallel and gestures shown in an LSS device in 60″ blackboard mode.

In this way, LSF allows the development of business applications with the ability to be deployed in different LSS platforms, in addition to contemplating design guidelines for that public. It also provides an implementation process that includes hardware and software, including the construction of an LSS device. Finally, the logical framework of the LSF is scalable, where each developer can create additional modules to the four already existing.

LSF has been validated through two applications, where each application used different logical framework modules and were tested in the LSS device built for that case. LSF and its documentation are available in Bitbucket to benefit other researchers or developers in LSS

[1] https://www.youtube.com/watch?v=X6GDMUQ72N8.

The device is 75.32% cheaper (see Table 2) compared with actual products in the market (compared with S/30,596.00 Microsoft Surface Hub, from Table 1), and the development time reduces 3 times due to the portability of the code.

Acknowledgements We thank Professor Alfredo Barrientos for his guidance, to Kamila Flores, Mariano Ponce, Ana Lucia Lira, Marco Florian and Juan Carlos Ramos for the ideas provided for their support in the graphic design, illustrations and interfaces of the applications and our University to provide the necessary resources to implement the LSS device.

References

1. Remy, C., Weiss, M., Ziefle, M., Borchers, J.: A pattern language for interactive tabletops in collaborative. In: Proceedings of the 15th European Conference on Pattern Languages of Programs (EuroPLoP '10). ACM Press, New York, NY, 1–48 (2010). http://dx.doi.org/10.1145/2328909.2328921
2. Gai, O., Regazzo, M.: Tactive, a framework for cross platform development of tabletop applications. In: Proceedings of the 10th International Conference on Web Information Systems and Technologies (WEBIST). SciTePress, Barcelona, 91–98 (2014). http://dx.doi.org/10.5220/0004857800910098
3. Martinez-Maldonado, R., Yacef, K., Kay, J.: TSCL: a conceptual model to inform understanding of collaborative learning processes at interactive tabletops. Int. J. Hum. Comput. Stud. **83C**, 62–82 (2015)
4. Microsoft Store: (2016). https://www.microsoft.com/en-us/store/b/surface-hub
5. Lee, J.H., Kim, H., Park, J.H.: SpreadView: a multi-touch based multiple contents visualization method composed of aligned layers. In: Held as Part of HCI International 2015, pp. 305–316. Los Ángeles (2015)
6. Scott, S.D., Grant, K.D., Mandryk, R.L.: System guidelines for co-located, collaborative work on a tabletop display. In: Proceedings of the 8th conference on European Conference on Computer Supported Cooperative Work (ECSCW'03). Kluwer Academic Publishers Norwell, Massachusetts, MA, pp. 159–178 (2003). http://dx.doi.org/10.1007/978-94-010-0068-0_9

Energy Model Based on Fluvial Rainfall for the Rural Population with Torrential Rain

Javier Perales, Gianpierre Zapata and Carlos Raymundo

Abstract In Latin America, the lack of electricity has been a serious problem for over several years. To overcome this lack of supply in electricity supply, hydraulic energy is now being used in a greater proportion to fulfill the electricity needs in the rural areas. Investigations have been conducted to assess the environmental conditions of these rural areas to optimize the functionality of turbines used for hydraulic energy generation. However, there are very few focused on turbines of less than 0.5 kW generation. The proposed study aims to analyze the positioning of the blades of the cross-flow turbines and designing an electric generation system for rural dwellings. A simulation of each evaluated design was performed, and the power generated from these turbines was calculated. The results show that the power outputs initially were high and stabilized at a value of approximately 180 W, hence satisfying the minimum demands of a rural house.

Keywords Renewable energy · Pluvial model · Micro-crossflow turbine

1 Introduction

Energy plays an important role in defining the quality of life in a country [1].

Lack of electricity arises due to the location of the user, meteorological characteristics of the area such as rainfall, or insufficient allocated budget. Currently, according to the World Bank, more than 10% of the world population does not have electricity in their homes. At the rural level, this percentage increases to 20% globally. People

J. Perales (✉)
Escuela de Ingeniería Industrial, Universidad Peruana de Ciencas Aplicadas (UPC), Lima, Peru
e-mail: u201311633@upc.edu.pe

G. Zapata · C. Raymundo
Dirección de Investigación, Universidad Peruana de Ciencas Aplicadas (UPC), Lima, Peru
e-mail: u201214895@upc.edu.pe

C. Raymundo
e-mail: Carlos.raymundo@upc.edu.pe

Y. Iano et al. (eds.), *Proceedings of the 4th Brazilian Technology Symposium (BTSym'18)*, Smart Innovation, Systems and Technologies 140,
https://doi.org/10.1007/978-3-030-16053-1_16

Table 1 Turbine classification by working principle

Turbine type	Fall classification		
	Low (<10 m)	Medium (10–50 m)	High (>50 m)
Impulse	Crossflow	Turgo Pelton Crossflow	Turgo Pelton
Reaction	Blade Kaplan Francis	Francis	

do not have an easy access to electric generators, and it emits polluting particles with a diameter of 2.5 microns in the proportion of 100 $\mu g/m^3$.

A significant number of researches have been dedicated in the field of hydropower generation that focused on different types of turbines, and some of these researches have explored the possibility of implementing them in rural areas in different parts of the world.

This article is organized as follows: In the second section, the literature on the hydraulic technologies used for power generation is presented. The proposed model is detailed in Sect. 3. The validation of the proposed model is explained in Sect. 4. And finally, conclusions and future research are discussed.

2 State of the Art

The consumption of renewable energy has been increasing globally, exceeding 18% of total consumption [2]. Hydraulic energy is the most widely used form of renewable energy with 2.38% of the total consumption. Hydraulic turbines are being designed to take the maximum advantage of this source. These turbines can transform wind energy into electrical energy that can be used for domestic and industrial applications.

There is a wide range of hydraulic turbines with different working conditions and requirements for optimal performance. These can be classified based on their working principle as impulse and reaction turbines. Within these 2 categories, there are some known turbines, such as pelton, blade, and francis, among others, as shown in Table 1.

Recently, Singh [3] proposed optimizing the design for a model that initially had a blade exit angle of 85°, which was decreased to 74° and the blade inlet angle was increased to 65°, obtaining a higher turbine efficiency (74%). Unlike the previous model, Ramos [4] proposed a new 5–blade turbine, compared it with 2 different measurement diameters (100 and 200 mm), and showed the variation in the power generated by the turbine with the largest diameter (from 100 to 9900 W). For obtaining greater power generation, Biner [5] designed a turbine with 2 passages, the first with 3 blades and second with 5, hence achieving an efficiency higher than 85%. Samora

[6] improved the design of a 5–blade turbine by decreasing the initial diameter and considering a greater thickness of the blades to prevent pressure effects, without modifying the exit angle of the blade, thus obtaining a power of only 330 W and an efficiency of 64%. Galindo [7] compared 5 turbines with a different number of blades (6, 12, 14, 16, and 18 blades) with an aim of finding the one with greater power and efficiency without modifying the blade characteristics, such as exit or inlet angles. After the simulation in the CFX software, the results showed that the ideal amount of blades in the turbine is 14, which obtained the highest power and efficiency. However, the 12–blade turbine has very close values and it could also be used.

These design parameters were increased in the model proposed by Zidonis [8] by considering 15 blades and including inlet angles with 2 methods where both manage to optimize the initial model. Then, to obtain the ideal number of cubes that the turbine should have, Zidonis [9] analyzed a turbine with 14 to 18 cubes, achieving a decrease in the number of cubes from 18 to 15 and still obtaining better results. On the contrary, Nigussie maintained the number of cubes as 3 but reduced the dimensions of passage to optimize the model productivity [10]. Additionally, Nigussie presented a design with percentage dimensions that can be calculated based on the diameter, relying on the Ansys software to confirm that turbine with a diameter of 500 mm has greater efficiency than the one with 400 mm.

The crossflow turbine is another type of impulse turbine, which unlike the pelton, can work in low falls and its blades have different geometry. Zanette proposed a model of 3 trapezoidal blades and showed design parameters, such as sweep angle, turbine radius, and height, blades, among others [11]. The model was designed to reduce tension in the blades and to improve turbine performance. Sammartano presented a completely different blade model with has no curvatures along the blade and with a shape very similar to that of a moon [12, 13]. Within its parameters, there are 2 radii that delimit this blade as well as the angles of attack (similar to the inlet angles in the blade turbines) and exit angles and includes the tangential velocity calculation. After the analysis, Sammartano modified the larger diameter and radius of the blade based on a 35–blade turbine, achieving an efficiency of 86%.

Although the turbines analyzed consider multiple variables, they do not consider rainfall as a source of generation nor evaluating low power turbines; therefore, the use of the crossflow turbine is recommended.

3 Contribution

3.1 Research Design

The research design comprised 4 main steps, as shown in Fig. 1.

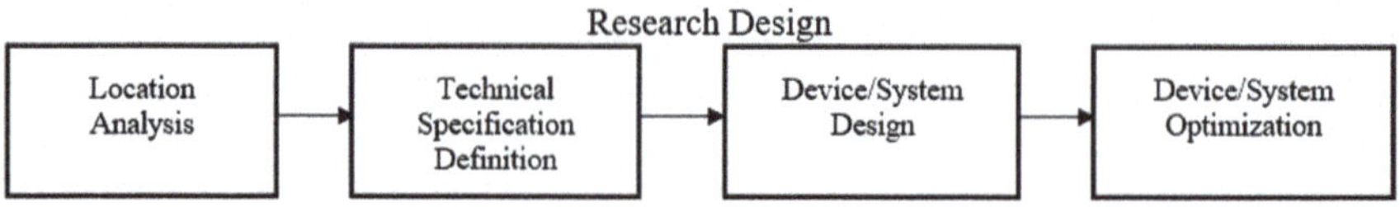

Fig. 1 Research design

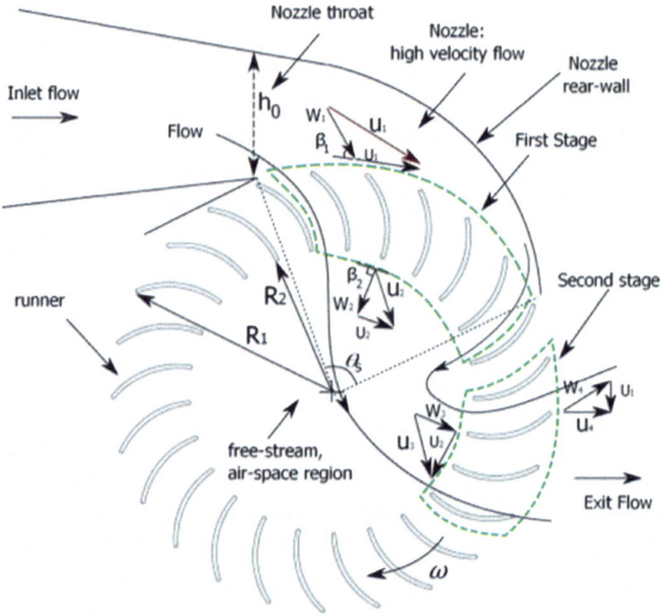

Fig. 2 Crossflow turbine

3.2 *Activities*

Location Analysis. In the first step, the characteristics of the location are identified, such as speed and height of the house, where ρ_A and ρ_W are the air and water drop densities, respectively, g is gravity, A is the cross-sectional area of the drop, and C is the drag coefficient. C value of 0.45 is assumed as Wong, in addition to a spherical ly shaped drop.

Technical Specifications Definition. Once both the speed and the height are known, a turbine is selected by considering the turbine in Fig. 2 used by Adhikari. Additionally, a 4-inch diameter polyvinyl chloride piping is used.

Device or System Design. It is important to analyze the parts while designing the system, starting with the 4-inch diameter gutters as well as the elbows and pipes of the same dimensions. The tank that will cover the turbine must also be sized. Giving more importance to the position of the blades, the initial case is shown in Fig. 3.

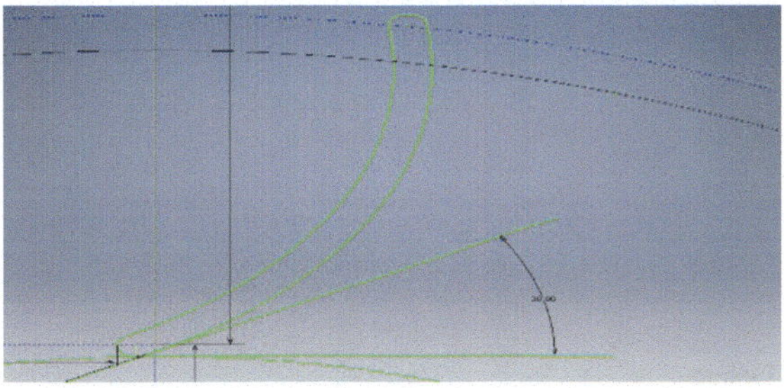

Fig. 3 Initial blade design

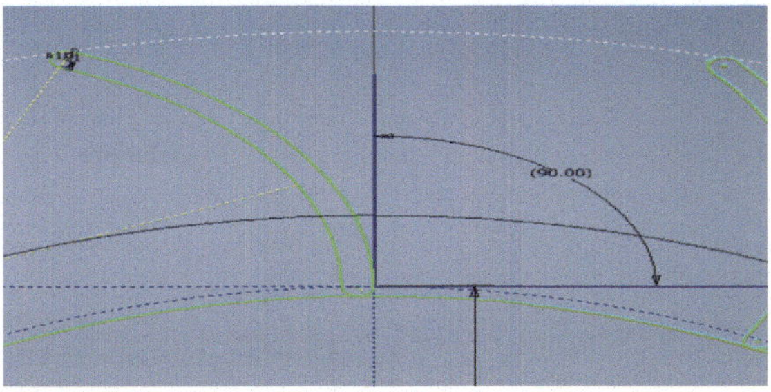

Fig. 4 New blade positioning

The Figure shows that the initial angle considered is 30°. The purpose is a better intake of rainfall.

Device or System Optimization. Once the first model or a base model has been designed, a modification to angle α is proposed, turning the blades to obtain greater power. The design is shown in Fig. 4.

3.3 Proposed Model

In the proposed model, gutters, piping, tank, turbine with generator, and battery, can be seen (Fig. 5). The operation is as follows: first, the rainwater that hits the roof of the house is collected using the gutters that are located on the edge on one of the sides of the roof. The water flows through the gutters and then falls through the pipes following the blue arrows until hitting and triggering the turbine located at the bottom. This hydraulic system will generate energy for the house, which will be stored in the battery for a maximum of 20 h.

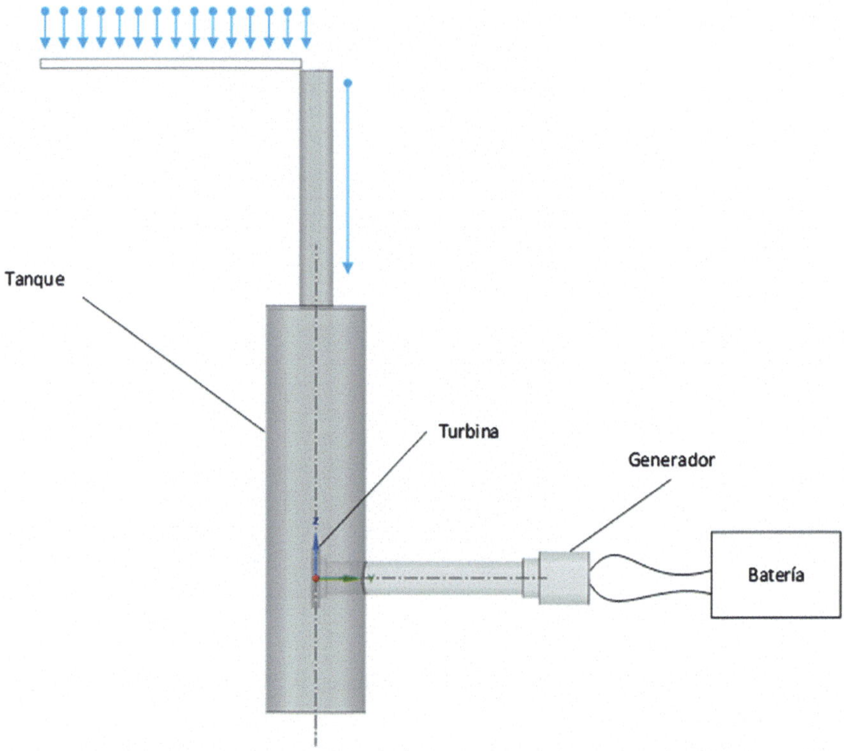

Fig. 5 Proposed energy model

4 Validity

4.1 Model and Situation for Simulation

The town of Quincemil was considered for the case study, which is located on the Peru–Brazil interoceanic highway, in the Cusco region, and faces difficulties while supplying electricity to houses that are far from the town center. This is due to transmission lines not covering the town entirely, in addition to having a relatively low allocated budget, making it unlikely for the new lines to be deployed for these houses. In many cases, these individuals are forced to use alternate sources, such as generators or candles, exposing themselves to contamination after combustion and release of micro particles [4]. Therefore, it is necessary to have systems outside the network that are supplied with renewable energy.

The main characteristic of the town is that it has the highest amount of rainfall in Peru with torrential rains that reaches 80 mm daily.

A similar model will be simulated with certain dimensions in the pipe and tank, in addition to its position, which will be supplied with this rain.

4.2 Simulation Model

One of the simulation models is shown in Fig. 6, showing the turbine blades, as well as the tank with the fall pipe in which the simulation will run.

The following parameters will be taken into account in the simulation:

- Fluid: Liquid water,
- Exit pressure: 1 atm,
- Speed (Inlet): 0.8 m/s (only for the simulation),
- Turbine angular velocity: 500 rpm,
- Average turbulence with 5% intensity is assumed, and
- Turbulence model: k-Epsilon escalable,

where t is the generated torque, N is the number of revolutions per minute in the turbine. The hydraulic power is not calculated because there is also rain falling from the sky.

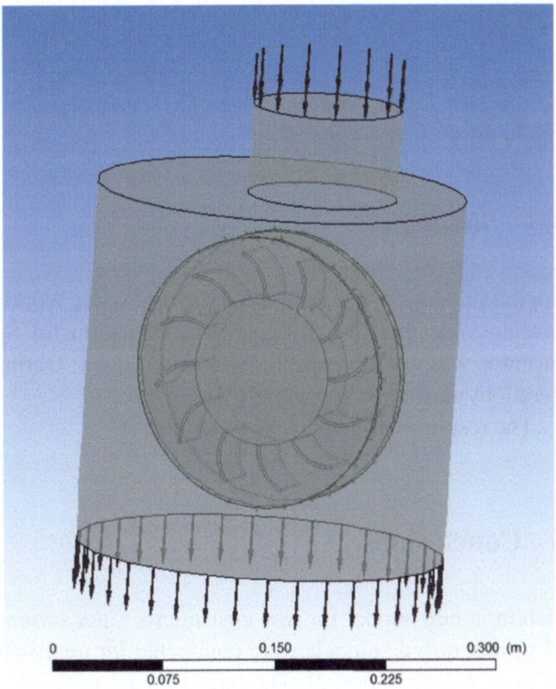

Fig. 6 Simulation model

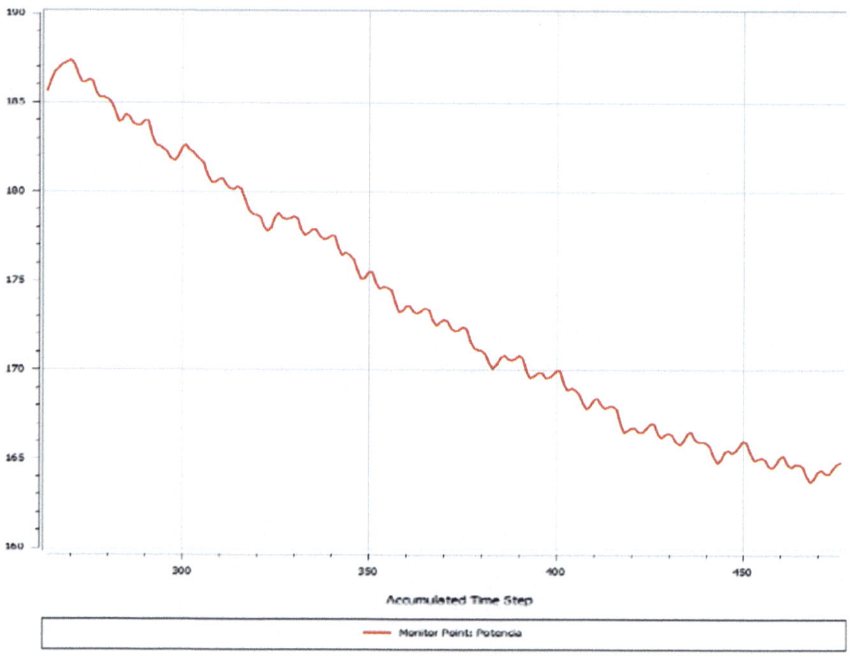

Fig. 7 Results from first design

4.3 *Simulation*

For the simulation, it was necessary to use Ansys Workbench software, and Autodesk Inventor was used for design, with adaptation for SpaceClaim software because Inventor was not compatible with Workbench. During the tests, 15 periods of 60 iterations each were considered.

The results obtained are shown in Fig. 7.

5 Conclusions

Herein, a new model for low-cost micro water turbine was proposed. Simulations of similar turbine models were conducted for micro-electricity applications in rural areas with heavy rainfall. The initial model with a 162.26 mm diameter was optimized by varying the position of the blades, taking system power as a criterion. The pipe dimensions were designed to meet the minimum required flow. The results show that the turbine with the highest power was the one that had an angle of 90°, exceeding the one with a 30° angle that better collected rain. During the simulation, some assumptions that may vary the results were made. However, the variation was minimal. The power variation will increase if turbines of greater dimension are considered.

The proposed system presented is cost-effective in comparison with transmission lines, with a return period of only 4 years considering that it would save approximately 18 dollars a month. Additionally, the budget for electrification (5400 dollars) can cover this investment of 4500 dollars.

The social impact of the power generation projects at the educational level of children is noteworthy as it will bring improvement in the quality of life in terms of water, internet, and electricity, among other services.

References

1. Singh, V.K., Singal, S.K.: Operation of hydro power plants-a review. Renew. Sustain. Energy Rev. **69**, 610–619 (2017). https://doi.org/10.1016/j.rser.2016.11.169
2. Banco Mundial, Acceso a la electricidad, sector rural. Retrieved from https://datos.bancomundial.org/indicator/EG.ELC.ACCS.RU.ZS
3. Singh, P., Nestmann, F.: Experimental optimization of a free vortex blade runner for micro hydro application. Exp. Thermal Fluid Sci. **33**(6), 991–1002 (2009). https://doi.org/10.1016/j.expthermflusci.2009.04.007
4. Ramos, H.M., Simão, M., Borga, A.: Experiments and CFD analyses for a new reaction micro-hydro blade with five blades. J. Energ. Eng. **139**(2), 109–117 (2013). https://doi.org/10.1061/(ASCE)EY.1943-7897.0000096
5. Ramos, H.M., Simão, M., Borga, A.: Experiments and CFD analyses for a new reaction micro-hydro blade with five blades. J. Energ. Eng. **139**(2), 109–117 (2013). https://doi.org/10.1061/(ASCE)EY.1943-7897.0000096
6. Samora, I., Hasmatuchi, V., Münch-Allignė, C., Franca, M.J., Schleiss, A.J., Ramos, H.M.: Experimental characterization of a five blade tubular blade turbine for pipe inline installation. Renew. Energy **95**, 356–366 (2016). https://doi.org/10.1016/j.renene.2016.04.023
7. Galindo-Luna, J., Garcia-Castrejon, J., Castro-Gomez, L., Urquiza-Beltran, G.: Mechanical power variation in a hydraulic microturbine regarding the number of blades. DYNA **92**(5), 503–506 (2017). https://doi.org/10.6036/8109
8. Židonis, A., Panagiotopoulos, A., Aggidis, G.A., Anagnostopoulos, J.S., Papantonis, D.E.: Parametric optimisation of two pelton turbine runner designs using CFD. J. Hydrodyn. **27**(3), 403–412 (2015). https://doi.org/10.1016/S1001-6058(15)60498-X
9. Židonis, A., Aggidis, G.A.: Pelton turbine: Identifying the optimum number of buckets using CFD. J. Hydrodyn. **28**(1), 75–83 (2016). https://doi.org/10.1016/S1001-6058(16)60609-1
10. Nigussie, T., Engeda, A., Dribssa, E.: Design, modeling, and CFD analysis of a micro hydro pelton turbine runner: for the case of selected site in Ethiopia. Int. J. Rotating Mach. (2017). https://doi.org/10.1155/2017/3030217
11. Kim, J.W., Jo, I.C., Park, J.H., Shin, Y., Chung, J.T.: Theoretical method of selecting number of buckets for the design and verification of a Pelton turbine. J. Hydraul. Res. **55**(5), 695–705 (2017). https://doi.org/10.1080/00221686.2017.1354933
12. Zanette, J., Imbault, D., Tourabi, A.: A design methodology for cross flow water turbines. Renew. Energy **35**(5), 997–1009 (2010). https://doi.org/10.1016/j.renene.2009.09.014
13. Arévalo, J., Quispe, G., Raymundo, C.: Sustainable energy modal for the production of biomass briquettes basad on rice huskin low-income agricultural areas in Peru. Energ. Procedia **141**, 138–145 (2017)

Remote Monitoring System of Physiological Parameters in Home Care

Marcus V. B. Franco, Thiago G. Terra, Theu S. Castro, Isadora B. Souto, Daniela O. H. Suzuki and Jefferson L. B. Marques

Abstract With the increasing number of individuals with chronic diseases and elderly people in Brazil, regular medical monitoring is necessary to monitor and preserve the physiological parameters. The present study developed a system of remote monitoring of body temperature, heart rate and oxygen saturation of the blood for the application in Home Care systems. The remote monitoring system developed has three stages that are able to collect the data, transmit this data via Wi-Fi to a cloud server and analyze them on a computer remotely. The designed system also notifies the user via his cell phone of the appropriate time and which measures to take according to a patient specific schedule and if the procedure was successful.

Keywords Physiological parameters · Remote monitoring system · Home care

M. V. B. Franco (✉) · T. G. Terra · T. S. Castro · I. B. Souto · D. O. H. Suzuki · J. L. B. Marques
Institute of Biomedical Engineering, Federal University of Santa Catarina,
Florianopolis-SC 88040-900, Brazil
e-mail: marc_vinici@hotmail.com

T. G. Terra
e-mail: tgterra@hotmail.com

T. S. Castro
e-mail: theu_sc@hotmail.com

I. B. Souto
e-mail: isa.souto.18@gmail.com

D. O. H. Suzuki
e-mail: daniela.suzuki@ufsc.br

J. L. B. Marques
e-mail: jlbmarques@gmail.com

Y. Iano et al. (eds.), *Proceedings of the 4th Brazilian Technology Symposium (BTSym'18)*, Smart Innovation, Systems and Technologies 140,
https://doi.org/10.1007/978-3-030-16053-1_17

1 Introduction

Due the improved quality of care and living conditions, life expectancy has increased in recent years [1]. According to the Brazilian Institute of Geography and Statistics (IBGE), in Brazil the number of elderly people rose by 50% in a decade [2]. This population often needs regular medical follow-up to prevent and/or control possible diseases [3].

In many cases of chronic patients who have already received primary care, whose treatment is not restricted to the hospital environment, Home Care is employed, which is a method that minimizes costs and complications associated with hospitalization, since the patient returns to his/her home, receiving a treatment closer to their social life, while releasing hospital beds for new patients [4].

The evaluation of the physiological parameters can identify basic needs of the patients, therefore showing a fast and efficient way to solve clinical problems [5]. Parameters such as body temperature, heart rate, blood oxygen saturation, among others, help in the identification of irregularities of the human body.

The remote monitoring of patients allows an early detection of changes in patients health conditions and enables earlier start of the therapy, with good results for several treatments and potential positive impact on clinical outcomes presented by patients [6].

Therefore, the objective of the present study is to develop a remote monitoring system within the context of a home care system, through the recording and analysis of physiological parameters such as body temperature, heart rate and blood oxygen saturation.

2 Materials and Methods

The remote monitoring system developed in this project has three stages that, acting together, are capable of collecting the data of a persons temperature, heart rate and oxygen saturation, transmit the data collected through the Wi-Fi network and perform data analysis in a computer. In addition, the developed system notifies the user at the appropriate time to make the data collections and if it has been successfully sent to the cloud service. For this, a prototype and an interface in the LabView were developed, as can be seen in Fig. 1.

2.1 Data Acquisition

In order to perform the data acquisition, two sensors connected to an Arduino Nano development platform (Fig. 1c) were used and a software to collect and send the data.

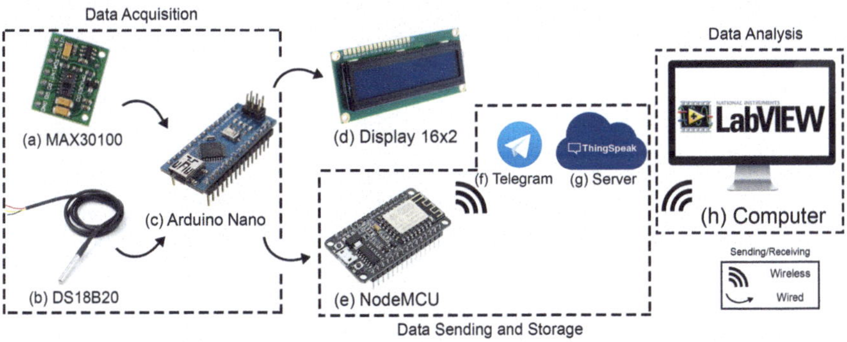

Fig. 1 Block diagram of the remote monitoring system developed showing the subparts in each stage

The sensor used to perform body temperature measurements was the DS18B20 (Fig. 1b) with a reading range of −55 to +125 °C and 0.5 °C of accuracy from −10 to +85 °C. This sensor communicates with the Arduino Nano through only one line of data.

For the physiological parameters of heart rate and blood oxygen saturation, the MAX30100 sensor (Fig. 1a) was used, which has two LEDs, one red light emitter and the other infrared light emitter combined with a photosensor. Oxygen saturated blood has a different light-absorbing capacity than unsaturated blood. The amount of red and infrared light absorbed by the blood can be used to measure the rate of oxygenated hemoglobin related to the total hemoglobin in the arterial blood. The heart rate determined from the range of the peaks of the pulse waves by means of the Photoplethysmography (PPG) technique [7]. This sensor reads the absorption levels for both light sources and stores them in a buffer that can be read via the Inter-Integrated Circuit - (I2C) communication protocol with the Arduino Nano.

In order to guide the user and present information regarding the data collected, a 16 × 2 display model JHD162 (Fig. 1d) was used. The instructions given are sent by Arduino Nano according to the software developed.

The operation of the developed software follows the flow chart in Fig. 2.

Initially all sensors and peripherals are started, where the first is the WiFi module that connects in a configured network and get the date and local time, which are shown in the display. After that, the routine turns in standby mode until the user clicks the start button that will start the measurement procedure of the sensors, where the user is instructed to positioning the sensors and so the data acquisition is performed.

For the temperature sensor 60 s of measurement has been defined. This time was estimated due the period needed the sensor to reach the body temperature. For MAX30100 sensor data is acquired for 20 s and the 10 first seconds are discarded due to the stabilization period of this sensor.

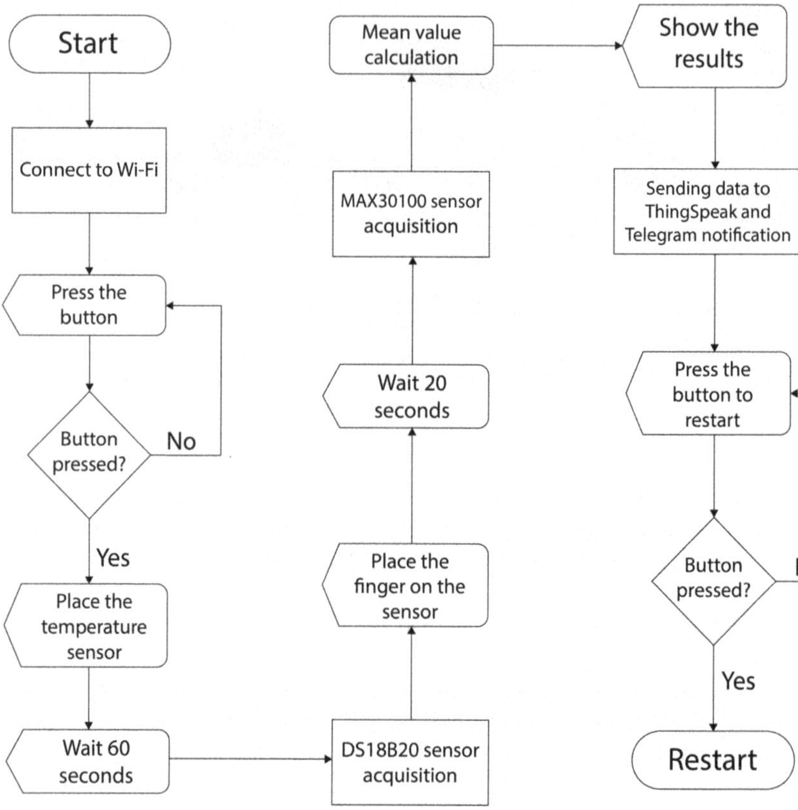

Fig. 2 Software flowchart developed showing the operation sequence

After the measurement stage of the sensors, duly showed in the display, the values of each parameter are averaged and three numerical values representing the physiological parameters described in this article are presented. After that, the data collected is sent via the Arduino Nano serial to the Wi-Fi module.

2.2 *Data Sending and Storage*

The NodeMCU module was used to remotely transmit date to a server (Fig. 1e). This module can provide a Wi-Fi connection easily, efficiently and at a low cost. Its operating voltage is 3.3 V and supports 802.1 b/g/n networks. The developed code connects the module to the local Wi-Fi network and sends the values received by the serial to the MathWorks ThingSpeak server through the Hypertext Transfer Protocol (HTTP) protocol.

The ThingSpeak enables the creation of private or public channels, which can receive different types of data from various sensors. Each channel has a write and read key and an identification number. The parameters measured by the system at the data acquisition stage were sent to a private channel with the HTTP protocol [8].

In addition, the Wi-Fi module is responsible for sending notifications to the user signaling when the user should perform data acquisition and confirming if the data was properly sent to the cloud server. To do this, a bot has been created in the Telegram application and defined notification messages that are sent automatically in certain periods according to the users schedule. Each bot has an identification number and a token key used when sending and receiving messages. The notifications are sent according to a scheduled time. In this study we used the 30-minutes period between each notification for notification validation [9].

2.3 Data Analysis

To visualize the data with more details a graphical interface was developed in the LabView software that shows the graphs of the 3 physiological parameters. In the interface was implemented the login fields of the ThingSpeak account to access the channels with the data and a field to determine the number of samples to be downloaded from the server. Also, two fields show the date-time, beginning, and the end of that data set.

The prototype was tested in academic laboratories in order to verify its functionality, simulating home environment with Wi-Fi network. In addition to the equipment developed measurement and data sending tests were performed.

3 Results

The data collected and sent to the ThingSpeak server can be viewed in the developed interface. To access the data stored on the server, you must log in to the interface and set the number of samples you want to view (Fig. 3).

In addition, it is possible to analyze the history of the collections performed for the parameters of heart rate, oxygen saturation and body temperature of the user. To do this, simply select the start date and end date of the desired history.

The implemented notification system provides different information to the user as: need for collecting data, sending status to the server and medical information.

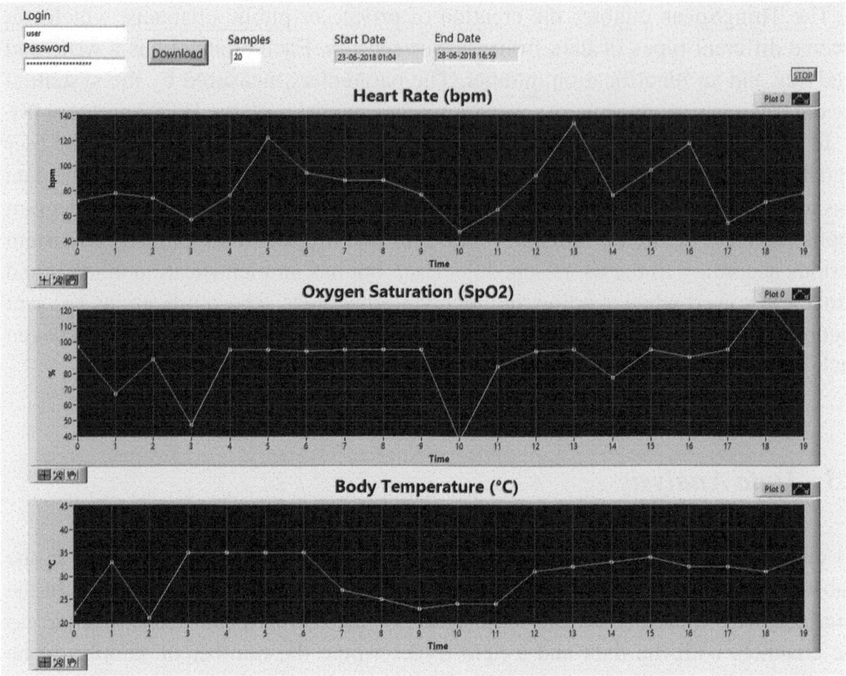

Fig. 3 Labview interface developed to analyze the data collected by the system

These notifications are generated following a schedule set by the health professional. Therefore, the Telegram is linked to the NodeMCU and plays a role of interface for the patient (Fig. 4).

After testing for system validation and circuit board design, the final version of the prototype was encapsulated to benefit the mobility and safety of the system (Fig. 5), an external 9–12 V DC with at least 300 mA power supply is necessary for energize this prototype.

4 Discussion

The main objective of the developed system is to provide access to the people who need the care of health professionals without necessarily having to be in care units preserving the autonomy and quality of life.

In the prototype developed in [10] a monitoring system was implemented with the same idea of transmitting data for storage on the ThingSpeak server. Although they did not use the same range of acquisition sensors, their results were equally satisfactory. As in the present study, the cited system has the characteristic of low cost aiming at the non-necessity of the patient being in a clinical environment. However,

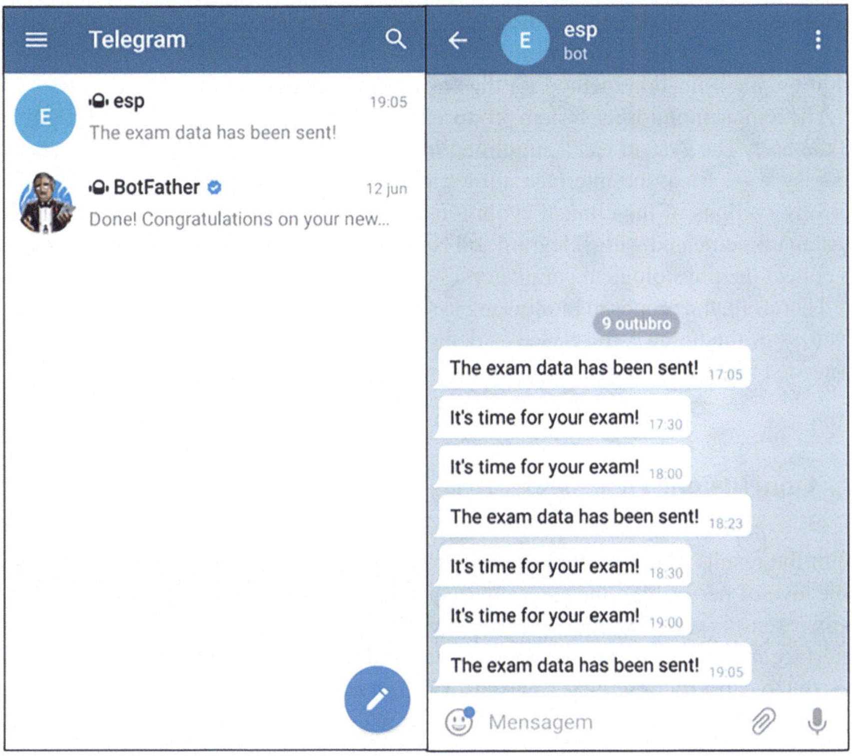

Fig. 4 Telegram bot notification system showing the interaction with the user

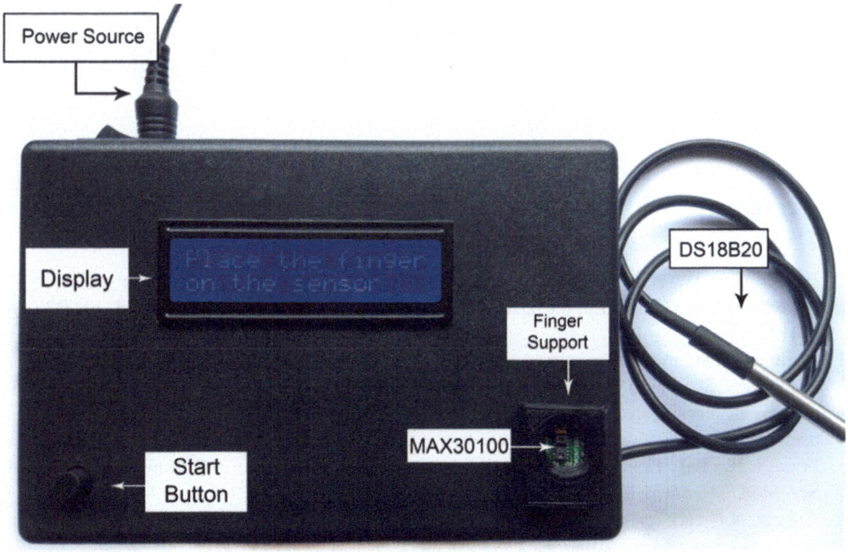

Fig. 5 Prototype of the monitoring system developed

the absence of a more intuitive interface does not offer practicality in the use, since this system proposes the analysis of the data in the own site of ThingSpeak. In addition, a notification method for the patient was not designed.

The remote monitoring system proposed in this article allows security and comfort to the user. The system has a simplified interface allowing the collection of data in an easy way. Also, the interface allows adjustments in the intervals of the graphs for observations of the clinical evolution of each user. In addition, the notification system implemented with Telegram and NodeMCU eliminates the need to remember to collect the physiological parameters.

Therefore, the proposed monitoring system may contribute to the increased effectiveness of the home care service as demonstrated in [11]. However, further testing is needed to evaluate the usability of the system in use over the time.

5 Conclusion

With the results obtained, it is possible to conclude that the present study met the objective of performing the stages of acquiring values obtained through microcontrolled sensors, sending the data to a cloud server and analyzing it in a graphical interface. Moreover, with the topology of the developed system it was possible to verify the capacity of adding new sensors of physiological parameters allowing the customization of its use according to the demands of the patient.

The developed system also enables the creation of a patient monitoring network separated by data acquisition units allowing the devices of this system to be easily scaled and increasing the reach of Home Care for remote monitoring. Therefore, with the developed system it is possible to validate the application of a remote system of measurement and analysis of physiological parameters within the context of Home Care.

References

1. Bourennane, W., et al.: Homecare monitoring system: A technical proposal for the safety of the elderly experimented in an Alzheimers care unit. In: ELSEVIER, IRBM 34, pp. 92–100 (2013)
2. Instituto Brasileiro de Geografia e Estatstica. IBGE: populacao brasileira envelhece em ritmo acelerado. So Paulo (2008). https://agenciadenoticias.ibge.gov.br/agencia-noticias/2013-agencia-de-noticias/releases/13577-asi-ibge-populacao-brasileira-envelhece-em-ritmo-acelerado.html
3. Fried, L.P., et al.: Frailty in older adults: evidence for a phenotype. J. Gerontol. Serie A Biol. Sci. Med. Sci. **56**(3), 146–56 (2001)
4. Amaral, N.N., et al.: Assistnci Domiciliar Sade (Home Health Care): Sua Histria e sua Relevncia para o Sistema de Sade Atual: Rev. Neurocincias **9**(3): 111–117 (2001)

5. Santos, E.S.F., Passos, V.C.S.: Procedimentos de verificao de sinais vitais e controles do cliente. In: Volpato ACB & Passos VCS(org). Tcnicas Bsicas de Enfermagem. Editora Martinari.4 ed, 480p (2015.)
6. Carvalho, M.A.S.: Um Sistema de Monitoramento Remoto de Pacientes usando Rede sem Fio. Dissertao (Mestrado em Cincia da Computao). Universidade Federal de Minas Gerais, Belo Horizonte (2005)
7. Chaithanya, M.K., Kishore, K.V.K., Srinivasulu, A.: Continues blood pressure measurement and data logging device with SMS alert. Int. J. Multimedia Ubiquitous Eng. **9**, 25–38 (2014)
8. Configure Accounts and Channels. https://www.mathworks.com/help/thingspeak/users.html
9. Bots: An Introduction for Developers. https://core.telegram.org/bots
10. Singh, A., Vishwakarma, A.D.: Real time ECG parameter identification and monitoring. Int. J. Recent Innov. Trends Comput. Commun. **4**(5) (2016)
11. Boonchieng, W., Boonchieng, E., Tuanrat, W., Khuntichot, C., Duangchaemkarn, K.: Integrative system of virtual electronic health record with online community-based health determinant data for home care service: MHealth development and usability test. In: IEEE Healthcare Innovations and Point of Care Technologies (HI-POCT). Bethesda, MD, pp. 5–8 (2017)

Online Control of a Robotic Manipulator by a Brain Computer Interface Based on SSVEP

Guilherme V. Vargas, Romeu Y. Takeda, Harlei M. A. Leite, Thiago B. S. Costa, Romis Attux and Sarah N. Carvalho

Abstract A Brain Computer Interface (BCI) provides a channel of communication and control amid the human brain and a computer or electronic device. In this paper we describe the implementation of all the main steps involved in the operation of a BCI system based on Steady State Visually Evoked Potential (SSVEP) paradigm. Also, we present a comparative study about the performance of our BCI employing two feature extraction techniques (Welch's Method and Discrete Fourier Transform), and considering the signal analysis window of two sizes (4 and 6 s). The results guided the conception of an online BCI-SSVEP. Our system was successfully controlled by two volunteers, who completed the task of moving a foam block with a robotic manipulator.

Keywords BCI · SSVEP · Brain signal processing · Welch's method · DFT · Robotic manipulator

1 Introduction

In Brazil, it is estimated that 6.2% of population has some kind of disability, among which are physical, auditory, visual and intellectual disability [6]. In this scenario, several studies have been carried out with the aim of developing assists for patients with some type of physical limitation [16], such as automatic wheelchair [16] and prosthetics [18]. A promising technology in this field is the Brain Computer Interface

G. V. Vargas (✉) · R. Y. Takeda · H. M. A. Leite · S. N. Carvalho
Federal University of Ouro Preto (UFOP), Ouro Preto, Brazil
e-mail: guilherme.vargas@aluno.ufop.edu.br

R. Y. Takeda
Federal University of Minas Gerais (UFMG), Belo Horizonte, Brazil

T. B. S. Costa · R. Attux
University of Campinas (UNICAMP), Campinas, Brazil

H. M. A. Leite · T. B. S. Costa · R. Attux · S. N. Carvalho
Brazilian Institute of Neuroscience and Neurotechnology (BRAINN), Campinas, Brazil

Y. Iano et al. (eds.), *Proceedings of the 4th Brazilian Technology Symposium (BTSym'18)*, Smart Innovation, Systems and Technologies 140,
https://doi.org/10.1007/978-3-030-16053-1_18

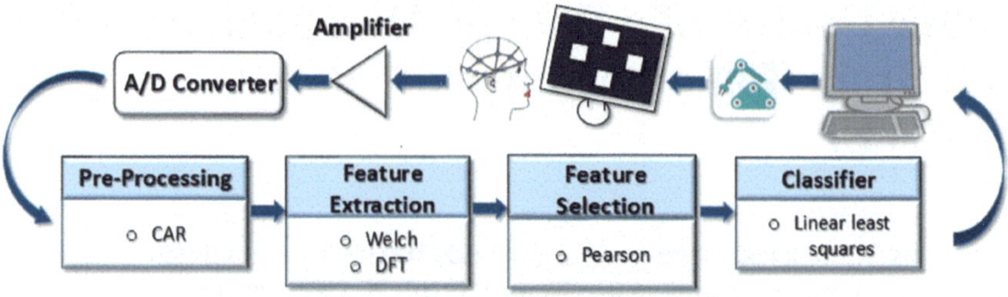

Fig. 1 Overview of a BCI-SSVEP system

(BCI) that allows a direct interaction amid the human brain and such equipment. A BCI establishes a non-muscular (or with minimal muscle interaction) communication or control channel [22].

There are several approaches to conceive a BCI [8] depending on the type of stimulation (internal or external) employed. The most usual are: motor imagery [19], task imagination [5], P300 [2] and Steady State Visually Evoked Potential (SSVEP) [12].

We implemented a BCI under SSVEP paradigm employing the following steps: brain signals acquisition through electroencephalography (EEG) [15]; signal amplification, filtering and digitization; digital signal processing (data segmentation and filtering, features extraction, features selection and linear classifier). The performance of the BCI-SSVEP system was evaluated in two steps, first using a database of four volunteers to test the digital signal processing techniques employed. Then, two volunteers controlled a robotic manipulator with four degrees of freedom by an online BCI-SSVEP (Fig. 1).

2 Methodology

2.1 Data Acquisition

The brain signals were registered by EEG of four male subjects with an average age of 28 ± 6.97 years. The dry electrodes were placed on the subjects' scalp at O1, O2, Oz, POz, Pz, PO4, PO3, PO8, PO7, P2, P1, Cz, C1, C2, CPz and FCz, according to the international 10-10 system [13], and the ground and reference were positioned on mastoids [23]. The subjects were informed about the experiment and signed the Ethics Committee of the University of Campinas (n. 791/2010) [14]. The visual stimulation corresponded of four scintillating squares at frequencies 6, 10, 12 and 15 Hz alternating between white and black on a black background. The volunteers focused theirs eyes on each stimulus for 12 s, with 8 replicates. The sampling rate employed was 256 Hz, utilizing the g.® SAHARAsys and the g.® USBamp biosignal amplifier [7].

2.2 Preprocessing

After the acquisition process, the collected signals were filtered by two analog filters, a notch filter (60 Hz) and a Butterworth bandpass filter (5–60 Hz). So, the brain signals were amplified and digitalized.

In order to improve the signal-to-noise ratio (SNR), the spatial filter Common Average Reference (CAR) [4] was employed to remove the artifacts from electrical source and body functions, such as eye blinking, deglutition and muscles in general. This filter realizes the subtraction of each signal from the average of all signals registered, as follows:

$$V_i^{CAR} = V_i^{electrode} - \frac{1}{n}\sum_{j=1}^{n} V_j^{electrode} \quad (1)$$

where n is the number of electrodes (channels) used, which in this case corresponds to 16 channels; $V_i^{electrode}$ is the potential measured at the electrode number i; and V_i^{CAR} is the potential of the electrode i after the application of the CAR spatial filtering.

After, the data set was segmented, employing rectangular windows of size of 4 and 6 s, in order to adjust the running time for the online application.

2.3 Features Extraction

The features extraction is responsible to compactly describe the signal information useful to identify and discriminate among the possible classes (visual stimulus, in our case) [3]. In this study, two spectral techniques were utilized: Welch's method [21] and the Discrete Fourier Transform (DFT) [11].

Welch's method This method applies the Fast Fourier Transform (FFT) to perform the estimation of the power spectral density (PSD) of the signal [21]. For this, the input signal x[n] is firstly divided into L smaller segments of size N with D points of overlapping. Subsequently, these segments are multiplied by a window function w[n] and the periodogram of each one is computed. So, the average of the periodograms are obtained [17], as follows:

$$\hat{S}(w) = \frac{1}{LNU}\sum_{i=1}^{L}\left|\sum_{n=1}^{N} w[n]x[n+iD]e^{-j\omega n}\right|^2 \quad (2)$$

where U is a constant given by:

$$U = \frac{1}{N}\sum_{n=1}^{N}|w[n]|^2 \quad (3)$$

In this study, Hamming windows of size 4 and 6 s were employed, estimating the PSD at frequencies 6, 10, 12 and 15 Hz with a step of 0.001 Hz and no overlapping.

Discrete Fourier Transform This approach consisted in transforming the time domain signal into the frequency domain using the FFT algorithm [17]. The FFT was computed in segments of 4 and 6 s of the signal and the features were the amplitude in frequencies 6, 10, 12, 15, 20 and 30 Hz.

2.4 *Feature Selection*

The features selection is useful to reduce the dimension of the database without losing system generalization, improving classifier performance and reducing the computer complexity [9]. In this study, to choose the relevant features for the classifier stage, the Pearson's filter technique was employed.

Pearson's Filters This method [10] performs the association amid an input vector X_i of features of determined channel i and its label Y, by computing the correlation coefficient in the form:

$$R(i) = \frac{cov(X_i, Y)}{\sqrt{var(X_i)var(Y)}} \tag{4}$$

This strategy first calculates $R(i)$ for $i = 1,\ldots,m$, where m is the number of features and, subsequently, ranks the channels using the criterion of the maximum values of $R(i)$.

2.5 *Linear Classification*

A classifier system interprets the features vector and generates a command for the application to perform the associated task [20].

In this study, a linear least squares classifier was employed to discriminate among the four visual stimuli based on features vector. This approach aims to minimize the mean square error between the desired values and the values obtained at the classifier output.

Considering X as the features matrix and y the labels vector, in the classifier training stage, the weights vector can be estimated by:

$$\hat{w} = \left(X^T X\right)^{-1} X^T y \tag{5}$$

After the definition of the separation hyperplane, the classifier output is given by:

$$Y = w^T X \tag{6}$$

For classifier training, 80% of the data from the features matrix was used and the rest 20% was employed for validation. The mutually exclusive partitions were randomly generated and the system performance was considered as the average of 100-cross-validation process.

2.6 *Robotic Manipulator*

The system was integrated with an acrylic robotic manipulator composed by four servomotors (Fig. 2). The communication amid the system and the manipulator was serial performed utilizing a MATLAB® Support Package for Arduino® toolbox.

In the start position, the servomotors 1, 2 and 3 were adjusted at 90° angulation each, so that the arm would be disposed in a central position relative to the support base, with medium lift and semi-open grip. Each visual stimulus (6, 10, 12 and 15 Hz) was associated with a task to be performed by the robotic manipulator:

- Rotation of 90 degrees in clockwise direction—6 Hz (servo 1);
- Rotation of 90 degrees in counterclockwise direction—10 Hz (servo 1);
- Claw opening and downward movement—12 Hz (servos 2 and 3);
- Claw shutting and upward movement—15 Hz (servos 2 and 3).

Despite the four servo motors available, only three of them were utilized, due to the possibilities of movements found related to the number of stimuli available in the experiment.

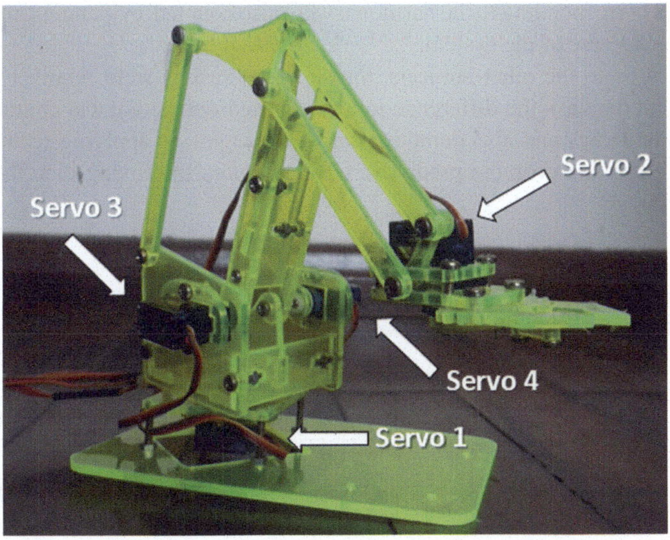

Fig. 2 Robotic manipulator employed

3 Results and Discussion

3.1 Offline Tests

The offline tests were directed to obtain the best parameters of digital signal processing module of BCI to be employed by the online system. The parameters considered were:

- Windows size (4 and 6 s): both windowing are suitable for the control of the robotic manipulator with mechanical inertia;
- Features extraction method (Welch's Method and DFT): possible techniques to be incorporated in an online system and of low computational cost.

Table 1 presents the results for the four subjects analyzed. We can see that for 4 s windowing, the Welch's Method provided hit rates slightly greater than the ones provided by the DFT method, except for the subject 3; whereas the appliance of windows of 6 s revealed certain discrepancy amid the two methods, evidencing the superiority of the DFT method in this scenario, mainly for subjects 2, 3 and 4. Plausible explanations support this discrepancy: the first is related to the amount of data available for system training and validation. With windows of 4 s, the system holds 96 inputs, of which 70 are destined for training and 26 for validation. In the case of windows of 6 s, 64 entries are available, 50 for training and 14 for validation. This difference of material for system training can be a determining factor for the drastic reduction of BCI accuracy rate in the scenario with windows of 6 s. Another relevant factor is associated with the evaluation of harmonic components by the DFT method. In this technique, besides the fundamental frequencies of stimulation, the frequencies corresponding to the harmonic were also counted, evidencing an increase of the accuracy rate of the system.

If compared to the other subjects, the subject 1 results were relatively inferior [1]. This fact displays the difference in terms of neurophysiological characteristics amid human beings and also demonstrates the necessity of applying feature selection techniques that promote predictive accuracy gain, improving then the system performance.

Table 1 BCI-SSVEP system accuracy rate for different scenarios

Accuracy rate (%)				
Subject	4 s		6 s	
	Welch	DTFT	Welch	DTFT
1	48.8	38.5	30.3	37.5
2	94.0	87.8	59.4	98.1
3	86.0	89.0	39.1	95.0
4	91.0	88.5	48.1	97.8

3.2 *Online Tests*

Based on offline tests results, the online tests employed windows of 6 s and the DFT method for feature extraction. The task to be performed was to move a foam cube from the ground level, moving it leftward or rightward and finally, returning the cube to the ground level, totalizing four movements. The robotic manipulator began in its initial position (arm disposed in a central position relative to the support base, with medium lift and semi-open grip).

As the volunteer needed to focus on the stimulus, he did not observe the movement of the robotic manipulator, so received sound feedback about the movement performed and an indication about the next command (stimulus to be focused on) to complete the task. The subjects 3 and 4 participated in this experiment. To complete the task it was necessary to focus on the stimuli in the following order: 15, 12, 10 and 15 Hz; this would take 24 s considering the window size set and taking into account that no error would occur.

The subject 3 completed the task in 42 s, with 7 commands (4 corrects and 3 system errors). The error always occurred between the stimulus 12 Hz interpreted as 15 Hz, in 3 consecutive occasions. The subject 4 needed 102 s (17 commands), and many confusions were observed, mainly in relation to frequencies 10 Hz with 15 Hz and 10 Hz with 12 Hz. Possible causes of this reduction in the quality of system performance can be raised, such as the individual's fatigue during the training and testing processes and the lack of visual feedback to improve understanding and motivation (Table 2).

Table 2 Online tests results

Time(s)	Frequency focused on (Hz)		Classifier response (Hz)		Frequency focused on (Hz)		Classifier response (Hz)		
	Sub. 3	Sub. 4	Sub. 3	Sub. 4	Sub. 3	Sub. 4	Sub. 3	Sub. 4	Time(s)
6	15	15	15	15		10		12	60
12	12	12	15	15		10		15	66
18	12	12	15	12		10		12	72
24	12	10	15	12		10		12	78
30	12	10	12	15		10		15	84
36	10	10	10	12		10		12	90
42	15	10	15	12		10		10	96
48		10		15		12		12	102
54		10		12					108

4 Conclusion

During the offline tests a comparative study was carried out between the feature extraction techniques, DFT and Welch's Method, considering analysis window of 4 and 6 s, in order to evaluate which provided a higher accuracy rate for our BCISSVEP system. The results showed that both feature extraction techniques are efficient, providing an acceptable performance (accuracy greater than 80%) for the control of an actual application. In the online tests, we opted to use windows of 6 s, adequate to the robotic manipulator dynamics and with a higher hit rate. The brain signal features were extracted using DFT and the feature selection stage, by Pearson's Coefficient, was included as a way to increase the predictive capacity of the system. The two volunteers were able to perform the proposed task of controlling the robotic manipulator, despite errors in identifying the desired command by the volunteer made by the system.

The proposed study revealed the real possibility of systems control by applying this emerging technology. System performance could be improved with visual feedback.

References

1. Allison, B.Z., Neuper, C.: Could anyone use a BCI? In: Braincomputer Interfaces, pp. 35–54. Springer, London (2010)
2. Bashore, T.R., van der Molen, M.W.: Discovery of the P300: a tribute. Biol. Psychol. **32**(2), 155171 (1991)
3. Bishop, C.M.: Neural networks for pattern recognition. Oxford university press (1995)
4. Carvalho, S.N., et al.: Comparative analysis of strategies for feature extraction and classification in SSVEP BCIs. Biomed. Sig. Process. Control **21**, 34–42 (2015)
5. Curran, E., et al.: Cognitive tasks for driving a brain computer interfacing system: a pilot study. IEEE Trans. Neural Syst. Rehabil. Eng. **12**(1), 4854 (2004)
6. da Saude, O.M.: CIF: classificacao internacional de funcionalidade, incapacidadee sade. Editora da Universidade de Sao Paulo, Sao Paulo (2003)
7. G.tec. G.tec Medical Engineering (2018). Available http://www.gtec.at/. Accessed Jan 2018
8. Graimann, B., Allison, B., Pfurtscheller, G.: BrainComputer interfaces: a gentle introduction. In: Brain-Computer Interfaces, p. 127. Springer, Heidelberg (2010)
9. Guyon, I., et al.: Feature Extraction: Foundations and Applications, v. 207. Springer, Heidelberg (2008)
10. Guyon, I., Elisseeff, A.: An introduction to variable and feature selection. J. Mach. Learn. Res. **3**, 11571182 (2003)
11. Haykin, S.S.: Adaptive Filter Theory. [S.l.]: Pearson Education India (2008)
12. Hermann, C.S.: Human EEG responses to 1–100 Hz flicker: resonance phenomena in visual cortex and their potential correlation to cognitive phenomena. Exp. Brain Res. **137**(3–4), 346353 (2001)
13. Jurcak, V., Tsuzuki, D., Dan, I.: 10/20, 10/10, and 10/5 systems revisited: their validity as relative head-surface-based positioning systems. Neuroimage **34**(4), 16001611 (2007). (Elsevier)
14. Leite, H.M.A., et al.: Analysis of user interaction with a BrainComputer interface based on steady-state visually evoked potentials: case study of a game. Comput. Intell. Neurosci. (2018)
15. Millan, J.R., Carmena, J.: Invasive or noninvasive: understanding BrainMachine Interface technology. In: IEEE Engineering in Medicine and Biology Magazine, Institute of Electrical and Electronics Engineers, vol. 29, n. EPFL-ARTICLE-150426, p. 1622 (2010)

16. Muller, S.M.T., Bastos, T.F., Filho, M.S. Proposal of a SSVEP-BCIto command a robotic wheelchair. J. Control Autom. Electr. Syst. **24**(1–2), 97105 (2013). ISSN 21953880
17. Oppenheim, A.: Discrete-Time Signal Processing. Pearson, Upper Saddle River (2010). ISBN 978-0-13-198842-2
18. Ortner, R., et al.: An SSVEP BCI to control a hand orthosis for persons with tetraplegia. IEEE Trans. Neural Syst. Rehab. Eng. **19**(1), 15 (2011)
19. Pfurtscheller, G., Flotzinger, D., Neuper, C.: Differentiation between finger, toe and tongue movement in man based on 40 Hz EEG. Electroencephalogr. Clin. Neurophysiol. **90**(6), 456–460 (1994)
20. Theodoridis, S., Koutroumbas, K.: Pattern Recognition. Elsevier (2009)
21. Welch, P.D.: The use of Fast Fourier Transform for the estimation of power spectra: a method based on time averaging over short, modified periodograms. IEEE Trans. Audio Electroacoust. **15**(2), 7073 (1967)
22. Wolpaw, J.R., et al.: Braincomputer Interfaces for communication and control. Clin. Neurophysiol. **113**(6), 767791 (2002) (Elsevier)
23. Wu, Z., Su, S.: A dynamic selection method for reference electrode in SSVEP-basedBCI. PLoS ONE **9**(8) (2014). ISSN 19326203

An Algorithm to Obtain the QRS Score Based on ECG Parameters Detection and Neural Networks for Confounder Classification

Julio Cabanillas, Gustavo Tello, Brandon Mercado, Guillermo Kemper, Mirko Zimic and Robert Gilman

Abstract The present work proposes an algorithm to calculate the QRS Score and the determination of confounders starting from Electrocardiographic (ECG) signals. The QRS Score is a parameter that indicates how big the scar is in the wall of the patient's myocardium; It is also helpful in determining how healthy the heart is. Said parameter is calculated from signal information such as time measurements, amplitude relationships and waveforms. The evaluation of the ECG signals is usually done by visual perception of the graph paper where it is printed as a result of the electrocardiogram examination. However, the reproducibility of this method is 60% and the repeatability is 66%. This definitely affects the accuracy of the score obtained and therefore the diagnosis of a disease. The proposed algorithm aims to reduce the subjectivity of the analysis and standardize the punctuations to be obtained. The algorithm is made up of processing stages that involve the conditioning of the signal using finite impulse response (FIR) filters, decontamination of confounders by neural networks, detection of the QRS complex, detection of times and amplitudes and finally obtaining the QRS score from a table of criteria. Finally, the proposed

J. Cabanillas (✉) · G. Tello · G. Kemper
Universidad Peruana de Ciencias Aplicadas, Av. Prolongación Primavera 2390, Santiago de Surco, Lima, Peru
e-mail: jujulioccu@gmail.com

G. Tello
e-mail: gtellop346@gmail.com

G. Kemper
e-mail: guillermo.kemper@gmail.com

B. Mercado · M. Zimic
Universidad Peruana Cayetano Heredia, Av. Honorio Delgado 430, San Martin de Porres, Lima, Peru
e-mail: brandonmercado4@gmail.com

M. Zimic
e-mail: mirko.zimic@upch.pe

R. Gilman
Johns Hopkins University, 3400 N. Charles Street, Baltimore, USA
e-mail: gilmanbob@gmail.com

Y. Iano et al. (eds.), *Proceedings of the 4th Brazilian Technology Symposium (BTSym'18)*, Smart Innovation, Systems and Technologies 140,
https://doi.org/10.1007/978-3-030-16053-1_19

algorithm obtained a reproducibility of 75% and a repeatability of 100% exceeding the performance of the specialist.

Keywords Confounder · ECG signals · Neural networks · Signal processing · Waveforms · Chagas · QRS score

1 Introduction

The blood supply system of the heart can be compromised during a cardiac event, leading to the scarring of the myocardial tissue. Such scarring occurs mainly due to the lack of blood flow, which can cause damage to the heart's capacity of electrical conduction [1]. In this context, the Selvester QRS Score can be used to determine the magnitude of tissue scarring based on the QRS complex because it presents the heart's electrical conduction characteristics. This method can evaluate effects of tissue scarring on ECG signals [2]. To obtain the QRS score, time and voltage of the waves present in the QRS complex are measured, and these measurements are interpreted for each ECG derivation under the criteria established by Selvester, Wagnder, and Hindman [3]. However, these criteria vary due to pre-existing anomalies called confounders, which affect the heart's electrical conduction, and their effects on the QRS score have been reported by Loring [4].

The QRS score is useful as an alternative diagnostic imaging modality used in clinical standards. Moreover, it has the potential as a tool for preventing diseases, such as Chagas disease or Wolf–Parkinson–White syndrome. However, its clinical use is limited due prolonged examination times and low repeatability and reproducibility owing to the limitation of specialists' precision.

Computational algorithms have been proposed in scientific literature with the aim to overcome these limitations. Xia et al. [5] and Bono et al. [6] have proposed computational methods to calculate the QRS score based on the derivations of ECG; both of which are useful as an initial framework for the presented research since they consider stages involved in the process as well as certain criteria, considerations, and relevant techniques.

However, Xia et al.'s study focuses solely on the analysis of ECG results of patients with a left bundle branch block (LBBB), while Bono et al.'s study is more detailed and presents more complex mathematical processes and tools.

Furthermore, Gautam and Giri [7] have proposed another procedure for the design and development of an artificial neural network. Their proposed procedure is rather interesting, which is also used as a reference for the presented work, with some changes and adjustments to calculate any confounders.

The proposed algorithm aims to increase the reproducibility of the QRS score to facilitate its use as a clinical diagnostic tool in conjunction with ECG. Moreover, we attempted to create a procedure that does not generate an excessive computational load to reduce examination time. The performance of the proposed algorithm was satisfactory, with 75% reproducibility and 100% repeatability, surpassing the specialist's results.

2 Description of the Proposed Algorithm

Figure 1 shows a block diagram of the proposed algorithm. Details of each processing stage are described in the following sections.

2.1 *Signal Improvement*

The aim at this stage was to condition the discrete signals $s_i(n)$ to increase precision in the extraction of characteristics. In this case, i indicates the signal number associated with a given derivation (Table 1).

Unwanted noise components are removed for signal improvement. For this purpose, a linear phase FIR low pass filter with a cutoff frequency of 0.12 (in standardized relative frequency) and a Barlett–Hanning window [3] is used. The use of the filter with a window system provided the best results in terms of lowering noise and unwanted distortion. The signals resulting from the filtering are defined as $y_i(n)$ (Fig. 2).

2.2 *Detection and Extraction of a Cardiac Period*

For the extraction of cardiac period, the QRS patterns of each period are highlighted and the P and T low-frequency waves in the $y_i(n)$ signals resulting from the improve-

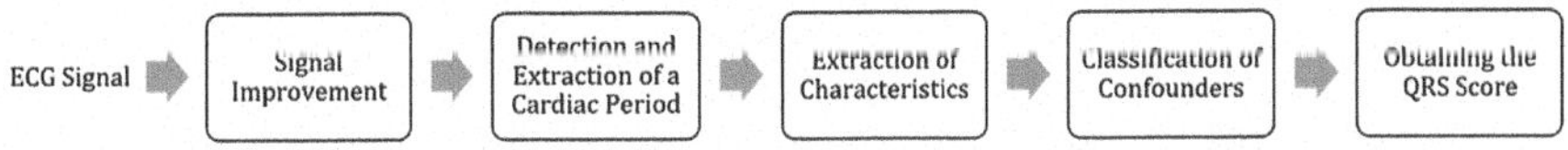

Fig. 1 Block diagram of the proposed algorithm

Table 1 Association of discrete signals

i	Lead
1	I
2	II
3	aVL
4	aVF
5	V1
6	V2
7	V3
8	V4
9	V5
10	V6

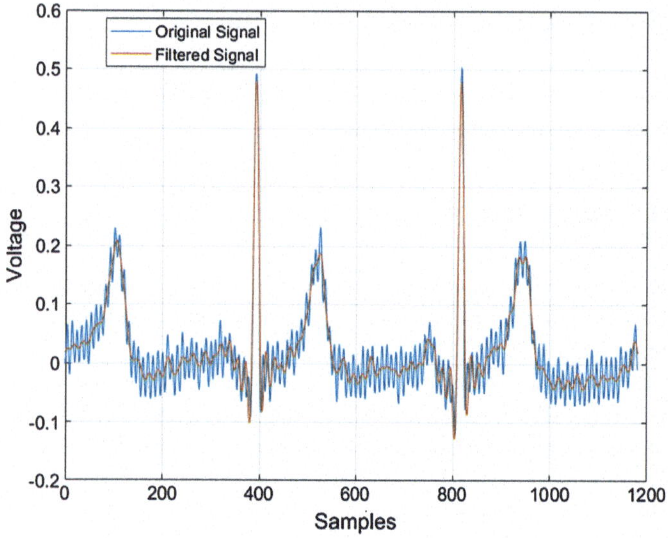

Fig. 2 Original signal and filtered signal

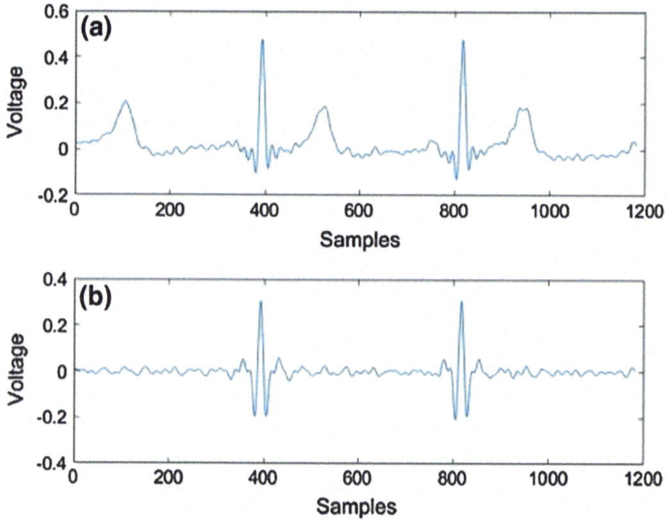

Fig. 3 **a** Filtered signal $y_i(n)$ and **b** Signal resulting from high-pass filter $y1_i(n)$

ment process are attenuated. For this purpose, a linear-phase high-pass FIR filter with a cut-off frequency of 0.04 (in normalized relative frequency) and a Barlett–Hanning window is used. The signals resulting from the process are defined as $y1_i(n)$ (Fig. 3). The following procedure is applied for the extraction of a cardiac period from $y1_i(n)$.

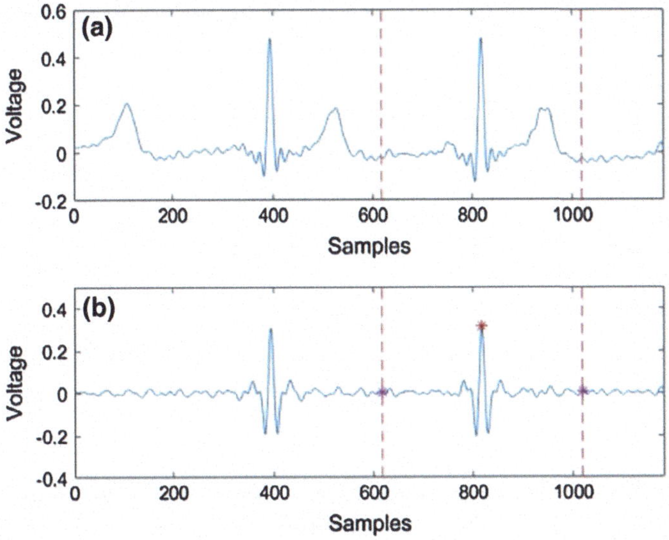

Fig. 4 **a** Filtered signal $y_i(n)$ and **b** Signal resulting from high-pass filter $y1_i(n)$ with extraction indices

Stage 1. The maximum value of $y1_i(n)$, corresponding to the position $n = n0_i$ (R-wave), is located.

Stage 2. A signal segment is extracted from the position $n1_i$ up to the position $n2_i$, producing the following:

$$n1_i - n0_i - round\left(\frac{0.5\,\text{bpm}}{60} \cdot T\right) \quad (1)$$

$$n2_i = n0_i + round\left(\frac{0.5\,\text{bpm}}{60} \cdot T\right) \quad (2)$$

where *bpm* is the heart rate in beats per minute and T is the sampling period. The *round* function provides the nearest integer value (Fig. 4).

Stage 3. The cardiac period is extracted from each signal $\boldsymbol{y}1_i.(n)$ which is defined as $p_i(n) = y1_i(n + n1_i)\ for\ n = 0, 1, \ldots, n2_i - n1_i$ (Fig. 5).

2.3 *Extraction of Characteristics*

The criteria to be considered at this stage have been established elsewhere [3]. The following procedure is used to extract the characteristics:

Stage 1. A linear phase low-pass FIR filter with a cut-off frequency of 0.08 (in standardized relative frequency) is applied to the $p_i(n)$ signals to attenuate the cardiac

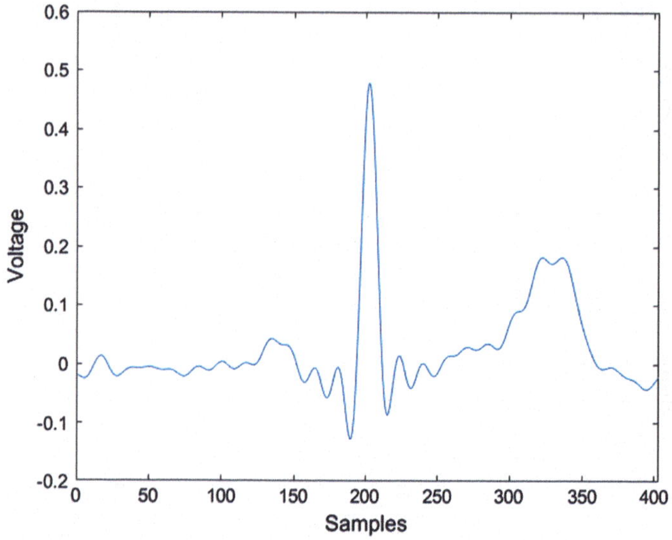

Fig. 5 Resulting signal $p_i(n)$

period waves that do not belong to the QRS complex and thus avoid errors in feature extraction. The signals resulting from this filtering are defined as $p_i^*(n)$.

Stage 2. The QRS complex is divided into three segments. For this purpose, two rates are defined, which apply to the signals of all derivations:

$$m0 = 1.5 \times round\left(\frac{QRSTime}{T}\right) \tag{3}$$

where $QRSTime$ is the total duration of the QRS complex, which is obtained from ECG data. Then, we extract segments of the signal of interest from the extracted QRS period. These segments are defined as follows:

$$p_i^{Q-R}(n) = p_i(n + n0_i - m0)\ para\ n = 0, 1, \ldots, m0 \tag{4}$$

$$p_i^{R-S}(n) = p_i(n + n0_i)\ para\ n = 0, 1, \ldots, m0 \tag{5}$$

$$p_i^{R}(n) = p_i\left(n + n0_i - \frac{m0}{2}\right)\ para\ n = 0, 1, \ldots, m0 \tag{6}$$

Stage 3. Based on the maximum value of $p_i^{Q-R}(n)$ located at $n = n0_i$ and defined as $p_i^{Q-R}(n0_i)$, we find the first local maximum by decreasing the value of n in each iteration. Once the position of this first maximum $(m2)$ is located, the second crossing by 0 (decreasing n) is found from this position. Once this last position $(m4)$

has been found, the minimum value of the signal in the section located between the two crossings is finally determined.

This value is defined as $p_i^{Q-R}(m3_i)$ and used in the calculation of the QRS score as the "zero reference."

Stage 4. For each signal *i*, we obtained the duration in time of segment Q, which is obtained from the following expression:

$$tQ_i = (m2_i - m4_i)T \tag{7}$$

Stage 5. For each signal *i*, the duration of the R segment is obtained. In this case, the maximum value of the signal p_i^R(n) is located, which is considered to be located at the position n = $m6_i$. Then from this position, n is decreased until the first crossing of the "zero reference" is found. This position is defined as $n = m5_i$.

The procedure is repeated but with the purpose of finding the first crossing of the "zero reference" from n = $m6_i$ by increasing the value of n. This position is defined as n = $m7_i$. The duration of the R segment is finally expressed as follows:

$$tR_i = (m7_i - m5_i)T \tag{8}$$

Stage 6. The minimum value of $p_i^{R-S}(n)$ is obtained, which is defined as $p_i^{R-S}(m8_i)$. The parameters obtained in this procedure are used later for the calculation of the QRS score.

2.4 Classification of Confounders

A self-organized map-type neural network is used for the classification of the confounder present in the electrocardiogram, which is characterized by no requirement an output vector or references in the training process; thus, it presents autonomous learning. The proposed classification procedure is as follows:

Stage 1. The amplitude frequency spectrum of the derivations $i = 1, 3, 5, 6, 9, 10$ is calculated for each ECG. In this case, fast Fourier transform (FFT) of 512 samples was used in the region of interest ranging from 0 to 1/2T Hz. This sample number in FFT allows for generating fewer inputs, which increases the speed of training without compromising the resolution of FFT. The obtained module spectra are defined as $AX_i^j(k)$, where *j* is the ECG number used in the conformation of the database (*j* = 0, 1,…, 899) and *k* = 0, 1,…, 511 is the frequency index. The database comprises 700 samples for the neural network training and 300 for the validation process.

Stage 2. The database of the spectra obtained in the previous stage is created. In this case, there are 6 frequency vectors (of 512 samples each) for each ECG.

Stage 3. The neural network is created according to the procedure described by Kohonen [8]. Figure 6 shows the block diagram of the network with inputs and outputs considered for detection.

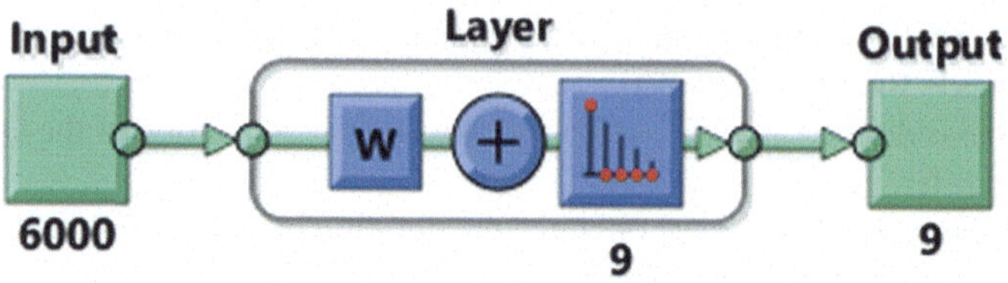

Fig. 6 Block diagram of the neural network

Table 2 Confounder identifiers

Confounder	Identifier
None	0
LVH	1
LAFB	2
RBBB	3
RBBB + LAFB	4
LBBB	5

Table 3 Variables used for calculation of the QRS score

Parameter	Description
$Sex(j)$	Patient's sex
$Age(j)$	Patient's age
$tQ_i(j)$	Q wave duration
$tR_i(j)$	R wave duration
$p_i^{Q-R}(m3_i)$	Maximum Q wave value
$p_i^{Q-R}(m0)$	Maximum R wave value
$p_i^{R-S}(m8_i)$	Maximum S wave value

Stage 4. The network training process begins. This process can take several minutes and is carried out without supervision.

Stage 5. Correct functioning is validated with the trained network. In case of excessive error, the network is retrained until the expected results are obtained. Table 2 shows the network output identifiers for each confounder.

2.5 Obtaining the QRS Score

Finally, the QRS score is calculated. For this purpose, the criteria of time, amplitude, and waveform parameters are applied for the electrodes considered according to the detected confounder, age, and sex, as described elsewhere [4]. Table 3 shows all variables used for calculating the QRS score.

Table 4 Score distribution for the QRS score

Lead	RBBB Criteria	Pts	LAFB Criteria	Pts	LAFB + RBBB Criteria	Pts	LVH Criteria	Pts	No Confounders Criteria	Pts
I	Q ≥ 30 ms	1	Q ≥ 30 ms	1	Q ≥ 30 ms	1	Q ≥ 30 ms	1	Q ≥ 30 ms	1
	R/Q ≤ 1	1	R/Q ≤ 1	1	R/Q ≤ 1	1	R/Q ≤ 1	1	R/Q ≤ 1	1
	R ≤ 0.2 mV		R ≤ 0.2 mV		R ≤ 0.2 mV		R ≤ 0.2 mV		R ≤ 0.2 mV	
II	Q ≥ 40 ms	2	Q ≥ 40 ms	2	Q ≥ 40 ms	2	Q ≥ 40 ms	2	Q ≥ 40 ms	2
	Q ≥ 30 ms	1	Q ≥ 30 ms	1	Q ≥ 30 ms	1	Q ≥ 30 ms	1	Q ≥ 30 ms	1
aVL	Q ≥ 30 ms	1	Q ≥ 40 ms	1	Q ≥ 40 ms	1	Q ≥ 40 ms	1	Q ≥ 30 ms	1
	R/Q ≤ 1	1	R/Q ≤ 1	1	R/Q ≤ 1	1	R/Q ≤ 1	1	R/Q ≤ 1	1
aVF	Q ≥ 50 ms	3	Q ≥ 50 ms	3	Q ≥ 50 ms	3	Q ≥ 60 ms	3	Q ≥ 50 ms	3
	Q ≥ 40 ms	2	Q ≥ 40 ms	2	Q ≥ 40 ms	2	Q ≥ 50 ms	2	Q ≥ 40 ms	2
	Q ≥ 30 ms	1	Q ≥ 30 ms	1	Q ≥ 30 ms	1	Q ≥ 40 ms	1	Q ≥ 30 ms	1
	R/Q ≤ 1	2	R/Q ≤ 1	2	R/Q ≤ 1	2	R/Q ≤ 1	2	R/Q ≤ 1	2
	R/Q ≤ 2	1	R/Q ≤ 2	1	R/Q ≤ 2	1	R/Q ≤ 2	1	R/Q ≤ 2	1
V1 Ant.	Q ≥ 50 ms	2			Q ≥ 50 ms	2	any QR			
	any Q	1	any QR	1	any Q	1	(or any Q if *)	1	any Q	1
	Init R ≤ 20 ms						Ntchinit40			
V1 Post.**			R/S ≥ 1	1		1	R/S ≥ 1	1	R/S ≥ 1	1
	Init R ≥ 60 ms	2	R ≥ 50 ms	2	Init R ≥ 60 ms	2	R ≥ 50 ms	2	R ≥ 50 ms	2
	Init R ≥ 1.5 mV		R ≥ 1mV		Init R ≥ 1.5 mV		R ≥ 1mV		R ≥ 1mV	
	Init R ≥ 50 ms	1	R ≥ 40 ms	1	Init R ≥ 50 ms	1	R ≥ 40 ms	1	R ≥ 40 ms	1
	Init R ≥ 1.0 mV		R ≥ 0.7 mV		Init R ≥ 1.0 mV		R ≥ 0.7 mV		R ≥ 0.7 mV	
			Q≤0.2&S≤0.2mV	1			Q≤0.2&S≤0.2 mV	1	Q≤0.2&S≤0.2mV	1
V2 Ant.	Q ≥ 50 ms	2			Q ≥ 50 ms	2				
	any Q	1	any QR	1	any Q	1	any QR	1	any Q	1
	R ≤ 10 ms		R ≤ 10 ms		R ≤ 10 ms		(or any Q if *)		R ≤ 10 ms	
	R ≤ 0.1mV		R ≤ 0.1mV		R ≤ 0.1mV		Ntchinit40		R ≤ 0.1mV	
V2 Post.**			R/S ≥ 1.5	1			R/S ≥ 1.5	1	R/S ≥ 1.5	1
	Init R ≥ 70 ms	2	R ≥ 60 ms	2	Init R ≥ 70 ms	2	R ≥ 60 ms	2	R ≥ 60 ms	2
	Init R ≥ 2.5 mV		R ≥ 2 mV		Init R ≥ 2.5 mV		R ≥ 2 mV		R ≥ 2 mV	
	Init R ≥ 50 ms	1	R ≥ 50 ms	1	Init R ≥ 50 ms	1	R ≥ 50 ms	1	R ≥ 50 ms	1
	Init R ≥ 2.0 mV		R ≥ 1.5 mV		Init R ≥ 2.0 mV		R ≥ 1.5 mV		R ≥ 1.5 mV	
			Q≤0.3&S≤0.3mV	1			Q≤0.3&S≤0.3mV	1	Q≤0.3&S≤0.3mV	1
V3	Q ≥ 30 ms	2	Q ≥ 30 ms	2	Q ≥ 30 ms	2	QR& (Q ≥ 30 ms)	2	Q ≥ 30 ms	2
	R ≤ 10 ms		R ≤ 10 ms		R ≤ 10 ms		Ntchinit40	1	R ≤ 10 ms	
	Q ≥ 20 ms	1	Q ≥ 20 ms	1	Q ≥ 20 ms	1	any QR		Q ≥ 20 ms	1
	R ≤ 20 ms		R ≤ 20 ms		R ≤ 20 ms		(or any Q if *)		R ≤ 20 ms	
V4	Q ≥ 20 ms	1	Q ≥ 20 ms	1	Q ≥ 20 ms	1	Q ≥ 20 ms	1	Q ≥ 20 ms	1
	R/Q ≤ 0.5	2	R/Q ≤ 0.5	2	R/Q ≤ 0.5	2	R/Q ≤ 0.5	2	R/Q ≤ 0.5	2
	R/S ≤ 0.5		R/S ≤ 0.5		R/S ≤ 0.5		R/S ≤ 0.5		R/S ≤ 0.5	
	R/Q ≤ 1	1	R/Q ≤ 1	1	R/Q ≤ 1	1	R/Q ≤ 1	1	R/Q ≤ 1	1
	R/S ≤ 1		R/S ≤ 1		R/S ≤ 1		R/S ≤ 1		R/S ≤ 1	
	R ≤ 0.5 mV		R ≤ 0.5 mV		R ≤ 0.5 mV		R ≤ 0.5 mV		R ≤ 0.5 mV	
	Ntchinit40		Ntchinit40		Ntchinit40		Ntchinit40		Ntchinit40	
V5	Q ≥ 30 ms	1	Q ≥ 30 ms	1	Q ≥ 30 ms	1	Q ≥ 30 ms	1	Q ≥ 30 ms	1
	R/Q ≤ 1	2	R/Q ≤ 1	2	R/Q ≤ 1	2	R/Q ≤ 1	2	R/Q ≤ 1	2
	R/S ≤ 1		R/S ≤ 1		R/S ≤ 1		R/S ≤ 1		R/S ≤ 1	
	R/Q ≤ 2	1	R/Q ≤ 2	1	R/Q ≤ 2	1	R/Q ≤ 2	1	R/Q ≤ 2	1
	R/S ≤ 2		R/S ≤ 1.5		R/S ≤ 1.5		R/S ≤ 2		R/S ≤ 2	
	R ≤ 0.6 mV		R ≤ 0.6 mV		R ≤ 0.6 mV		R ≤ 0.6 mV		R ≤ 0.6 mV	
	Ntchinit40		Ntchinit40		Ntchinit40		Ntchinit40		Ntchinit40	
V6	Q ≥ 30 ms	1	Q ≥ 30 ms	1	Q ≥ 30 ms	1	Q ≥ 30 ms	1	Q ≥ 30 ms	1
	R/Q ≤ 1	2	R/Q ≤ 1	2	R/Q ≤ 1	2	R/Q ≤ 1	2	R/Q ≤ 1	2
	R/S ≤ 1		R/S ≤ 1		R/S ≤ 1		R/S ≤ 1		R/S ≤ 1	
	R/Q ≤ 3	1	R/Q ≤ 3	1	R/Q ≤ 3	1	R/Q ≤ 3	1	R/Q ≤ 3	1
	R/S ≤ 3		R/S ≤ 2		R/S ≤ 2		R/S ≤ 3		R/S ≤ 3	
	R ≤ 0.6 mV		R ≤ 0.6 mV		R ≤ 0.6 mV		R ≤ 0.6 mV		R ≤ 0.6 mV	
	Ntchinit40		Ntchinit40		Ntchinit40		Ntchinit40		Ntchinit40	
Total	Points		Points		Points		Points		Points	
	%LV infarct (3 x #pts)	___	% LV infarct (3 x #pts)	___	%LV infarct (3 x #pts)	___	% LV infarct (3 x #pts)	___	%LV infarct (3 x #pts)	___

Lead	LBBB Criteria	Pts
I	any Q	1
	R/Q ≤ 1	2
	R/S ≤ 1	
	R/Q ≤ 1.5	1
	R/S ≤ 1.5	
II	Q ≥ 40 ms	2
	Q ≥ 30 ms	1
	R/Q ≤ 0.5	1
	R/S ≤ 0.5	
aVL	Q ≥ 50 ms	2
	Q ≥ 40 ms	1
	R/S ≤ 0.5	2
	R/Q ≤ 0.5	
	R/S ≤ 1	1
	R/Q ≤ 1	
aVF	Q ≥ 50 ms	2
	Q ≥ 40 ms	1
	R/Q ≤ 0.5	1
	R/S ≤ 0.5	
V1 Ant.***	Nchinit40	1
	R ≥ 0.3 mV	2
	R ≥ 30 ms	
	R ≥ 0.2 mV	1
	R ≥ 20 ms	
V1 Post	S/S' ≥ 2.0	3
	S/S' ≥ 1.5	2
	S/S' ≥ 1.25	1
V2 Ant.***	Nchinit40	1
	R ≥ 0.4 mV	2
	R ≥ 30 ms	
	R ≥ 0.3 mV	1
	R ≥ 20 ms	
V2 Post	S/S' ≥ 2.5	3
	S/S' ≥ 2.0	2
	S/S' ≥ 1.5	1
V5	any Q	1
	R/R' ≥ 2	2
	R/R' ≥ 1	1
	R/S ≤ 2	
	R ≤ 0.5 mV	1
V6	Q ≥ 20 ms	1
	R/R' ≥ 2	2
	R/R' ≥ 1	1
	R/S ≤ 2	
	R ≤ 0.6 mV	1
Total	Points	
	%LV infarct (3 * #pts)	___

A score is obtained according to the values of each parameter and the presence of confounders. Then the scores are accumulated for each derivation, and they are finally added to obtain the final QRS score. Table 4 shows the distribution of assigned scores for each derivation and those corresponding to each confounder.

3 Results

To evaluate the proposed algorithm, a part of the database was provided by Dr. Robert H. Gilman, which was created during the project "Predictors of Cardiomyopathy Progression in A Chagas Disease Cohort in Bolivia" developed by Brandon Mercado [6].

We analyzed 905 12-electrode ECGs that had the following confounders: LAFB, RBBB, LAFB + RBBB, LVH, and LBBB (validated by the specialist).

Table 5 presents the performance results that correspond to the artificial neural network to detect for the presence of confounders and the total number of confounders present in the analyzed database. It is evident that classification is more successful in the detection of confounders when there are more samples to train the neural network.

In contrast, for the calculation of the QRS Score, we considered a coincidence criterion of ±1.5 points difference between the total QRS score obtained by the proposed algorithm and that obtained by the specialist.

Table 6 shows the number of coincidences that were obtained from the proposed algorithm against what the specialist obtained by a visual inspection.

The degree of coincidence was measured using Cohen's Kappa index [9]. In this case, a value of 0.772 was obtained (where 1.0 represents a perfect coincidence), which is approximately equivalent to a reproducibility level of 75% considering the error of ±1.5. This level of agreement is considered satisfactory. However, it can be significantly improved with more training samples for the neural network.

Table 5 Results of confounder detection

Confounder	Total presences in the database	Correct classification (%)
None	736	98.70
LVH	51	74.51
LAFB	40	80.02
RBBB	38	65.79
RBBB + LAFB	27	74.07
LBBB	13	69.23

Table 6 Results of the calculation of the QRS score

Specialist			
Algorithm	YES	NO	Marginal
YES	**790**	**24**	814
NO	**15**	**76**	91
Marginal	805	100	905

References

1. Selvester, R., Wagner, J., Rubin, H.: Quantitation of myocardial infarct size and location by electrocardiogram and vectorcardiogram. In: Snellen, H.A., Hemker, H.C., Hugenholtz, P.G., Van Bemmel, J.H. (eds.) Quantitation in Cardiology. Boerhaave Series for Postgraduate Medical Education (Proceedings of the Boerhaave Courses Organized by the Faculty of Medicine, University of Leiden The Netherlands), vol. 8. Springer, Dordrecht (1971)
2. Barrett, K., Barman, S., Boitano, S., Brooks, H.: Fisiología médica de Ganong. McGraw-Hill Interamericana Editores S.A., Mexico (2010)
3. Selvester, R., Wagner, G., Hindman, N.: The Selvester QRS scoring system for estimating myocardial infarct size. The development and application of the system. Arch. Intern. Med. **145**, 1877–1881 (1985)
4. Loring, Z., Chelliah, S., Selvester, R., Wagner, G., Strauss, D.: A detailed guide for quantification of myocardical scar with the Selvester QRS score in the presence of electrocardiogram cofounders. J. Electrocardiol. **44**, 548 (2011)
5. Xia, X., Wieslander, B., Strauss, D.G., Wagner, G.S., Zareba, W., Moss, A.J., Couderc, J.-P.: Automatic QRS Selvester scoring system in patients with left bundle branch block. Europace **18**(2), 308–314 (2015)
6. Bono, V., Mazomenos, E.B., Chen, T., Rosengarten, J.A., Acharyya, A., Maharatna, K., Curzen, N.: Development of an automated updated Selvester QRS scoring system using SWT-based QRS fractionation detection and classification. IEEE J. Biomed. Health Inform. **18**(1), 193–204 (2014)
7. Gautam, M.K., Giri, V.K.: A Neural network approach and wavelet analysis for ECG classification. In: IEEE International Conference on Engineering and Technology (ICETECH) (2016)
8. Kohonen, T.: Self-organized formation of topologically correct feature maps. Biol. Cybern. **43**, 59–69 (1982)
9. Cohen, J.: A coefficient of afreement for nominal scales. Educ. Psychol. Meas. **20**, 37–46 (1960)

Comparison Between DMD and Prony Methodologies Applied to Small Signals Angular Stability

Zulmar Soares Machado Jr. and Gabriel de Vasconcelos Eng

Abstract The power systems operation is in current evolution as new technologies are added and influence the dynamics of the system. Increased monitoring of the electric variables to be observed as a way to prevent oscillations and unstable transient effects also yields on the analysis of a lot of data. This paper describes two methodologies for assessment of the angular stability of small signals, using data obtained through measurements. A case with noise on measurement, and another without, will be used for evaluations of the methodologies. An acknowledged computational tool will be used to allow the comparative analysis and the results discussed.

Keywords Angular stability · Dynamic mode decomposition · Prony method · Small signals

1 Introduction

Power systems can be subjected to a condition that yields the occurrence of small signals caused by the transient unbalance between the generation and load of electric energy. Usually, the power flow is unidirectional, i.e. from power plants to the big loading center (cities) and, transferred by a power transmission system. However, in front of new technologies for distributed generation and metering devices being implemented to the power systems, there is a need to integrate them so that the system can maintain itself in stable operation, starting the concept of smart grids.

With the increase of generation's availability, the reliability seeks to ensure electrical energy to the consumers and may obtain reductions in tariff prices. On the other hand, if the installation of the power electronics can cause harmonic phenomena on the quality of power transfer, its combination with Phasor Measurement Units (PMUs), which carry on accurately monitoring over the system, can improve the electrical grid topology and allow an appropriate preventive maintenance [1]. Through

Z. S. Machado Jr. (✉) · G. de Vasconcelos Eng
Federal University of Itajuba, P.O. Box 50, Itajuba, MG, Brazil
e-mail: zulmar@unifei.edu.br

Y. Iano et al. (eds.), *Proceedings of the 4th Brazilian Technology Symposium (BTSym'18)*, Smart Innovation, Systems and Technologies 140,
https://doi.org/10.1007/978-3-030-16053-1_20

the installation of PMU to monitor low frequency oscillations in large power systems is possible to achieve satisfactory results [2–5].

The Prony method is the most usual tool in signal analysis applied to power systems [6–9]. However, the following disadvantages of this methodology are: the analysis of only one signal at a time; the need to previously estimate the system order. The first disadvantage may cause inaccuracy of results since there are multiple inputs and outputs to be considered within an electrical system. The second disadvantage consists of a hard task to estimate the system order. One needs knowledge of the dynamic phenomena and the noise characteristics associated with metering devices. Such automatic methodologies were proposed in [7], along with alternative versions with multiple signals are presented in [10].

In the scientific literature there is a robust methodology to the estimation of oscillation modes, known as dynamic mode decomposition (DMD) [11, 12]. Initially, this methodology was applied to identify dynamic modes in fluid mechanics, where its robustness was presented [1, 2, 5]. The DMD works with multiple signals, requiring only that the data is equally sampled in time.

Other benefits of DMD methodology are its possibilities to work with noise signals [13]. Computational implementation based on optimization problems also was studied with DMD [14–16]. Although the optimization improved the results, it was checked that the sparsity characteristic would benefit the identification of the most dominant dynamic modes [17, 18]. For power system modeling, this paper presents a comparison between the Prony and DMD methodologies applied on angular stability to small signals in an IEEE test system.

This paper is structured as follows: In Sect. 2 a conventional methodology of the Prony is shown. Section 3 evolves the theory of DMD. Section 4 show the results obtained through the case studies on the test system. Finally, the conclusions of this research are presented in Sect. 5.

2 Prony Method

The Prony method is a methodology that allows the estimation of frequency, damping, amplitude and phasing angle of the modal component responses [8]. Consider a time invariant dynamic linear system, which receive as input an impulse function in $t = t_0$, that yields an initial condition for the state variable $x(t_0) = x_0$. Thus, at the instant that the signal is applied to the input, it is turned off without others inputs [6]. The output variables of the system will oscillate until a new state is reached, as shown by the state space modeling in (1).

$$\dot{x}_{nx1} = A_{nxn}x_{nx1} \tag{1}$$

In (1), x is the vector of the n-states of a dynamic system. From the solution of (1), the correspondent dynamic system time response is presented in (2).

$$y(t) = \sum_{i=1}^{n} c_i \exp(\lambda_i t) = \sum_{i=1}^{n} z_i \quad (2)$$

In (2), λ_i is i-th eigenvalue in the complex domain. Usually, the equations of the state space modeling of the system are unknown, which is necessary to sample at each Δt intervals (units of time), a point of $y(t)$, that results in (3).

$$y[k] = y(k\Delta t) = \sum_{i=1}^{n} z_i^k; \quad k = 1, 2, \ldots, N \quad (3)$$

The goal of the Prony method consists in reconstruct the sampled signal in (3) by a linear prediction model as presented in (4).

$$\hat{y}[n+k] = \hat{a}_1 y[n-1+k] + \hat{a}_2 y[n-2+k] + \cdots + \hat{a}_n y[k] \quad (4)$$

Building this model for all N realized samples in signal, the matrix formulation (5) of the prediction linear problem can be calculated.

$$\begin{bmatrix} y[n] & y[n-1] & \ldots & y[1] \\ y[n+1] & y[n] & \ldots & y[2] \\ \vdots & \vdots & \ddots & \vdots \\ y[N-1] & y[N-2] & \ldots & y[N-n] \end{bmatrix} \begin{bmatrix} \hat{a}_1 \\ \hat{a}_2 \\ \vdots \\ \hat{a}_n \end{bmatrix} = \begin{bmatrix} y[n+1] \\ y[n+2] \\ \vdots \\ y[N] \end{bmatrix} \quad (5)$$

The solution of (5) yields in the coefficients of the characteristic polynomial, shown in (6), which contains information about the discrete eigenvalues of the system.

$$\hat{z}^n - \hat{a}_1 \hat{z}^{n-1} - \hat{a}_2 \hat{z}^{n-2} - \cdots - \hat{a}_n \hat{z}^0 = 0 \quad (6)$$

With the calculation of the roots that satisfy (6), it will be necessary to convert them to continuous time domain, using (7), which will be possible in order to estimate the parameters of the original signals.

$$\hat{\lambda}_i = \ln(\hat{z}_i)/\Delta t \quad (7)$$

Therefore, by solving (5), (6) and (7), the Prony method will yield a pole estimation for the dynamic characteristics of the system.

3 Dynamic Mode Decomposition

Dynamic mode decomposition (DMD) is a methodology of modal identification which simultaneously analyzes a set of signals under the condition that the sample at each Δt interval (units of time) is spaced in equally [5]. Consider a set of M signals that contains a $N+1$ sample each one, ordained on matrix Y. as in (8).

$$Y = \begin{bmatrix} \begin{bmatrix} y_0 \; y_1 & \cdots & y_i & \cdots & y_N \end{bmatrix}_1 \\ \begin{bmatrix} y_0 \; y_1 & \cdots & y_i & \cdots & y_N \end{bmatrix}_2 \\ \vdots \\ \begin{bmatrix} y_0 \; y_1 & \cdots & y_i & \cdots & y_N \end{bmatrix}_M \end{bmatrix} \tag{8}$$

In (8), y_i is i-th sample of the signal to be analyzed. Then, consider Y separated into two matrices with M rows and N columns, as can be seen in (9) and (10).

$$Y0_{MxN} = \begin{bmatrix} \begin{bmatrix} y_0 \; y_1 & \cdots & y_i & \cdots & y_{N-1} \end{bmatrix}_1 \\ \begin{bmatrix} y_0 \; y_1 & \cdots & y_i & \cdots & y_{N-1} \end{bmatrix}_2 \\ \vdots \\ \begin{bmatrix} y_0 \; y_1 & \cdots & y_i & \cdots & y_{N-1} \end{bmatrix}_M \end{bmatrix} \tag{9}$$

$$Y1_{MxN} = \begin{bmatrix} \begin{bmatrix} y_1 \; y_2 & \cdots & y_i & \cdots & y_N \end{bmatrix}_1 \\ \begin{bmatrix} y_1 \; y_2 & \cdots & y_i & \cdots & y_N \end{bmatrix}_2 \\ \vdots \\ \begin{bmatrix} y_1 \; y_2 & \cdots & y_i & \cdots & y_N \end{bmatrix}_M \end{bmatrix} \tag{10}$$

Assuming that a linear and constant mapping matrix that contains information about the dynamic of the system exists [1], then (9) and (10) can be combined as shown in (11).

$$Y1_{MxN} \approx A_{MxM} Y0_{MxN} \tag{11}$$

In situations where there is a large amount of data to be considered with the DMD, applying the Eigen decomposition directly over matrix A may be computationally expensive and can yield inaccurate results [12]. An alternative to spending less time consuming can be realized with an approximation of the mapping matrix using the singular value decomposition (SVD) of Y_0, as shown in (12).

$$Y0_{MxN} = U_{MxM} \sum_{MxN} (V_{NxN})^T \tag{12}$$

The matrix U is orthonormal matrix that contains the left eigenvectors. V^T is also an orthonormal matrix for the right eigenvectors and Σ is a diagonal matrix with N singular values $\{\sigma_1 \; \sigma_2 \ldots \; \sigma_n \ldots \; \sigma_N\}$. However, the matrix Σ can also be built with numerical eigenvalues, even if the measurement presents either kind of noise. In this study a tolerance for the SVD was considered, such that σ_n divided by σ_1 will allow for the estimation of an upper limit for the system order.

Considering the order n ($n < N$) and replacing (12) in (11), it is possible to calculate an alternative formulation for Y_1, shown in (13).

$$Y_{1_{MxN}} = A_{MxM} U_{Mxn} \sum_{nxn} (V_{Nxn})^T \quad (13)$$

By taking the mapping matrix in (13), and knowing that the matrices U and V are orthonormal, i.e., $U^{-1} = U^T$ and $V^{-1} = V^T$, so it is possible to obtain an approximation for A, as in (14).

$$A_{MxM} = Y_{1_{MxN}} V_{Nxn} \sum_{nxn}^{-1} (U_{Mxn})^T \quad (14)$$

Finally, an approximation for the mapping matrix $\tilde{A}$ can be calculated multiplying to the left by U^T, and to the right by U, where $U^T U$ is equal to an I_n (identity matrix). This operation is regarded as a robust implementation because it combines the DMD with an Eigen system identification methodology known as own orthogonal decomposition (OOD). In [12], there is a description between the DMD and OOD, and an expression of $\tilde{A}$ is shown in (15).

$$\tilde{A}_{nxn} = (U_{Mxn})^T Y_{1_{MxN}} V_{Nxn} \sum_{nxn}^{-1} \quad (15)$$

The Eigen decomposition of $\tilde{A}$ yields the discrete modes of the modeling which can be transformed into continuous modes of time domain through of (7). Thus, the DMD is calculated applying the following terms (8)–(15).

4 Results

The methodologies will be applied in an electrical power system known as New England, which is usually considered in stability studies [19, 20]. The system has 39 buses, 34 transmission lines, 12 transformers and 10 generators (More details in [21]).

The assessment of small signals on angular stability is established by the analysis of the equilibrium point of the system. Under this condition, the set of differential equations can be linearized and the eigenvalues can be calculated. Whereas if all eigenvalues have a negative real part, the system presents dynamic stability.

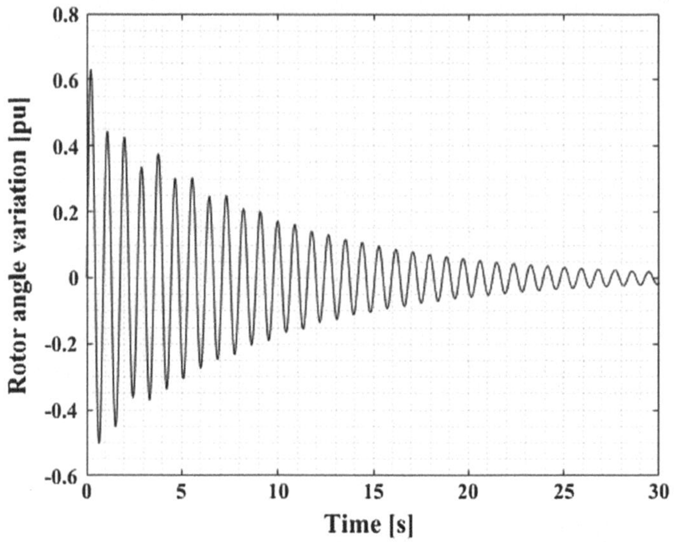

Fig. 1 Rotor angle at bus 30 (noiseless)

A comparative analysis will be realized between the methodologies, calculating all eigenvalues of the system through of the PacDyn software [22] and 65 eigenvalues were found and considered as a reference. The PacDyn is capable of building a time invariant linear dynamic system through of a state space modeling as shown in (16).

$$\begin{cases} \dot{x} = Ax + Bu \\ y = Cx + Dy \end{cases} \tag{16}$$

Usually the dynamic system is unknown (16). However, known inputs will be applied to the system, resulting in output curves which will have estimated modes through of the Prony and DMD methodologies.

4.1 Case 1—Response to Impulse

In this case, mechanical power of the generators was considered as an input in (16) while the angular velocity and rotor angle were considered as output signals. With impulse function inputs, 180 curves were yielding that were sampled in 2000 points and distributed in 30 s. In Fig. 1, the response of the rotor angle in bus 30 is shown through of the impulse applied on the mechanical power of the generator connected in the same bus.

The DMD was applied with multiple signals, which have an estimated 35 eigenvalues from which 30 matching with the 65 referent eigenvalues. However, in the

Prony method, the algorithm needed to be applied 180 times when only one signal is considered at each time. Through all of the calculated Prony simulations, 1080 eigenvalues were determined of which only 22 are very close to the reference. However, 19 presented a positive real part.

A comparison between the estimated eigenvalues from Prony and DMD methodologies is presented in Fig. 2 as well as the eigenvalues reference from PacDyn. In Table 1, the 9 reference modes called as electromechanical of the New England system are presented, and the comparison with the respective modes estimated with both methodologies.

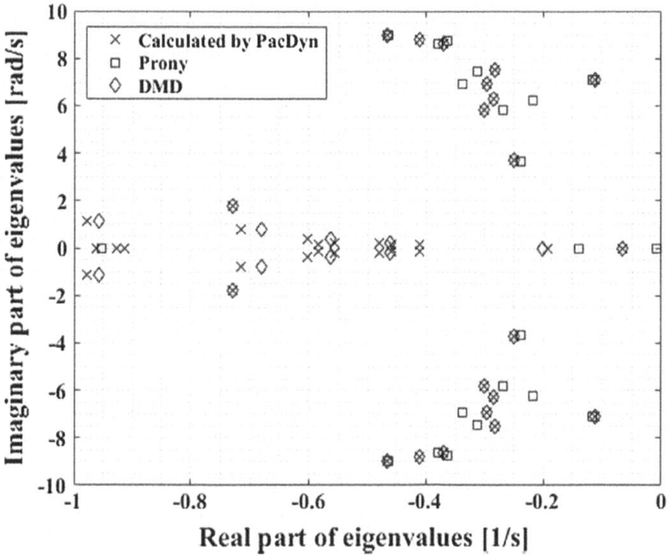

Fig. 2 Comparative Eigenvalues Analysis—Case 1

Table 1 Electromechanical Modes—Case 1

PacDyn	Prony	DMD
−0.4672 ± j8.9644	−0.4635 ± j8.9808	−0.4672 ± j8.9645
−0.4118 ± j8.7786	−0.3627 ± j8.7517	−0.4118 ± j8.7786
−0.3704 ± j8.6111	−0.3802 ± j8.6133	−0.3704 ± j8.6111
−0.2817 ± j7.5367	−0.3125 ± j7.4713	−0.2817 ± j7.5367
−0.1118 ± j7.0950	−0.1156 ± j7.0889	−0.1118 ± j7.0950
−0.2968 ± j6.9562	−0.3367 ± j6.9169	−0.2968 ± j6.9562
−0.2834 ± j6.2816	−0.2181 ± j6.2539	−0.2834 ± j6.2816
−0.3009 ± j5.7921	−0.2686 ± j5.8398	−0.3009 ± j5.7921
−0.2491 ± j3.6862	−0.2387 ± j3.6612	−0.2491 ± j3.6862

Through Table 1 and Fig. 2, it is possible to verify that the DMD methodology was capable of identify all of the electromechanical modes of the system with accuracy. However, the Prony methodology does not identify all modes with the same precision.

4.2 Case 2—Noise in Measurements

The case 2 considerations are the same as those from case 1, but the difference is in the noise added in the output signals. More severe noise was applied in order to test the accuracy of the methodologies. Figure 3 represents the same signal shown in Fig. 1, but with noise in its measurements.

For case 2, the Prony estimated 1080 eigenvalues, but only 12 were close to the reference. Otherwise, the DMD estimated 37 eigenvalues with 30 close to the reference. However, no eigenvalues were found with a positive real part in either of the methodologies. A comparison between the estimated eigenvalues from Prony and DMD methodologies is presented in Fig. 4 along with the eigenvalues reference from PacDyn.

The DMD methodology was estimated with accuracy for all of called electromechanical modes whereas Prony did not identify any mode. The reason for this result is explained by the difficulty of estimating the order of the system in which there is noise in the measurements.

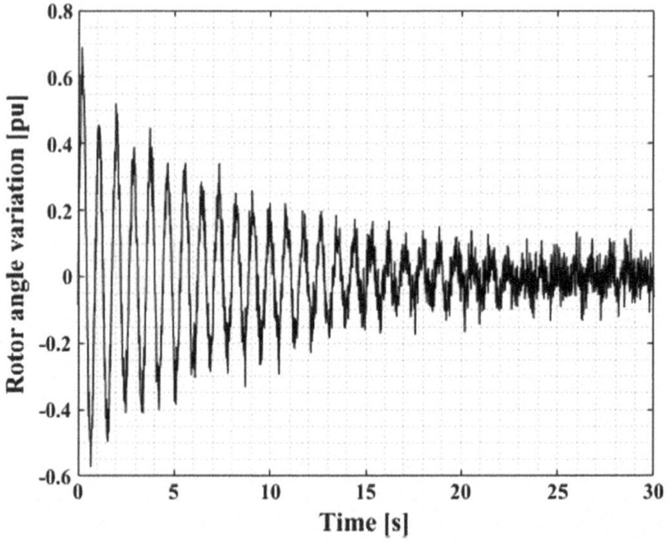

Fig. 3 Rotor angle at bus 30, with noise

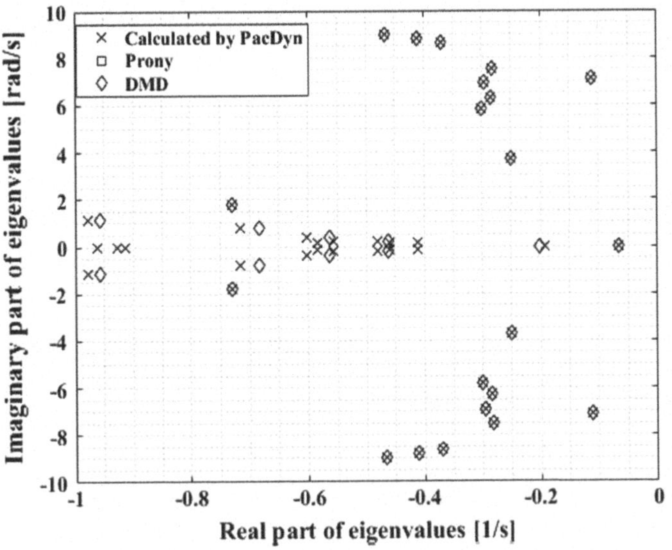

Fig. 4 Comparative Eigenvalues Analysis—Case 2

5 Conclusion

This work has evaluated two methodologies for assessment of small signals angular stability. It was observed that both methods presented acceptable estimates that captured the dynamics of the system. However, the DMD methodology presented highlighted results when compared to Prony methodology.

An important feature of DMD methodology was the ability to estimate close distinct eigenvalues to PacDyn, and the identification of electromechanical modes, showing a high performance even when there is noise in the measurements.

Further tests should be performed to improve the methodologies considered, especially with regard to the analysis of PMU signals.

Acknowledgements The authors thank CEPEL (Electrical Energy Center Research) for helping with the sourcing of an academic version of the PacDyn computer program.

References

1. Saldaña, A.E., Barocio, E., Messina, A.R., Ramos, J.J., Segundo, R.J., Tinajero, G.A.: Monitoring harmonic distortion in microgrids using dynamic mode decomposition. In: 2017 IEEE PES General Meeting, pp. 1–5. Chicago, IL (2017). https://doi.org/10.1109/pesgm.2017.8274696
2. Barocio, E., Pal, B.C., Thornhill, N.F., Messina, A.R.: A dynamic mode decomposition framework for global power system oscillation analysis. IEEE Trans. Power Syst. **30**(6), 2902–2912 (2015). https://doi.org/10.1109/TPWRS.2014.2368078

3. Hauer, J.F., Trudnowski, D.J., DeSteese, J.G.: A perspective on WAMS analysis tools for tracking of oscillatory dynamics. In: 2007 IEEE PES General Meeting, pp. 1–10. Tampa, FL (2007). https://doi.org/10.1109/pes.2007.386186
4. Power System Dynamic Performance Committee, Task Force on Identification of Electromechanical Modes: Identification of electromechanical modes in power systems, IEEE/PES, Special Publication TP462 (2012)
5. Mohapatra, S., Overbye, T.J.: Fast modal identification, monitoring, and visualization for large-scale power systems using dynamic mode decomposition. In: 2016 Power Systems Computation Conference (PSCC), pp. 1–7. Genoa (2016)
6. Hauer, J.F., Demeure, C.J., Scharf, L.L.: Initial results in prony analysis of power system response signals. IEEE Trans. Power Syst. **5**(1), 80–89 (1990). https://doi.org/10.1109/59.49090
7. Zhou, N., Pierre, J., Trudnowski, D.: A stepwise regression method for estimating dominant electromechanical modes. In: 2012 IEEE PES General Meeting, pp. 1–1. San Diego, CA (2012). https://doi.org/10.1109/pesgm.2012.6344635
8. Zhou, N., Pierre, J., Trudnowski, D.: Some considerations in using prony analysis to estimate electromechanical modes. In: 2013 IEEE PES General Meeting, pp. 1–5. Vancouver, BC (2013). https://doi.org/10.1109/pesmg.2013.6672888
9. Khan, F.Z., Nagaraja, R., Khincha, H.P.: Improved fault location computation using prony analysis for short duration fault. In: 2014 Eighteenth National Power Systems Conference (NPSC), pp. 1–5. Guwahati (2014)
10. Ordonez, C.A., Chamorro, H.R., Quintero, J., Leelaruji, R., Peng, J.C., Nordström L.: Prony-based on-line oscillation detection with real PMU information. In: 2016 IEEE Colombian Conference on Robotics and Automation (CCRA), pp. 1–5. Bogota (2016). https://doi.org/10.1109/ccra.2016.7811401
11. Schmid, P., Sesterhenn, J.: Dynamic mode decomposition of numerical and experimental data. In: Bulletin of the American Physical Society, vol. 53 (2008)
12. Schmid, P.J.: Dynamic mode decomposition of numerical and experimental data. J. Fluid Mech. **656**, 5–28 (2010)
13. Dawson, S.T.M., Hemati, M.S., Williams, M.O., Rowley, C.W.: Characterizing and correcting for the effect of sensor noise in the dynamic mode decomposition. In: Experiments in Fluids, vol. 57, no. 3, p. 42 (2016)
14. Wang, S., Qing, X.: Nuclear norm regularised dynamic mode decomposition. IET Signal Process. **10**(6), 626–632 (2016)
15. Héas, P., Herzet, C.: Optimal low-rank dynamic mode decomposition. In: 2017 IEEE International Conference on Acoustics, Speech and Signal Processing (ICASSP), pp. 4456–4460. New Orleans, LA (2017). https://doi.org/10.1109/icassp.2017.7952999
16. Takeishi, N., Kawahara, Y., Yairi, T.: Sparse nonnegative dynamic mode decomposition. In: 2017 IEEE International Conference on Image Processing (ICIP), pp. 2682–2686. Beijing (2017). https://doi.org/10.1109/icip.2017.8296769
17. Jovanović, M.R., Schmid, P.J., Nichols, J.W.: Sparsity-promoting dynamic mode decomposition. In: Physics of Fluids, vol. 26, no. 2, p. 024103 (2014)
18. Annoni, J., Seiler, P., Jovanović, M.R.: Sparsity-promoting dynamic mode decomposition for systems with inputs. In: 2016 IEEE 55th Conference on Decision and Control (CDC), pp. 6506–6511. Las Vegas, NV (2016). https://doi.org/10.1109/cdc.2016.7799270
19. Martins, N., Lima, L.T.G., Pinto, H.J.C.: Computing dominant poles of power system transfer functions. IEEE Trans. Power Syst. **11**(1), 162–170 (1996). https://doi.org/10.1109/59.486093
20. Athay, T., Podmore, R., Virmani, S.: A practical method for the direct analysis of transient stability. In: IEEE Transactions on Power Apparatus and Systems, vol. PAS-98, no. 2, pp. 573–584 (1979). https://doi.org/10.1109/tpas.1979.319407
21. Pai, M.A.: Energy Function Analysis for Power System Stability. Kluwer Academic Publishers, Boston, MA (1989)
22. Eletrobras CEPEL. PacDyn 9.8.3 - User's Manual, CEPEL/DRE - Department of Electrical Networks (2017)

Rheological Analysis of Blood Flow in the Bifurcation of Carotid Artery with OpenFOAM

Marcus Vinicius Paes Carvalho, Raquel Jahara Lobosco and Guilherme Barbosa Lopes Júnior

Abstract **Purpose**: Evaluate the influence of rheological models for blood flow in the carotid artery bifurcation. **Methods**: The geometry (developed in SolidWorks) was based on the mean dimensions of several angiograms of adults and children. Newtonian and non-Newtonian (Casson, Carreau-Yasuda and Power Law models) hypothesis were admitted for modelling the rheological behavior of the blood. The parameters of the equations that govern the rheological behavior were taken from blood itself. The flow simulation was performed using open source software OpenFOAM, with the solvers nonNewtonianIcoFoam (non-Newtonian hypothesis) and icoFoam (Newtonian hypothesis). Both use the PISO algorithm. **Results**: In all models, the fluid presents a "detachment" in the non-divider wall. This behavior is more noted in the Newtonian hypothesis, which also presents "C" shaped profiles. **Conclusions**: The Newtonian behavior diverges from the non-Newtonian and the results demonstrate that there is a good concordance between the non-Newtonian models. For a mid-sized artery such as the carotid artery, the non-Newtonian hypothesis is more indicated. There are no significant differences between the Casson, Power Law and Carreau-Yasuda models.

Keywords Hemodynamics · OpenFOAM · Rheology

M. V. P. Carvalho (✉)
Universidade Federal do Rio de Janeiro, Rio de Janeiro, RJ 21941-901, Brazil
e-mail: mvpaescarvalho@peb.ufrj.br

R. J. Lobosco
Universidade Federal do Rio de Janeiro, Macaé, RJ 27930-560, Brazil
e-mail: raquelobosco@gmail.com

G. B. L. Júnior
Universidade Federal de Pernambuco, Recife, PE 50670-901, Brazil
e-mail: guilherme.lopes@ufpe.br

Y. Iano et al. (eds.), *Proceedings of the 4th Brazilian Technology Symposium (BTSym'18)*, Smart Innovation, Systems and Technologies 140,
https://doi.org/10.1007/978-3-030-16053-1_21

1 Introduction

The study of fluid biomechanics is very important since 65% of the human body is composed of water. The design of any coronary implant depends on the study of fluid mechanics, especially those in direct contact with blood, such as stents and heart valves. According to the American Heart Association, in 2011, more than 82,500,000 people in the United States suffered from at least one cardiovascular disease. Most of these cases are associated with hypertension (approximately 74 million) or coronary heart disease (approximately 18 million). In addition, cardiovascular diseases are the cause of one in three deaths in the US, that is, 1.5 times greater than cancer [1]. As an example, atherosclerosis and thrombosis can be cited, which triggers are influenced by hemodynamics. Local transfer of atherosclerotic plasma and cellular components are regulated by the flow altered by recirculation. Low and oscillatory shear stress values on arterial walls modify their permeability to blood components. Other factors influencing local hemodynamics are pulse pressure (difference between systolic and diastolic pressures), flow geometry and rheological properties of blood. Therefore, the use of the correct model is important for this representation, since it will help to describe the dynamics of fluids in the cardiovascular system. Therefore, it will be possible to design better prostheses and improve clinical and surgical procedures [2]. Displayed equations are centered and set on a separate line. According to clinical observations, regions of bifurcations and curvatures are more sensitive to pathological changes. In particular the carotid sinus (region where the carotid artery bifurcation occurs), since it is the most affected by atherosclerotic lesions [3]. Thereby, it is very important to study hemodynamics for a better understanding of atherosclerosis and its relationship with flow [4]. It is also important to mention that it is extremely difficult to carry out in vivo measurements of certain parameters with precision, such as shear stress. So, the numerical simulation becomes a great ally in the development of this area [4].

Due to the complexity of blood rheology, there is no unanimous model to represent its viscosity. The most used fluid models for this purpose are the Power Law, Casson and Carreau-Yasuda [2]. Some researches argue that in large arteries, the hypothesis of Newtonian behavior for blood is a good assumption [3, 5]. Other studies, however, argue that although non-Newtonian behavior is more present in capillaries and small vessels, the shear stress in vessel walls is highly related to atherosclerosis, so it would be important to investigate non-Newtonian behavior in major arteries [4]. On this divergence, it is worth mentioning that Perktold et al. [3] used parameters of a blood analog fluid for the non-Newtonian Casson model, which does not guarantee their reliability. Boyd et al. [4], reused these parameters and compared them with three models: Newtonian, Carreau-Yasuda (non-Newtonian) model and Casson model (non-Newtonian), from results obtained from blood itself, in a study carried out by Abraham et al. [6]. The study shows that there are significant differences between blood and blood analog fluid properties. Thus, the adjustment of the parameters for the rheological models was more relevant than the choice of the model itself. The work Gijsen et al. [7] also defends the non-Newtonian assumption. This feature was included using the Carreau-Yasuda model. Nevertheless, more hemodynamic studies on cardiovascular diseases and a more accurate rheological model of blood are needed [2]. The non-Newtonian Power Law, Casson and Carreau-Yasuda models are defined by Eqs. 1, 2 and 3, respectively:

$$\eta = k\dot{\gamma}^{(n-1)} \tag{1}$$

$$\eta = \left(\sqrt{\frac{\tau_0}{\dot{\gamma}}} + \sqrt{m}\right)^2 \tag{2}$$

$$\eta = \eta_\infty + (\eta_0 - \eta_\infty)\left[1 + (\lambda\dot{\gamma})^a\right]^{\left(\frac{n-1}{a}\right)} \tag{3}$$

2 Methods

For the development of a numerical CFD simulation, it is necessary to create a geometry and the generation of its mesh. After this procedure, the appropriate solver is chosen. The geometry used was developed in the software SolidWorks that was later imported into Salome to generate the mesh of 152,631 tetrahedral elements. Numerical simulations were performed using OpenFOAM. For Newtonian hypothesis simulation, the icoFoam solver was used and for the non-Newtonian hypothesis the nonNewtonianIcoFoam solver was used. Both use the PISO algorithm for the matching between pressure and velocity.

It is known that small changes in the geometry can produce different results in the flow behaviour in bifurcations. The choice of a geometry of a specific individual or of an average of several individuals is an exhaustive discussion [8]. This work decided to make use of a model used by several works such as Bharadvaj et al. [8, 9], Ku et al. [10], Gijsen et al. [7], which was produced from fifty-seven angiograms of twenty-two adults between thirty-four and seventy-seven years and sixty-seven angiograms of children under eighteen years.

The velocity measurements were performed for $t = 1s$, in steady state in five positions in the internal carotid along the y and z axes, distant 4 mm along the x′ axis, as can be seen in Fig. 1.

The geometry has one input (common carotid) and two outputs (internal and external carotid). At the entrance, the boundary condition used was a parabolic velocity profile and maximum velocity $v = 0.09$ m/s [10]. For all simulations, a stress-free condition was established in the two exits and a non-slip condition in the walls.

The parameters used by Abrahamet al. [6] are valid for the Carreau-Yasuda model and are shown in Table 1.

For the Casson and Power Law models, the parameters were obtained by curve fitting in MATLAB. The fitting of Casson and Power Law models to the curve generated by the Carreau-Yasuda equation using the data in Table 1 produced the curves shown in Fig. 2.

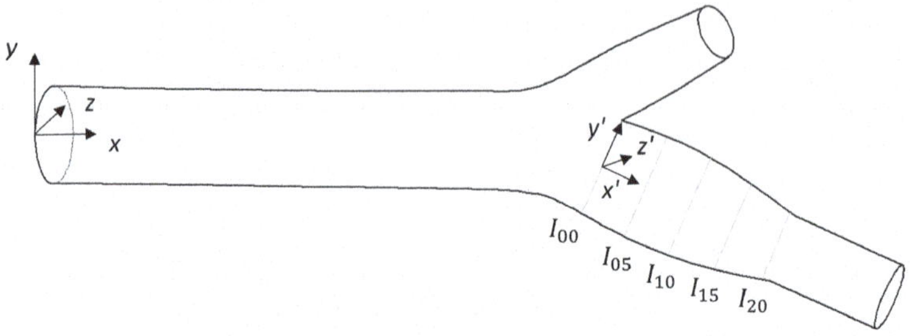

Fig. 1 Geometry of the carotid artery with the sites of velocity measurement. The sections I_{00}, I_{05}, I_{10}, I_{15} and I_{20} are separated by 4 mm

Table 1 Values of Eq. (3) for blood

Parameters	Values
η_0	1.6×10^{-1} Pa s
η_∞	3.5×10^{-3} Pa s
λ	0.110 s
a	0.640
n	0.2128

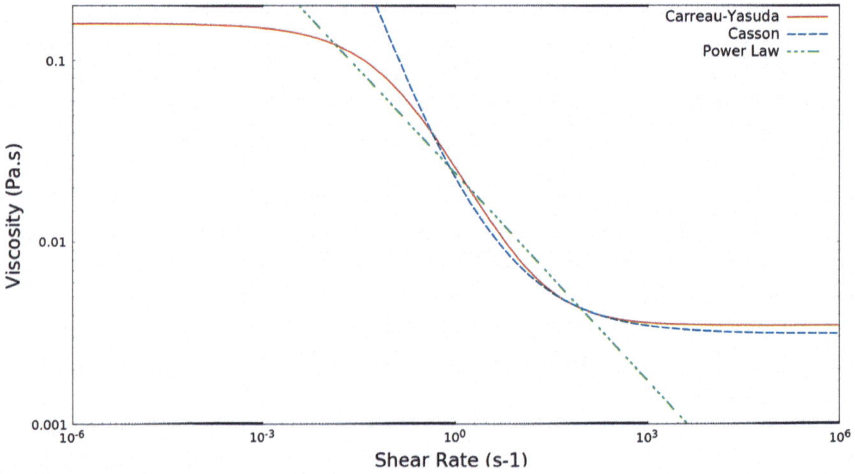

Fig. 2 Fit of the Casson and Power Law models to the curve generated by Eq. (3) with the values presented in Table 1

The Power Law model has $k = 0.024$ Pa s and $n = 0.62$ and the Casson model has $\tau_0 = 0.0090$ Pa and $m = 0.0031$ Pa s. The viscosities were limited to $\eta_0 = 1.6 \times 10^{-1}$ Pa s and $\eta_\infty = 3.5 \times 10^{-3}$ Pa s.

It can be seen that at high shear rates, the blood behaves like a Newtonian fluid (constant viscosity). It is believed that when the shear rates approach zero, the blood would behave as a solid because of the agglomeration of the red blood cells [1]. These are characteristics of a Bingham plastic fluid.

3 Results

Figures 3 and 4 show the flow profiles for the sections I_{00}, I_{05}, I_{10}, I_{15} and I_{20} along the y and z axes, respectively, for Newtonian and non-Newtonian hypothesis using the Casson, Power Law and Carreau-Yasuda.

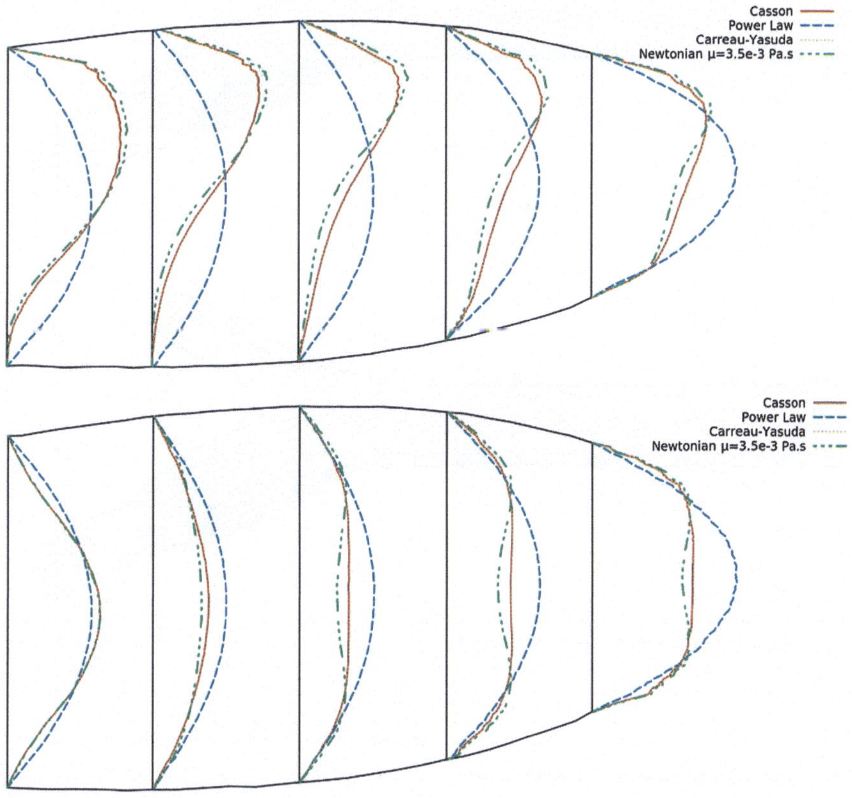

Fig. 3 Velocity profiles with respect to the y′ and z′ axis, respectively, for the Newtonian hypothesis and non-Newtonian models (Casson, Power Law and Carreau-Yasuda)

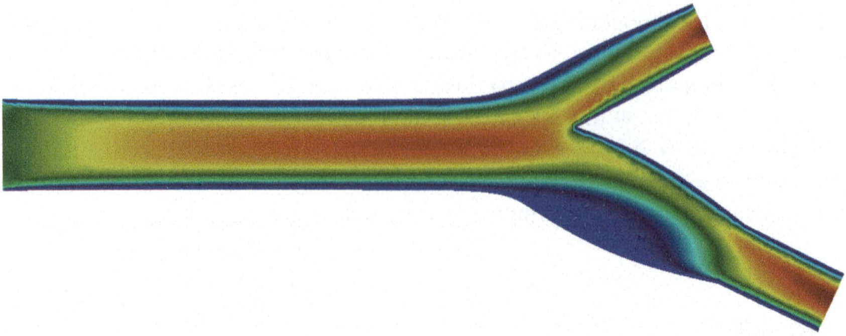

Fig. 4 Illustration of the velocity field for the Newtonian hypothesis

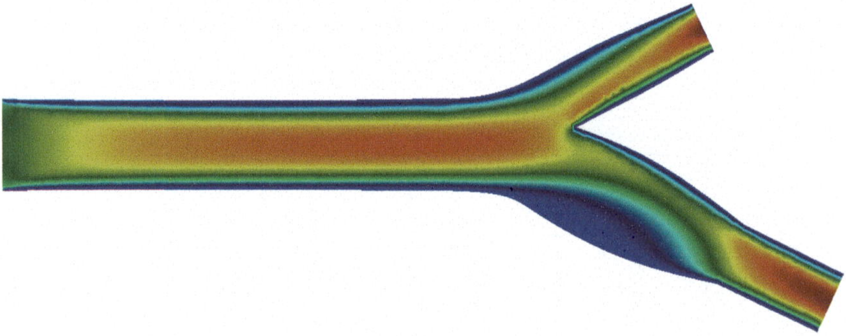

Fig. 5 Illustration of the velocity field for the Casson model

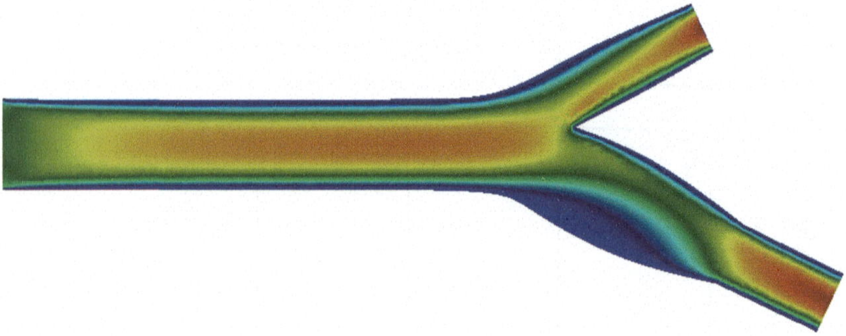

Fig. 6 Illustration of the velocity field for the Power Law model

The velocity field for the different models is shown by a slice in the xy plane of the geometry in Figs. 4, 5, 6 and 7.

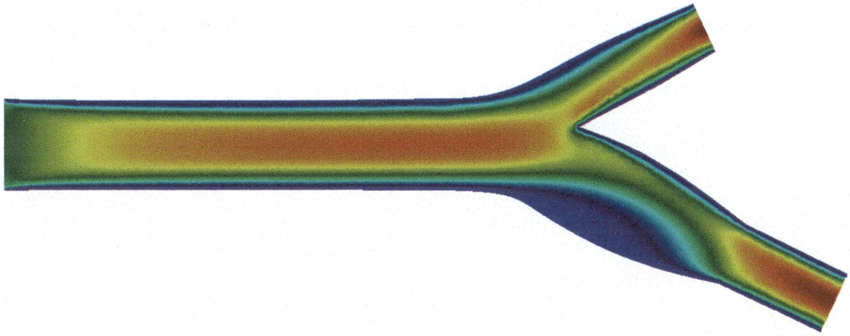

Fig. 7 Illustration of the velocity field for the Carreau-Yasuda model

Fig. 8 Velocity magnitude scale in m/s

4 Discussion

The Casson, Carreau-Yasuda and Power Law models show good agreement in the results. The differences in velocity magnitudes between the divider and non-divider walls are the largest when $\mu = 3.5 \times 10^{-3}$ Pa s for the Newtonian hypothesis. In the z-axis, Fig. 5, the difference of the velocity magnitudes at the center of the flow become significant when the Newtonian hypothesis with constant viscosity is adopted. Parabolic profiles (in "C" format) are predominant. It can be noticed that, for non-Newtonian models, the axial velocity profile is flat, common to pseudoplastic fluids [7] (Fig. 8).

Figures 4, 5, 6, and 7 illustrate the longitudinal sections of the arterial bifurcation computational domain. It is possible to note the similarity between the non-Newtonian models. The "detachment" of the fluid in the non-divider wall is more pronounced for the Newtonian flow.

5 Conclusion

For a mid-sized vessel such as the carotid artery, it is possible to conclude that there are significant differences between the Newtonian and non-Newtonian hypothesis for blood modeling, which demonstrates agreement with the results of Gijsen et al. [7], Boyd et al. [4]. Among the non-Newtonian models analyzed in this study (Casson, Carreau-Yasuda and Power Law), all presented similar results.

It is important to emphasize that any analysis cannot be considered definitive, since simplifications such as constant velocity in the entrance and absence of movement in the walls of the vessel were adopted.

Acknowledgements Some of this work benefited from the support of Rio de Janeiro Federal University computer resources at the Thermal and Fluid Mechanics Laboratory. The research was part of the Advanced Computational Fluid Dynamics project financed by Foundation for Research Support of Rio de Janeiro State—FAPERJ.
The authors would also like to thank CAPES for financing support through scholarship as part of Science Without Borders program.

References

1. Rubenstein, D., Yin, W., Frame, M.D.: Biofluid Mechanics: An Introduction to Fluid Mechanics, Macrocirculation, and Microcirculation. Academic Press (2015)
2. Shibeshi, S.S., Collins, W.E.: The rheology of blood flow in a branched arterial system. Appl. Rheol. **15**(6), 398 (2005) (Lappersdorf, Germany: Online)
3. Perktold, K., Resch, M., Peter, R.O.: Three-Dimensional Numerical Analysis of Pulsatile Flow and Wall Shear Stress in the Carotid Artery Bifurcation. J. Biomech. **24**(6), 409–420 (1991)
4. Boyd, J., Buick, J.M., Green, S.: Analysis of the Casson and Carreau-Yasuda non-Newtonian blood models in steady and oscillatory flows using the lattice Boltzmann method. Phys. Fluids **19**(9), 093103 (2007)
5. Wu, W.-T., et al.: A numerical study of blood flow using mixture theory. Int. J. Eng. Sci. **76**, 56–72 (2014)
6. Abraham, F., Behr, M., Heinkenschloss, M.: Shape optimization in steady blood flow: a numerical study of non-Newtonian effects. Comput. Methods Biomech. Biomed. Eng. **8**(2), 127–137 (2005)
7. Gijsen, F.J.H., van de Vosse, F.N., Janssen, J.D.: The influence of the non-Newtonian properties of blood on the flow in large arteries: steady flow in a carotid bifurcation model. J. Biomech. **32**(6), 601–608 (1999)
8. Bharadvaj, B.K., Mabon, R.F., Giddens, D.P.: Steady flow in a model of the human carotid bifurcation. Part II—laser-Doppler anemometer measurements. J. Biomech. **15**(5), 363–378 (1982)
9. Bharadvaj, B.K., Mabon, R.F., Giddens, D.P.: Steady flow in a model of the human carotid bifurcation. Part I—flow visualization. J. Biomech. **15**(5), 349–362 (1982)
10. Ku, D.N., et al.: Hemodynamics of the normal human carotid bifurcation: in vitro and in vivo studies. Ultrasound Med. Biol. **11**(1), 13–26 (1985)

Performance Measurements of Sobel Edge Detection Implemented in an FPGA Architecture and Software

Tercio Naoki Sato, Gabriel Pedro Krupa and Leonardo Breseghello Zoccal

Abstract This paper presents an implementation of Sobel edge detection in static images for the FPGA, described in Verilog, and for the respective software: Mathematica, MATLAB, OpenCV and in-house C++ written program. Comparisons in performance were made between the different applications. The results concluded that the FPGA was faster.

Keywords Edge detection · Digital image processing · FPGA

1 Introduction

Field-Programmable Gate Arrays (FPGAs) are composed of interconnected element arrays. In general, these interconnections between the elements can be programmed via hardware description language (HDL) through computer-aided design (CAD) tools. Thereby combinational or sequential logic functions can be implemented by the user [1]. The basic structure [2] of an FPGA includes a lookup table (LUT), which performs a logical operation; a flip-flop, which stores the LUT results; the wires, which connect one element to another and input/output blocks, which obtain the data. Note that FPGAs haven't had gate arrays inside of it because this was an alternative to integrated circuits, that had gate arrays in the middle of 1980s [3]. Also, the term "field programmable" has been used due to the possibility of programming after the manufacturing and configuration in the field [4, 5].

T. N. Sato (✉) · G. Pedro Krupa · L. B. Zoccal
Systems Engineering and Information Technology Institute/Microelectronics Group, Federal University of Itajubá, 1303 BPS Ave, Pinheirinho, Post Box 50, Itajubá, MG 37500 903, Brazil
e-mail: sato.tercio@gmail.com

G. Pedro Krupa
e-mail: gabrielkrupa23@gmail.com

L. B. Zoccal
e-mail: lbzoccal@unifei.edu.br

Y. Iano et al. (eds.), *Proceedings of the 4th Brazilian Technology Symposium (BTSym'18)*, Smart Innovation, Systems and Technologies 140,
https://doi.org/10.1007/978-3-030-16053-1_22

Besides the development of the FPGAs, the field of image processing continues significant. One of the local image preprocessing techniques used to detect changes in light intensity is called edge detection [6]. As stated by [7], the objective of edge detection "is to return a binary image where a nonzero value denotes the presence of an edge in the image". According to [8], an edge is characterized as "a border between two regions, each of which has approximately uniform brightness", and additionally [9] suggests that an edge "is characterized by its height, slope angle and horizontal coordinate of the slope midpoint". Therefore "Edge detection is one of the most important and widely studied aspects of image processing" [10]. According to [11] digital image processing "has been widely used in a range of industrial, commercial, civilian and military applications for over two decades. And edge detection is always the first step in many image processing algorithms because it significantly reduces the amount of data and filters out useless information". Also, edge detection is computationally demanding and inherently parallelizable [12]. As FPGAs have fine-grained reconfigurable architectures that allow any operation to be done in hardware, algorithms for video and image processing have better results when applied to such architectures [13]. Due to its high computing power, FPGAs are used in a great variety of applications, being able to, in some cases, substitute multiple digital signal processors (DSPs) and as they present low power consumption, FPGAs can even be implemented in battery powered devices [14].

For these reasons, an FPGA was chosen to perform Sobel edge detection in this work, which was based on the "ECE536 final project" from Cornell University [15], written in Verilog, and the goal was to measure the performance of the Sobel edge detection algorithm when implemented in an FPGA architecture. Performance metrics in terms of execution time are repeatedly compared with results obtained with a few software, including (a) MATLAB; (b) Mathematica; (c) OpenCV, and (d) an in-house software written in C++ that does not make use of image processing libraries.

The following sections describe in more detail the different tools used to assess and evaluate potential benefits of applying FPGA architectures for image processing.

2 Methodology

In this section, the algorithm for Sobel edge detection and the tools used to develop the applications are listed and explained.

2.1 Altera DE2 Board

For this project, Altera DE2 Development and Education board were utilized. It can be considered a System on a Programmable Chip (SoPC) [16]. The main components used include Altera Cyclone II FPGA 2C35, USB Blaster, 512- Kbyte Static Random Access Memory (SRAM), 8-Mbyte Single Data Rate Synchronous Dynamic RAM

(SDRAM) chip, 4- Mbyte Flash memory, 18 interrupters, 18 light emitting diodes (LED), 9 red light emitting diodes, 27-MHz Oscillator, 50- MHz Oscillator, Video Graphic Array Digital to Analogic Converter (VGA DAC) and a VGA Video port as detailed in its manual [17].

2.2 *Sobel Algorithm*

The Sobel algorithm is derivative which means that, for a given image, the gradients of light intensity in the x and y directions are approximated using convolution masks, or kernels. For each pixel in the image, the horizontal and vertical changes are calculated, and the magnitude of the resultant gradient is approximated by the square root of the sum of the squared values for both directions. Because there is no significant change in the magnitude of the gradient, the absolute sum of the squared root of the gradients in the x and y directions can be used for a less computationally costly equation [18].

2.3 *Tools*

Quartus II, Signal Tap and DE2 Control Panel. Quartus II is a software used in a project of SoPC. It has resources to accomplish all stages of an FPGA design [19]. SignalTap II Logic Analyzer is a debugging tool that comes with Quartus II. It takes and reveals signal response in real time, and can choose the beginning, the number and the nature of the signal [19]. DE2 Control Panel is a software that simplifies the users access to the peripheral components of the board. It permits command of each input/output device by instantiating a controller in the Cyclone II FPGA [17].
MATLAB. MATLAB is a high-level matrix-based language software that engineers and scientists can program intuitively [20]. In this project, the function edge () [21] was applied for Sobel edge detection.
Mathematica. Mathematica is a software which allows a wide range of technical computing applications [22]. As MATLAB, the Sobel method was utilized as well [23].
OpenCV. OpenCV is an open-source library for computer vision and image processing applications that contains hundreds of algorithms and methods for image manipulation [24]. The code used in this work for comparison was provided by [25].
GIMP. GIMP is an open-source free image editor that allows images to be created and saved in a variety of formats. GIMP was used to save pixel data from images in array format ".data" so that these could be processed. The resultant arrays from edge detection where opened and visualized through GIMPs interface.
C++. Using the C++ programming language, a software was developed to implement Sobel algorithm. The current version receives an image in array format, processes it and returns an output array that can be converted to any image format. To validate

the functionality of the implemented tool, GIMP image editor was used to visualize the output image and the library "<chrono>" to measure execution time [26].

2.4 Simulation

As cited before, this work is based on the "ECE536 final project" from Cornell University. The FPGA model used here is the Cyclone II EP2C35F672C6 for Altera DE2 board. The original DE2 2MP CMOS camera module was replaced by a flash memory reader module, thus the edge detection could be performed over a single image. After being changed, the 640 × 480 bitmap original image portrayed in Fig. 1 [27], was converted to a 640 × 480 raw grayscale image as depicted in Fig. 2 via the Terasic Image Converter program [17].

Fig. 1 Original Image 640 × 480

Fig. 2 Gray Image 640 × 480

The grayscale image, Fig. 2, was stored in the flash memory, using the DE2 Control Panel application. Next, the project SRAM Object File (.sof) was loaded in the FPGA. After that, the Sobel edge detection algorithm was used to convert the grayscale image, displaying the outcome image in the LCD monitor. Before calculating the execution time of the edge detection algorithm, a counter module was implemented in the algorithm to count the number of clock cycles. For each positive edge of a 50-MHz clock rate signal, the value stored in the counter was incremented by one. The algorithm assigns the zero value to the counter when it reaches a negative edge of the Sobel edge detection signal. Subsequently, during the computation of Sobel edge detection, the execution time was measured by coupling the counter block in the SignalTap II Logic Analyzer software. Similarly, the execution time of MATLAB, Mathematica, OpenCV, and the in-house C++ code were measured for performance evaluation.

The resulting execution times of each application were compared with the execution time of the equivalent FPGA implementation as described next:

Execution time for MATLAB. The MATLAB in-built function edge() [21], used to perform edge detection in MATLAB, calls the sub-function imshow() [28], which shows the processed image on the screen. The execution time of imshow() was subtracted from the edge() function to measure the execution time when only the edge detection algorithm is used.

Execution time measurement for Mathematica. Timing[] command [29] was used to measure the execution time spent by the processor to perform edge detection with Mathematica. The command ClearSystemCache[] [30] was used to avoid increased time by past iterations.

Execution time measurement for OpenCV. The functions getTickCount() [31], which returns the number of clock cycles after a relevant event, and getTickFrequency() [31], which returns the number of clock-cycles per second, were used to measure the execution time of the Sobel implementation in OpenCV [25].

Execution time measurement for C++ software. The time library "<chrono>" [26] was used to measure the execution time of the block of code corresponding to the edge detection algorithm, the resulting time was then obtained in milliseconds.

3 Results

The provided OpenCV software [25] was executed on a desktop with Windows 10 Home Edition and 4.00 GHz Intel® Core™ i7-4790K CPU. The intention was to show that, even for a highly optimized code executed in a faster processor, the FPGA would surpass regarding performance. All the remaining software were executed on a laptop with Windows 10 Home Edition and 2.20 GHz Intel® Core™ i5-5200U CPU. The frequency of the FPGA Sobel edge detector block was 50 MHz. The results of the edge detection are showed below through the Figs. 3, 4, 5, 6 and 7. Difference between the images could be due to noise, thresholding, different designs or false negatives [18]. Table 1 contains the execution time in milliseconds. The data revealed

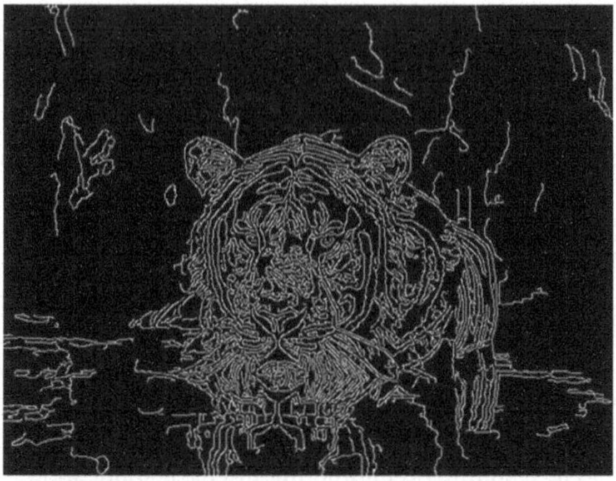

Fig. 3 Mathematica edge detection

Fig. 4 GIMP edge detection

Fig. 5 MATLAB edge detection

Fig. 6 C++ edge detection

Fig. 7 FPGA edge detection

that the FPGA scored the lowest execution time and is up to 16.435 (1643.5%) times speedup compared to MATLAB, 26.060 (2606.0%) times to Mathematica, 10.593 (1059.3%) times to the C++ application and 3.240 (324.0%) times to OpenCV. Moreover, examining the first table in [13], it can be observed that for 512 $\times$ 512 and 1024 $\times$ 1024 image resolutions, the execution time for FPGA was 1.10 ms and 4.37 ms, respectively. Therefore, the execution time for the FPGA in Table 1 is consistent with [13] because the value of 1.439 ms is in the range from 1.10 to 4.37 ms for an intermediate image resolution of 640 $\times$ 480.

Table 1 All execution times

Hardware/Software	Execution Time (ms)
FPGA	1.439
MATLAB	23.65
Mathematica	37.5
C++	15.244
OpenCV[a]	4.663

[a]Executed on 4.00-GHz Intel® Core™ i7 CPU

4 Conclusion

The present work had demonstrated that FPGA applications, which involve digital processing, are indeed faster and less computationally expensive than similar implementations performed with different software such as, MATLAB, OpenCV, Mathematica and an in-house C++ software that does not use image processing libraries.

References

1. Ashenden, P.J.: Digital Design. An Embedded Systems Approach Using Verilog, Morgan Kaufmann (2007)
2. Xilinx Inc.: SDAccel Environment Optimization Guide UG1207 v2017.2 (2017)
3. Vahid, F.: Digital Design. Wiley (2011)
4. Intel Corporation: What is an fpga?. https://www.intel.com/content/www/us/en/products/programmable/fpga/new-to-fpgas/resource-center/overview.html. Last Accessed 06 Sep 2018
5. Moore, A.: FPGAs For Dummies. John Wiley & Sons Inc (2014)
6. Sonka, M., Hlavac, V., Boyle, R.: Image Processing, Analysis and Machine Vision, 3rd edn. Thomson (2008)
7. Prince, S.J.D.: Computer Vision. Models, Learning, and Inference, 1st edn. Cambridge University Press (2012)
8. Horn, B.K.: Robot Vision. MIT Electrical Engineering and Computer Science, mit press edn. (1986)
9. Pratt, W.K.: Introduction to Digital Image Processing. CRC Press (2013)
10. Solomon, C., Breckon, T.: Fundamentals of Digital Image Processing: A Practical Approach with Examples in Matlab, 1 edn. Wiley-Blackwell (2011)
11. Li, X., Jiang, J., Fan, Q.: An improved real-time hardware architecture for canny edge detection based on fpga. In: Intelligent Control and Information Processing (ICICIP), 2012 Third International Conference on, pp. 445–449. IEEE (2012)
12. Nita, I., Costachioiu, T., Lazarescu, V., Seceleanu, T.: Multiprocessor real time edge detection using fpga ip cores. In: Intelligent Computer Communication and Processing (ICCP), 2011 IEEE International Conference on, pp. 331–334. IEEE (2011)
13. Possa, P.R., Mahmoudi, S.A., Harb, N., Valderrama, C., Manneback, P.: A multi-resolution fpga-based architecture for real-time edge and corner detection. IEEE Trans. Comput. **63**(10), 2376–2388 (2014)
14. Stavinov, E.: 100 Power Tips for FPGA Designers (2011)

15. Chin, A., Chan, G.: Real-time edge detection. In: Course ece 5760 advanced microcontroller design and system-on-chip. https://people.ece.cornell.edu/land/courses/ece5760/FinalProjects/f2007/hc454_gtc32/hc454_gtc32/index.html. Last Accessed 27 Nov 2017
16. Hamblen, J.O., Hall, T.S.: Using system-on-a-programmable-chip technology to design embedded systems. Int. J. Comput. Appl. **13**(3), 142–152 (2006)
17. Altera Corporation: DE2 Development and Education Board User Manual Version 1.41 (2007)
18. Parker, J.R.: Algorithms for Image Processing and Computer Vision, 2 edn. Wiley Publishing (2011)
19. Corporation, Altera: Introduction to the Quartus II Software Version **10.0** (2010)
20. MathWorks: What is an matlab?. https://www.mathworks.com/discovery/what-is-matlab.html. Last Accessed 06 Sep 2018
21. MathWorks: Detecting a cell using image segmentation. https://www.mathworks.com/help/images/detecting-a-cell-using-image-segmentation.html. Last Accessed 28 Sep 2018
22. Wolfram Research: Wolfram mathematica. https://www.wolfram.com/mathematica/. Last Accessed 06 Sep 2018
23. Wolfram Research: Edgedetect—Wolfram Language Detection. https://reference.wolfram.com/language/ref/EdgeDetect.html. Last Accessed 28 Sep 2018
24. OPENCV: Opencv library. https://opencv.org/. Last Accessed 06 Sep 2018
25. OPENCV: Opencv: Sobel Derivatives. https://docs.opencv.org/master/d2/d2c/tutorial_sobel_derivatives.html. Last Accessed 27 Nov 2017
26. <chrono> - c++ reference. http://www.cplusplus.com/reference/chrono/. Last Accessed 28 Sep 2018
27. ALTERA CORPORATION: CD-ROM: DE2 System Version 1.6 (2007)
28. MathWorks: Display image—matlab imshow. https://www.mathworks.com/help/images/ref/imshow.html. Last Accessed 28 Sep 2018
29. Wolfram Research: Timing—wolfram language detection. https://reference.wolfram.com/language/ref/Timing.html. Last Accessed 28 Sep 2018
30. Wolfram Research: Clearsystemcache—wolfram language detection. https://reference.wolfram.com/language/ref/ClearSystemCache.html. Last Accessed 28 Sep 2018
31. OPENCV: Opencv: Performance measurement and improvement techniques. https://docs.opencv.org/3.3.1/dc/d71/tutorial_py_optimization.html. Last Accessed 28 Sep 2018

Geometric Characterization of Yagi Antenna Using Digital Image Processing

Heyul Chavez, Hugo Hernandez, Luciano Prado, Guillermo Kemper and Christian del Carpio

Abstract This work proposes a system capable of obtaining the geometric measurements of a Yagi antenna using digital image processing. The objective is to obtain the measurements automatically in 3D using the CST STUDIO software, which is a program specializing in electromagnetic simulations in 3D. The antenna images are acquired by means of a depth sensor and an RGB sensor. By applying image processing techniques such as thresholding, morphology, filtering, and segmentation, the image obtained by the RGB sensor will identify the parts that are comprised by the antenna and their measurements in pixels. Through the image obtained with the depth sensor, it will be possible to transform said measurements from pixels into a unit of length. To obtain these measurements with the least possible error, the system processes 30 samples and averages them to obtain the final measurements. With this process, it was possible to obtain a measurement error of 0.16 cm.

Keywords Yagi antenna · Digital image processing · Geometric characterization · Telecommunication

H. Chavez · G. Kemper · C. del Carpio (✉)
Universidad Peruana de Ciencias Aplicadas (UPC), Lima, Peru
e-mail: cdelcarpiod@gmail.com

H. Chavez
e-mail: heyul.chavez@gmail.com

G. Kemper
e-mail: guillermo.kemper@gmail.com

H. Hernandez · L. Prado
Universidade Estadual de Campinas (UNICAMP), Campinas-SP, Brazil
e-mail: hugoehf@gmail.com

L. Prado
e-mail: luciano.prado@icloud.com

Y. Iano et al. (eds.), *Proceedings of the 4th Brazilian Technology Symposium (BTSym'18)*, Smart Innovation, Systems and Technologies 140,
https://doi.org/10.1007/978-3-030-16053-1_23

1 Introduction

In recent years, antenna manufacturing has increased greatly due to the demand that currently exists in the area of telecommunications. To guarantee the correct functioning of the antennae, its physical dimensions must be validated as correct in accordance to its design. This validation guarantees its maximum performance.

The geometrical measurements of an antenna are important because any change or modification may directly affect the radiation pattern for which it was de-signed. No information has been found thus far about a system that automatically obtains geometric measurements from an antenna. The proposed system uses image processing to obtain the 3D model of a Yagi antenna. The programming language used for the implementation of the algorithms was C++ [1], with an OpenCV artificial vision library [2].

2 Description of the Proposed System

The block diagram of the proposed system is shown in Fig. 1. The processing stages, from image capturing to obtaining the corresponding measurements, will be described in the following sections.

2.1 Image Acquisition

The Intel® RealSense™ Depth Camera D435 [3], which contains a depth sensor and an RGB sensor, was used to acquire the images. The resolution used was 1280 × 720 pixels. Two images are taken to start the process: one image obtained with the RGB sensor, defined as $I_C(x, y)$, and one image obtained with the depth sensor, defined as $I_D(x, y)$, as shown in Fig. 2a, b, respectively.

Likewise, 3 components of the antenna will be identified, as shown in Fig. 3.

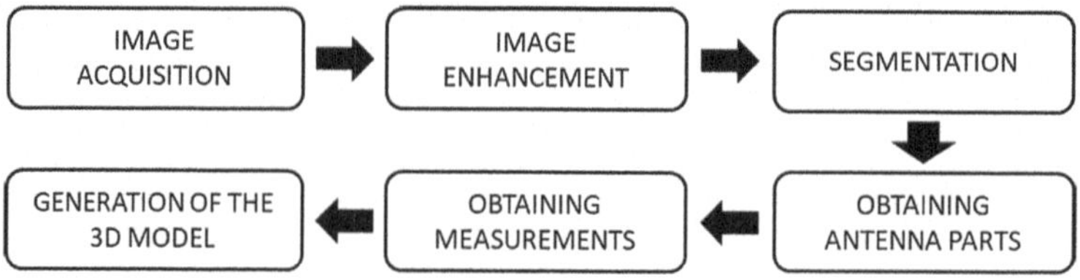

Fig. 1 A block diagram of the proposed method

2.2 Image Enhancement

For this stage, the first step is to separate the three primary components of the image $I_C(x, y)$, i.e., the R component, expressed as $I_{CR}(x, y)$, the G component, expressed as $I_{CG}(x, y)$, and the B component, expressed as $I_{CB}(x, y)$.

Next, a grayscale image is obtained using Eq. 1, $I_G(x, y)$. The resulting image is shown in Fig. 4a.

$$I_G(\mathrm{x}, \mathrm{y}) = 0.299 \cdot I_R(x, y) + 0.587 \cdot I_G(x, y) + 0.114 \cdot I_B(x, y) \quad (1)$$

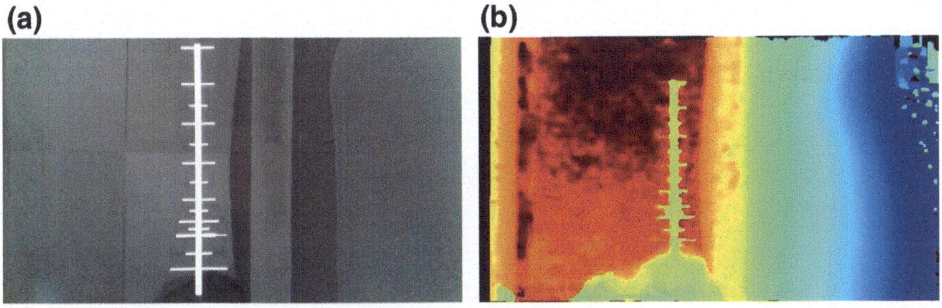

Fig. 2 **a** RGB image. **b** Depth image

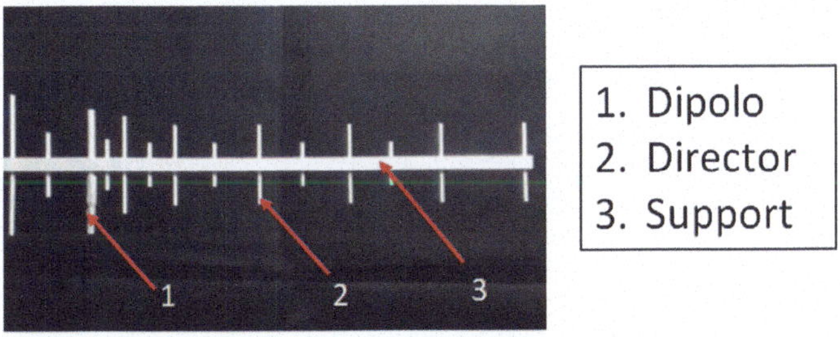

Fig. 3 Parts of the antenna

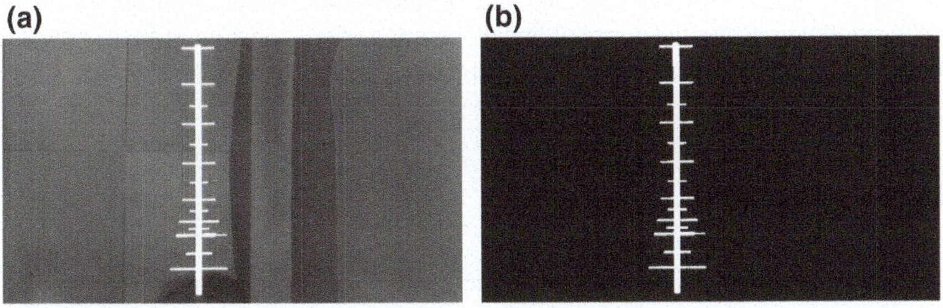

Fig. 4 **a** Grayscale image $I_G(\mathrm{x}, \mathrm{y})$. **b** Binarized image $I_B(x, y)$

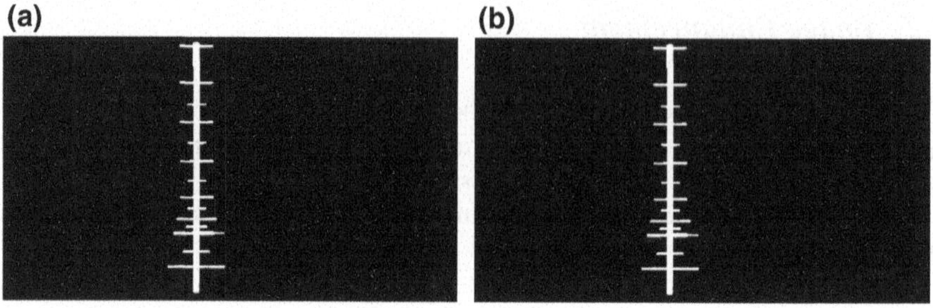

Fig. 5 **a** Image with opening. **b** Image with closing

Then, we proceed to binarize the image by applying Eq. 2. The result of this process is shown in Fig. 4b.

$$I_B(x, y) = \begin{cases} 1\ , & I_G(x, y) > \mu_0 \\ 0\ , & otherwise \end{cases} \tag{2}$$

where μ_0 is the adaptive threshold obtained by applying the Otsu method [4].

Then, the morphological opening and closing operation [5] is applied to provide greater robustness to the image. This is obtained from Eqs. 4 and 5. The structural element is defined in Eq. 3

$$EE = \begin{matrix} 1\ 1\ 1 \\ 1\ 1\ 1 \\ 1\ 1\ 1 \end{matrix} \tag{3}$$

$$I_{o1} = (I_o \ominus EE) \oplus EE \tag{4}$$

$$I_{o2} = (I_{o1} \oplus EE) \ominus EE \tag{5}$$

where I_{o1} and I_{o2} constitute the matrices that represent images $I_{o1}(x, y)$ and $I_{o2}(x, y)$, respectively. The $\oplus$ symbol indicates the morphological operation of dilation, and the $\ominus$ symbol indicates the morphological operation of erosion. Figure 5a, b show the images resulting from the morphological process of opening and closing respectively.

2.3 *Segmentation*

Segmentation of the antenna was performed with the following procedure. Assume the object of interest to be a white pixel.

Step 1: Starting in the first row, the image is scanned from left to right and from top to bottom until the first white pixel is found. The position of said pixel will allow us to obtain the x_1 row.
Step 2: Starting in the last row, the image is scanned from left to right and from bottom to top until the first white pixel is found. The position of said pixel will allow us to obtain the x_2 row.
Step 3: Starting in the first column, the image is scanned from top to bottom and from left to right until the first white pixel is found. The position of this pixel will allow us to obtain the y_1 column.
Step 4: Starting in the last column, the image is scanned from top to bottom and from right to left until the first white pixel is found. The position of said pixel will allow us to obtain the y_2 column.

Equation 6 displays the image obtained from this process.

$$I_{ROI}(x, y) = I_{o2}(x_1 : x_2, y_1 : y_2) \tag{6}$$

The result of this segmentation process is shown in Fig. 6a.

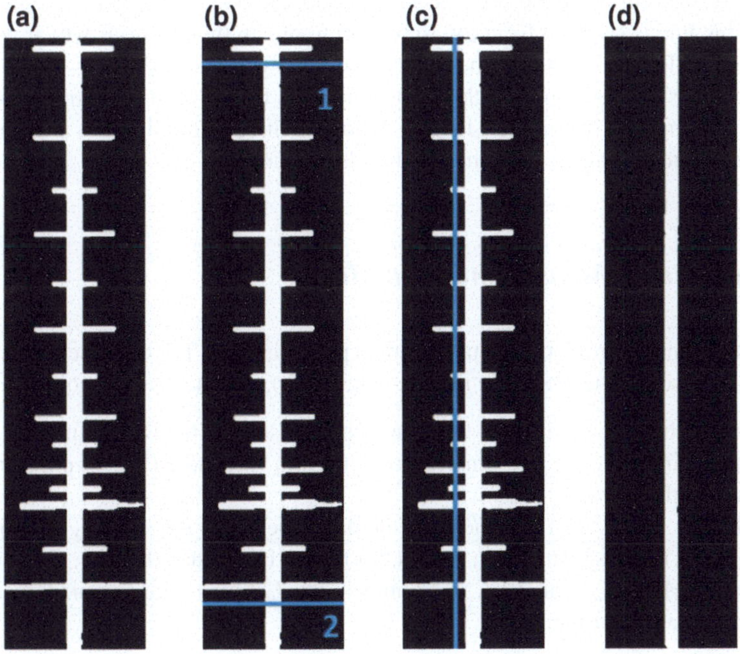

Fig. 6 **a** Segmented image. **b** Support Identification. **c** Identification of directors. **d** Side image

2.4 Obtaining Antenna Parts

Supposing the object of interest to be a white pixel. The following steps are followed to find the main support of the antenna:

Step 1: Obtain the position of two rows, an upper row and a lower row, where a dipole is not present, as shown in Fig. 6b.
Step 2: Scan the upper row found from left to right until finding a white pixel. From this pixel, we obtain the c_1 column value. Continue scanning the row until no more white pixels are found. From this last pixel, we obtain the c_2 column value. The difference of c_2 and c_1 will be the width a_1 of the antenna support.
Step 3: Repeat step 2 with the lower light blue row from Fig. 6b. Here, we obtain width a_2. The final thickness is obtained from averaging $a_1 y a_2$.
Step 4: The $A(x_a, y_a)$ point is defined as midpoint a_1 and the $B(x_b, y_b)$ point as $a\text{midpoint}_2$.
Step 5: To obtain the thickness of the support profile, follow the same steps 2, 3 and 4 for the image shown in Fig. 6d.

The following steps are performed to find the directors and dipole:

Step 1: A line is drawn parallel to the antenna support as shown in Fig. 6c (light blue line). This line is scanned from top to bottom to obtain the positions which will be used to calculate the widths of the corresponding directors and dipole.
Step 2: Once the position of each director/dipole is found, each of them is scanned from left to right to obtain their length. All of the director/dipole rows are scanned horizontally to guarantee obtaining their ends properly.

2.5 Obtaining Measurements in Meters

All measurements previously obtained are in pixel distance. The depth sensor is used to obtain these measurements in meters. (Figure 2b). The steps are as follows:

Step 1: Synchronize the images from the depth sensor and the RGB sensor.
Step 2: Obtain the x, y, z coordinates from the $A(x_a, y_a)$ and $B(x_b, y_b)$ points in the $I_D(x, y)$ depth image.
Step 3: The $d(A_m, B_m)$ distance is calculated between the $A_m(x_a, y_a, z_a)$ and $B_m(x_b, y_b, z_b)$ points in meters in the $I_D(x, y)$ depth image, and the $d(A, B)$ distance between the $A(x_a, y_a)$ and $B(x_b, y_b)$ points in the $I_{o2}(x, y)$ image in pixel distance using Eqs. 7 and 8, respectively.

$$d(A_m, B_m) = \sqrt{(x_b - x_a)^2 + (y_b - y_a)^2 + (z_b - z_a)^2} \tag{7}$$

$$d(A, B) = \sqrt{(x_b - x_a)^2 + (y_b - y_a)^2} \tag{8}$$

Step 4: With this reference, we obtain the $F_c(x, y)$ conversion factor between the distance in meters and the distance in pixels by applying Eq. 9.

$$F_c = \frac{d(A_m, B_m)}{d(A, B)} \tag{9}$$

Step 5: Obtain the distance of the support between the highest and lowest point in pixels (see Fig. 6a), and multiply it by the *Fc*. to obtain the distance in meters.
Step 6: The same procedure as in the previous step is used to find the measurements in meters of the directors.

2.6 Generation of the 3d Model—Connection with CST STUDIO SUITE

CST STUDIO SUITE [6] is a software for designing, simulating and optimizing electromagnetic systems. To allow this software to interpret the data obtained from the antenna, data must be exported in a format supported by the software. For this

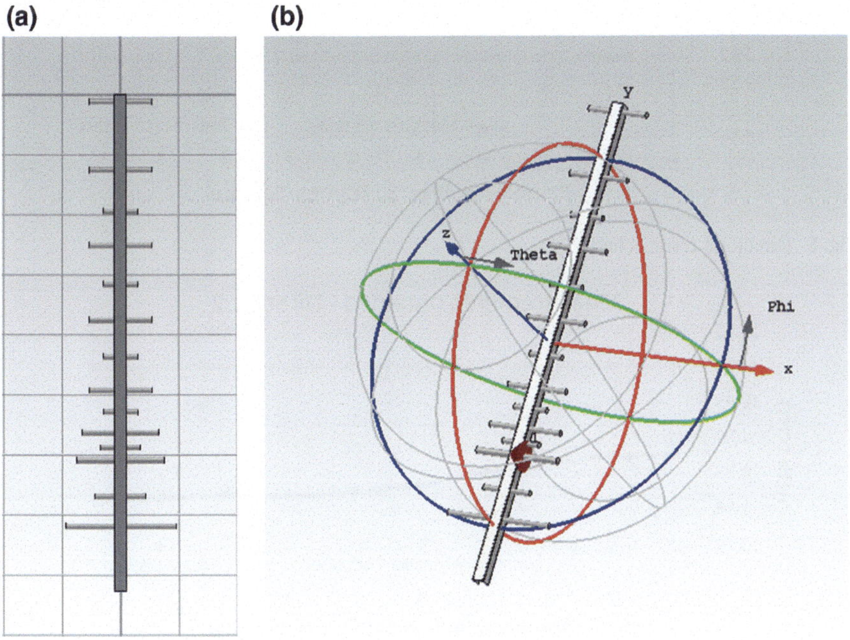

Fig. 7 Digital 3D model generation of the antenna in CST STUDIO SUITE. **a** Front view. **b** 3D view

reason, the *.bas* extension was selected, which is an extension used for source code files written in the BASIC programming language. Therefore, all the measurements obtained from the antenna are entered in a.*bas* file, in addition to the configuration parameters, so that CST STUDIO can interpret the data. Figure 7a shows the front view and Fig. 7b the 3D view of the antenna generated in the CST STUDIO.

3 Results

The measurements obtained by the system are shown in Figs. 8 and 9. The measurements of two directors will be taken as reference. For Director 1, the average error obtained is 0.08031 cm, with a standard deviation of 0.01437 cm. For Director 2, the average error obtained is −0.15855 cm, with a standard deviation of 0.02436 cm.

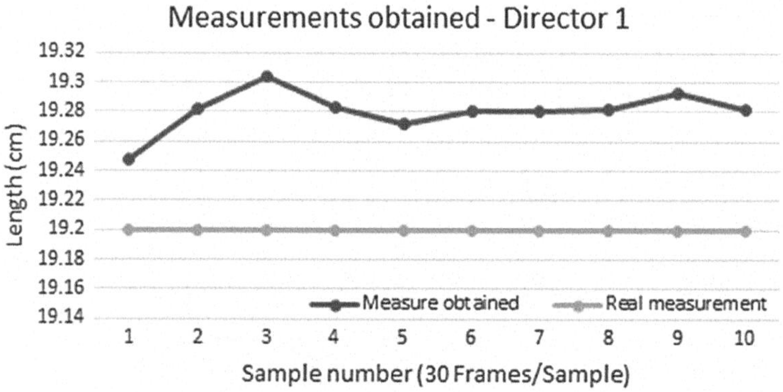

Fig. 8 Director 1 results chart

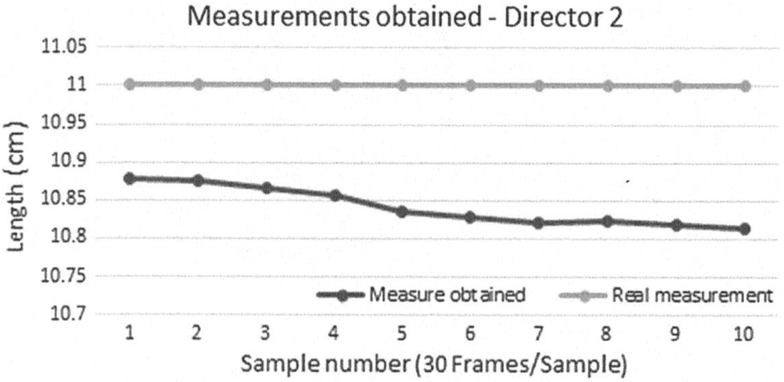

Fig. 9 Director 2 results chart

4 Conclusions

- The proposed system obtained the geometrical measurements with a maximum error rate of 0.16 cm. This error could be ameliorated by increasing the resolution of the RGB and depth sensors.
- However, obtaining measurements from a larger number of frames does not guarantee greater precision. In fact, this study showed that the results from 30 frames were identical to those obtained from 400 frames.
- In addition, we were able to generate the 3D digital model of the antenna using the CST STUDIO software, which may later be used to generate an electromagnetic analysis of the antenna.

References

1. Lenguaje de programación C++—http://www.cplusplus.com. Ultimo acceso 30 Sep 2018
2. Biblioteca libre de visión artificial OpenCV 3.4.2—https://opencv.org/. Ultimo acceso 30 Sep 2018
3. Intel® RealSense™ Depth Camera D435. https://www.intel.com/content/dam/support/us/en/documents/emerging-technologies/intel-realsense-technology/Intel-RealSense-D400-Series-Datasheet.pdf, Ultimo acceso 30 Sep 2018
4. Otsu, N.: A threshold selection method from gray level histograms. IEEE Trans. Syst. Man Cybernet. **9**, 62–66 (1979)
5. Gonzalez, R.C, Woods, R.E.: Digital Image Processing Second Edition, Pearson Education, New Jersey (2002). ISBN 978-0201180756
6. Computer Simulation Technology: CST—https://www.cst.com/. Ultimo acceso 30 Sep 2018

Study of an Air Navigation System Based on Pilot Hypoxia Level

Raffael Fonseca de Mattos and Rogério Bastos Quirino

Abstract This paper presents a preliminary study of the development of an air navigation safety system based on one of the major physiological problems of the navigational pilot. This proposal primarily aims at the life safety of pilots and crew as well as the integrity of aircraft. In this context, the work brings a search of synaptic response patterns to different levels of hypoxia and, based on these, discuss actions of control of the aircraft.

Keywords Hypoxia · Synapse · Safety

1 Introduction

From the beginning, Man lived in an environment characterized by various physical parameters necessary for the optimal functioning of his organism. Within this variety of parameters, the concentration of oxygen in the atmospheric air is present.

Even with various organic limitations when exposed to hostile environments, Man dared to leave his habitat, seeking the new. Upon reaching the airspace, the vital importance of oxygen was investigated, and with that, it looked for methods to connect to the terrestrial surface.

The first reference to altitude sickness, known also as hypoxia, that is, oxygen reduction offered to tissues by blood [1], occurred in 1590, by the Spanish Jesuit priest José Acosta, in Peru. From the seventeenth century, the search for understanding the evil of the heights began. In 1650, the German physicist Otto Von Guericke developed an air pump that allowed the simulation of an environment with high altitudes. In 1735, the Spanish scientist Antônio de Ulloa described the clinical picture of hypoxia

R. F. de Mattos (✉) · R. B. Quirino
University of the State of Mato Grosso, Sinop, Brazil
e-mail: mattos.aero@gmail.com

R. B. Quirino
e-mail: rb_quirino@hotmail.com

Y. Iano et al. (eds.), *Proceedings of the 4th Brazilian Technology Symposium (BTSym'18)*, Smart Innovation, Systems and Technologies 140,
https://doi.org/10.1007/978-3-030-16053-1_24

in Peru, citing some characteristics such as tiredness, vertigo, lipothymia, dyspnea and pyrexia.

With technological advancement, Man finally reached the heavens, but his greatest physiological problem remained hypoxia. In view of the great sensitivity of the nervous system and the complexity of the piloting, the absence of oxygen or the reduction of its supply to the tissues reduces the efficiency of the operations carried out, even the simplest, putting at risk the lives of the pilots and crew involved [1].

Hypoxia has been a major concern for aviation. It can deteriorate execution or even produce complete incapacity and threats increase as altitude rises. As altitude increases and barometric pressure decreases, less air will be available per unit volume. Since oxygen is constant at 20.95% air, there is, in turn, less oxygen per unit volume. While an individual's lung volume is approximately constant, the amount of inspired oxygen available with each breath decreases with increasing altitude.

For example, the atmospheric pressure at sea level is about 760 mmHg, proceeding at an alveolar oxygen pressure of approximately 103 mmHg. In contrast, the altitude of about 14,000 feet, the atmospheric pressure decreases about 447 mmHg and the alveolar oxygen pressure decreases to approximately 48 mmHg [1]. Meanwhile, body oxygen wear at altitude continues the same in the face of reduced oxygen. Therefore, individuals exposed to altitude envisage the battle to function under conditions of reduced accessible oxygen. As altitude increases further with additional decreases in available oxygen, the challenge becomes even greater [1].

In view of the information described above the intention of the project is to make a study of an air navigation safety system based on the level of hypoxia of the pilot, to carry out commands automatically in case of loss of consciousness on the part of the pilot, also analyzing the state of evolution within the framework of hypoxia of the same.

2 Historic

Hypoxia is recognized as a significant physiological threat at altitude. The crew is periodically trained to recognize hypoxic symptoms using a hypobaric chamber at simulated altitudes of 25,000 feet or more. A 2003 study in Australia used the following data for analysis: type of aircraft, number of people on board (POB), number of POB hypoxic, fatalities, whether the victims were trained or untrained as the symptoms were recognized as hypoxia, symptoms experienced, altitude at which the incident occurred and probable cause. Outcome: In the period studied, there were 27 reports of hypoxia, involving 29 troops. In two cases the consciousness was lost, one of them fatality. Most incidents (85.1%) were in combat aircraft or training with crews using oxygen equipment routinely.

Most of the symptoms occurred between 10,000 and 19,000 feet. The most common cause of hypoxia (63%) in these aircraft was due to degradation of the mass regulator, or restriction of water. Decompression was not presented as hypoxia.

Ten involved had subtle effects due to a cognitive incompatibility with the feeling of relief. The majority (75.8%) of these episodes, the crew itself recognized, thus giving great value and importance to the training of hypoxia [2].

Between 2008 and 2010, fourteen physiological events related to hypoxia were registered as a result of the embedded oxygen generation system. The Lockheed Martin F-22 Raptor of the United States Air Force (USAF) at Langley Air Force Base stopped flying in 2011 for 142 days [3].

In 2005, the Boing 737–300 of the Cypriot company Helios Airways hit a mountain northeast of Athens. One hundred and twenty one passengers and crew were killed and the cause was because the aircraft had not been pressurized causing everyone present to be unconscious due to hypoxia. For lack of fuel, the flight hit the Greek mountains, after almost three hours of flight.

In May and June 2017, F-35As fighters were canceled at Luke Air Force Base due to five incidents related to hypoxia.

3 Fundamentation

Oxygen is one of the most important components needed for the maintenance of life. The inadequate amount of oxygen to the tissues is called hypoxia. Man is highly sensitive and vulnerable to the effects of oxygen deprivation, and severe hypoxia often results in major bodily functions problems, leading to death in some cases [4].

When the human being goes to the air space, it is necessary that he take with him a part of the atmosphere that surrounds him to the level of the sea. This equates to a mini habitat that allows it to survive and overcome physiological barriers that otherwise could not be overcome [1].

3.1 Hypoxia

Hypoxia can be defined as a low concentration of oxygen in the tissues. This low concentration is capable of causing various flaws in the physiological functions of humans. It has an insidious installation feature [5].

Hypoxia has four variations: hypoxic hypoxia (hypobaric), hypemic hypoxia (anemic), stagnant hypoxia (circulation deficit) and histotoxic hypoxia (poisoning) [5].

Hypoxic hypoxia (hypobaric). Poor supply of oxygen will result in lack of oxygen—O_2 in the tissues. Insufficient oxygenation of the blood in the lungs, and may be caused by the reduction of oxygen partial pressure in the inspired air (altitude), airway obstruction (asthma) or ventilation/perfusion defect (pneumonia). It represents the most common form of hypoxia found at altitude [5].

Hypemic hypoxia (anemic). Deficiency of transporting O_2 to tissues through the blood. The most common causes are: significant blood loss, carbon monoxide poisoning, anemia, generalized or localized circulatory deficiency (peripheral, cerebral, coronary vessels) [5].

Stagnant hypoxia (deficit in circulation). Due to poor circulation, O_2 deficiency occurs in the body. They can also be caused by G-charges, blood vessel occlusion, arterial spasm, venous congestion, and the period of positive breathing.

Histotoxic hypoxia (poisoning). The inability of tissues to use O_2 available. Reduction of the cellular metabolic capacity for the use of oxygen. Caused by: carbon monoxide poisoning, cyanide, alcohol ingestion and narcotics [5].

As man ascended at ever higher altitudes, using aviation, there was a need to understand the effects of altitude and low pressures on the human body. Tables 1 and 2 below show the barometric and oxygen pressures at different altitudes.

Tables 1 and 2 above show barometric pressures at different altitudes. At sea level, i.e. 0/0, the barometric pressure is 760 mmHg. At 15250 m altitude, the barometric pressure is only 87 mmHg. This reduction of barometric pressure motivates all hypoxia problems at high altitudes. Therefore the partial pressure of the oxygen (PO_2) reduces proportionally with the decrease of the barometric pressure.

From 3657 m, one may note some effects related to hypoxia for people breathing air. They are: drowsiness, lassitude, muscular and mental fatigue, sometimes

Table 1 Effects of acute exposure to low atmospheric pressures on concentrations in alveolar gas and natural oxygen saturation

Breathing Air					
Altitude (m/ft)	Barometric pressure (mmHg)	PO_2 in the air (mmHg)	PCO_2 in the alveoli (mmHg)	PO_2 in the alveoli (mmHg)	Saturation of arterial oxygen(%)
0/0	760	159	40	104	97
3050/10,000	523	110	36	67	90
6100/20,000	349	73	24	40	73
9150/30,000	226	47	24	18	24
12,200/40,000	141	29	–	–	–
15,250/50,000	87	18	–	–	–

Table 2 Effects of acute exposure to low atmospheric pressures on concentrations in alveolar gas and blood oxygen saturation

Breathing pure oxygen				
Altitude (m/ft)	Barometric pressure (mmHg)	PCO_2 in the alveoli (mmHg)	PO_2 in the alveoli (mmHg)	Saturation of arterial oxygen (%)
0/0	760	40	673	100
3050/10,000	523	40	436	100
6100/20,000	349	40	262	100
9150/30,000	226	40	139	99
12,200/40,000	141	36	58	84
15,250/50,000	87	24	16	15

headache, occasionally nausea and sometimes euphoria. The reactions progressively evolve when it is above 5486 m, causing the person to have muscle tremors and seizures. The last stage for non-acclimated persons would be above 7010 m, which would cause coma and soon after death [5].

Decreased mental proficiency is one of the most important effects of hypoxia, since it hinders memory, movement, and judgment. For example, if an unacclimated pilot remains at 4572 m over a period of 1 h, his mental proficiency will be reduced by about 50% from normal [5].

3.2 *Hypobaric Chamber*

The hypobaric chamber training has been used as a way of demonstrating the effects of hypoxia on pilots. Hypobaric chamber training demonstrates rapid decompression by breathing 100% oxygen from a demand regulator and mask. Once at altitude, this mask is removed and the ambient air is breathed until hypoxic symptoms are demonstrated. The results of the present analysis show that hypoxia is more likely to occur in training or in the combat environment where mask removal is a more natural and potentially dangerous maneuver. In this aircraft, hypoxia occurs in most cases with the mask [2].

In Brazil, there is only one Hypobaric Chamber focused on this purpose. It is present at the Institute of Aerospace Medical Brigadeiro Roberto Teixeira—IMAE, located in the complex of the University of the Air Force—UNIFA, Rio de Janeiro-RJ (Fig. 1).

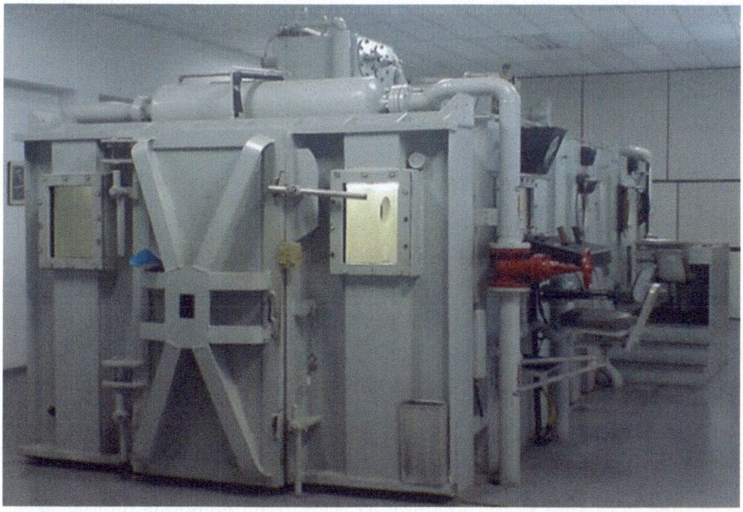

Fig. 1 Hypobaric Chamber, IMAE-RJ

Table 3 Cell rest potential

Cell restore potential	
Type	Resting potential (mV)
Muscle cell	Approximately −90 mV
Nervous cell	Between −80 and 40 mV
Epithelial cell	Approximately −50 mV

3.3 Physiological Data

For the study will be used the signals from the synapses, since the lack of oxygen in the tissues and blood will cause a direct reaction to the process of synapse formation and thus will not provide the signal for muscle activation. The research aims to compare these two variables (hypoxia and synapses) and thus create a standard coding to understand how the lack of oxygen reacts in the body at the cellular level.

3.4 The Cell Restore Potential

The value of the membrane resting potential varies from cell to cell. The table below shows the value of the resting potential of some cell types (Table 3).

Because of this potential difference in cells, a flow of electrical current between cells may occur. If he had not, the skeletal muscles could not contract, nor could the nerves transmit impulses. All this happens because the inside of the cell is not in ionic equilibrium with the extracellular medium, because if it were, there would be no membrane potential. When ions are predominantly extracellular, the cell is not in equilibrium with the extracellular medium, while others will be predominantly intracellular [6].

3.5 Sodium-Potassium Pump

Na/K-ATPase pumps are cell membrane proteins that, in the presence of the energy released by the breakdown of ATP in ADP + P, send three sodium ions out of the cell and two potassium ions in. They work to restore the original concentrations when sodium inlet and potassium outlet occur by physiological process of the cell or to restore the original concentrations when sodium leaks into the cell and potassium out. It is driven by ATP [6].

Certain physiological conditions alter the electrical potential of the membrane. The changes are determined by the passage of ions through the membrane. The passage of ions is related to the opening of membrane channels which, at rest, are closed. The opening can be by alteration in the spatial configuration of the same or

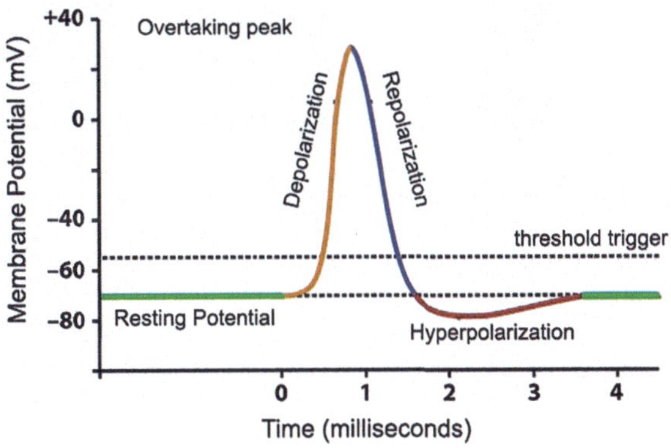

Fig. 2 Membrane potential phases

several physical conditions (electric field relative to the membrane DDP, pH change, mechanical stress on the membrane, temperature change, action of various chemical substances and etc.).

3.6 *Action Potential*

Soon after a stimulus has triggered the action potential, it propagates through the axon without losing its amplitude (the PA is a regenerable and active process of the cell, which has a signal with a few milliseconds, the sodium input and the successive output of potassium sufficient to produce a voltage variation of approximately 100 mV).

To migrate information from one point to another in the nervous system, it is necessary that the action potential be conducted along the axon. A PA initiated at one end of an axon will propagate in a single direction, not returning through it. Because of this they are characterized as unidirectional and called orthodromic conduction. As the axonal membrane is excitable, the AP will propagate without decay. The speed that the AP propagates along the axon will depend on the depolarization and the shape that it is projected ahead of the PA [7] (Fig. 2).

4 Methodology

The data acquisition test will be performed in Hypobaric Chamber. Only pilots and crew will be tested. The test is done superficially to the skin, through electrodes, requiring no invasive maneuver.

To acquire the signal will be used a signal acquisition system. Data acquisition system is the device responsible not only for the acquisition of data, but also the manipulation of such data if necessary. The system also known as Data Acquisition—DAQ [8]. In order to have control of a data acquisition system, a Personal

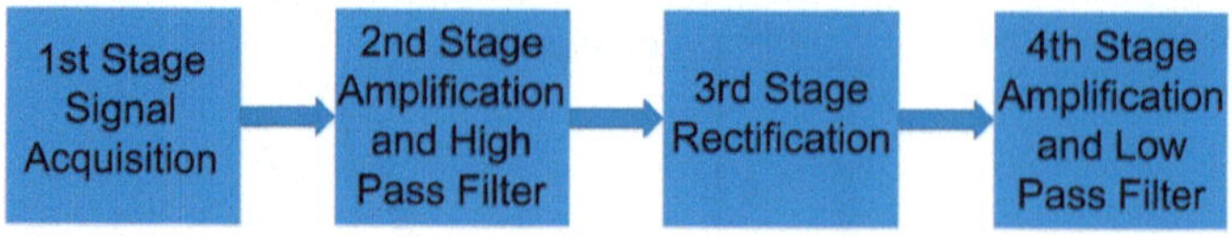

Fig. 3 Stages of data acquisition

Fig. 4 Operational Amplifier (AOP)

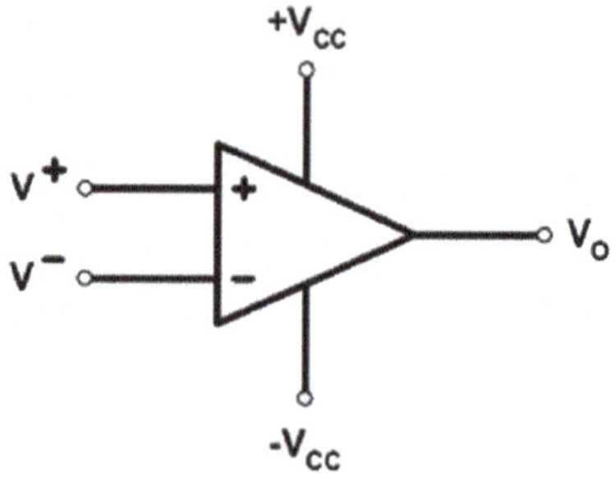

Computer (PC) is present, and can be organized by several blocks of hardware. The circuit for data acquisition will be based on 4 stages, shown in the figure below: (Fig. 3).

The first stage will be responsible for measuring the pilot's electrical impulses, which are used for muscle activation. An instrumentation amplifier will be used to acquire and compare the acquired signals. As the acquired signal is very small and any noise can be part of the signal response. Because of this, it is necessary to use an instrumentation AOP so that its gain is accurate and that it has good common-mode rejection (CMRR) response. CMRR is an important characteristic of a differential connection. Every signal that is common to the two inputs will have its amplification reduced. In this way, the amplifier will amplify the signal that really matters and reject the common signals. As the noise is generally equal in both inputs, the tendency is the attenuation of this unwanted signal [9] (Fig. 4).

The second stage will receive the signal acquired, and because it is very small, it is necessary that amplification of the same occurs. An inverter amplifier will be used to perform this amplification. To avoid DC error displacement, the circuit will have an AC coupling capacitor and a high pass filter. The filter will also aid in the removal of low frequency noise (Fig. 5).

After the signal has been amplified and filtered, the third stage will be the rectification of this signal. The rectifier used will be full wave active. Its use is present in order to work only with a positive signal. Once the signal is rectified, the signal will be routed to a filter for transforming the AC voltage into a DC voltage; this stage is the preparation of the signal for the input of a microcontroller (Fig. 6).

The last stage of data acquisition is given by an amplifier and a low—pass filter, whose purpose is to deliver a smoother signal to the microcontroller. However, since this is an active filter, the output signal will be reversed. Due to this a further inverting amplifier is used in order to correct the phase of the output signal of the filter PB

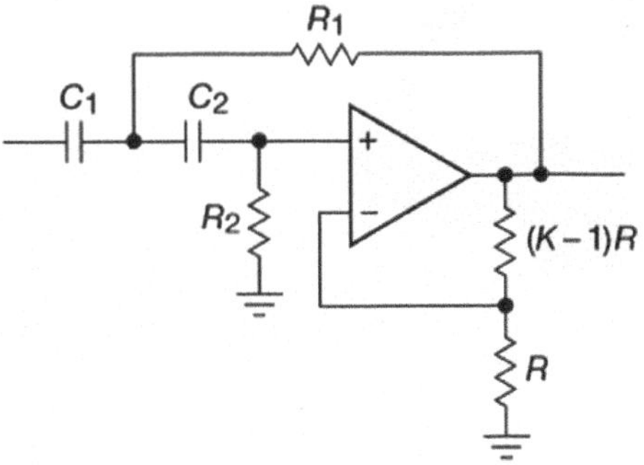

Fig. 5 Active high pass filter

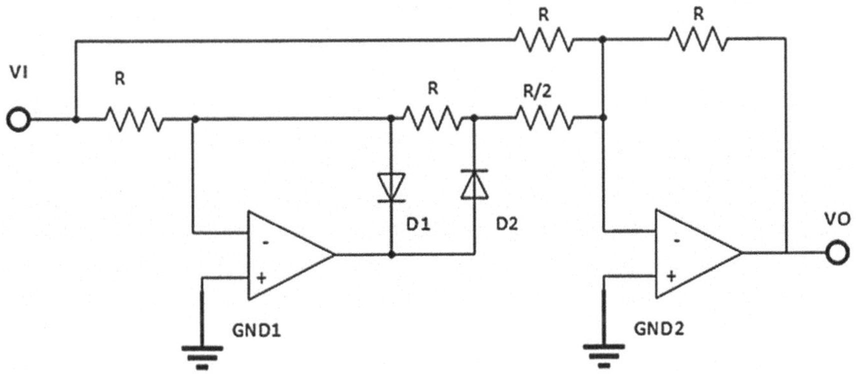

Fig. 6 Active full wave rectification

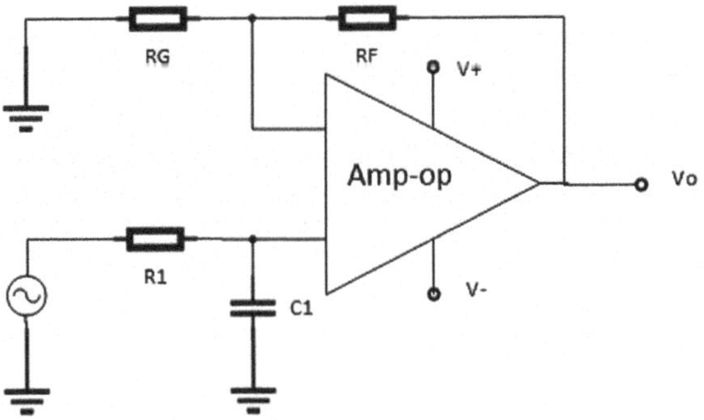

Fig. 7 Low pass filter active

and with this also to allow if necessary the gain increase. From the fourth stage, the signal is ready to be recorded and sent to the microprocessor (Figs. 7 and 8).

With the data, it will be possible to feed a database and program experimental actions in a flight simulator, stipulating maneuvers and procedures in case of loss of consciousness on the part of the pilot.

When the system is installed on aircraft, it will be intermediated by the Automatic Flight Control System, commonly called Autopilot. This system used for decades

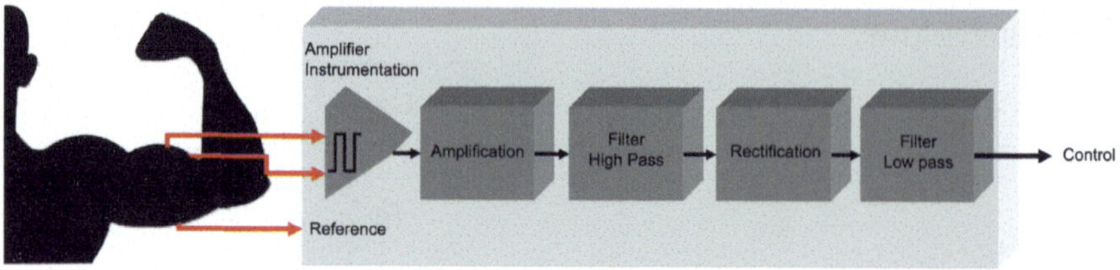

Fig. 8 Acquisition system

helps the pilot during in-flight operations. As systems advance, aircraft already have automatic pilots performing complex activities, such as automatic flight, performing virtually every aspect of the flight. Through it is possible to tune in radios, fly on standby circuits and even perform automatic landing functions [10].

The Automatic Flight Control System uses the ARINC 429 bus as the standard for data communication. There is also the possibility of using the Ethernet type bus, both of which traffic digital signals.

The projection of the Automatic Flight System aims to fulfill the duties of the pilot automatically [10]. Although the complexity of the systems varies according to the aircraft, all autopilots have the same basic functional elements. For all of this to occur, the system must have one or more processor units, known as the autopilot computer.

This computer receives information through the main autopilot control panel through the digital data bus. The computer sends signals to the outputs of the servomechanisms to control the aircraft and this system is fed back to confirm the position of the control surface.

As mentioned above, the automatic flight system must perform all the activities of a human pilot, including getting acquainted with the navigational routes, continuously monitor the parameters of the aircraft, where the Air Navigation Safety System Based on Hypoxia level would be part, analyze the current situation and make decisions about possible corrections and perform the replacement of control surfaces [10].

5 Final Remarks

This work, which was still in the experimental phase, was judged by the Research Ethics Committee (CEP) and approved by Opinion No. 2,832,244. It is currently in the process of using the Hypobaric Chamber at the Institute of Aerospace Medical Brigadeiro Roberto Teixeira (IMAE) to perform the various analysis procedures in a hypoxia situation.

The project's perspective is that in the future, it will be able to correlate with the aircraft's Automatic Flight Control Systems, enabling practical application and further enhancing flight safety (Fig. 9).

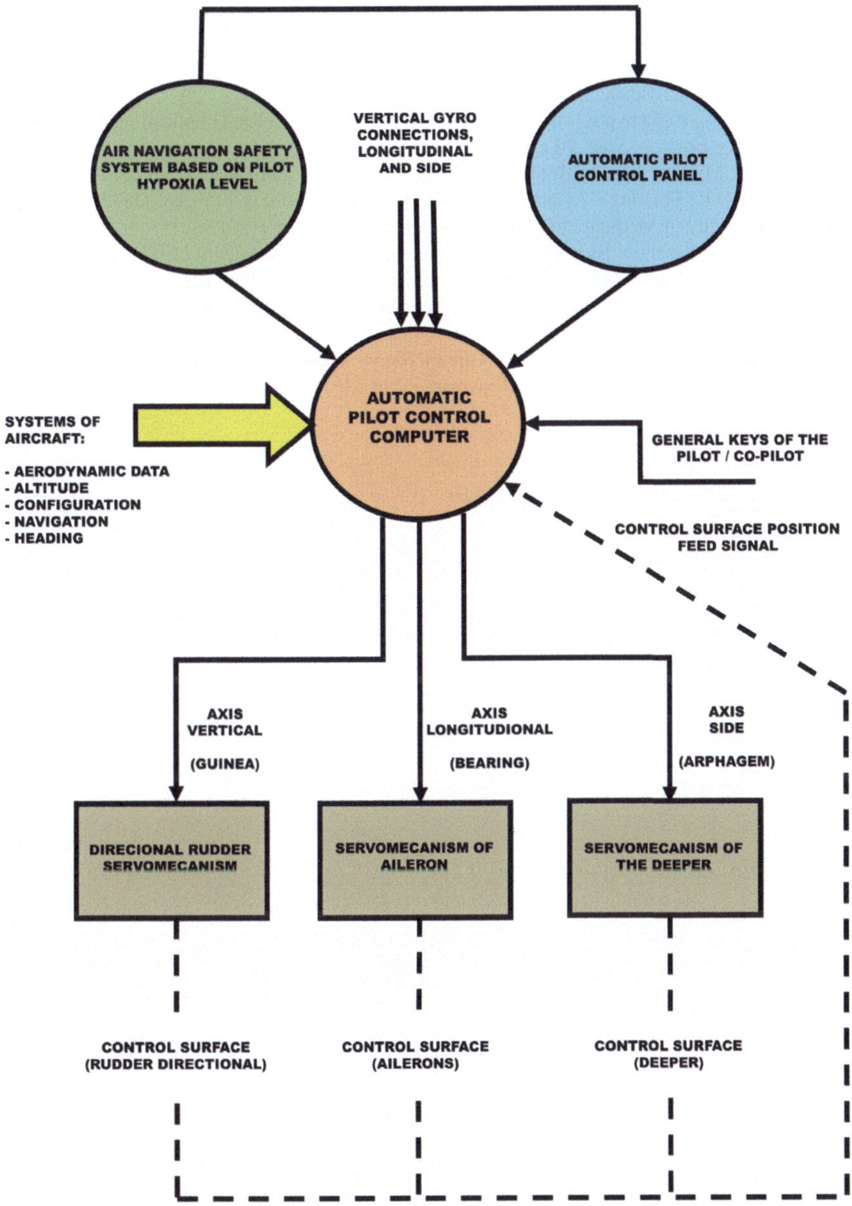

Fig. 9 Scheme of association between systems

References

1. Temporal, W.: Aerospace Medicine. Lights, Rio de Janeiro (2005)
2. CABLE, G.: In-flight hypoxia incidents in military aircraft: causes and implications for training. Aviat. Space Environ. Med. **74** (2003)
3. Klotzel E.: Military pilots suffer from hypoxia, a long story. Aero Magazine (2017)
4. Gradwell, D.P.: Ernsting's Aviation and Space Medicine. Taylor & Francis Ltda (2016)
5. Hall, G.: Treaty of Medical Physiology. Elseiver, Rio de Janeiro (2011)
6. Abramov, M.: Essential Biophysics. Guanabara Koogan, Rio de Janeiro (2011)
7. Amabis, J.M., Martho, G.R.: Concepts of Biology. Modern, São Paulo (2001)
8. Park, J.: Practical Data Acquisition for Instrumentation and Control Systems. Elsevier, Burlington (2003)
9. Boylestad, R.: Electronic Devices and Circuit Theory. Prentice Hall, São Paulo (2013)
10. Eismin, T.K.: Aircraft Electronics. Bookman, Porto Alegre (2016)

Estimate of Three-Phase Distribution Transformer Losses Through Artificial Neural Networks

Daniel Sousa da Silva, Daniel Guzman del Rio, Wheidima Carneiro de Melo, Israel Gondres Torné and Lennon Brandão Freitas do Nascimento

Abstract This article presents a study of a neural network applied to estimate losses in core and winding of three-phase distribution network. The architecture of neural network used was the topology Multiple Layer Perceptron and training algorithm Levenberg-Marquard that use non-linear methods. From collate of data is create a database with all selected attribute for subsequently be used on simulation of artificial neural networks. All samples of learning are collected from transformers electric tests of automated software of routine testing used in diverse transformers industry and concessionaire of electric energy. The test stage represents 15% of samples, this part represents not supervision stage of training process where is possible observe ANN behavior after training stage (70% of samples) and validation (15% of samples). The evaluation of neural network was made by tools Mean Square Error, Linear Correlation Coefficient and graphic analyzer of cross-validation process. And in the training process obtain accuracy of 80 and 96% of data samples test of a transformer industry.

Keywords Losses · Neural network · Levenberg-Marquard · Three-phase distribution transformer

1 Introduction

The transformer is fundamental to the electric power system because of its presence in three supply stages of electric energy: Generation, transmission, and distribution. [1]. In Brazil, a reflection of its importance is in the Interconnected National System (SIN) that corresponds to the voltages of 230 to 750 kV, there as principal functions: an electric energy transmission generated by power plants for the load centers, of a course a integration between various elements of electric system for guarantee stability and reliability to grid [2].

D. S. da Silva (✉) · D. G. del Rio · W. C. de Melo · I. G. Torné · L. B. F. do Nascimento
Universidade Do Estado Do Amazonas, Manaus, Amazonas 1200, Brazil
e-mail: dass.eng@uea.edu.br

Y. Iano et al. (eds.), *Proceedings of the 4th Brazilian Technology Symposium (BTSym'18)*, Smart Innovation, Systems and Technologies 140,
https://doi.org/10.1007/978-3-030-16053-1_25

Indeed, the transformer to be a great promoter of electric energy transmission and distribution, its indispensable your demand on the electric energy market, including for the next 15 years, the Brazilian consumption of electricity in the grid is expected at the rate of 3.6% per year until 2032 [3]. Therefore, in a competitive scenario is necessary analysis the losses associated to transformer that it is in the order of millions of Reais annually [4, 5] and the transformer industries as consequence tend to produce in a manner compatible, with quality, with low cost, diversified items and delivered to a certain period [6].

In this article, it is proposed through the applications of artificial neural networks (ANN) in transformers, estimate three-phase distribution transformers (TDT) losses, because a reduction of transformers losses by estimating will enable a reduction of distribution grid cost e more agility to traditional methods used in the branch industries. Instead with a typical engineer of simulation where it is necessary a great effort create a learning model and learn the data structure from the training data [7].

Thus, the proposal is to estimate TDT losses by using ANN, the network architecture is composed by topology Multiple Layer Perceptron (MLP) learning algorithm Levenberg Marquard (LM). The evaluation of results founds, it is used the Mean Square Error (MSE), Linear Correlation Coefficient (R), graphic analyze and also the comparison with industry results.

2 Three-Phase Transformers Losses

The losses in the core when obtained in the transformer's industry, traditionally can be determinate with the use of predetermined curves (W/kg) by electric characteristics of sheet and empirical correction values [8]. However, the losses in the core are difficult to calculate with accuracy and also calculated results are very different of values obtained in tests [9].

As mentioned by [1, 8], Eq. 1 represents the losses in core traditionally calculated in industries. Where kp is the empirical correction factor, M is the core weight in kg and p is the specific losses of a sheet in W/kg.

$$Po = kp.p.M \tag{1}$$

In meantime, Eq. 2 can be considered for the losses in the windings of High Voltage (HT) and Low Voltage (LT) of TDT.

$$Pcc = \rho.\frac{i}{l.s} \tag{2}$$

where ρ is the resistivity of the material, i is the current, l is the length of wire and s is the section of wire.

3 Artificial Neural Networks

The ANN is intelligent systems compost of parallel operation systems inspired by biological nervous systems [10]. The great advantage of using this technique is a reconstruction of input in a useful output [11]. This process occurs by adjusting of synaptic weights is used in the training algorithm.

The ANN applications are too many in different areas of knowledge, but primarily focusing on problems related to classification and regression [12].

There too many topologies possible for RNA, however, a more commonly used for prediction problems is the Multiple Layer Perceptron (MLP) because it's possible a wide range for diverse problems as your structure is compost at least 3 layers: input, hidden and output [11].

The algorithm of learning normally used with topology MLP, it is backpropagation by gradient descending (BGD) that use optimization methods for optimization of synaptic weights. However, a diverse modification is made to better algorithm convergence and accuracy as backpropagation Levenberg-Marquard (BLM).

4 Methodology

The methodology proposal pretends to relate through methods and tools the input sets of ANN with ANN output by means of the learning algorithm to be possible to estimate TDT losses [10]. Therefore, the methods propose efficient means for evaluation of ANN and tools to the evaluation of quantitative manner the proposed method in function of your datasets [13].

So, the methodology uses the following steps: analyses of relevant parameters of TD losses, analyses of ANN parameters for losses prediction, analyses of training samples and simulation, analyses of ANN performance and evaluation of ANN application to predicate transformer losses.

5 Implementation

The first step is to select parameters to ANN inputs, it is selected two sets for TDT losses, thus they are constituted of losses in core and winding. Initially, it is relating attributes for losses in core according to Table 1 designed conformer [14, 15], however, certain different constructive characteristics of the core are necessary modifications, because core in [14, 15] is wound, but on this article is stacked.

The parameters selected in Table 1 correspond to core dimensions (ATRC1-ATRC3), characteristic of silicon sheet (ATRC4-ATRC6) and rate of a volt per coil of TDT (ATRC7). Among the main aspects correspondents to losses in the core, beyond the core dimensions is important emphasize on construction of table the characteris-

Table 1 The parameter for ANN of losses in the core

Symbol	Description	Symbol	Description
ATRC1	Diameter	ATRW1	Material resistivity
ATRC2	Height of windows core	ATRW2	Length of high voltage wire
ATRC3	Width of windows core	ATRW3	Length of low voltage wire
ATRC4	Max. 1.5T magnetics losses in 60 Hz	ATRW4	Current of high voltage
ATRC5	Max. 1.7T magnetics losses in 60 Hz	ATRW5	Current of low voltage
ATRC6	Thickness of silicon sheet	ATRW6	Wire cross-section of high voltage
ATRC7	Volt/coil	ATRW7	Wire cross-section of low voltage

tic of silicon sheet, thus the conventional methods to calculate losses in core is made by curve W/kg of sheet in function of magnetic induction that was affected by shift due to ATRC4 and ATRC5 of Table 1 [14, 15]. In the meantime, the input parameters for losses in the winding used the Eq. 2.

According to input parameters of ANN to losses in the core and winding, on account of being various input and the problem be a prediction was used the following architecture: MLP and backpropagation by gradient descending (BDG). Since that the objective of ANN is a prediction, it was realized a linear regression for model creation [16]. However, was used a variation BLM of BDG as long as BLM for ANN that owns hundreds of weights acquire faster convergence, higher precision in training process and lower MSE [14]. Therefore, ANN topology is initially defined with one hidden layer (initially 20 hidden neurons) and the output layer (1 neuron), respectively, with activation function hyperbolic tangent e linear.

6 Analyses of Results

During the process was analyzed diverse alteration of initially established for optimization of results. The two networks established, respectively, prediction of losses in winding and core obtained satisfactory results that were an evaluation by means of MSE, R and graphic analyze of performance curve of cross-validation. Figure 1 represents a performance of ANN prediction losses in the winding, it is possible that MSE obtains typical behavior, this means decrescent behavior until low values and find stable values. With this, obtain a value of 9.58E−05 to MSE. Also by Fig. 1, the

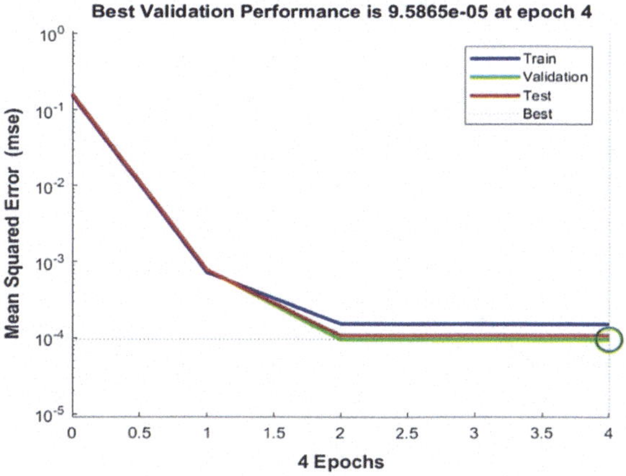

Fig. 1 ANN performance of losses in the windings with 20 neurons

Table 2 MSE and R per neurons of ANN losses in the winding

Neurons	MSE	R-Train	R-Valida.	R-Test	R-All
5	1.7E−04	0.9628	0.9348	0.9419	0.9562
10	1.3E−04	0.9560	0.9606	0.9532	0.9562
15	1.2E−04	0.9563	0.9615	0.9524	0.9543
20	9.58E−05	0.9512	0.9801	0.9619	0.9562
25	1.0E−04	0.9530	0.9658	0.9505	0.9545
30	1.2E−04	0.9583	0.9477	0.8865	0.9406

test set has satisfactory behavior, despite presents a low difference in relation to the training set, however, symbolize a minimum rate of underfitting.

According to Fig. 2, all sets present values R closer to 1, therefore the relation of input set data corresponds to output values that are losses in the winding. In addition, by means of R may be affirmed that ANN obtains adequate precision, thus a regression straight positions itself close of an ideal line of inclination of 45°.

As a consequence of satisfying network result, it is need to verify if the neurons numbers are ideal for ANN. Thus, was elaborate a table for this analyze. The results of Table 2 were satisfactory and because of curve graphic analyze of performance, best results correspond to the number of 15 neurons in a hidden layer. Figure 3 represents the performance of the network with 15 neurons. Its possible to observe in Fig. 3 there is no case of underfitting or overfitting, so this fit of the curve is appropriate to the ANN losses in the winding.

Figure 4 represents results found of ANN of the core losses prediction. With this, similar analyses made for ANN of core losses in winding prediction, the ANN of losses in the core results corresponds to Fig. 4 and optimize result was found with 10 neurons in a hidden layer. In addition, it is possible to observe MSE and R in Table 3.

Established the optimum topologies of ANN, it was possible to establish the accuracy of each network from samples test results, respectively, losses in windings

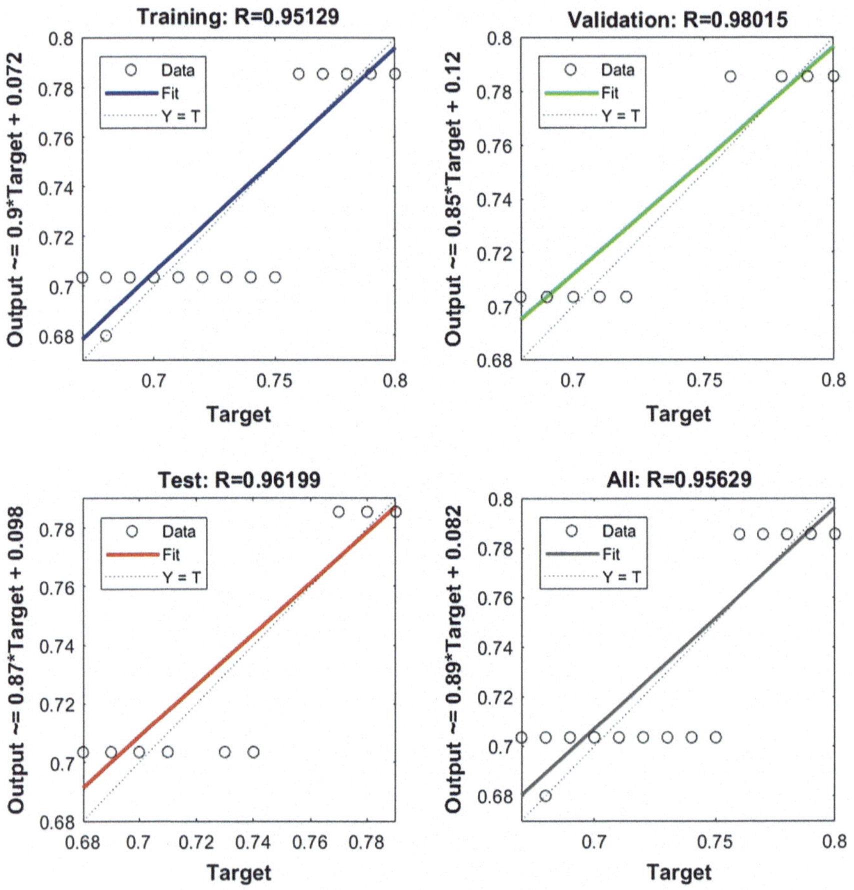

Fig. 2 ANN R of losses in the windings with 20 neurons

Table 3 MSE and R per neurons of ANN losses in the core

Neurons	MSE	R-Train	R-Valida.	R-Test	R-All
5	7.1E−04	0.9823	0.9766	0.9794	0.9811
10	7.3E−04	0.9820	0.9770	0.9801	0.9809
15	9.7E−04	0.9818	0.9716	0.97618	0.9793
20	8.9E−04	0.9831	0.9737	0.9795	0.9811
25	1.8E−03	0.9814	0.9447	0.9843	0.9763
30	6.3E−04	0.9814	0.9810	0.9767	0.9806

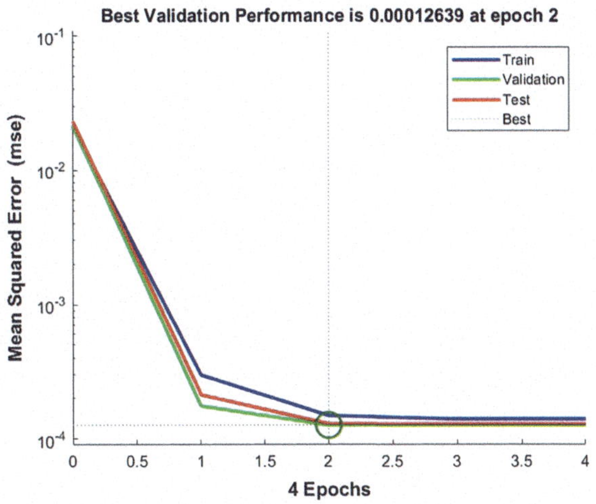

Fig. 3 ANN performance of losses in windings with 15 neurons

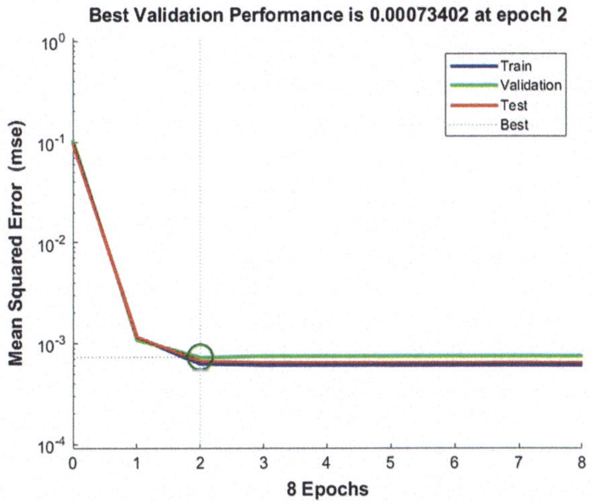

Fig. 4 ANN performance of losses in the core with 10 neurons

and in the core, 80.15 and 96.05%. This performance could be considered by R higher because it is shown that the estimated data are closer of expected.

After established the number of optimum neurons of each ANN, realized a simulation with real values of industry. The objective of the simulation was estimate losses of some transformer of 75 kVA of class D and E and was compared with yours test results after produced (Table 4). Each sample was rebooted the weights to guarantee the ANN reach the convergence regardless of initialized synaptic weight. The best result happened on first sample of losses in the core with error of 0.086% and worst result happened on second sample of losses in the windings with error of 3.927%.

Table 4 Results of samples test class D and E

N°	Losses in the core			Losses in the windings		
	Real	Est.	Err.(%)	Real	Est.	Err(%)
1	291	290.75	0.086	1054.6	1096.10	3.927
2	292	288.45	1.212	1041.3	1095.80	5.233
3	292	290.02	0.678	1054.6	1095.72	3.898
4	291	291.67	0.229	1054.6	1095.88	3.914
5	291	292.60	0.549	1054.6	1096.37	3.960
6	291	290.64	0.124	1054.6	1095.80	3.960
7	289	291.25	0.778	1054.6	1092.48	3.591
8	290	292.60	0.896	1054.6	1095.60	3.888
9	191	198.78	4.073	974.15	975.50	0.097
10	195	198.77	0.115	974.15	976.50	0.241

7 Conclusion

According to the described methodology, it was possible to elaborate a study to estimate the losses of TDT by of the ANN with the LM training algorithm. From this, it was verified that it was possible to relate electrical characteristics of the TDT with losses in the core and to propose a model for RNA.

Thus, it was possible to obtain a linear relation between the inputs of ANN with the outputs (losses) a correlation coefficient of 95 and 98%, respectively, losses in the core and verifying that the choice of suitable parameters for the proposed model.

The results achieved using the cross-validation were satisfactory, where accuracy was obtained in 80% of the test samples for winding losses and 96% for core losses. And though the simulation realized to caparison to industry values, it was to possible to provide more confidence to ANN modeling with best results of error 0086% and worst result of with error of 3.927%. Finally, the proposed model was adequate despite the complexity due to the various factors that comprise TDT.

References

1. Ries, W.: Transformadores: Fundamentos para projeto e cálculo. EDIPCCRS (2007)
2. Empresa de Pesquisa Energética. Transmissão de Energia Elétrica: EPE (2017)
3. Empresa de Pesquisa Energética. Caderno de Demanda de Eletricidade N1: EPE (2018)
4. Gradiz, E.: Eficiência Energética em Transformadores de Distribuição: Procobre (2009)
5. Kundu, M., Jadhav, S., Bagdia, K.: Technical loss reduction through active repair of distribution transformers: results from the field. In: International Conference on Power Systems—ICPS (2017). https://doi.org/10.1109/icpes.2017.8387304
6. Bessa, J.: Alteração de um layout funcional motivado pelos fundamentos da manufatura. Holos – Federal Institute of Rio Grande do Norte (2018). https://doi.org/10.15628/holos.2015.1556
7. Mayr, A., Meyer, A., Seerfired, J.: Potentials of machine learning in electric drives production using the example of contacting processes and selective magnet assembly. In: International

Conference on Industrial Engineering and Operations Management (2017). https://doi.org/10.11909/edpc.2017.8328166
8. Wojtkun, J., Stachowiak, D.: The influence of core geometry on no-load of medium power transformers. In: International Interdisciplinary PhD Workshop —IIPhDW (2018). https://doi.org/10.1109/iiphdw.2018.8388339
9. Souza, K., Castro, T.N., Pereira, T.M., Pontes, R.S.T, Braga, A.P.S.: Prediction of core losses on a three-phase transformer using neural networks. In: IEEE International Symposium on Circuits and Systems—ISCAS. https://doi.org/10.1109/iscas.2011.5937763
10. George, L.P., Dhas, J.E.R., Stheesh, M.: Forecasting of machining quality using predictive neutal networks. In: International Conference on Control, Instrumation, Communication and Computational Technologies—ICCICCT (2015). https://doi.org/10.1109/iccicct.2015.7475276
11. Tumer, A.E., Ederbali, S: Prediction of wavewater treatment plant perfomance using multilinear regression and artificial neural networks. In: International Symposium of Innovations in Intelligent Systems and Application—INISTA (2015). https://doi.org/10.1109/inista.2015.7276742
12. Keong, K.C., Mustafa, M., Mohammad, A.J., Sulaiman, M.H., Abdullah, N.R.H., Samad, R., Perbrianti, D.: Levenberg—Marquardt Prediction for Sungai Isap Residence. In: Conference on Systems, Process and Control—ICSPC (2016). https://doi.org/10.1109/spc.2016.7920722
13. Yadav, S., Shukla, S.: Analysis of k-fold cross-validation over hold-out validation on colossal datasets for quality classification. In: 6th International Conference on Advanced Computing (2016). https://doi.org/10.1109/iacc.2016.25
14. Yadav, A.K., Azeem, A., Sinhg, A., Rahi, O.P.: Application research based on artificial neural networks (ANN) to predict No Load Loss for transformers's design. In: International Conference on Communication Systems and Network Technologies (2011). https://doi.org/10.1109/csnt.2011.45
15. Georgilakis, P.S., Hatziargyriou, N.D., Doulamis, A.D., Doulamis, N.D., Kollias, S.D.: A Neural Network Framework for Predicting Transformer Core Losses. In: Proceeding of the 21st International Conference on Power Industry Computer Applications Connecting Utilities PICA99 To the Millennium and Beyond (1999). https://doi.org/10.1109/pica.1999.779511
16. Izzah, A., Sari, Y.A., Widyastuti, R., Cinderatama, T.A.: Mobile app for stock prediction using Improved Multiple Linear Regression. In: International Conference on Sustainable Information Engineering and Technology—SIET (2017). https://doi.org/10.1109/siet.2017.8304126

Wheelchair Simulator Controlled by a Brain-Computer Interface Based on Steady State Visually Evoked Potentials

Maycon Miranda, Guilherme V. Vargas, Sarah N. Carvalho, Dalton Arantes and Harlei M. A. Leite

Abstract Once unthinkable, it is now possible to record and interpret brain signals to generate application commands through a Brain-Computer Interface (BCI). This innovative technology of human-computer interaction brings hope to people with severe motor disabilities to improve their quality of life and social insertion. This study aims to analyze the feasibility of using a BCI based on Steady State Visually Evoked Potentials (SSVEP) in the control of a wheelchair. For this purpose, a 3D simulator of a wheelchair was developed and the user interaction was analyzed. The results show the feasibility of controlling the wheelchair simulator by BCI-SSVEP. However, it is still necessary to analyze the interaction performance in a real environment and to make improvements in the BCI and in the control system to guarantee a safer and more efficient experience to the user.

Keywords Brain-computer interface · SSVEP · Wheelchair simulator

1 Introduction

With the exponential evolution of technology, applications previously restricted to science-fiction movies have now become a reality, revolutionizing everything around

M. Miranda (✉) · G. V. Vargas · S. N. Carvalho · H. M. A. Leite
Federal University of Ouro Preto, João Monlevade, Brazil
e-mail: maycon.miranda@aluno.ufop.edu.br

G. V. Vargas
e-mail: guilherme.vargas@aluno.ufop.edu.br

S. N. Carvalho
e-mail: sarah@ufop.edu.br

H. M. A. Leite
e-mail: harlei@ufop.edu.br

D. Arantes
University of Campinas, Campinas, Brazil
e-mail: dalton@decom.fee.unicamp.br

Y. Iano et al. (eds.), *Proceedings of the 4th Brazilian Technology Symposium (BTSym'18)*, Smart Innovation, Systems and Technologies 140,
https://doi.org/10.1007/978-3-030-16053-1_26

us. Brain Computer Interface (BCI) system is such an example of "futuristic" technology that attracts the curiosity of people, since it brings to reality the dream of direct communication between user and computer through brain signals.

Since the development of the first BCI system by Dr. Grey Walter in 1964 [4], who implanted an electrode directly into the brain of a patient during neurosurgery, allowing him to shift slides in a projector, to the current BCI systems which are applied in the control of prostheses and electronic games, many challenges have been conquered. However, this interaction technology is still restricted to research laboratories, being the motivation for countless research projects in different areas. These efforts range from basic understanding of the human brain, in medicine, through computation, physics and engineering, in the improvement of acquisition techniques and algorithms for brain signal processing.

BCI systems have been primarily devoted to applications for users with disabilities that cannot manipulate the conventional interaction media, such as mouse, keyboard and joysticks. The wheelchair controlled by BCI systems is certainly an example of application that may be crucial for these users. However, it is a critical application, since for example it involves the physical integrity of the user in collision situations. Therefore, as BCI systems are still under development, with unexpected behavior mainly due to failures in brain signal interpretation, the control of a wheelchair by BCI is still a challenge. In the present study we present a wheelchair simulator and analyze the viability of its control by a BCI-SSVEP system, without the need of building a real automatic wheelchair and exposing the user to risks in the preliminary tests.

1.1 Brain-Computer Interface

Brain-Computer Interface is a system that acquires brain signals and interprets them to generate control signals. A BCI usually consists of five steps: (i) signal acquisition, (ii) digital analog conversion, (iii) feature extraction, (iv) feature translation and (v) device output, as shown in Fig. 1.

The signal acquisition consists basically in the recording of the electrical activity of the brain through electrodes. The acquired signal is amplified, filtered, digitized and sent to the signal processing module, which performs feature extraction that will be related to application commands.

1.2 BCI Based on SSVEP

A BCI system can be exogenous or endogenous. Exogenous BCI depends on the user's brain response caused by external stimuli, while endogenous BCI does not require any kind of external stimulus, depending only on the user's ability to control his/her cognitive activity through concentration.

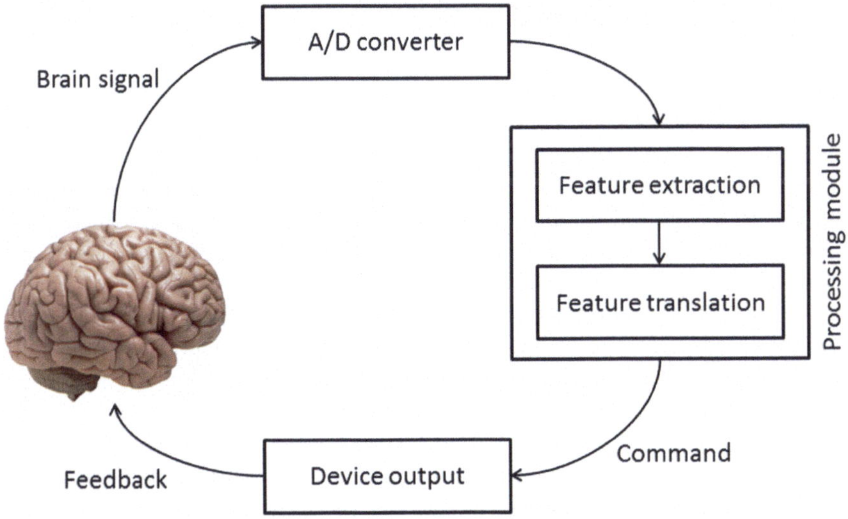

Fig. 1 A typical BCI system

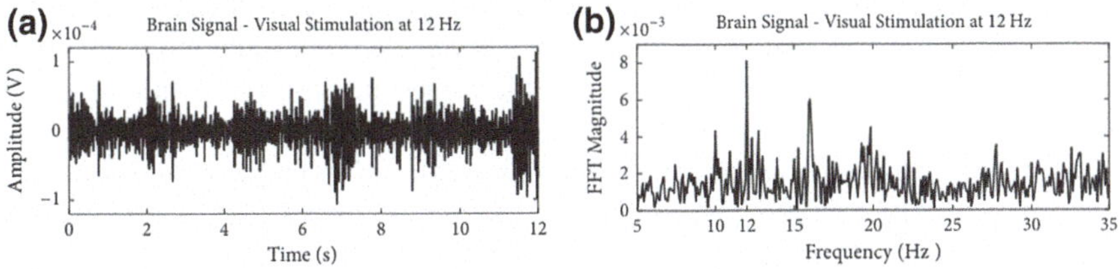

Fig. 2 SSVEP response at 12 Hz. **a** time domain and **b** frequency domain [5]

In the present study we work with the exogenous BCI based on Steady State Visually Evoked Potentials. This approach explores the evoked potential that emerges in the visual cortex of a subject exposed to a visual stimulus that flickers on a given frequency.

Figure 2 shows a typical brain signal recorded by electroencephalography (EEG) in the time and frequency domains for a user exposed to a visual stimulus flickering at 12 Hz. In both cases it is possible to note the presence of Gaussian and impulsive noises due to artifacts and interferences. Also, in the frequency domain we can observe a power spectral density peak at the evoked frequency (12 Hz). The signal was registered at Oz position (standard 10–10) during 12 s.

Generally, BCI-SSVEP has an interface with several stimuli, each scintillating at a well-known frequency and associated with a given command to be sent to the application.

2 Wheelchair Controlled by BCI

BCI-controlled wheelchair research has not yet been matured sufficiently to be commercially available. There are still several challenges to be overcome, all of which have to be addressed before ensuring the security and reliability of these systems.

Leeb et al. [6] demonstrated that a tetraplegic can control a wheelchair in virtual reality (VR) by brain waves. The brain signal was acquired through EEG and the control was obtained from the motor imagery (MI). The participants of this study achieved a good performance, motivating additional studies in a real environment.

Rebsamen et al. [7] developed a wheelchair controlled by a BCI system that detects P300 waves. Movement occurs when the user selects the destination they want from a list of predefined locations. The experiments were performed with healthy users. The results demonstrated the efficiency of this strategy.

Zhang et al. [8] proposed that the control of wheelchair controlled by BCI system should be aided by automated navigation techniques. Its control system consists of selecting waypoints in the environment. The selection is performed by motor imagery (MI) or P300. This approach reduced mental fatigue significantly.

The greatest challenge in wheelchair control using BCI systems is the system hit rate, since countless variables can affect performance, such as user distraction, electromagnetic waves that cause interference to the electrodes and the acquisition equipments, physiological artifacts, among others. Despite the effort to minimize the impact of these variables, there is still no solution that always correctly identifies the user's intention. However, based on the progress achieved so far, there will certainly be a time when BCI systems will be robust enough to control critical applications such as automatic wheelchairs.

2.1 Wheelchair Simulator

Due to the instability of BCI systems, we present a wheelchair simulator controlled by BCI-SSVEP that avoids collision through a set of proximity sensors. The objective of the simulator is to train new users without being exposed to the risks of a training involving a real wheelchair.

The simulator was developed in Unity3D and offers a first-person view, making the experience closer to the real one. The movement of the wheelchair occurs through three visual stimuli in the square format that flicker at the frequencies 6 Hz (turn left), 10 Hz (turn right) and 15 Hz (up) displayed in the Heads-Up Display (HUD) of the simulator, as shown in Fig. 3a.

The user's goal is to control the wheelchair, starting from the blue cube at the bottom of the square to the orange cylinder, located at the top of the square, as shown in Fig. 3b. The region in gray represents the hunt, where it is possible to move around. The blue region in the center of the square, as well as the green areas, are

Fig. 3 **a** Wheelchair simulator. **b** Top view of the scenario used in the simulation

not accessible to the wheelchair. Finally, the brown rectangles represent the banks of the square.

When the user focuses on the left or right stimulus, the wheelchair rotates 15° to the respective side and when the user focuses on the up stimulus, the wheelchair advances 2 meters. Movements are performed only if the sensors fixed in the wheelchair do not indicate any obstacles. Figure 4 shows the position and reach monitored by the sensors. The front sensors that reach 2 meters are only activated when the user wants to move forward. The other sensors are used to verify collision, when the user wishes to rotate the wheelchair.

The online BCI-SSVEP used was developed by a group of researchers from the School of Electrical and Computer Engineering (FEEC) at the State University of Campinas (UNICAMP) [2]. The brain signal was recorded at 256 Hz by EEG using 16 dry electrodes and a biopotential amplifier (gUSBamp). The digital signal processing consisted of a spatial filtering by the CAR (Common Average Reference) method [3]. The features were extracted by Fast Fourier Transform (FFT) using windows of 4 s and discriminated by a linear classifier based on the least squares method [1].

3 Results

A healthy male participant with no previous BCI application control experience participated in the experiment. The participant was adequately informed about the research and the experimental protocol and signed the consent form approved by the Ethics Committee of the University of Campinas (n. 791/2010).

Figure 5 shows the path traveled by the user (yellow line), starting from the initial position delimited by the square in the lower region of the scenario, to the final position delimited by the cylinder at the top of the scenario, in two attempts. Additionally, the blue line projects the ideal path using the fewest steps to achieve the goal.

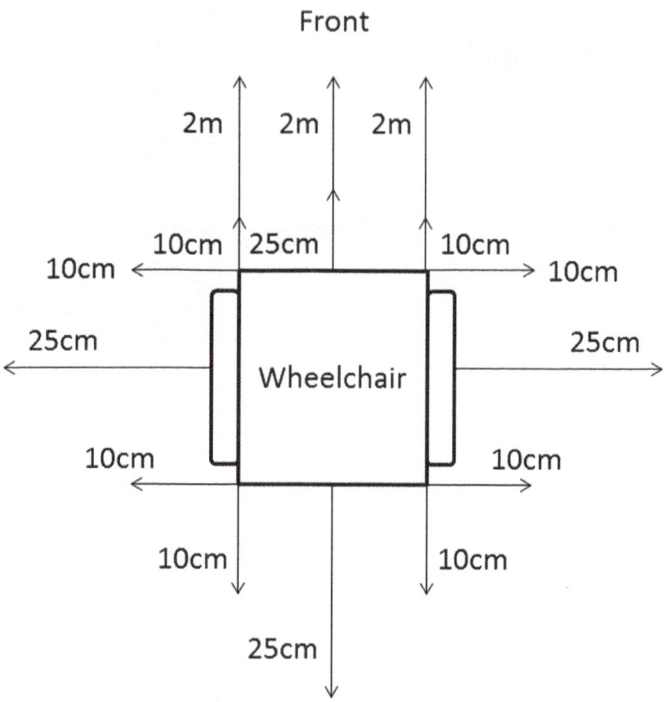

Fig. 4 Position of proximity sensors

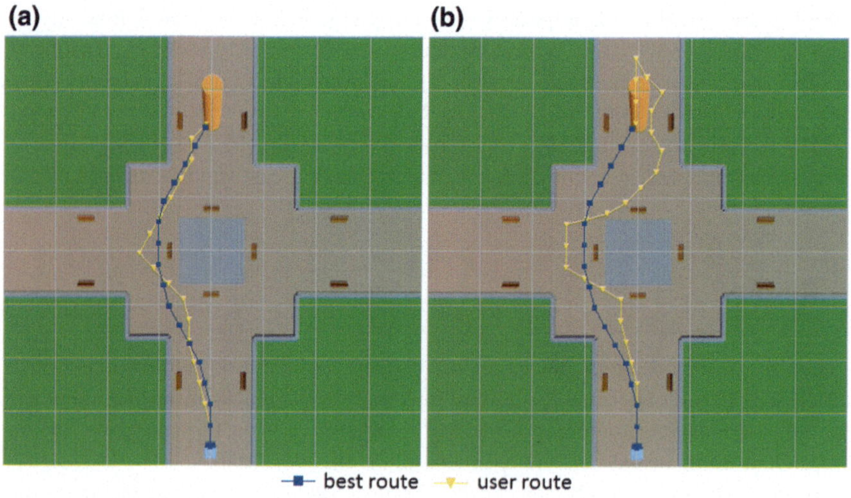

Fig. 5 Path traversed by the wheelchair. **a** First attempt and **b** Second attempt

In the first attempt, Fig. 5a, the user issued 59 commands through the BCI system, while on the second attempt, in Fig. 5b, the user issued 124 commands due to errors in the trajectory caused probably by the loss of concentration of the volunteer or due to errors in the classification system.

Through this experiment it was possible to observe that the misinterpretation of user intention can influence considerably the experience of wheelchair use. At times when the chair moved in a direction not desired by the user, it was possible to visualize signs of frustration, through limb movements and deep breathing. In a real scenario, therefore, in which the chair is controlled by a disabled person, failures in the interpretation of the user's intention may expose the user to danger, both due to accidents involving the wheelchair and increased stress.

Obstacle sensors have shown promise in increasing the level of safety of the wheelchair. Even in situations where the chair moved in a direction not desired by the user, at no time there was collision with elements of the scenario. However, additional studies involving a real chair should be performed, since the sensors inaccuracies are non-existent in the simulation, not corresponding to real sensors.

The signal processing techniques configured in the BCI system proved to be satisfactory. A larger study involving new signal processing techniques is desirable. However, it is important to note that the configuration of the BCI system used in the experiment of this work is widely known and has been used in other works, such as [1, 2].

4 Conclusion

In the experiments carried out in the present work, it was possible to observe that the control of a wheelchair by means of a BCI system based on SSVEP is possible, although new achievements are still necessary in order to improve control robustness and efficiency.

Despite the numerous research works that seek to improve the accuracy of the BCI systems, they are still in the development phase, being very sensitive to numerous environmental variables and also to variables resulting from the physiological system. These variables can degrade performance considerably and therefore hinder the quality of interaction.

The popularization of BCI systems, as a form of human-computer interaction, still depends on evolutions that would make it possible to identify and correct possible misinterpretations of the brain signal, therefore avoiding the generation of unwanted commands that impairs communication efficiency.

Finally, this work shows that the current state of BCI systems already allows a demonstration of the potential of this interaction medium. However, there is still a long way forward with challenges to be overcome before they leave the research laboratories and enter the life of people who need this type of interaction device.

References

1. Carvalho, S.N., et al.: Comparative analysis of strategies for feature extraction and classification in SSVEP BCIs. Biomed. Signal Process. Control **21**, 34–42 (2015)
2. Costa, T.B.S., Carvalho, S.N., Uribe, L.F.S., Ferrari, R., Souza, R.S., Soriano, D.C., Castellano, G., Attux, R.R.F., Romano J.M.T., Cardozo, E.: Online brain-computer interface based on steady-state visually evoked potentials. In: 2nd Brainn Congress, Campinas. Journal of Epilepsy and Clinical Neurophysiology vol. 21. pp. 75–75 (2015)
3. Garcia-Molina, G., ZHU, D.: Optimal spatial filtering for the steady state visual evoked potential: BCI application. In: Neural Engineering (NER), 5th International IEEE/EMBS Conference on. IEEE, pp. 156–160 (2011)
4. Graimann, B., Allison, B., Pfurtscheller, G.: Brain-Computer Interfaces: A Gentle Introduction. The Frontiers Collection. Springer-Verlag, Berlin Heidelberg (2010)
5. Leite, H.M.A., Carvalho, S.N., Costa, T.B.S., Attux, R., Hornung, H.H., Arantes, D.S.: Analysis of user interaction with a brain-computer interface based on steady-state visually evoked potentials: case study of a game. Comput. Intell. Neurosci. **2018**, 10 (2018)
6. Leeb, R., Friedman, D., Müller-Putz, G.R., Scherer, R., Slater, M., Pfurtscheller, G.: Self-paced (Asynchronous) BCI control of a wheelchair in virtual environments: a case study with a tetraplegic. Comput. Intell. Neurosci. **2007,** 8 (2007)
7. Rebsamen, B., Guan, C., Zhang, H., Wang, C., Teo, C., Ang, M.H., Burdet, E.: A brain controlled wheelchair to navigate in familiar environments. IEEE Trans. Neural Syst. Rehabil. Eng. **18**(6), 590–598 (2010)
8. Zhang, R., Li, Y., Yan, Y., Zhang, H., Wu, S., Yu, T., Gu, Z.: Control of a wheelchair in an indoor environment based on a brain-computer interface and automated navigation. IEEE Trans. Neural Syst. Rehabil. Eng. **24**(1), 128–139 (2016)

A Static Hand Gesture Recognition for Peruvian Sign Language Using Digital Image Processing and Deep Learning

Cristian Lazo, Zaid Sanchez and Christian del Carpio

Abstract The work consists in recognizing the gestures of the alphabet in Peruvian sign language using techniques of digital image processing and a model of Deep Learning (CNN). Image processing techniques are used for segmentation and tracking of the hand of the person making the gestures. Once the image of the segmented hand is used, a CNN classification model is used to be able to recognize the gesture. The image processing and CNN algorithms were implemented in the Python programming language. The database used was 23,000 images divided into 70% for training, 15% for testing and 15% for validation. Likewise, said data corresponds to 1000 images for each non-mobile gesture of the alphabet. The results obtained for the precision of the classifier were 99.89, 99.88 and 99.85% for the data of training, test and validation respectively. In the case of the Log Loss parameter, 0.0132, 0.0036, and 0.0107 were obtained for the training, testing and validation data, respectively.

Keywords Static hand gesture · Digital image processing · Deep learning · Python

1 Introduction

Sign language communication is an important means of communication that allows people with hearing difficulties to be understood. Inclusion through technological advancement is the main motivation of this project to facilitate daily tasks and even in the world of work for those with a disability. A clear example of this problem is that of a deaf worker who must communicate face to face with his client. The sign language may differ in each country, for this reason it is intended to generate a

C. Lazo (✉) · Z. Sanchez · C. del Carpio
Universidad Nacional de Ingeniería, Lima, Peru
e-mail: mecatronico.lazo@gmail.com

Z. Sanchez
e-mail: zse.mu10@gmail.com

C. del Carpio
e-mail: cdelcarpiod@gmail.com

Y. Iano et al. (eds.), *Proceedings of the 4th Brazilian Technology Symposium (BTSym'18)*, Smart Innovation, Systems and Technologies 140,
https://doi.org/10.1007/978-3-030-16053-1_27

system of recognition of the Peruvian Sign Language, and be able to help people to communicate easily.

The work consists of two parts, the first part is the image processing that locates and segments the hand for later classification and the second part is a classifying model based on a Deep Learning technique, convolutional neural networks (CNN), which is used in the recognition of non-mobile gestures. The algorithms were implemented in Python [1] and the OpenCV [2] and Keras [3] libraries were used.

2 Proposed Method

The block diagram of the proposed algorithm is showed in Fig. 1.

2.1 Image Acquisition

The image is obtained through a webcam, in a RGB color model, with a resolution of 720 × 480 pixels.

2.2 Background Substraction

First, the reference image of what the background will be (without the object of interest) is obtained and 5 s later (time for the person to raise their hand), another image is taken with the object of interest (hand). The reference image is expressed as $I_1(x, y)$, and the image with the object of interest is represented as $I_2(x, y)$. In

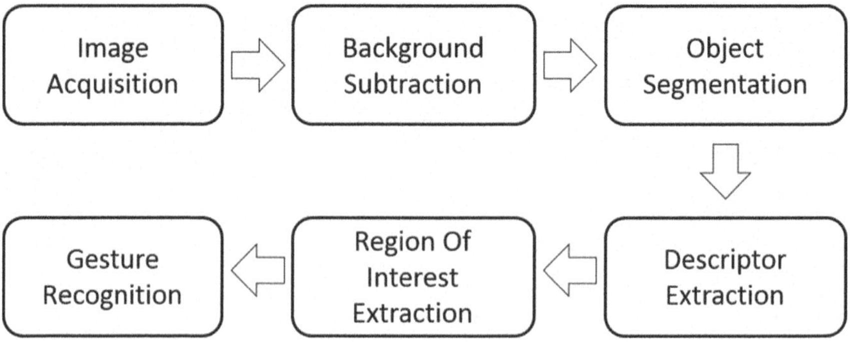

Fig. 1 Proposed method flowchart

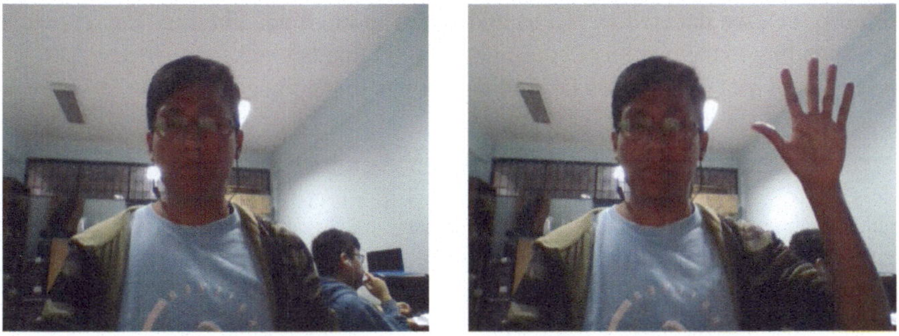

(a) Reference Image $I_1(x, y)$

(b) Image with the object of interest $I_2(x, y)$

Fig. 2 Image acquisition

Fig. 3 Image $I_3(x, y)$

Fig. 2a the background image is shown and in Fig. 2b the image with the object of interest is shown.

Secondly, in order to eliminate the background, each of the pixels of the 2 images and their 8 close neighbors (Eq. 1) are compare. When they are the same, that pixel becomes zero otherwise, it stays with its original values. This technique is very effective because the background image is static while the one with the object of interest is differentiated by the hand and its movements. The result of this operation is shown in Fig. 3.

$$I_3(x, y) = \begin{cases} I_2(x, y) & N_8(p) \neq N_8(q) \\ 0 & N_8(p) = N_8(q) \end{cases} \tag{1}$$

In Eq. (1), we denoted as $N_8(p)$ and $N_8(q)$ to the 8 neighboring (Eq. 2) pixels corresponding to the pixel p of coordinates (x, y) that belongs to the image $I_1(x, y)$ and the pixel q of coordinates (x, y) that belongs to the image $I_2(x, y)$.

$$\begin{aligned}&(x-1, y-1), (x-1, y), (x-1, y+1), (x, y-1),\\&(x, y+1), (x+1, y-1), (x+1, y), (x+1, y+1)\end{aligned} \tag{2}$$

2.3 Object Segmentation

In the previous process much of the background of the image was removed, however, thj9imnhk b1ykokkis removal is not total, because it is only possible if both images $I_1(x, y)$ and $I_2(x, y)$, are in the same position and there are no new objects.

In order to completely segment the object of interest, the S component of the HSV color model is obtained, for this, the image $I_3(x, y)$ is transformed from the RGB color model to the HSV color model (image $I_4(x, y)$).

The first step to move from one color model to another is to separate the primary components of $I_3(x, y)$. In this sense, component R is expressed as $I_{3R}(x, y)$, component G is expressed as $I_{3G}(x, y)$ and component B is expressed as $I_{3B}(x, y)$. Using equations Eqs. (3) and (4) the S component of the HSV color model is obtained. Figure 4 shows image $I_{4s}(x, y)$.

$$\begin{aligned}MIN &= \min(I_{3R}(x, y), I_{3G}(x, y), I_{3B}(x, y))\\MAX &= \max(I_{3R}(x, y), I_{3G}(x, y), I_{3B}(x, y))\end{aligned} \tag{3}$$

$$I_{4S}(x, y) = \begin{cases} 1 - \frac{\text{MIN}}{\text{MAX}}, & otherwise \\ 0, & MAX = 0 \end{cases} \tag{4}$$

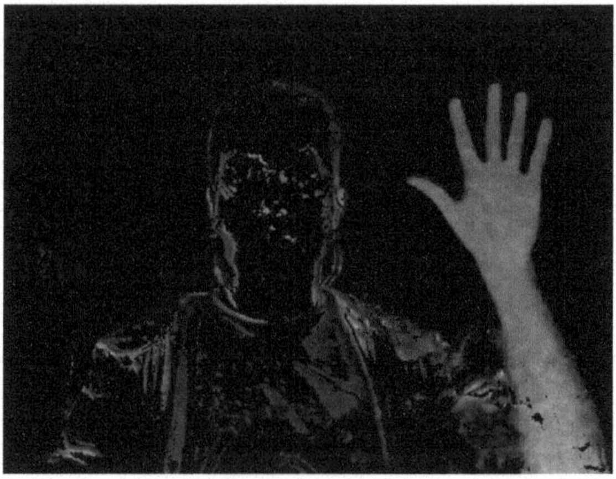

Fig. 4 Component S of the image, $I_{4S}(x, y)$

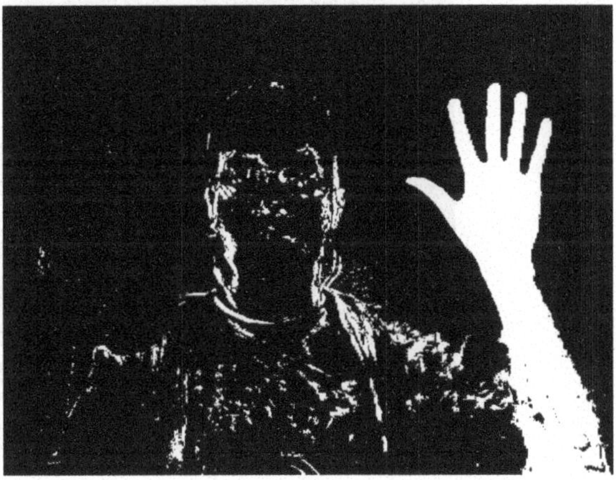

Fig. 5 Binary image $I_B(x, y)$

Once the *S* component of the HSV color model is obtained, Eq. 5 is applied to obtain the binary image $I_B(x, y)$ as shown in Fig. 5. The threshold values were obtained from the multiple tests that were performed.

$$I_B(x, y) = \begin{cases} 1, & 0.24 \leq I_{4S}(x, y) \leq 1 \\ 0, & otherwise \end{cases} \tag{5}$$

In order to provide solidity to the segmented object, the morphological opening process Eq. 7 is applied, with a structural element *EE* shown in Eq. 6.

$$EE = \begin{matrix} 0 & 1 & 0 \\ 1 & 1 & 1 \\ 0 & 1 & 0 \end{matrix} \tag{6}$$

$$I_{B1} = I_B \circ EE \tag{7}$$

where I_B and I_{B1} constitute the matrices that represent the images $I_B(x, y)$ and $I_{B1}(x, y)$ respectively. The symbol "∘" indicates the operation of the morphological opening process [4]. The result of this process is shown in Fig. 6.

2.4 Descriptor Extraction

Once the image $I_{B1}(x, y)$ is obtained, the descriptors are calculated to identify the hand, this is achieved thanks to its unique characteristics such as fingers and palm. First, the entire contour of the image is extracted and the thinnest areas are searched to characterize the fingers and the largest area to identify the palm. Once the fingers

Fig. 6 Image after the morphological process I_{B1}

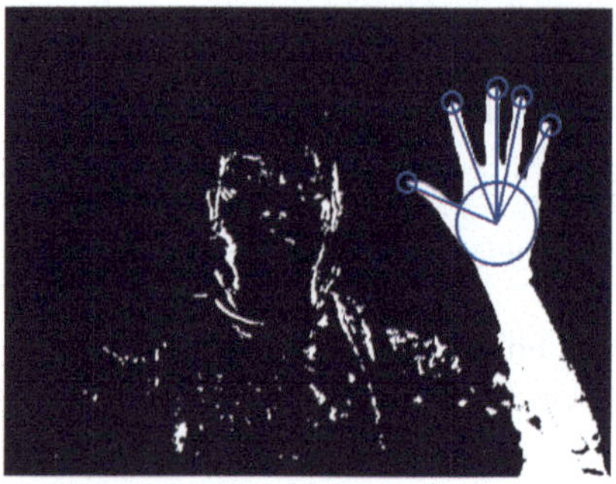

(a) Descriptors of the hand in the binarized image

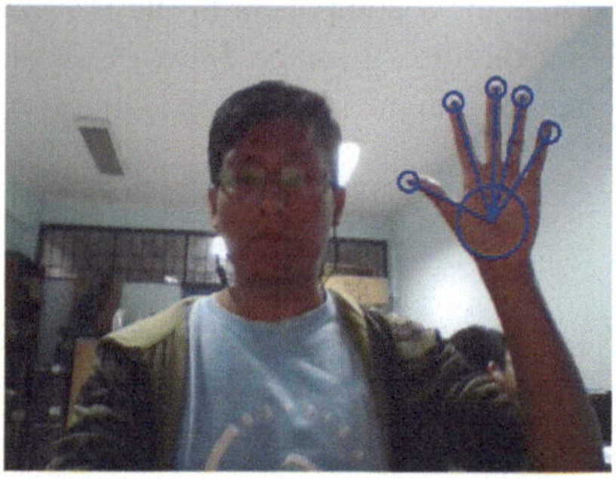

(b) Descriptors in the image in RGB.

Fig. 7 Calculation of descriptors

and palm are identified, markers are placed on each of them. Circle for the palm and lines for the fingers. The result is shown in Fig. 7.

2.5 Region Of Interest Extraction

Once the object of interest (hand) has been identified within the image, the object is extracted. In this case, the hand is tracked using the descriptors obtained in the previous step to identify it and thus only obtain this area. The area obtained is shown in Fig. 8. The image of the extracted hand is defined as $I_M(x, y)$.

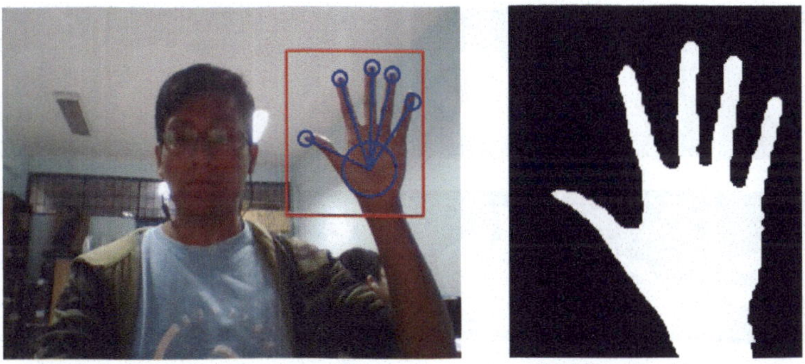

(a) Selection of the region of interest (b) Binarized image of the region of interest.

Fig. 8 Obtention of the area of interest

3 Gesture Recognition

In this stage, the first step is to resize by interpolation [4] the image $I_M(x, y)$, to an image $I_{M1}(x, y)$ of 128 × 128 pixels. Next, a Convolutional Network (CNN) was used for the gesture recognition process, the network model used is shown in Fig. 9.

The CNN architecture consists of 2 convolutional layers, 2 maximum grouping layers where the output maps of the last layer are sampled at 1 pixel per map, resulting in a 1 × 30,752 feature (fully connected layer) to then get the 1 × 23 classification vector and finally use an activation function (softmax) to classify. The total data used consists of 23,000 images, corresponding to 23 non-mobile signs, divided into training group (70%), test group (15%) and validation group (15%) randomly selected. The signs are shown in Fig. 10.

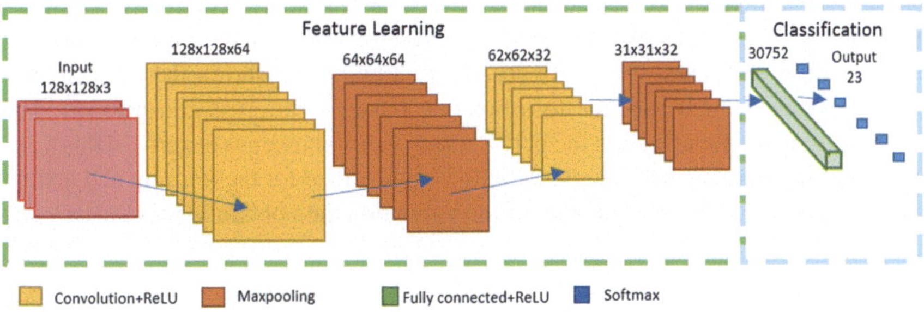

Fig. 9 Configuration of the architecture used in the classification algorithm

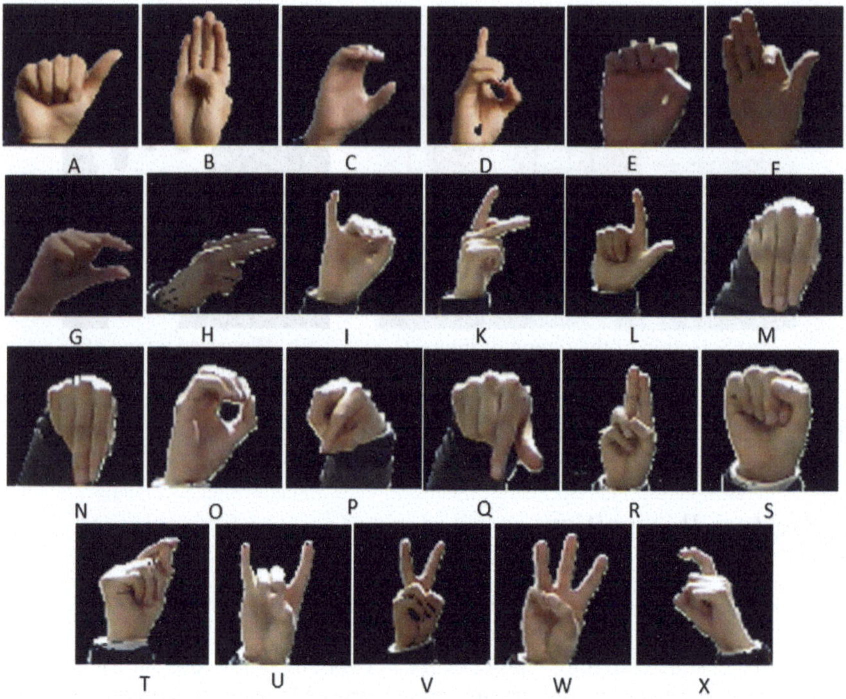

Fig. 10 Worked signs of the Peruvian sign language alphabet

4 Results

The results obtained for the proposed recognition algorithm for the validation data (3450 images) are shown in Table 1, which indicates the number of hits and errors by sign. For the validation data an accuracy of 99.85% was achieved. Figure 11a shows the accuracy obtained by the classifier in 10 training epochs for training and testing data, reaching 99.89 and 99.88% respectively. Figure 11b shows the performance of the classifier, which uses the loss function (categorical cross entropy) [5]. The classifier is most efficient when the performance index converges to zero. In this case, for 10 training periods, 0.0132 and 0.0036 were obtained for the training and testing data, respectively. The performance of the validation data obtained was 0.0107.

5 Conclusions

The proposed algorithm is robust enough to learn 23 non-mobile gestures with an error rate of 0.15%.

Table 1 Results of recognized signs

Sign	Number of correct detections	Number of incorrect detections	Sign	Number of correct detections	Number of incorrect detections
A	132	0	N	144	0
B	146	0	O	165	0
C	171	0	P	154	0
D	139	0	Q	140	0
E	150	0	R	150	1
F	147	1	S	137	0
G	130	0	T	129	0
H	141	0	U	174	1
I	154	0	V	161	0
K	155	0	W	159	2
L	166	0	X	139	0
M	162	0	–	–	–

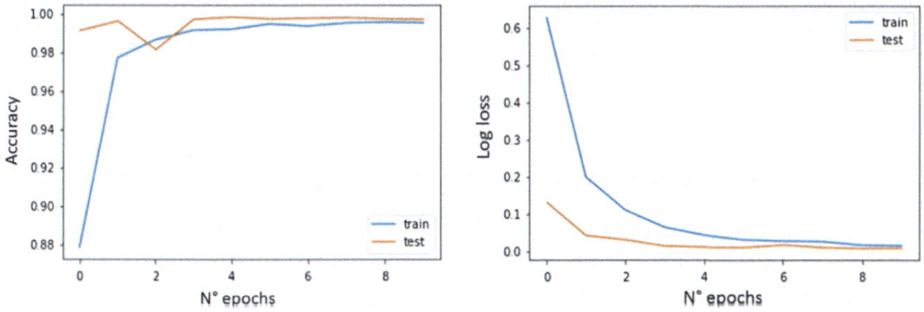

(a) Classifier Accuracy (b) Classifier performance

Fig. 11 Classifier training

The joint use of digital image processing and convolutional neural networks improves classifier performance. Also, the use of a large database helps in the accuracy of the convolutional neural network.

Finally, the use of free software was taken into consideration so that the developed system is accessible and has no restrictions on use.

References

1. Python Software Foundation: Python 2.x Docs, https://www.python.org/downloads/. Last Accessed 27 Feb 2018
2. Mordvintsev A., Abid, K.: OpenCV-Python Tutorials. https://opencv-pythontutroals.readthedocs.io/en/latest/py_tutorials/py_tutorials.html. Last Accessed 30 Sep 2018
3. Keras: The Python Deep Learning library. https://keras.io/. Last Accessed 30 Sep 2018
4. Gonzalez, R.C, Woods, R.E.: Digital Image Processing Second Edition, Pearson Education, New Jersey (2002). ISBN 978-0201180756
5. Goodfellow, I., Bengio, Y., Courville, A.: Deep Learning, MIT Press (2016)

Performance Comparison between a Single-Phase Bidirectional Converter Connected to the Grid Using PI and Fuzzy Controllers

Reginey Azevedo Barbosa and Durval de Almeida Souza

Abstract This paper compares the control of a bidirectional converter connected to the grid using two different controllers: the conventional PI controller and a fuzzy controller. The approach is done by detailing the design of the fuzzy controller, indicating the inputs and outputs, pertinence functions, and inference rules. The performance of the controllers is evaluated based on simulations using Simulink/Matlab, so that the output current of the drive follows a given reference. This reference goes through changes, thus the transient behavior for each case can be studied. In the conclusion of this paper a quantitative and comparative comparison is done, in which the observed advantages of each controller analyzed are highlighted.

Keywords Bidirectional converter · Fuzzy · PI controller · Microgrid

1 Introduction

The climatic variations have worried environmentalists and stimulated technological advances which aim at preserving the environment and replacing technologies that affect the ecosystem. In this context, investments in technologies involving clean and renewable generation that imply lower environmental costs have been growing considerably, with high performance in response to requests from the energy supply system [1].

With the strengthening of research in the renewable energy sources field, the concept of distributed generation emerged. This term exemplifies a smaller form of generation compared to traditional (hydroelectric) power plants and can be connected at various points in the electrical system, either in transmission or distribution

R. Azevedo Barbosa · D. de Almeida Souza
Universidade Federal da Bahia, Salvador, BA, Brazil
e-mail: reginey.azevedo@ifba.edu.br

D. de Almeida Souza (✉)
Instituto Federal da Bahia, Lauro de Freitas/Irecê, BA, Brazil
e-mail: durval@ifba.edu.br

Y. Iano et al. (eds.), *Proceedings of the 4th Brazilian Technology Symposium (BTSym'18)*, Smart Innovation, Systems and Technologies 140,
https://doi.org/10.1007/978-3-030-16053-1_28

[1]. In light of this, the concept of microgrids emerges as a way of inserting the distributed generation into the existing electrical system, while promoting the control and management of locally generated energy.

The microgrids are composed by generating units and loads of electronic components that treat the energy generated and consumed. They also have tools that can control all the data generated on the electrical grid, in order to adapt the variables involved such as bus voltage, voltage, output frequency and current, power flow and other parameters depending on the operations topology and form. Thus, this paper has as main focus the study of the bidirectional converter control in the single phase, which interfaces with the utility grid. There are international standards that establish tolerable minimum values for signal disturbance, such as harmonic voltage variation and distortion of the output wave, which should serve as parameters for performance analysis of the system and the designed controller. According to [2], a Proportional Integral Controller (PI) can fulfill some requirements established, as response speed and overshoot, however in some situations imposed to the microgrid, such as a non-linear load, this controller presents a slow and unsatisfactory response.

2 Microgrid

Microgrid is understood by the grouping of loads and energy sources which form a subsystem in a self-sufficient way and without damaging the grid [3]. A very common microgrid is usually composed of a renewable source, the MPPT converter, the DC bus and a bidirectional converter for grid integration, as shown in Fig. 1.

The MPPT tracks the maximum power point to improve the efficiency of the PV panel in order to extract the maximum power from the panels. The DC bus is where the DC loads are connected and the bidirectional converter is connected to the distribution grid, which must control and manage the energy consumed and injected into the grid.

The integration between the DC bus and the main grid is carried out by means of a bidirectional converter that has the function of regulating the voltage of the DC bus through the control of the power flow. The voltage of the bus varies in a range that

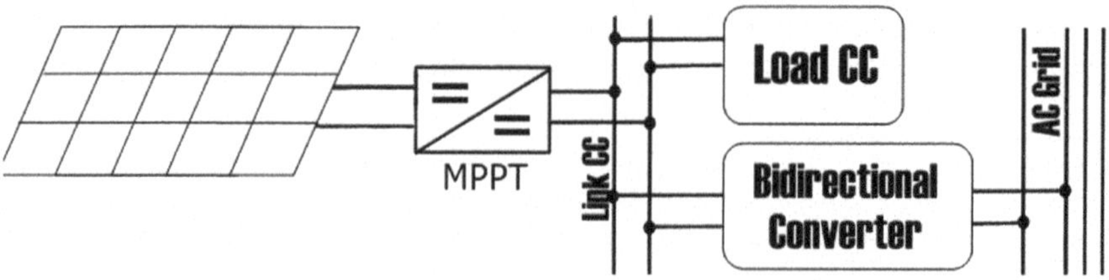

Fig. 1 Typical topology of a single microgrid

allows the flow control according to the need of consumption and generation, doing the energy management [4].

Studies focused on integrations about microgrids deal with the following aspects: equipment architecture design for the system optimization; power flow control; protection and quality of energy and system-oriented modeling of power converters.

The need for control to make it possible to connect an energy generation system to the utility grid is prominent, since it needs to be in sync with the network to be connected, respecting the maximum tolerances allowed by the Brazilian Electricity Regulatory Agency. With this, the scientific community concentrates research on converter topologies and control algorithms to ensure energy quality, efficiency and reliability.

2.1 Microgrid Operation Mode

The microgrids can operate either interconnected with the medium voltage distribution network or isolated from the medium voltage network [4]. The operation form depends on the situation or planning in which the microgrid was given, and may be presented in the following ways:

- Normal mode: Microgrid is interconnected to the conventional grid, feeds local loads and can inject or consume power from the grid to which it is connected, exporting the surplus produced or absorbing the deficit needed to supply the entire load.
- Emergency mode: it happens if some disturbance or upstream fault occurs and the protection system acts to open the circuit, or even in a planned situation that leaves the microgrid running isolated from the conventional network (islanding).

2.2 Microgrids Topologies

The topologies of the microgrids can present different equipment depending on their type, which can be differentiated as direct current (DC) and alternating current (AC) microgrids.

In general, the microgrid concept can be applied to the residential power system at low power level (10–100 kW) [6]. Figure 2 (left) shows an AC microgrid (or AC nanogrid) architecture. These microgrids combined with computerized meters, communications and remote control could become blocks that come to build a smart grid.

Since the microgrids are dynamically separated from the rest of the grid, the advantages presented by the DC microgrid promote the idea that in the future the construction of electrical systems could be based on microgrids like these, as shown in Fig. 2 (right). In comparison to an AC microgrid, a DC microgrid offers many

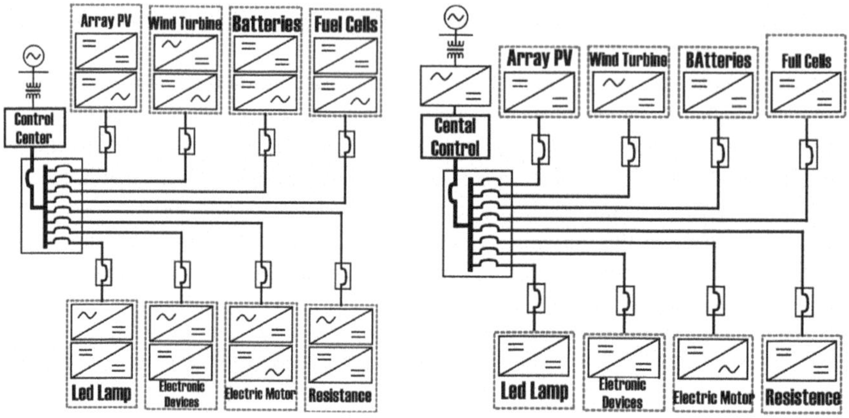

Fig. 2 Topology microgrid AC (left) and CC (right)

advantages, starting with fewer power converters, greater overall system efficiency and easier interface of renewable energy sources for a DC system. There aren't concerns about grid frequency values and reactive power issues, reducing AC losses. Furthermore, the electronic equipments, LED lighting and variable speed drive motors can be more conveniently powered by DC sources [6].

There are several factors that enhance the use of DC systems rather than AC systems. Its possible to mention:

- Many forms of renewable energy generation are already suitable for this system, such as photovoltaics, fuel cells and energy storage system with batteries;
- Most residential and building loads are DC ones;
- The integration of electric vehicles into the system will increase consumption through battery chargers in the form of direct current;
- The DC system becomes more efficient because there is no reactive energy present;
- The system as a whole has a smaller number of power converters, avoiding losses in the AC-DC and DC-AC conversion; Each component of the microgrid has its function which they must perform satisfactorily in order to improve the characteristics of the microgrid and the system altogether. Efficiency, reliability and performance depend on the performance of each of these components.

2.3 Control of the Microgrid Power Elements

The importance of microgrid management and control is justified by the need to maintain power balance between generating sources, energy storage devices and loads, as well as adding features that optimize the system. Microgrids must have power flow control and the bus voltage must be kept within a certain range so that it operates in a stable way without compromising the loads or even the system in which it is connected [3].

3 Studied System

The system to be evaluated consists of a complete bridge converter, capable of draining the current flow from the DC link to the grid, and in the opposite direction. The DC link was modeled as a fixed DC current source at 400 V, while the power grid is an alternating source of 127 V, 60 Hz. There is an RLC output filter, which also models the intrinsic resistance of the components. Figure 3 illustrates the model converter in Simulink's SimPowerSystems Toolbox. It is intended to evaluate the tracking capability of a reference sinusoid given by a conventional PI-type controller and a Fuzzy controller, the results obtained are compared through simulation using Matlab/Simulink. There is a voltage sensor in the network from which the sinusoidal reference current will be extracted so that it will be in phase with the voltage (or 180° out of phase). The current sensor in the inductor measures the output current and compares it with the reference sinusoid, from where the error which is processed by the controller is calculated and then the decision making is given by the designed control rules.

In order to study the behavior of the controller during transient periods, the reference voltage value will be constantly modified. These modifications are strategic and will be described in the Results and Discussions topic.

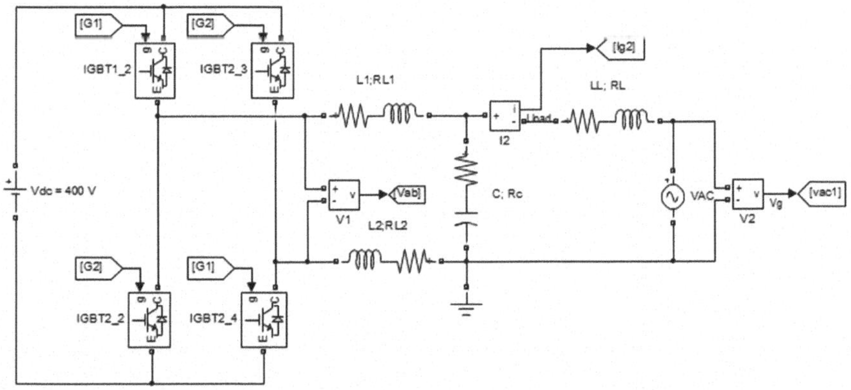

Fig. 3 Modeled converter in SimPowerSystem

4 Projected Fuzzy Controller

In this section, the proposed controller will be described. The controller is based on the artificial intelligence technique of fuzzy controllers. The controller has as input the error values of the inductor current in relation to the reference current and its rate of change. These two inputs are fuzzyfied, transforming each of them into seven triangular memberships, named: NG, NM, NP, NU, PP, PM, PG, according to Fig. 4.

The membership functions can receive values from zero to one and are distributed within a range of −1 to 1. The rule basis is made with the logic "If Error and VarErr then D" in which the output D, control signal which goes to the Pulse Width Modulation (PWM), is related to the mentioned inputs. The elaborate rule basis is shown in Table 1.

The output signal is defined in a similar way to the inputs, containing seven functions of triangular pertinence, with names resembling those of input variables, but distributed in the range of −1.2 to +1.2. The defuzzification occurs through the bisect method, available in the Matlab ToolBox.

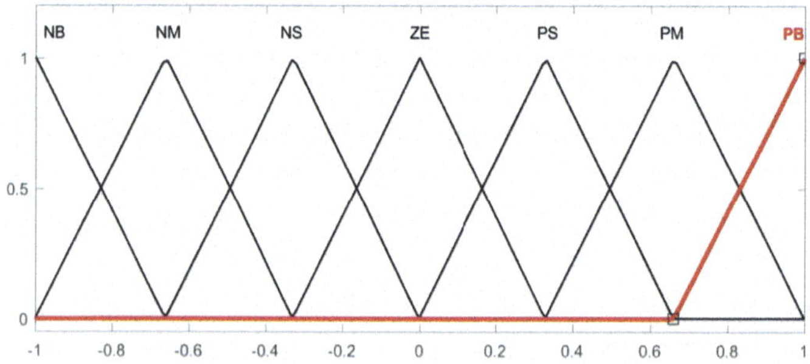

Fig. 4 Assignment functions assigned to inputs

Table 1 Rulles base of fuzzy controller

		Error						
		NG	NM	NP	NT	PP	PM	PG
ΔError	NG	NG	NG	NG	NM	PP	PM	PG
	NM	NG	NG	NG	NP	PP	PM	PG
	NP	NG	NG	NG	NP	PM	PM	PG
	NT	NG	NG	NM	NT	PM	PG	PG
	PP	NG	NM	NM	PP	PG	PG	PG
	PM	NG	NM	NP	PP	PG	PG	PG
	PG	NG	NM	NP	PM	PG	PG	PG

5 Results and Discussions

The simulations were performed with the purpose of evaluating the transient response for two types of controllers: fuzzy and conventional PI. In order to evaluate the transient behavior of the output current with reference changes, strategic points that could present different responses were chosen, such as the crest, the valley, the zero and some intermediate points of the reference sinusoid.

In the first moment, only the behavior of the inverter in steady state was analyzed for the two types of controllers designed. Then, the controller was tested when there is a change in the drives power flow. Finally, the simulation was carried as a means to analyze the converters behavior when there is variation of the reference intensity, as described in the sequence.

5.1 PI Controller

The first phase of the study consisted in evaluating the performance of the PI controller on steady-state. In order to do so, it was tested the case in which there would be no load variation in the converter, so ability of this controller to trace the reference sinusoid could be observed. In this way, the behavior of the error in permanent and transient state was observed when a reference signal in phase with the voltage was applied. Figure 5 shows the result obtained for such simulation.

Its possible to notice that the output has an initial transient state that goes up to about 2 ms, when it reaches the regime considered permanent. The transient regime presents a high error, which starts negative at 3.86A, but reaches the steady-state without oscillating. On the other hand, the steady-state error is very small, around 0.05 A and oscillates symmetrically around zero during all simulation time, as shown in Fig. 5.

For the test with load flow inversion, the simulation was done as previously mentioned, so that there could be observed several possible situations. Figure 8 shows the

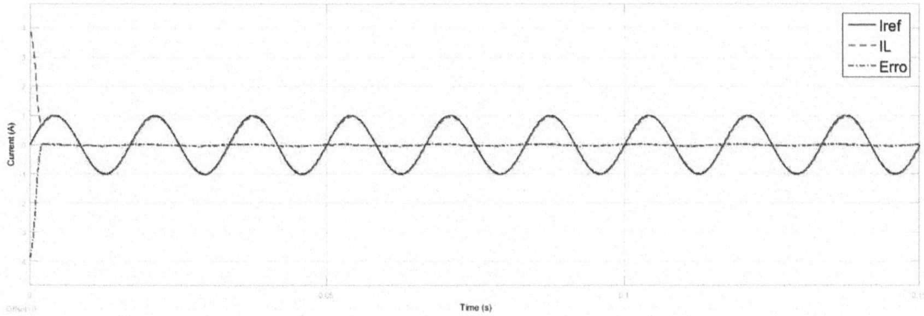

Fig. 5 Simulation for continuous regime analysis, PI control

response of these situations to the PI controller. Observing the figure, it can be seen that the characteristics of the transient are relative to the position where the power flux inversion in the sinusoid occurred. When this transition is at zero the error is very small, about 0.124 A and a transient duration time is 1.5 ms. However, the error in the transient state is greater when the power flux inversion occurs at the crest or in the valley, reaching a value of 2.02 A, presenting a transient duration time of about 3.5 ms. The graph of the error is shown in Fig. 6.

The next test was done by increasing and reducing the power demand, and similarly, the analyses were done in several conditions, some of which are more significant here. Table 2 shows the variations of the correlation in percentage in relation to the previous value and the time in which the current was changed.

Figure 7 shows the reference current, Iref, and the output current in the inductor, ILg. In this figure it can be observed the behavior of the current in the inductor, which

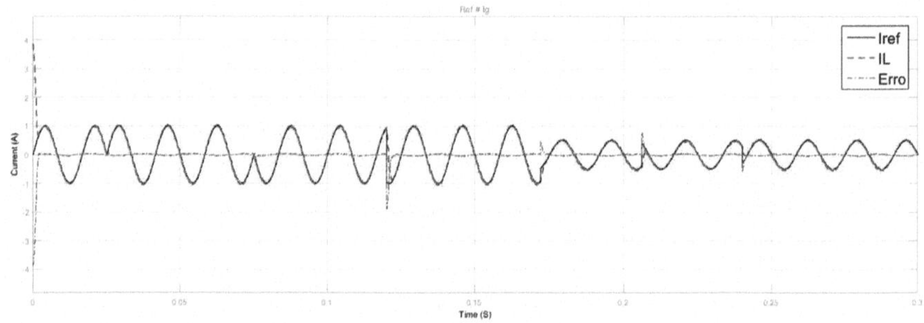

Fig. 6 Simulation for flow inversion analysis, PI control

Table 2 Variation of reference current and instant of variation

Instant	1	2	3	4	5	6
Time (ms)	25	75	120	172	206	240
Variation (%)	+25	−50	+25	−50	100	−90

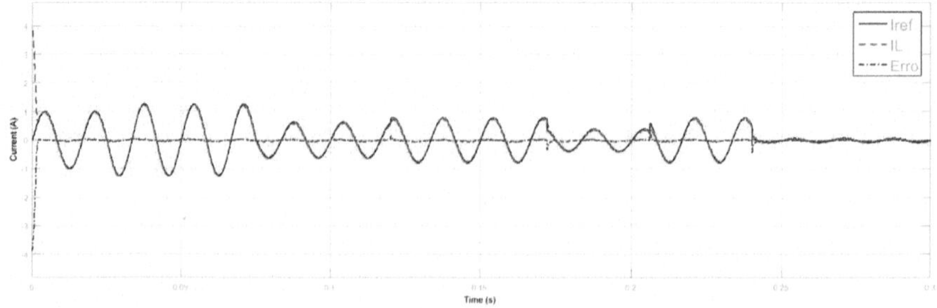

Fig. 7 Simulation for load variation analysis, PI control

can follow the given reference. When the intensity of the current changes to zero, the effect is practically imperceptible, presenting a very low error, and its mixed up with the steady state error. The largest observed error was 0.4 A at time 0,24 s, in which there was a reduction in reference current at a sinusoid descent. The transitory state of this point lasted around 2 ms. The other errors were below 0.33 A with duration below 1.7 ms. Figure 7 shows the graph of the error for such simulation.

5.2 *Fuzzy Controller*

The first analysis, as in the case of the PI controller, was done in order to evaluate the controller performance. The converter was then tested using the fuzzy controller with sinusoidal current reference, constant amplitude and in phase with the mains voltage. Figure 8 shows the result obtained for such a simulation, from which the transient state and the steady-state error are observed.

It can be seen that the output has a transient state that goes up to 5.28 ms until reaching the steady-state. The transient state presents a high error, which starts negative from 3.86 A, rises to a positive value of 1.75 A, and fluctuates decreasing until it reaches the permanent regime, at 5.28 ms. The observed error in the steady state is very low, around 0.035 A, at the beginning of the permanent state and continues to decrease reaching values below 0.02 A, as observed in Fig. 8. It is interesting to notice that the steady-state error is always negative.

The next test is the inversion of the current flow, that is, in the transition from a situation in which the microgrid presents energy generated by the sources greater than the energy consumed by the loads. For this purpose, the simulation was done where the load flow occurred in several different points, as for the PI controller, the same criteria were adopted for the fuzzy controller, changing the load flow in the same instants of time described for the PI controller.

Figure 9 shows the system response when subjected to these power flow inversions by adopting the fuzzy controller. From the figure it can be observed that, just as in the case of the PI controller, the transient differs depending on the point where the load

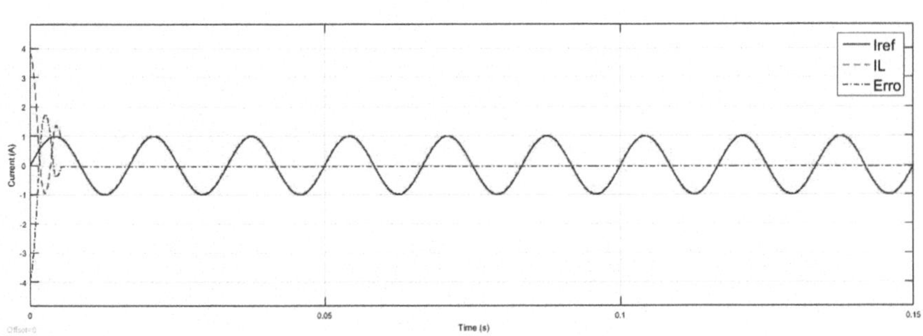

Fig. 8 Simulation for continuous regime analysis, Fuzzy control

flow changes. The zero transition of the sinusoidal reference continues to present a low error, while the largest error is observed in reference wave crest and valley. The greatest error observed in the transient was 2.02 A, with a duration of 4 ms. The error graph is illustrated in Fig. 9.

The other transient state times observed were below 4 ms for the direction change of the power flow.

The last test was performed by increasing and reducing the power demand, now using the fuzzy controller with the same current variations at the same time instants adopted for the PI controller shown in Table 2.

Figure 10 shows the reference current, Iref, and the output current at the inductor, ILg, obtained for the fuzzy controller for increasing and reducing the reference current. In this graph we can observe the behavior of the current in the inductor, which can also follow the given reference. The output current response is almost insensitive when the increase or reduction of the current occurs at zero of the reference, presenting a very low error, and mixing up with the steady state error. The largest error observed in this case was 0.33 A at the same point as the largest error occurred using the PI controller, where there is a reduction in the reference current at a descent of the sine. This points transitory state lasted around 9 ms. The other errors were below 0.3 A with duration under 6ms. Figure 10 shows the graph of the error for such simulation.

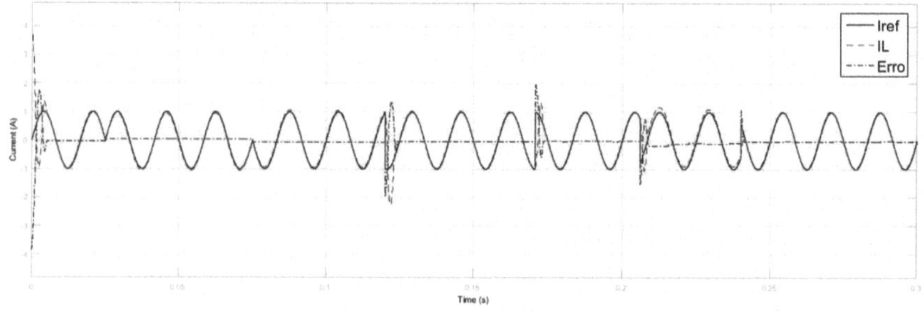

Fig. 9 Simulation for flow inversion analysis, Fuzzy control

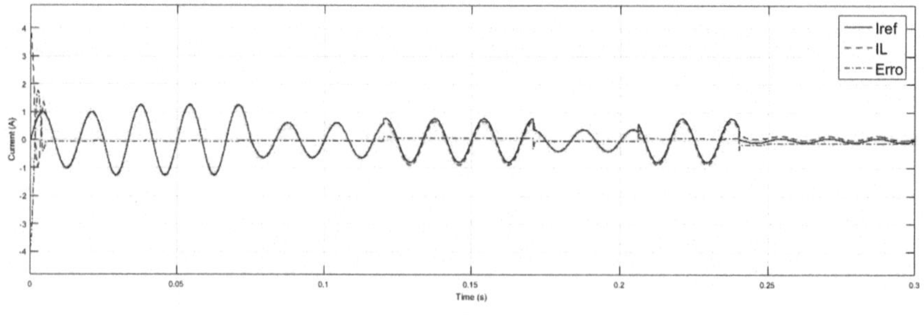

Fig. 10 Simulation for load variation analysis, Fuzzy control

6 Conclusion

The insertion of the microgrids requires mastery of all the technologies involved to make them increasingly reliable and efficient. Thus, it is necessary to develop research that covers the whole scope of this subject, evolved in the matter of converter topologies, controllers design and control strategies. The traditional controller is simple to implement and has good results, but it is necessary to develop other controllers that allow a more precise control of the variables involved. This paper describes the performance of a fuzzy controller by comparing it with traditional PI controller performance. The basic rules and memberships associated to the inputs and outputs of the controller are described. Observing the results, interesting conclusions can be drawn from the controller designed in relation to the PI controller. In general, it is well known that the PI controller has lower time in the transient state than for the fuzzy controller, while the fuzzy controller overrides the PI controller in the steady state error condition, presenting smaller error for all situations. In the first test, the fuzzy controller presented transient oscillations while the PI controller had better performance. The error in the transient state of the fuzzy controller was better only in the third test case, which was slightly smaller than the PI controllers, but showed longer accommodation time in the same situation, as in all other cases. Finally, it is concluded that the PI controller has better transient performance while the fuzzy controller outperforms the PI controller in steady state.

References

1. Marinho, G.S.P., Jnior, D.P.: Qualidade da Energia Eltrica de uma Indstria Sucroalcooleira Apresentada como uma Microrrede
2. Chang, E.C., Cheng, H.L., Liao, K.Y.: Design of adaptive fuzzy sliding-mode PI control for full-bridge inverters. In: Fuzzy Theory and Its Applications (iFUZZY), 2014 International Conference on, pp. 144-147. IEEE (2014)
3. Tahim, A.P.N.: Controle de microrredes de distribuio de energia eltrica em corrente contnua (2015)
4. Ribeiro, M.L.: Estratgia de Reposio de Servio utilizando Micro-Redes. Faculdade de Engenharia da Universidade do Porto, Porto, Dissertao realizada no mbito do Mestrado Integrado em Engenharia Electrotcnica e de Computadores (2010)
5. Wu, T.F., Kuo, C.L., Sun, K.H., Chen, Y.K., Chang, Y.R., Lee, Y.D.: Integration and operation of a single-phase bidirectional inverter with two buck/boost MPPTs for DC-distribution applications. IEEE Trans. Power Electron. **28**(11), 5098–5106 (2013)
6. Boroyevich, D., Cvetkovi, I., Dong, D., Burgos, R., Wang, F., Lee, F.: Future electronic power distribution systems a contemplative view. In: Optimization of Electrical and Electronic Equipment (OPTIM), 2010 12th International Conference on, pp. 1369–1380. IEEE (2010)
7. de Almeida Souza, D.: Mejora de la Eficiencia en los Generadores Empleados en Parques Elicos Utilizando Controladores Fuzzy Adaptativos. Zaragoza (2010)

8. Wu, T.F., Kuo, C.L., Lin, L.C., Chen, Y.K.: DC-bus voltage regulation for a DC distribution system with a single-phase bidirectional inverter. IEEE J. Emerg. Sel. Top. Power Electron. **4**(1), 210–220 (2016)
9. Bevrani, H., Shokoohi, S.: An intelligent droop control for simultaneous voltage and frequency regulation in islanded microgrids. IEEE Trans. Smart Grid **4**(3), 1505–1513 (2013)
10. Dang, D.Q., Choi, Y.S., Choi, H.H., Jung, J.W.: Experimental validation of a fuzzy adaptive voltage controller for three-phase PWM inverter of a standalone DG unit. IEEE Trans. Ind. Inf. **11**(3), 632–641 (2015)

10-Steps Method to Extract the I-V Curve of Resonant Tunneling Diode Based on Experimental Data Preserving Physical Parameters

Mateus H. R. Faria, Rafael V. T. da Nobrega and Ulysses R. Duarte

Abstract In this paper we establish a methodology for obtaining the I-V characteristic of fabricated and characterized resonant tunneling diodes (RTDs) based on experimental data available in the literature. In addition, a scaling factor is proposed, in order to taking account the reduction in the dimensionality between the reservoir and the active region. Hence, the proposed method allows preserving the physical meaning of the major parameters used in the theoretical model developed by Schulman et al. [1], while providing an acceptable fitting, at least, for the negative differential resistance (NDR) behavior of the I-V curve. The validation of the methodology is obtain by satisfactorily contrasting the simulated curve with the experimental data presented in Schulman et al. [1], for a RTD of the $In_{0.53}Ga_{0.47}As/AlAs$ system. Finally, we extend the model proposed for another two RTDs of the GaAs/AlAs system manufactured and characterized by Yang et al. [2] and Alkeev et al. [3]. The simulated results show good agreement regarding experimental peak and valley currents, peak and valley voltages and peak-to-valley ratio (PVR), with a minimum concordance of 94.5%. Thus, this method can be use in the design of RTD, aiming its application in oscillator circuits.

Keywords Quantum device · RTD · I-V characteristic

M. H. R. Faria · R. V. T. da Nobrega · U. R. Duarte (✉)
Federal Institute of Minas Gerais, Academic Area of Electrical Engineering, Formiga 35570-000, Brazil
e-mail: ulysses.rondina@ifmg.edu.br

M. H. R. Faria
e-mail: mateusherculanofaria@gmail.com

R. V. T. da Nobrega
e-mail: rafael.nobrega@ifmg.edu.br

Y. Iano et al. (eds.), *Proceedings of the 4th Brazilian Technology Symposium (BTSym'18)*, Smart Innovation, Systems and Technologies 140,
https://doi.org/10.1007/978-3-030-16053-1_29

1 Introduction

Since the beginning of the 21st century, there has been a graceful increase in the development of devices employing composite semiconductors, exhibiting distinct designs and capabilities. All this was possible due to emergence of the first quantum device: the diode tunnel or Esaki diode [4]. The effect of negative differential resistance (NDR) found on the diode has originate a new class of devices, which electronic conduction along the semiconductor system is ruled by quantum tunneling.

The improvement of semiconductor epitaxial manufacturing and growth techniques have allowed the design of quantum devices based on semiconductor heterostructures, which were first proposed by Tsu and Esaki in 1973 [5], originating the resonant tunneling diode (RTD). In generally, a resonant diode exhibit a negative differential region in the I-V characteristic due to the physical phenomenon known as resonant tunneling, which guarantees the oscillation behavior of the device.

In 1983, Sollner et al. [6] presented experimental results of a GaAs/AlAs RTDs, which was theoretical capable of to operate on frequencies higher than 2.5 THz at room temperature. Thus, this work established the insertion of RTD into the class of high-speed devices, aiming its application on telecommunication systems.

Since the RTD can be used in the designing of oscillator circuits, Oshima et al. in 2016 [7] presented the manufacture of a wireless transmitter operating on data rate of 34 Gbit/s based on InGaAs/AlAs RTD system. Later, in 2017, Tavares et al. [8] investigated the optical modulation characteristics based on an oscillator circuit, which was constituted by an InGaAs/AlAs RTD system. This circuit was used in the hardware implementation for applications in telecommunications systems known as Radio over Fiber (RoF). Thus, due to relevance of RTD in high-speed communication systems, the development and/or improvement of models capable of to reproduce the experimental I-V curve of such resonant diode can be considered an essential tool in the designing of future devices.

In this paper we propose a method to extract the I-V curve of manufactured RTDs, exploiting information of the physical parameters of the semiconductor materials which constitutes the device. Also, we propose a scale factor α in order to take account the difference between the emitters dimensionality (3D) and the quantum well (2D). The main goal of this work is to establish a methodology using the formalism developed by Schulman et al. [1] at starting point, which preserves the physical meaning of the major parameters used in theoretical model, while guarantying a reasonable fitting, at least, in the NDR behavior of the RTD.

For this purpose, we contrast, first, our simulation results to a I-V curve of a RTD manufactured on Ref. [1]. After validating the methodology, we extend the analysis to another two RTD devices [2, 3], in order to verify the limit of our model. Once again, a minimum concordance of 94.5% between simulated results and experimental data is obtained in regard to PVR values, although the model exhibits failure in fitting the conventional diode behavior (I-V region after valley voltage). Nevertheless, we believe this methodology can be applied in the designing of oscillator circuits in telecommunications systems based on RTDs.

2 Theoretical Fundamentals

The development of epitaxial growth techniques in 1970 has allowed the manufacturing of devices based on heterostructures (superlattices) of groups III-V, such as AlGaAs/GaAs, InGaAs/AlAs, GaAs/AlAs systems, among others. Exploiting this experimental tool, Tsu and Esaki in 1973 [5] were the pioneer to propose and evaluate the resonant tunneling of electrical carriers through a semiconductor heterostructure, originating the RTD device.

Generally, a resonant tunneling diode is composed by a two-terminal device, which the active region is constituted of two potential barriers separated by a quantum well. The active region is interconnected to the emitter and collector terminals [9], as depicted in Fig. 1a. For a specific range of applied voltage between the terminals, the energetic profile established in the active region allow the electron leaving the emitter to propagate through the double barrier of potential by means of the physical phenomenon known as resonant tunneling, reaching the collector in the opposite side *quasi-instantaneously*. In this way, the I-V curve of RTDs exhibit a negative differential region due to tunneling effect contribution along the device, as shown on Fig. 1b.

In this context, Tsu and Esaki in 1973 [5] developed and presented an expression to obtain the current density, $J(V)$, as a function of the applied voltage in the RTD, supported by the effective mass approximation,

$$J(V) = \frac{qm^* k_B T}{2\pi^2 \hbar^3} \int_0^\infty \mathbf{T}(E, V) \ln\left[\frac{1 + e^{(E_F - E)/(k_B T)}}{1 + e^{(E_F - E - qV)/(k_B T)}}\right] \mathrm{d}E, \quad (1)$$

where q is the elementary charge, m^* is the effective mass of the semiconductor, k_B is the Boltzmann constant, T is the operation temperature of the device, $\hbar$ is the reduced Planck constant, $\mathbf{T}(E, V)$ is the transmission coefficient and E_F is the Fermi level.

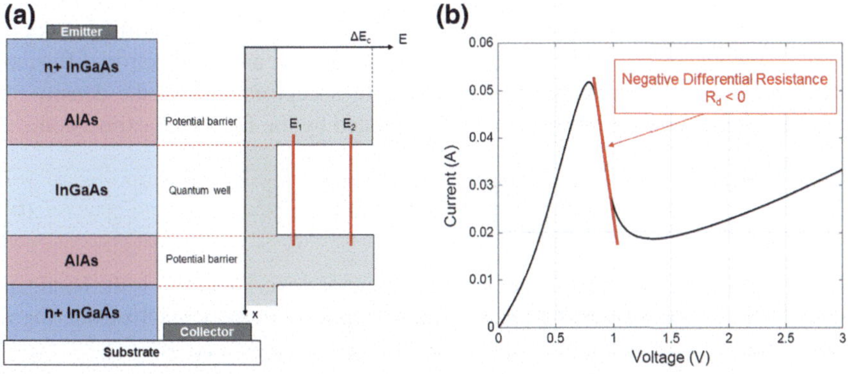

Fig. 1 **a** Schematic representation of an InGaAs/AlAs RTD, also showing the double barrier structure separated by a quantum well. **b** Typical I-V curve of an RTD, evidencing the NDR

According to the Eq. (1), $J(V)$ does not present a closed analytical solution. Therefore, in order to obtain an analytical expression for current density, Schulman et al. [1] have proposed a Lorentzian approximation for the transmission coefficient of RTD, considering that the energy values close to resonance (E_r) could be approximated to $E \approx E_r - qV/2$. This approach have allowed the extraction of the RTD current density as a function of the applied voltage, analytically expressed by,

$$J_1(V) = A \ln \left[\frac{1 + e^{(B-C+n_1 V)q/(k_B T)}}{1 + e^{(B-C-n_1 V)q/(k_B T)}} \right] \left[\frac{\pi}{2} + \tan^{-1} \left(\frac{C - n_1 V}{D} \right) \right], \qquad (2)$$

where,

$$A = \alpha \frac{qm^* k_B T \Gamma}{4\pi^2 \hbar^3}, B = E_F, C = E_r, D = \frac{\Gamma}{2}, \qquad (3)$$

Γ is the resonance width and n_1 is an adjustment parameter proposed by Schulman et al. [1]. In respect to the parameter A, we proposed in this work a scaling factor α, in order to adjust the current density presented by Schulman et al. [1], which does not take into account the difference between the emitters dimensionality (3D) and the quantum well (2D) [10, 11].

Equation (2) reproduces the current density of the first peak and the NDR of the RTD, however, it does not take into account the valley current, which is the term due to the tunneling of other energy levels and inelastic scattering present in the barriers of potential [1]. Then, for voltages higher than the peak voltage, the current density J_2 must be added,

$$J_2(V) = H \left(e^{n_2 V q/(k_B T)} - 1 \right), \qquad (4)$$

where n_2 is an adjustment parameter and H is the saturation current density, which can be obtained by the following expression,

$$H = qn_i^2 \left(\frac{D_n}{N_a L_n} + \frac{D_p}{N_d L_p} \right), \qquad (5)$$

where n_i is the intrinsic concentration of carriers, D_n and D_p are the diffusion coefficient of the electrons and holes, respectively, L_n and L_p are the electron diffusion length and holes, respectively, N_a the density of acceptors and N_d the density of donors [12]. The concentration n_i can be calculated by the following expression,

$$n_i = 2 \left(\frac{k_B T}{2\pi \hbar^2} \right)^{3/2} \left(m^* m_h \right)^{3/4} e^{-\frac{E_g}{2k_B T}}, \qquad (6)$$

where m_h is the mass of the hole in the semiconductor material and E_g is the bandgap energy [12]. The parameter B represents the Fermi level in the RTD. Considering a device which N_d is much higher than n_i, the E_F can be calculated as,

$$E_F = E_C - k_B T \ln\left(\frac{N_C}{N_d}\right), \tag{7}$$

where E_C is the conduction band and N_C is the effective density of states in the conduction band [12], which is obtained by the following expression,

$$N_C = 2\left(\frac{m^* k_B T}{2\pi\hbar^2}\right)^{3/2}. \tag{8}$$

The parameter C in Eq. (3) represents the resonance energy of the first level in the quantum well. In order to obtain this parameter, we used the Transfer-Matrix Method (TMM), a formalism already developed in a previous work [13].

Consequently, the expression of the total current density as a function of the applied voltage of an RTD can be obtained by adding Eqs. (2) and (4),

$$J(V) = J_1(V) + J_2(V). \tag{9}$$

The main goal of presenting the physical meaning of the parameters proposed by Schulman et al. [1] is the lack of technical information to obtain the simulated I-V curve from RTDs manufactured and available in the literature. In this way, the physical expressions necessary and steps to obtain the I-V characteristic from the constituent parameters of the RTD semiconductor materials were raised here. Thus, in the next section the methodology proposed in this work is presented.

3 Methodology

We describe bellow the proposed methodology to simulate the I-V curve based on experimental data of a manufactured RTD reported in literature:

1. Simulation input: the values of the physical constants: q, m_0 (free electron mass), $\hbar$ and k_B, as well as the operating temperature of the device, T;
2. Analyze the structure of the RTD and identify the semiconductor materials used in its manufacturing: active region, emitter and collector;
3. Search in the literature for the values of the physical parameters of the emitter and collector (reservoirs): effective mass of electron and hole and bandgap energy;
4. Replacing the information obtained from steps 1–3 in Eqs. (6) and (8) yields the values of n_i and N_C, respectively;
5. The value of N_d can be provided by the experimental data of the RTD, otherwise, the value of N_d can be obtained by means of Eq. (7), if the reference makes the Fermi energy available;
6. In the Eq. (7) it was considered that N_d is much larger than n_i, so N_a will be approximately n_i;

7. In order to obtain the value of saturation current density, H, the information of steps 1–3 and 6 must be replaced in Eq. (5), as well as the semiconductor parameters of the reservoir (emitter) should be investigated: D_n, D_p, L_n and L_p. As a note, these physical parameters must be limited to a superior value found in the literature, which depends of each semiconductor material;
8. After these steps, we can obtain the parameters A, B, C and D shown in Eq. (3);
9. Replacing the values found in Eq. (9), we have the expression of the RTD current density as a function of the applied voltage. The curve needs to be fitting due to the existence of adjustment parameters (empirical evaluation);
10. The parameters n_1 and n_2 are adjusted according to the experimental curve provided in the reference. The scale factor α is used to correct the dimensionality difference between emitter and quantum well. Lastly, Γ is a small value of the order of tens of meV. Depending on the reference, the value of Γ is supplied, otherwise it must be set.

Finally, we validated the methodology proposed in this work by applying it to three experimental characterized devices: InGaAs/AlAs RTD by Schulman et al. [1], GaAs/AlAs RTD by Yang et al. [2] and GaAs/AlAs RTD by Alkeev et al. [3]. The results and discussions are presented in the next section.

4 Results

In order to validate the methodology to simulate the I-V curves proposed in this work, we used the experimental data of a RTD device characterized by Schulman et al. [1] as the starting point.

According to step 1, it is necessary to include the values of the physical constants in the simulation: we set the operating temperature T to 300 K. From step 2, the semiconductor materials of the device should be identified: the active region is constitute by an $In_{0.53}Ga_{0.47}As/AlAs$ semiconductor system and the emitter and collector are composed by $In_{0.53}Ga_{0.47}As$. Applying step 3, the physical parameter of the reservoirs must be found in the literature: $m^* = 0.042m_0$, $m_h = 0.051m_0$ and $E_g = 0.74$ eV [14, 15]. Substituting these values into Eqs. (6) and (8) leads to: $n_i = 1.52 \times 10^{17}\ m^{-3}$ and $N_C = 2.16 \times 10^{23}\ m^{-3}$.

The Fermi level provided by Schulman et al. [1] is 35 meV. Hence, the value of N_d can be calculated by Eq. (7) according to step 5, using the bottom of the conduction band, E_C, as reference. The N_d value obtained was $8.36 \times 10^{23}\ m^{-3}$. The value of N_a is of the order of n_i (step 6), with $N_a = 1.52 \times 10^{17}\ m^{-3}$.

In order to obtain the value of H based on physical considerations it is necessary to search for the values of D_n, D_p, L_n and L_p reported in the literature. Based on references [14], the values for $In_{0.53}Ga_{0.47}As$ considered were: $D_n = 181.0\ cm^2/s$, $D_p = 6.2\ cm^2/s$, $L_n = 0.312\ \mu m$ and $L_p = 1.080\ \mu m$. As the value of H was provided by Schulman et al., these parameter values needed to be adjusted in order to produce the saturation density current value equal to the one presented in Ref.

[1]. It should be pointed out that the parameters D_n, D_p, L_n and L_p chosen are within the validity range for the $In_{0.53}Ga_{0.47}As$ [14]. Thus, with Eq. (5) we calculate $H = 1.411 \times 10^{-1}\ A/cm^2$.

Schulman et al. [1] present the value of parameter D as 5.2 meV. According to Eq. (3) we obtained $\Gamma = 2D = 10.4$ meV. These are the values used in the simulation of the I-V curve, since the Γ is of the order of ten meV [16].

The parameter C is obtained through the implementation of the Transfer-Matrix Method [13]. Considering the following characteristics of the RTD device identified in step 2: height of the potential barrier (offset) $V_0 = 1.0$ eV, length of the barriers (AlAs) $L_b = 26$ Å, quantum well length ($In_{0.53}Ga_{0.47}As$) $L_w = 48$ Å, effective mass of the barrier (AlAs) $m_b = 0.150m_0$ and effective mass of the well ($In_{0.53}Ga_{0.47}As$) $m_w = 0.042m_0$. Using these values we obtained the energy of the first peak within the quantum well (E_r) equal to 0.1447 eV, which agrees 98.3% with the value presented by Schulman et al. [1].

The parameter B has already been defined for the calculation of N_d in step 5, and it is equal to the Fermi level, i.e., $E_F = 35$ meV [1]. Finally, using Eq. (3) the value of parameter A can be expressed as a function of the proposed scale factor α, $A = (9.137 \times 10^4)\alpha\ A/cm^2$.

In order to perform the fitting of the simulated I-V curve based on experimental data, the adjusting parameters were set (optimized) to $\alpha = 0.107$, $n_1 = 0.115$ and $n_2 = 0.120$. The simulated curve obtained using the proposed methodology is shown on Fig. 2.

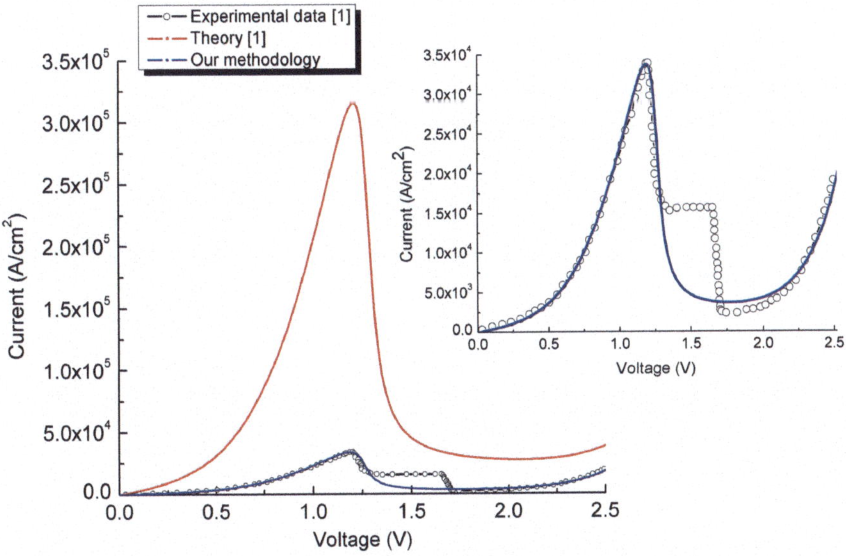

Fig. 2 Comparison between the model (theory) and the experimental RTD curve provided by Schulman et al. [1] and the simulated I-V curve using the methodology proposed in this work

According to the results, it can be noticed that the theoretical model presented by Schulman et al. [1] does not match the experimental curve of the characterized device. This is because the value of Γ used in parameter D is not the same as that used to obtain the parameter A on Ref. [1]. Thus, we suggest the insertion of a scale factor α to adjust the curve.

As it can be seen from the inset of Fig. 2, an excellent agreement is obtained between the profile of the curve produce by our methodology and the experimental data.

Once validate our proposal, we seek to identify the robustness of our methodology. For this purpose, we extend the analysis for other two RTD devices (GaAs/AlAs semiconductor active region) manufactured and characterized by Yang et al. [2] and Alkeev et al. [3]. The both physical parameters reported on literature and those calculated following the steps 1–10 are summarized on Table 1.

The simulated I-V curves employing our methodology based on experimental data from Yang et al. [2] and Alkeev et al. [3] are depicted of Fig. 3. For the RTD manufactured by Yang et al., i.e., Fig. 3a, we obtained $I_p = 1.107 \times 10^5\ A/cm^2$, $V_p = 1.17$ V, $I_v = 3.048 \times 10^4\ A/cm^2$ and $V_v = 1.94$ V. Comparing the simulated results with the experimental data, concordances of 99.2%, 97.1%, 94.5%, 95.1% were achieved in regard to I_p, V_p, I_v and V_v, respectively. The peak-to-valley ratio (PVR) value simulated is 3.6, which agrees in 94.6% with the experimental value.

Regarding the comparison between the simulated I-V curve and the experimental data of the RTD manufactured by Alkeev et al. [3], i.e, Fig. 3b, the following values were extracted: $I_p = 3.245\ kA/cm^2$, $V_p = 1.16$ V, $I_v = 1.123\ kA/cm^2$ and $V_v = 1.42$ V. Comparing the simulated values with the experimental data, we

Table 1 Values of physical parameters used to simulate the I-V curves

Calculated for Ref. [2]	Calculated for Ref. [3]	Literature
$n_i = 2.18 \times 10^{12}\ m^{-3}$	$n_i = 2.18 \times 10^{12}\ m^{-3}$	$m^* = 0.067m_0$ [14, 15]
$N_C = 4.35 \times 10^{23}\ m^{-3}$	$N_C = 4.35 \times 10^{23}\ m^{-3}$	$m_h = 0.510m_0$ [14, 15]
$E_F = 39.43$ meV	$E_F = 39.43$ meV	$E_g = 1.424$ eV [14, 15]
$N_a = 2.18 \times 10^{12}\ m^{-3}$	$N_a = 2.18 \times 10^{12}\ m^{-3}$	$N_d = 2.00 \times 10^{24}\ m^{-3}$ [2, 3]
$H = 6.99 \times 10^{-6}\ A/cm^2$	$H = 6.99 \times 10^{-6}\ A/cm^2$	$V_0 = 1.07$ eV [15]
$E_r = 0.135$ eV	$E_r = 0.132$ eV	$m_w = 0.067m_0$ [14, 15]
$A = (3.87 \times 10^5)\alpha\ A/cm^2$	$A = (2.10 \times 10^5)\alpha\ A/cm^2$	$m_b = 0.150m_0$ [14, 15]
$\alpha = 9.20 \times 10^{-2}$	$\alpha = 4.35 \times 10^{-3}$	$L_b = 1.7$ nm [2, 3]
$\Gamma = 27.6$ meV	$\Gamma = 15.0$ meV	$L_w = 4.0$ nm [2]
$n_1 = 0.107$	$n_1 = 0.105$	$L_w = 4.5$ nm [3]
$n_2 = 0.255$	$n_2 = 0.317$	
$D_n = 200\ cm^2/s$	$D_n = 200\ cm^2/s$	
$D_p = 9.2\ cm^2/s$	$D_p = 9.2\ cm^2/s$	
$L_n = 0.1\ \mu$m	$L_n = 0.1\ \mu$m	
$L_p = 0.4\ \mu$m	$L_p = 0.4\ \mu$m	

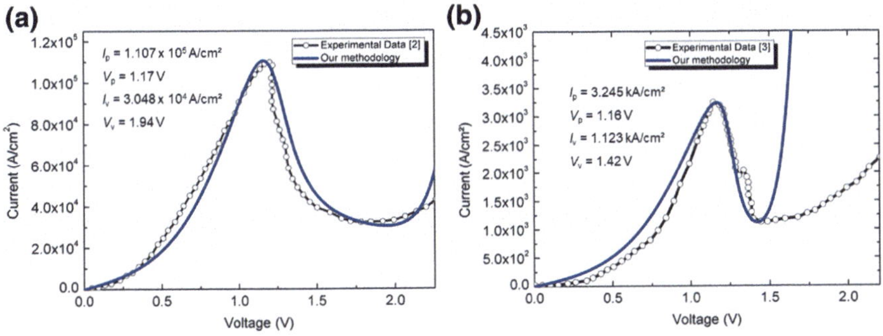

Fig. 3 Comparison between the simulated I-V curve and the experimental data of the GaAs/AlAs RTD device manufactured and characterized by: **a** Yang et al. [2], **b** Alkeev et al. [3]. Acronyms: peak current (I_p), peak voltage (V_p), valley current (I_v) and valley voltage (V_v)

obtained a concordance of 99.9%, 98.2%, 99.3%, 98.4% in relation to I_p, V_p, I_v and V_v, respectively. The PVR value calculate was 2.89, which agrees with the experimental value in 99.6%.

Fig. 3b shows that the proposed methodology is valid for modeling the negative differential resistance behavior in the RTDs, i.e, the first peak region of the current-voltage curve. However, the model fails for voltages higher than the valley voltage, since the RTD curve suffers from impairments effects related to resistance (R_s) and inductance (L_s) in series, as well as the appearance of capacitance (C_0) in parallel to the device. These parasitic effects imply in a pronounceable linear profile of the I-V curve, after the valley voltage region. On the other hand, the methodology proposed does not consider the presence of these parasitic effects in the electrical characterization of the devices, resulting in ideal parabolic behavior [17]. As a future work, we will include these effects of R_s, L_s, and C_0 on the simulated I-V curves.

5 Conclusion

In this work we proposed a methodology constituted of 10 steps to simulate I-V curves based on experimental data of a RTD, in order to preserve the physical meaning of the major parameters used in the theoretical model. Also, we have proposed a scale factor α which take accounts the difference between the emitters dimensionality (3D) and the quantum well (2D), which guarantees a better fitting in the NDR behavior of the RTD.

First, we have validated our proposal to the I-V curve of a RTD manufactured by Schulman et al. [1]. After validated the model, we have explored the robustness of our methodology, extending the analysis to RTD devices manufactured by Yang et al. and Alkeev et al. Once again, the results produced by the simulation achieved a concordance of 94.6% and 99.6 % in regard to PVR values for Refs. [2] and [3], respectively. Summarizing, the theoretical model developed in this work can satisfactory fit the negative differential resistance (NDR) behavior of the RTD.

A drawback identified in this proposal is the model failure in fitting the behavior of I-V curve for values higher than valley voltage (diode standard curve). It can be explained due to absence of parasitic effects on the model, such as resistance, capacitance and inductance. Nevertheless, according to the results, we believe this methodology could be used in the designing of RTD, aiming its application on oscillation circuits.

References

1. Schulman, J., Santos, H., Chow, D.: Physics-based RTD current-voltage equation. IEEE Electron Device Lett. **17**(5), 220–222 (1996)
2. Yang, L., Draving, S., Mars, D., Tan, M.: A 50 GHz broad-band monolithic GaAs/AlAs resonant tunneling diode trigger circuit. IEEE J. Solid-State Circuits **29**(5), 585–595 (1994)
3. Alkeev, N., Averin, S., Dorofeev, A., Gladysheva, N., Torgashin, M.: GaAs/AlAs resonant-tunneling diode for subharmonic mixers. Russian Microelectron. **39**(5), 331–339 (2010)
4. Esaki, L.: New phenomenon in narrow germanium pn junctions. Phys. Rev. **109**(2), 603–604 (1958)
5. Tsu, R., Esaki, L.: Tunneling in a finite superlattice. App. Phys. Lett. **22**(11), 562–564 (1973)
6. Sollner, T., Goodhue, W., Tannenwald, P., Parker, C., Peck, D.: Resonant tunneling through quantum wells at frequencies up to 2.5 THz. App. Phys. Lett. **43**(6), 588–590 (1983)
7. Oshima, N., Hashimoto, K., Suzuki, S., Asada, M.: Wireless data transmission of 34 Gbit/s at a 500-GHz range using resonant-tunneling-diode terahertz oscillator. Electron. Lett. **52**(22), 1897–1898 (2016)
8. Tavares, J., Pessoa, L., Figueiredo, J., Salgado, H.: Analysis of resonant tunneling diode oscillators under optical modulation. In: 19th International Conference on Transparent Optical Networks (ICTON), pp. 1–4. Girona (2017)
9. Mizuta, H., Tanoue, T.: The Physics and Applications of Resonant Tunneling Diodes. Cambridge University Press, Cambridge (1995)
10. Askerov, B.: Electron Transport Phenomena in Semiconductors. World Scientific, Singapore (1994)
11. Schulman, J.: Extension of Tsu-Esaki model for effective mass effects in resonant tunneling. Appl. Phys. Lett. **72**(22), 2829–2831 (1998)
12. Sze, S., Lee, M.: Semiconductor Devices: Physics and Technology, 3rd edn. Wiley, New York (2012)
13. Ragi, R., Nobrega, R., Romero, M.: Modeling of peak voltage and current of nanowire resonant tunneling devices: case study on InAs/InP double-barrier heterostructures. Int. J. Numer. Model. Electron. Networks Devices Fields **26**(5), 506–517 (2013)
14. Levinstein, M., Rumyantsevand, S., Shur, M.: Handbook Series on Semiconductor Parameters. vol. 1, 2. World Scientific, London (1999)
15. Vurgaftman, I., Meyer, J.: Band parameters for III-V compound semiconductors and their alloys. J. Appl. Phys. **89**(11), 5815–5875 (2001)
16. Ozbay, E., Bloom, D., Chow, D., Schulman, J.: 1.7-ps, microwave, integrated-circuit-compatible InAs/AlSb resonant tunneling diodes. IEEE Electron Device Lett. **14**(8), 400–402 (1993)
17. Rei, J., Foot, J., Rodrigues, G., Figueiredo, J.: Modelling of optoelectronic circuits based on resonant tunneling diodes. In: Proceedings of SPIE 10453 Third International Conference on Applications of Optics and Photonics, 1045308 (2017)

Efficient Optimization of Photocatalytic Process for Azo Dyes

Gustavo Duran Iga, André Luis de Castro Peixoto and Ademir Geraldo Cavallari Costalonga

Abstract The need of fresh water in our time has become a problem of important concern. Restore water for new uses makes purification an essential process to achieve a desired degree of quality. This need depends on each different purpose, such as purification of drinking water or recycle water for industrial need. Due to the toxic characteristics of non-biodegradable organic pollutants, new treatment technologies must be studied and Advanced Oxidation Processes (AOP's) come up as suitable for degradation of a wide range of organic contaminants in polluted water. The optimization of photocatalysis is an important issue to turn this process viable. Organic pollutants in effluents threats the environment in several ways, in this context, textile dyes are the major responsible for the colored wastewaters disposed at effluents. This work presents the experimental results of utilizing HNO_3 and $NaNO_3$ mixture in aqueous solution for the optimization of the photocatalytic activity. 97% of the dye was degraded in 16 min. Thus, the mechanism of reaction is discussed about a new source of hydroxyl radicals which reacts with the pollutant, as it happens in troposphere.

Keywords Photocatalysis · Titanium dioxide · Optimization

1 Introduction

The need for fresh water in our time has become a problem of important concern. The water is not only for industrial purposes, but also for domestic uses many freshwater

G. D. Iga · A. L. de Castro Peixoto · A. G. C. Costalonga (✉)
IFSP - Instituto Federal de Educação, Ciência e Tecnologia de São Paulo Câmpus Capivari, Avenida Doutor Ênio Pires de Camargo, 2971, CEP: 13360-000 Capivari, SP, Brazil
e-mail: ademirgcc@ifsp.edu.br

G. D. Iga
e-mail: gustavoiga@hotmail.com

A. L. de Castro Peixoto
e-mail: alcpeixoto@ifsp.edu.br

Y. Iano et al. (eds.), *Proceedings of the 4th Brazilian Technology Symposium (BTSym'18)*, Smart Innovation, Systems and Technologies 140,
https://doi.org/10.1007/978-3-030-16053-1_30

sources have been polluted due to the industrial activity. Many different chemicals are discharged into the aquatic environment. Some of them are biodegradable and nontoxic, however, others are.

Restore water for new uses makes purification an essential process to achieve a desired degree of quality. This need depends on each different purpose, such as purification of drinking water or recycle water for industrial needs.

Due to the toxic characteristics of non-biodegradable organic pollutants, new treatment technologies must be studied and Advanced Oxidation Processes (AOP's) come up as suitable for degradation of a wide range of organic contaminants in polluted water.

The optimization of photocatalysis is an important issue to turn the advanced oxidation process (AOP) viable. Organic pollutants in effluents threats the environment in several ways, in this context, textile dyes are the major responsible for the colored wastewaters disposed at effluents. The dyeing process does not provide the perfect dye fixation, about 20% of total world production of this compound contaminates the environment [1].

To work with this, the AOP's has become an important method to degrade organic pollutants in water. Heterogeneous photocatalysis is an example of AOP that is showing good results for the oxidation of these pollutants. TiO_2 is potentially the best catalyst that the researchers chose for this reaction, due to its high efficiency, physical and chemical stability, low cost and low toxicity.

This process has many variables that affect directly the efficiency of the reaction. Several parameters that influence the photocatalytic degradation of dyes such as pH, the initial concentration of dyes, photocatalyst concentration, reaction temperature, and light intensity.

One of them is the utilization of different inorganic ions that influence the photocatalytic activity of TiO_2 and photocatalytic degradation [2]. Reduction of nitrate ions is known to occur at heterogeneous photocatalysis in the presence of oxygen. Nevertheless, these ions compete with oxygen for photogenerated holes and nitrate is considered an inert species in TiO_2 photocatalytic suspension systems [3].

As nitrate ions only slightly influence the photocatalytic degradation [4], this work presents the experimental results of utilizing HNO_3 and $NaNO_3$ in aqueous solution for the optimization of the photocatalytic activity. Methyl orange, an azo dye, was chosen as an organic compound to be degraded.

2 Materials and Methods

The method applied in the experiments were adapted from Iga et al. [5] in a flow system. A peristaltic bomb pumps the fluid from a 500 mL vessel through silicone hoses, to the tubular reactor. An Hg low-pressure germicide 8 W UV lamp irradiates the fluid, in contact with the lamp, until the reactor is filled with the solution, then flows back to the vessel. The photoreactor material is polyvinyl chloride and has two fluid passages, one at each extremity.

There are many methods for the photocatalysis of organic molecules by TiO_2. This study uses the semiconductor in aqueous suspension at the concentration of 0.10 mg/L. To maintain the pH around 2 and generate conditions for the formation of OH radicals, a mixture of $HNO_3/NaNO_3$ was made. The organic compound concentration for degradation is about 8 mg/L. The temperature was controlled by a hydrostatic bath at 40 °C.

The hydroxyl radical formation was evaluated by the degradation of the organic dye methyl orange. Measurements were performed using UV-Vis spectrophotometer in the range of 200–600 nm. The mineralization of the dye solution was monitored at a wavelength of 510 nm (absorption spectrum maximum). The interval of samples was after every 15 min at the 1-hour experiments and every 4 min at the 16 min tests.

3 Results

The results of this work were fascinating. Iga et al. [5] degrade the samples with a buffer mixture of HCl/NaCl in 1-hour experiments, so we did the same experimental plan. However, at 15 min of the process, the dye was almost entirely degraded. Chloride ions are known to compete for the hydroxyl radicals [2], so perhaps switching the buffer mixture could accelerate the process.

HNO_3 with $NaNO_3$, as Sharcko [6] report is his work, at pH = 2 generate the compound HONO, a hydroxyl generator from nitrate anion. We reason that the photolysis of organic and nitrogen compounds generate a considerable new rate of hydroxyl radicals that oxidizes the dye faster. Reaction 1 summarizes the phenomenon.

$$\cdot OH + NO^{\cdot} \rightarrow HONO \xrightarrow{hv} \cdot OH + NO^{\cdot} \quad (1)$$

Folli et al. [7] investigate the reaction on TiO_2 surface under UV irradiation with NO(g) and shows a significant result for what could be happening at our system. Both H_2O and O_2 could be reacting at the surface of the catalyst generating ·OH radicals, or the NO generated by the nitrate mixture, could react with adsorbed ·OH forming nitric acid. This reaction is the truly sink of hydroxyl radicals at troposphere, however, this compound can be photochemically decomposed back to its components and generate more ·OH, degrading the azo dye.

In attempt to register the reaction progress, the experiment time was decreased to 16 min. This result was performed with the addition of 0.10 mg of TiO_2 in aqueous suspension, supposing that would have a synergy between the nitrous acid for the generation of ·OH. Kontos et al. [8] studied the degradation of NO(g) by TiO_2 films and indicates that NO concentration level is the limiting factor that determines the photocatalytic oxidation yields. For the analysis of this theory were performed tests with and without the semiconductor. Figure 1 shows the degradation at a short period of time, with or absence of titanium in the reaction.

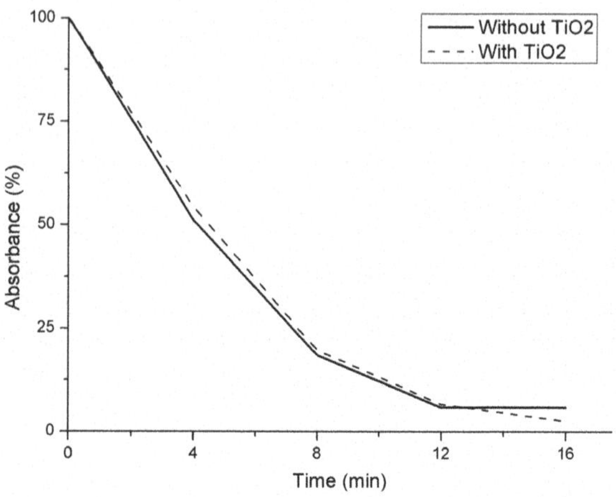

Fig. 1 Degradation of Methyl orange of experiments with and without TiO_2

There is no much difference between the two results, the addition of the catalyst makes the reaction slightly faster, but do not confirm the synergy of HONO hydroxyl generators and the semiconductor, both in acid aqueous solution. The pH didn't vary in the process.

Still, the degradation rate is about 97% at 16 min. This optimization is so important for the advanced oxidative process, knowing that this method hasn't experimented before in literature.

4 Conclusion

Azo dye methyl Orange had a degradation rate of 97% in 16 min of the essay. The solution of $HNO_3/NaNO_3$ acts with great influence at the time of degradation, not only maintaining the pH in an acid medium, but theoretically generating ·OH for the mineralization of the dye. The real mechanism is discussed, but no real NO or HONO formation is evaluated, for the next works or for the academy, this could be an important issue to be studied. AOP's optimized processes could be easily the next revolutionary large-scale treatment tool for effluents around the world, however, the byproducts must be also analyzed.

References

1. Konstantinou, I.K., Albanis, T.A.: TiO_2-assisted photocatalytic degradation of azo dyes in aqueous solution: kinetic and mechanistic investigations: a review. Appl. Catal. B Environ. **49**, 1–14 (2004)
2. Krivec, M., et al.: The nature of chlorine-inhibition of photocatalytic degradation of dichloroacetic acid in a TiO_2-based microreactor. Phys. Chem. Chem. Phys. **16**, 14867–14873 (2014)
3. Parrino, F., Livraghi, S., Giamello, E., Palmisano, L.: The Existence of nitrate radicals in irradiated TiO_2 Aqueous suspensions in the presence of nitrate ions. Angew. Chemie Int. Ed. **57**, 10702–10706 (2018)
4. Chen, H.Y., Zahraa, O., Bouchy, M.: Inhibition of the adsorption and photocatalytic degradation of an organic contaminant in an aqueous suspension of TiO_2 by inorganic ions. J. Photochem. Photobiol. A Chem. **108**, 37–44 (1997)
5. Iga G.D., de Castro Peixoto A.L., Costalonga A.G.C.: Optimization of photocatalytic degradation of methyl orange using TiO_2. In: Iano Y., Arthur R., Saotome O., Vieira Estrela V., Loschi H. (eds) Proceedings of the 3rd Brazilian Technology Symposium. BTSym 2017. Springer, Cham
6. Scharko N.K., Berke A.E., Raff J.D. Release of nitrous acid and nitrogen dioxide from nitrate photolysis in acidic aqueous solutions. Environ. Sci. Technol. 11991–12001 (2014)
7. Folli A., Campbell SB., Anderson J.A., MacPhee D.E.: Role of TiO_2 surface hydration on NO oxidation photo-activity. Photochem. Photobiol. A Chem. 85–93 (2011)
8. Kontos A.G., Katsanaki A., Likodimos V., et al.: Continuous flow photocatalytic oxidation of nitrogen oxides over anodized nanotubular titania films. Chem. Eng. 151–157 (2012)

Stress-Strain Analysis and Optimization of a Parking Duplicator Hanging Grid

Lucas Amaral Costa and Alessandro Corrêa Victorino

Abstract The increasing fleet of automobiles in metropolitan centers challenges humanity for engineering solutions that allow parking multiple vehicles in confined spaces. A possible solution to this problem is in the verticalization of the spaces through the development of high-tech parking systems capable of parking multiple vehicles in limited spaces. A previous paper presented an innovative parking duplicator device launching a possible solution to the stated challenge. The current discussion presents an optimization study of the dimensions of the elevating platform of the parking duplicator device in the prior academic work. In this study, the search was for enhancing the equivalent stress at minimum weight and this was achieved by means of stress-strain analysis and CAD/CAM technics. First, a conceptual design was developed and the design characteristics of the elevating platform, i.e., a grid were defined. Secondly, the structural analysis of the grid was performed in the software MASTAN2®. The obtained results of the stress-strain analysis were used while analytically dimensioning the grid and so determining the optimal thickness of the elements that constitute it. The solutions were refined with the finite element method. Finally, a virtual prototype and a detailed project of the grid was developed on the software SOLIDWORKS® v. 2018. The obtained structure was a simple and inexpensive engineering solution for the elevator platform optimizing the parking system. The methodology applied in the project development of the present discussion demonstrated to be efficient and so it can be applied to the development of virtually any mechanical design.

Keywords Stress-strain analysis · Finite element method · Parking duplicator · Grid

---

L. A. Costa (✉) · A. C. Victorino
Universidade Federal de Minas Gerais, Belo Horizonte, Brazil
e-mail: llucasac@ufmg.br

A. C. Victorino
e-mail: avictorino@ufmg.br

A. C. Victorino
Universitè de Technologie de Compiègne, Compiègne, France

Y. Iano et al. (eds.), *Proceedings of the 4th Brazilian Technology Symposium (BTSym'18)*, Smart Innovation, Systems and Technologies 140,
https://doi.org/10.1007/978-3-030-16053-1_31

1 Introduction

Researching data collected from the housing market since the year 1930, the Polytechnic School of the University of São Paulo (Poli) presented the study [1] showing that about 25%, i.e. one quarter, of the total area built in the metropolis São Paulo is used as parking slots. The increasing fleet of automobiles in metropolitan centers has brought the necessity of parking multiple vehicles in reduced spaces. Offering possible solutions to this parking problem, some parking duplicating devices have been proposed, for example, the patents PI 9906103-1A of 11/18/1999, PI 0002843-6A of 6/14/2000 and US 5110250A of 12/18/1990 present a drive formed by electro-hydraulic or electro-mechanical assemblies. Patents PI 9906103-1A of 11/18/1999 and PI 0002843-6A of 06/14/2000 for example, have a smart design but doesn't present the energy savings provided by the drive with a movable pulley. Offering another possible solution to this parking problem, the present research group has launched a parking duplicator device which was presented in a prior paper [2]. A patent application has been registered under the number BR1020170118142 and an innovative system was launched combining effective characteristics found in previous devices with new attributes. For example, in the launched device the elevating platform on which the vehicles park in order to be elevated is formed by a grid that hangs without support in one of its corners. Figure 1 shows the device and its operation. First, the superior vehicle parks over the elevating platform and the electric hoist elevates it through a steel cable and a movable pulley. Then the inferior vehicle parks under the upper one. It's important to notice that for the superior vehicle to leave, the inferior vehicle must be removed first. Another interesting observation to be done is the fact that the elevating platform is hanging without support on one of

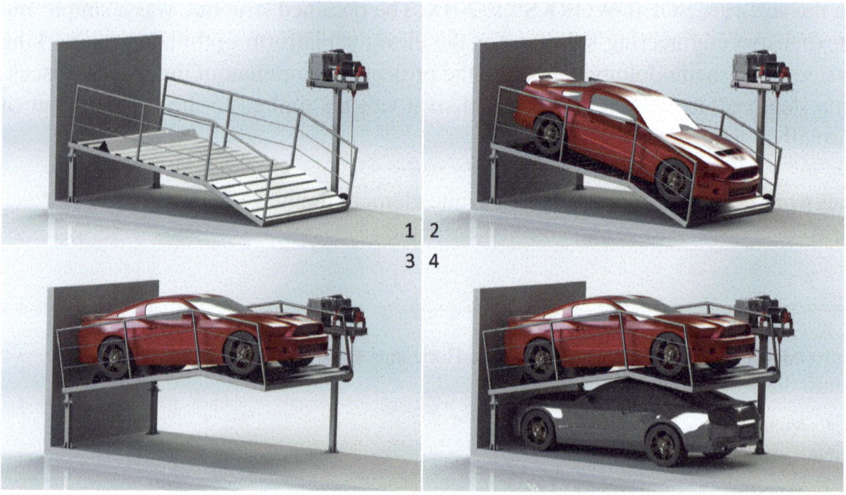

Fig. 1 Operation of the Parking duplicator with focus on the hanging elevating platform

its corners and thus this hanging platform is under a severe bending moment and requires proper dimensioning.

The present discussion presents an optimization study of the dimensions of the elevating platform, i.e., the grid of the parking duplicator device in the previous paper [2] aiming to enhance the equivalent stress at minimum weight by means of stress-strain analysis and CAD/CAM technics.

2 Materials and Methods

The mechanical project methodology [3, 4] adopted on the development of the grid includes the following steps: A. Definition of the Design and Its Parameters; B. Development of the Conceptual Design; C. Analysis of Main Parts; D. Finite Element Analysis and Refining of the Calculations; and E. Elaboration of a Virtual Prototype and Detailed Project.

2.1 Definition of the Design and Its Parameters

The design concept [5] of the elevating platform was primarily developed based on the maximum load conditions applied over the elevating platform and its applicability, regarding domestic or industrial use. So the design characteristics of the platform such as its dimensions and maximum supported load were defined based on the Honda Civic i-VTEC Sport 2.0 and on the observation of suchlike devices available on the market. In this step, it was defined that the maximum weight of a vehicle elevated in the platform would be 2000 kg. To support all the required loads, it was also determined that the architecture of the platform would be of a grid or frame. Thus, parameters such as the shape, disposition, dimensions and the cross-section of the beams that constitute the grid were determined. Different cross-sections were evaluated such as the I, H and rectangular sections. The rectangular section was selected. The limitations of the project were considered as well as similar existing platforms were observed giving an idea of the architecture.

2.2 Development of the Conceptual Design

Once the design and its parameters were determined a sketch of the platform was manually draw on paper. Simple and quick analytical stress-strain calculations took place in this step with rough assumptions giving a previous idea of important dimensions and of the project's economic viability. Then a conceptual design of the grid was developed in the CAD software SOLIDWORKS® v. 2018.

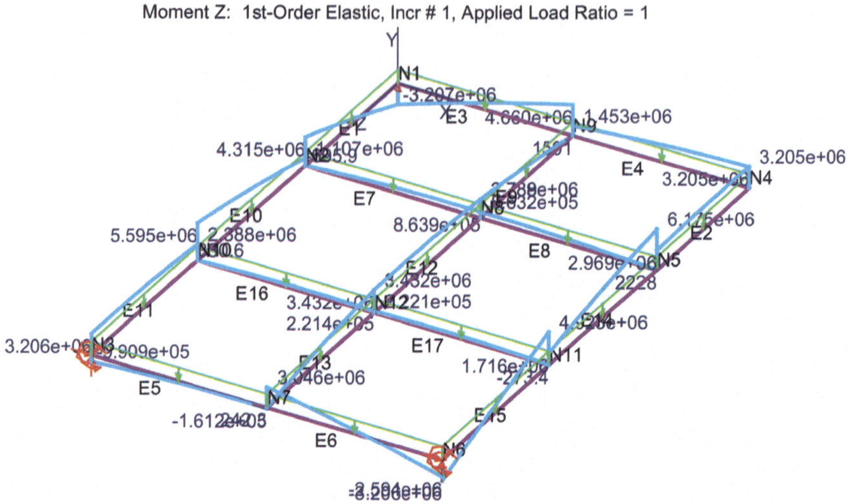

Fig. 2 Grid structural analysis in the software MASTAN2® exhibiting the bending moment

2.3 *Analysis of Main Parts*

In this step, the stress-strain analysis was performed determining the minimum thickness of the rectangular structural profile that constitutes the grid. The load was calculated as transmitted from the elevated vehicle to the folded sheet metal over which the upper vehicle is parked and so, finally transmitted to the grid as an uniformly distributed load. The load condition primarily subjects the grid to shear stress, twisting and bending moments. According to [6] these strains acting on the grid were calculated in the structural analysis software MASTAN2® as shown in Fig. 2.

The designed grid has shown to be relatively well resistant to the shear stress and the twisting moment. The bending moment is the limiting strain acting on it. Once the grid is dimensioned to resist the bending moment, the conditions required by the shear stress and the torsion moment are simultaneously attended. Thus, as described in [7, 8], the grid was primarily dimensioned to resist the bending moment through (1) where M_f symbolizes the bending moment, ω_{x0} represents the resistance module and σ_f symbolizes the permissible bending stress of the steel.

$$M_f = \omega_{x0\,x}\sigma_f \tag{1}$$

The thickness of the rectangular profile of the beams that form the grid is chosen according to the resistance modulus of the profile, which is selected to guarantee that the structural element won't suffer plastic deformation. If the selected beam has a resistance module higher than the one calculated through (1), there will only be elastic deformations and the mechanical component won't fail. The bending moment was calculated in the software MASTAN2® as aforementioned and the permissible

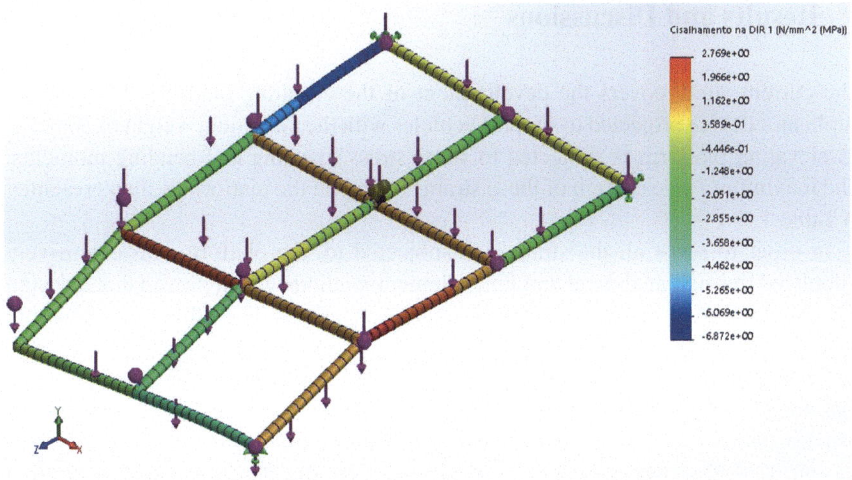

Fig. 3 FEM simulation on the software SOLIDWORKS® Simulation showing the shear stress acting on the grid

bending stress is the yield strength of the carbon ASTM A36 steel divided per a safety coefficient.

2.4 Finite Element Analysis and Refining of the Calculations

The analytic solution found in the previous step was refined in this step by means of the Finite Element Method (FEM) resulting in a lighter and thus cheaper structural profile that still resists to the required load conditions. The software SOLIDWORKS® Simulation was used as shown in Fig. 3.

A variety of commercial rectangular profile's thickness was simulated and the thinner one that attended the load conditions was then selected.

2.5 Development of a Virtual Prototype and Detailed Project

Once all the grid parameters were well defined, a virtual prototype was finally developed for the grid on the CAD software SOLIDWORKS® v. 2018 and a detailed project was elaborated containing all the dimensions and assembly informations.

3 Results and Discussions

The current study covers the development of the elevating platform of a parking duplicator device projected to support vehicles with the maximum weight of 2000 kg. Its elevating platform is subjected to shear stress, twisting and bending moments. The maximum value of each of these strains acting on the platform is then presented in Table 1.

In order to resist all the strain it is subjected to, the platform was extensively calculated through analytical and finite element methods which proved that a design of a structural steel frame, i.e., a grid as exhibited in Fig. 4 is an efficient architecture for it.

Different cross-sections were evaluated for the grid design. The I and H cross-sections have been shown to be very efficient sections typically used in steel construction, but as the rectangular cross-section is more suitable for domestic use, it was chosen. Table 2 shows the technical specifications of the elevating platform where it is interesting to notice that the platform is made of ASTM A36 steel and it weighs 337.73 kg, which corresponds to 16.9% of the weight it is dimensioned to support - 2000 kg. The thickness of the rectangular profile is 5 mm.

Table 1 Maximum strains acting on the grid

Parameter	Value
Max. shear stress	3.55 N/mm^2
Max. twisting moment	3208 Nm
Max. bending moment	6175 Nm

Fig. 4 Grid of the elevating platform

Table 2 Technical Details of the Grid

Parameter	Value
Max. possible weight	2000 kg
Material	ASTM A36 steel
Weight	337.73 kg
Structural profile of the beams	Rectangular
Thickness of the structural profile	5 mm

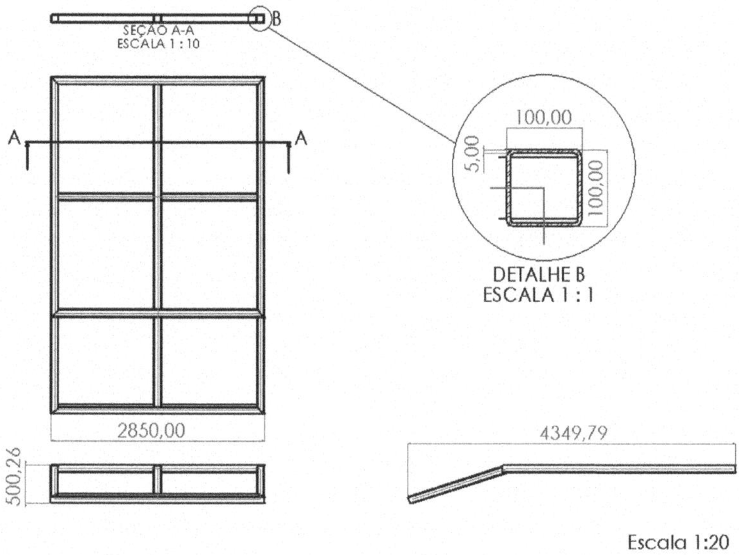

Fig. 5 Important dimensions of the grid

It's important to consider that the elevating platform is formed by the grid and a folded sheet metal that goes over the grid. As it can be viewed in Fig. 1, the folded sheet can support bending moment as well as other strains applied on the structure, but the grid is the element analyzed, dimensioned and projected to properly resist to all the required strains in a way that doesn't leave the folded sheet under these load conditions. Figure 5 shows important dimension of the grid.

It's also interesting to notice that the platform is formed by two planes in an angle reducing the linear space occupied by the elevated vehicle. This attribute was idealized in the step "A. Definition of the Design and Its Parameters" as described in the topic "2. Materials and Methods". It's also interesting to notice that the methodology described in this same topic is not so linear as it appears to be. It's dynamic. The steps are performed and reperformed as a previous or posterior step brings a solution that allows improvement of an already accomplished step. The very interaction with the project itself makes new ideas to arise. The methodology applied in the present study has shown to be an efficient methodology to develop the mechanical project of the present discussion and so it can be adopted on the development of virtually any mechanical project.

Considering these aspects, the obtained result is a good engineering solution compared to the similar equipment found on the market, so far patented or in the state-of-the-art as referred in [2]. References [9, 10] show suchlike studies of platforms engineered with structural sandwich plate systems where the equivalent stress was enhanced at minimum weight offering other attractive design solutions. Besides this good equivalent stress to weight relation, an architecture that spares space when employed in the parking system and that had a simpler assembly was also sought in the present discussion finally leading to the solution presented in the Fig. 4. The project will also conform to national and international standards aiming for security and sustainability, i.e., ABNT, NR-10, NR-12, NR-17, ISO 9000 and ISO 14000.

The development of the grid was an important step in the overall development of the parking duplicator device. The next step in the development of the duplicator will be adopting it with an assistive technology that will help the driver to successfully accomplish the task of parking. The assistance will help the driver, e.g., guiding to the best position on the parking slot and signaling if a possible collision is imminent.

4 Conclusion

A high-tech solution to save space when parking vehicles has been proven to be necessary and in a former paper, an innovative design of a parking duplicator device was launched as one possible solution to the presented parking problem. Then an optimization of the elevating platform of this parking duplicator was sought at the beginning of the current study and so it was performed. The designed grid provides an efficient mechanical solution for the elevating platform of the duplicator having been extensively evaluated in its maximal loading conditions in order to be dimensioned in its limit parameters minimizing the manufacturing costs and simplifying the assembly. In view of these results, we conclude that the project conforms to the expected specifications allowing the development of a product that can be widely used in urban centers conceiving wide parking possibilities. The methodology applied in the project development of the present discussion demonstrated to be efficient and so it can be applied to the development of virtually any mechanical design.

Acknowledgements The authors are gratified for the support of Programa de Pós-Graduação da Universidade Federal de Minas Gerais, Coordenação de Aperfeiçoamento de Pessoal de Nível Superior (CAPES), Fundação de Amparo à Pesquisa do Estado de Minas Gerais (FAPEMIG), Conselho Nacional de Desenvolvimento Científico e Tecnológico (CNPq) and Fundação de Desenvolvimento da Pesquisa (Fundep) all which had a crucial role in the development of the present academic work.

References

1. Leite Júnior, H.F., Alencar, C.T., John, V.M.: Evolution of Space for the Automobiles in Relation to the Total Built Area of the Buildings from São Paulo," São Paulo: 11ª Conferência Internacional da LARES, in press (2011)
2. Costa, L.A., Ferreira, M.F.S., Ferreira, A.G., Teixeira, L.F., Costa, J.M.: Development of a Parking Duplicator Mechanical Device Driven with Pulley, Proceedings, ISSN 2447-8326. V.1, in press BTSym (2017)
3. Ulrich, K.T., Eppinger, S.D.: Product Design and Development, 5th edn. McGraw-Hill Education (2011)
4. Dym, C.L., Little, P.: Engineering Design: A Project Based Introduction, 4th edn. Wiley, Hoboken (2003)
5. Tidd, J., Bessant, J., Pavitti, K.: Gestão da Inovação, 5th edn. Bookman, Porto Alegre (2015)
6. Zienkiewicz, O.C., Taylor, R.L.: The Finite Element Method, vol. 1, 5th edn. The Basis, Butterworth-Heinemann, Oxford, Massachusetts (2000)
7. Hibbeler, R.C.: Structural Analysis, 4th edn. Prentice-Hall, Englewood Cliffs, New Jersey (1998)
8. Timoshenko, S.P., Young, D.H.: Theory of Structures, 2nd edn. McGraw-Hill, New York (1965)
9. Tale, M.R., Sontakke, K.R., Satav, P.K.: Design Analysis and Optimization of Industrial Lift-platform Base on Composite Structural Sandwich Plate, IJRAT, E-ISSN 2321-9637, in press
10. Gopichand, A., Krishnaiah, G., Krishnaveni, D., Diwakar Reddy, V.: Numerical Simulation of Steelsandwich Plate System (SPS) Floor, IJIRSET, ISSN 2319-8753, in press

Physico-Chemical Analysis of Biodiesel Obtained from Blends of Virgin Castor Oil and Biodiesel Obtained from Mixture of Oil

Daniela Helena Pelegrine Guimarães, Lucas Romano de Aguirre, Bruna Caroline Gonçalves and Maria Thereza de Moraes Gomes Rosa

Abstract Castor oil is one of the most important potential feedstocks for biodiesel production, but the product obtained has problems within the specific density and viscosity specifications. A wat to improve these properties is to produce biodiesel form blends of castor biodiesel oil with other biodiesel oils. On the other hand, the recycling of waste oils and fats to obtain biodiesel can help in the economic development of several regions in Brazil. This work includes producing biodiesel from blends of virgin castor oil and waste frying oil. Biodiesel production from blends of virgin castor oil (VCO) and waste cooking oil (WCO) from ethylic route and alkaline catalysis was studied in this work. Physico-chemical characterization of VCO (**B1**), WCO (**B2**) and three different blends (**B3**: 50%VCO + 50%WCO; **B4**: 25%VCO + 75%WCO; **B5**: 75%VCO + 25%WCO) were evaluated. Initial blend was characterized by acidity level, free fatty acid, saponification index, and density. Iodine and cetane index, glycerin, absence of triglycerides and rheological behavior were determined for obtained biodiesel. The results indicated that the acidity level dropped considerably after the transesterification, however only biodiesel from B1, B4 and B5 showed acidity index in accordance with the standard parameters. Nevertheless, B1 viscosity and cetane number lied outside the legislation. All samples were within the specified standard for iodine index and density. Despite the fact that B4 and B5 results were within the legislation for biodiesel, B4 would be a better alternative due to higher quantity of WCO.

Keywords Biodiesel · Oilseeds · Castor beans · Transesterification · Waste cooking oil

D. H. P. Guimarães (✉) · L. R. de Aguirre · B. C. Gonçalves
Lorena Engineering School (EEL-USP), Lorena, Brazil
e-mail: dhguima@usp.br

M. T. de Moraes Gomes Rosa
Mackenzie Presbyterian University, Campinas, Brazil

Y. Iano et al. (eds.), *Proceedings of the 4th Brazilian Technology Symposium (BTSym'18)*, Smart Innovation, Systems and Technologies 140,
https://doi.org/10.1007/978-3-030-16053-1_32

1 Introduction

The oil crisis, in the 70s and 80s, has given rise to several studies concerning the pyrolysis of triglycerides. With the risk of depletion of some energy sources, especially fossil energy with no prospects of renewal, there is a motivation for the development of technologies that allow the use of renewable energy sources [1].

In addition, the growing concern with environmental issues should influence the decisions of the leaders regarding the possibilities of alternative energy sources use. In this regard, fossil fuels such as diesel oil and gasoline are the most criticized. These fuels belong to the category of non-biodegradable fuels. Moreover, pollutants, such as CO_2, are released into the atmosphere during their burn, besides the possibility of sulfur oxides emission. Once the planet cannot assimilate all CO_2 released in the long term, the accumulation of this and other gases cause the greenhouse effect increasing [2–5].

Biodiesel is a vegetable oils and fats derived fuel, commonly obtained by transesterification using a catalyst agent. It is considerate a renewable and biodegradable fuel, with low emission of polluting compounds into the atmosphere. Therefore, biodiesel has been seen as the fuel of the future. Indeed, biodiesel totally or partially replaces petroleum diesel fuel in automotive diesel engines (trucks, tractors, vans, cars, etc.) or stationary (generators of electricity, heat, etc.) and can be used pure or mixed with diesel in different proportions [6–8].

Biodiesel can be produced from any source of fatty acids, however not all of them make the process industrially feasible. The use of waste oils and fats such as frying oil, refining lees, sewage grease, vegetable oils or animal fats, which are out of legislation standards and useless for another ends is promising, avoiding improper disposal [9]. The recycling of waste oils and fats for biodiesel obtaining can cooperate with the economic development of several regions of Brazil. Moreover, the consumption of biodiesel and its blends can reduce petrol dependence, as well as generate alternative jobs in less favorable geographic areas to other economic activities [10, 11].

Nowadays, issues associated with the environment have worsened significantly, in view of the disorderly increase in population and the subsequent generation of large amounts of complex recycling waste. At present, to ensure survival on the planet, every effort should be made in order to recycle most of the generated waste, including the oil used in frying. This material can be transformed in biodiesel by transesterification and reused as fuel. Considering that after successive re-use in frying the oil has its fatty acid composition and physico-chemical characteristics altered, it becomesinadequate for food processing even when subjected to purification [12, 13].

In this study, the biodiesel was obtained from five different initial blends (**B1**: VCO;**B2**: WCO; **B3**: 50%VCO + 50%WCO; **B4**: 25%VCO + 75%WCO; **B5**: 75%VCO + 25%WCO), and the physical, chemical and rheological properties of the biodiesel produced from the blending ratios between castor and waste frying oils were analyzed.

2 Materials and Methodology

2.1 *Raw Material for Biodiesel Obtaining*

The raw material utilized in the transesterification reaction for biodiesel obtaining was prepared from a blend of VCO and WCO in different proportions (Table 1). WFO samples were provided by an event company located in the city of São Paulo, which prepares snacks by immersion in heated soybean oil. The WCO was filtered in two steps. Larger particles and food residues were removed through a pre-filter system consisting of a steel wool pad attached to a nylon sieve, while smaller particles were vacuum-filtered with Whatman filter paper no 2. The resulting filtrate was vacuum-dried for 90 min at 100 °C. The experiment were performed in triplicate.

2.2 *Transesterification Reaction for Biodiesel Obtaining from the Blends*

Transesterification was carried out in a magnetic stirrer with heating (Model 753A, FISATOM), where 100 mL of each formulation (B1, B2, B3, B4 and B5) was heated to 55 °C. After that, 33 mL ethyl alcohol and 1 g KOH (catalyst) were added to the reactional mixture with stirring in order to keep system uniform temperature (55 °C) and homogeneity for 80 min. After that, the mixture was cooled down at room temperature and distilled under reduced pressure in order to remove ethyl alcohol excess. The resulting ester/glycerin mixture was transferred to a separating funnel and kept standing for 24 h, yielding two phases: biodiesel (upper phase) and glycerin

Table 1 Blend composition used as raw material for biodiesel obtaining and physico-chemical parameters

Blend	Composition	Acidity level (mg_{KOH}/g_{oil})	Free fatty acid(%)	Saponification index (mg_{KOH}/g_{oil})	Density (g/cm^3)
B1	100% VCO	0.24 ± 0.18	1.52 ± 0.42	215.03 ± 0.56	0.96 ± 0.02
B2	100% WFO	1.70 ± 0.20	6.44 ± 0.77	328.99 ± 23.54	0.92 ± 0.01
B3	50% VCO + 50% WFO	1.11 ± 0.40	4.46 ± 0.39	311.92 ± 21.71	0.94 ± 0.01
B4	25% VCO + 75% WFO	1.82 ± 0.02	4.76 ± 0.40	296.15 ± 16.33	0.93 ± 0.01
B5	75% VCO + 25% WFO	1.12 ± 0.01	3.90 ± 1.19	298.32 ± 14.17	0.95 ± 0.01

VCO virgin castor oil; *WFO* waste frying oil

(lower phase). After glycerin removal, biodiesel was washed and heated at 100 °C for 15 min for water and residual ethyl alcohol removal.

2.3 Calculations

The yield of each biodiesel sample was calculated according to Eq. (1):

$$\text{Yield}\ (\%) = \left(\frac{V_{\text{biodiesel}}}{V_{\text{initial}}}\right)100 \tag{1}$$

where:

$V_{\text{biodiesel}}$ Volume of biodiesel obtained;
V_{initial} Volumeof reactional mixture (oil blend + ethyl alcohol).

2.4 Physico-Chemical Characterization

The initial blend was characterized according to acidity level (ASTM D664), saponification index (ASTM D1387) and density(ASTM D1298). The resulting biodiesel was characterized with respect to the acidity level (ASTM D664), iodine (ASTM D1959) andcetane index (ASTM D4737), density (ASTM D1298), glycerin [14]. The absence of triglycerides was carried outby means of proton nuclear magnetic resonance (NMR). The NMR spectra recorded on a spectrometer (Varian, Mercury-300 MHz). The rheological behavior of the blends and resulting biodiesel (at 40 °C) was analyzed in a rotating rheometer (Brookfield, LV—DVII).

3 Results and Discussion

Table 1 shows thephysico-chemical parameters of the tested blends, including average and standard deviation. According to Van Gerpen et al. [15] 3 is the maximum acidity level for industrial oil nº 1. Moreover, an alkali-catalyzed transesterification reactioncan be carried outin cases where the oil showsacidity level of less than 3%, in order to assure the efficiency of the conversion and avoid soap formation. Considering the results, all the blends showed acidity within the limit for classification as industrial oil, allowing the alkalinecatalytic route for the transesterification reaction.

Virgin castor oil (B1) showed lower acidity level (0.24 mg_{KOH}/g_{oil}) and free fatty acids content (1.52%) when compared to WCO (B2) (1.70 mg_{KOH}/g_{oil} and 6.44%, respectively) (Table 1). Moreover, the acidity level was higher in the blend with 75% WCO (B4/1.82 mg_{KOH}/g_{oil}), which also showed higher amount of free fat acids

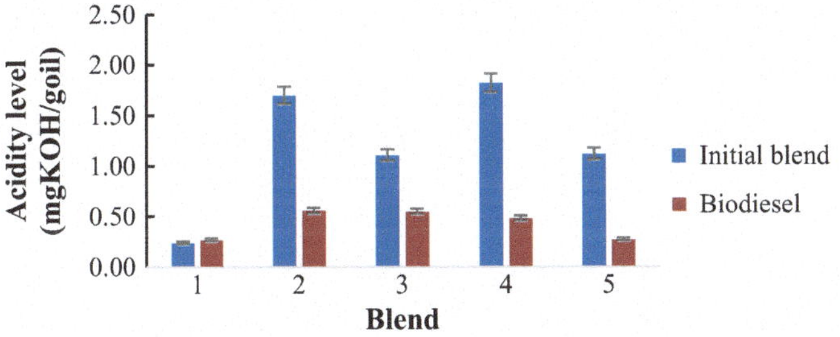

Fig. 1 Acidity index for initial blend and biodiesel obtained after the transesterification

(4.76%). All samples showed an acidity level inferior than the maximum allowed by the legislation (2%) [16] even though they had been exposed to adverse conditions of the frying process which involves heat and light exposure. The higher WCO acidity level is explained by the break down and release of the triglycerides chains whereas free fatty acids are produced by oxidation reactions which are intensified during the frying process [17].

The saponification index (Table 1) resulted in higher values than those found in the literature, for all blends. The first quality oil should have a saponification index ranging from 177 to 187 mg_{KOH}/g, however these values are established for refined oils and not virgin oils. The VCO (B1) density is consistent with Lôbo, Ferreia and Cruz [18] who observed density values ranging from 0.93 to 0.98 g/cm^3. However, according to the Brazilian ANP 07/2008 standard, the density specification for use in engines is within the range 0.85–0.90 g/cm^3. Valente et al. [19] observed this density specification in blends with ca. 70% biodiesel from waste cooking oil or from castor oil in diesel fuel No. 2.

Biodiesel yieldandphysico-chemical parameters are presented in Table 2. Transesterification yield ranged from 28.46% (B1) to 59.23% (B4). The lower reaction yield observed in B1 is attributed to the alkaline catalyst reaction with VCOfree fatty acids, whichresulted in soap. Yields higher than 50% of transesterification were observed for the other blends (B4 > B2 > B5 > B3).

Except for B1, the acidity level of the blends drastically reduced after the transesterification process (Fig. 1). According to Anvisa [16] the maximum acidity level for biodiesel is 0.50 mgKOH/goil. As shown in Table 2, B1, B4 and B5 are in accordance with the legislation. On the other hand, B2 and B3 showed acidity level slightly above the specification (0.56 and 0.55, respectively). All the samples showed iodine index within the specified standard in a maximum value of 120 g_{I2}/g_{oil} [18], as well as the density [20].

High amount of free glycerin was observed in blends composed of high VCO content (B1, B3 and B5). Itcan be explained by the low sensibility of the method. This author has analyzed the oligomerization of castor bean biodiesel and observed dimers formation after the reaction between the hydroxyl group of ricinoleic acid

Table 2 Volumetric yield of the transesterification reaction and biodiesel physico-chemical parameters

Blend	Composition	TransterificationYield (%)	Aciditylevel (mg_{KOH}/g_{oil})	Iodine Index (g_{I2}/g_{oil})	Density (g/cm^3)	Freeglycerin (%)	Cetanenumber	Conversion rate[a] (%)
B1	100% VCO	28.46	0.27 ± 0.01	110.86 ± 2.32	0.95 ± 0.01	0.64 ± 0.01	24.46	40.08
B2	100% WCO	56.15	0.56 ± 0.01	99.05 ± 0.98	0.92 ± 0.01	0.02 ± 0.01	33.39	23.82
B3	50% VCO + 50% WCO	52.31	0.55 ± 0.02	77.99 ± 0.13	0.92 ± 0.01	0.28 ± 0.04	31.90	61.49
B4	25% VCO + 75% WCO	59.21	0.48 ± 0.01	93.01 ± 4.81	0.92 ± 0,01	0.03 ± 0.03	30.99	32.68
B5	75% VCO + 25% WCO	53.08	0.27 ± 0.01	76.30 ± 3.49	0.94 ± 0.09	0.33 ± 0.01	31.19	40.53

[a]Obtained from spectrum analysis (NMR)

Table 3 Rheological parameters, cetane number and conversion rate of different biodiesel formulations

Biodiesel samples	Power			Casson		
	K	n	r^2	K_{OC}	K	r^2
B1	0.95 ± 0.13	0.98 ± 0.01	0.99	0.03 ± 0.00	0.76 ± 0.07	0.99
B2	0.22 ± 0.11	0.99 ± 0.02	0.99	0.02 ± 0.00	0.47 ± 0.12	0.99
B3	0.21 ± 0.19	0.97 ± 0.03	0.99	0.09 ± 0.00	0.43 ± 0.05	0.99
B4	0.39 ± 0.12	1.01 ± 0.03	0.99	0.02 ± 0.00	0.63 ± 0.14	0.99
B5	0.28 ± 0.15	0.98 ± 0.04	0.99	0.06 ± 0.00	0.50 ± 0.18	0.99

K consistency coefficient, *n* behavior index, r^2 correlation coefficient, K_{OC} Casson yield stress

andcarboxyl group of the esters. These dimers are detected as diglycerides resulting in values above the expectation of the process.

B1 was the only formulation that presented a cetane number which lie outside the allowed range (30–45) (Table 2). Lôbo, Ferreia and Cruz [20] reported that the cetane number increases with the length of the non-branched carbon chain. Ricinoleic is the predominant fatty acid in VOC and pursuit a short branched chain, therefore a low cetane number for B1 was expected. The results found herein for cetane index is below the cetane levels reported by Valente et al. [19]. These authors observed that the oil source and cooking process have some impact on cetane number due to the effect on the waste cooking oil molecular structure.

There was no significant difference between the results for proton nuclear magnetic resonance for B1 and B5 conversion rate (Table 2). However, increased amounts of WCO in the blend led to lower conversion rates. The conversion was greater when B3 was used as raw material for biodiesel production. A possible explanation for this would be the fact that the NMR methodology was unable to separate and quantify the individual sterols [18].

All biodiesel samples showed a rheological behavior of a Newtonian fluid, since the values for behavior index (n) Casson yield stress (K_{OC}) are close to 1 and 0, respectively (Table 3). Thus, the rheogram can be represented by a straight line through the origin, represented in Fig. 2. According to the ANP specification (ANP resolution 42, 2004), the kinematic viscosity of biodiesel must be 5.0 mm^2/s at 40 °C. In this way, except for B1, the biodiesel kinematic viscosity values are within the normalization.

4 Conclusions

The significant reduction in the kinematic viscosity of biodiesel, obtained when used portions of WCO, demonstrated the great feasibility of reusingthe oil discarded from frying for the production of that fuel.

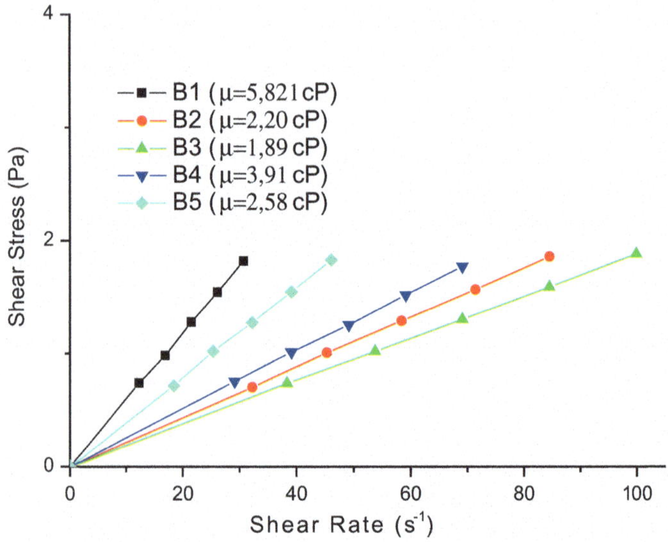

Fig. 2 Reograms of biodiesel formulations

Considering the physico-chemical characteristics, VOC showed the lower transesterification rate. The blend composed by 50% VOC + 50% WCO (B3) would be the best formulation for biodiesel obtaining, showing high transesterification yield (52.31%), iodine index and density within the standards in addition to higher conversion rate (61.49%). However, B3 showed an acidity level slightly above the allowed by Anvisa (0.55 mgKOH/goil). In this way, B4 (25% VOC + 75% WFO) would be an alternative of formulation for biodiesel obtaining, once all parameters are within the legislation. Moreover, higher amounts of WCO incorporated into the blend make biofuel attractive with respect to oil reuse, in addition to reducing the total cost of the process and avoiding the wrong disposal of waste to the environment.

Acknowledgements The authors would like to thank Fapesp - Brazil (Process number 2014/27341-5) and the Lorena Engineering School (EEL/USP).

References

1. Martín, M., Grossmann, I.E.: On the synthesis of sustainable biorefineries. Ind Eng. Chem. Res **52**, 3044–3083 (2012)
2. Sharma, Y.C., Singh, B., Upadhyay, S.N.: Advancements in development and characterization of biodiesel: a review. Fuel **87**(12), 2355–2737 (2008)
3. Basha, A.S., Gopal, K.R., Jebaraj, S.: A review on biodiesel production. Combust Emissions Performance Renew Sustain. Rev. **13**, 1628–1639 (2009)
4. Fitzpatrick, M., Champagne, P., Cunnigham, M.F., Whitney, R.A.: A biorrefinary processing perspective: treatment lignocellulosic materials for the production of value-add products.

Bioresour. Technol. **101**(23), 8915–8922 (2010)
5. Dogaris, I., Mamma, D., Kekos, D.: Biotechnological production of ethanol from renewable resources by Neurosporacrassa: an alternative to conventional yeast fermentations? Appl. Microbiol. Biotechnol. **97**, 1457–1473 (2013)
6. Costa Neto, P.R., Rossi, L.F.S., Produção de Biocombustível Alternativo ao Óleo Diesel Através da Transesterificação de Óleo de Soja Usado em Frituras, Química Nova, **23**(4), 531–537 (2000)
7. Ferrari, A.R., Oliveira, V.S., Seabio, A.: Biodieselde Soja: Taxa de Conversão em Ésteres Etílicos. Caracterização Físico-Química e Consumo em Gerador De Energia, Química Nova **28**(1), 19–23 (2005)
8. Francisco, E.C., Franco, T.T., Maroneze, M.M., Zepka, L.Q., Lopes, E.J.: ThirdGeneration biodiesel production from microalgae. RevistaCiência Rural **45**(2), 349–355 (2005)
9. Godos, I., Vargas, V.A., Blanco, S, Gonzáles, M.C.G., Soto, R., García-Encina, P.A., Becares, E., Munhoz, R. A.: Comparative evaluation of microalgae for the degradation of piggery wastewater under photosynthetic oxynenation. Bioresour. Technol. **101**(14), 5150–5158 (2009)
10. Demirbras, A., Demirbras, M.F.: Importance of Algae as a source of biodiesel. Energy Convers. Manag. **52**, 163–170 (2011)
11. Silva, L.T., Gouveia, L., ReisS.A.: Integrated microbial processes for biofuels and high value-added products: the way to improve the cost effectiveness of biofuel production. Appl. Microbiol. Biotechnol. **98**, 1043–1053 (2014)
12. Brennan, L., Owende, P.: Biofuels form microalgae: a review of technologies for production, processing and extractions of biofuels and co-products. Renew. Sustain. Energy Rev. **14**(2), 557–579 (2010)
13. Xia, C., Zhang, J., Hu, B.: A new cultivation method for microbial oil production: cell pelletization and lipid accumulation by mucorcircinelloides. Biotechnol. Biofuels **4**, 4–15 (2011)
14. Spudeit, D.A.: Determinação de parâmetros de qualidade do biodiesel utilizando espectrofotometria UV/VIS. https://repositorio.ufsc.br/bitstream/handle/123456789/99781/Daniel.pdf?sequence=1 (2009)
15. Van Gerpen, J., Shanks, B., Pruszko, R., Clements, D., Knothe, G.: Biodiesel Production Technology, Ed. National Renewable Energy Laboratory, Colorado, USA, p. 100 (2004)
16. ANVISA Homepage.Consulta Pública n° 85, de 13 de dezembro de 2004, D.O.U de 17/12/2004, http://www.anvisa.gov.br/alimentos/informes/11_051004.htm. 2015/09/10
17. Silva, T.A.R., Neto, W.B.: Estudo da Redução da Acidez do Óleo Residual para a Produção de Biodiesel Utilizando Planejamento Fatorial Fracionado. Revista Virtual Química **5**(5), 828–839 (2013)
18. Lôbo, I.P., Ferreira, S.L., Cruz, R.S.: Biodiesel: Parâmetros de Qualidade e Métodos Analíticos. Quim. Nova **32**(6), 1596–1608 (2009)
19. Valente, O.S., Pasa, V.M.D., Belchior, C.R.P., Sodré, J.R.: Physical–chemical properties of waste cooking oil biodiesel and castor oil biodiesel blends. Fuel **90**(4), 1700–1702 (2011)
20. Moretto, E., Fett, R., Tecnologia de Óleos e Gorduras Vegetais na Indústria de Alimentos, Ed. Varela, S. Paulo, Brazil, 120 p (1998)

Sensorial and Rheological Parameters of Cured Cheese Produced with Goat Milk

Daniela Helena Pelegrine Guimarães, Débora Barros and Maria Thereza de Moraes Gomes Rosa

Abstract The goat breeding has increased the participation on the agricultural scenario, where goat's milk and its derivatives have conquered and maintained the new markets. The present work aims to analyze sensorial and rheological parameters of cured cheese produced with goat milk, by acid coagulation. The same product was prepared in a conventional manner, which is, with cow's milk. The products were analyzed with regard to the physical, chemical, texture and sensorial properties. The results presented in sensorial tests demonstrate that only for the texture, the products presented no significant difference, although rheological tests have found the penetration of about 37% higher in the cheese prepared from goat's milk. The other sensorial attributes showed significant differences between the two products, where the cheese made from goat's milk had slight preference, by the tasters.

Keywords Dairy products · Goat breeding · Texture · Sensorial attributes

1 Introduction

Milk is a complex blend, consisting by an emulsion composed of lipids and a colloidal dispersion of proteins, lactose, minerals, vitamins, enzymes and certain organic compounds [1]. The opaque color of milk is mainly due to the dispersion of proteins and calcium salts [2]. Even, at the global level, cow milk remains the most abundant and the most consumed by humans, in the last years, an increase on goat milk production has occurred in many developing countries, indicating an interest increasing on this activity, which plays a key role in livestock production [3, 4]. The success in the

D. H. P. Guimarães (✉)
Lorena Engineering School (EEL-USP), Lorena, Brazil
e-mail: dhguima@usp.br

D. Barros
Taubate University (UNITAU), Taubaté, Brazil

M. T. de Moraes Gomes Rosa
Mackenzie Presbyterian University, Campinas, Brazil

Y. Iano et al. (eds.), *Proceedings of the 4th Brazilian Technology Symposium (BTSym'18)*, Smart Innovation, Systems and Technologies 140,
https://doi.org/10.1007/978-3-030-16053-1_33

market and growing consumer interest became the goat milk profitable and popular as alternatives to bovine milk [5].

The advantage in consuming goat milk is its lesser particles of fat, facilitating digestion process, also recommended on allergic [6–8]. In the past two decades, the volume of goat milk production increased dramatically and worldwide production of goat milk has been risen by approximately 60% between 1993 and 2013 [9].

In this context, the large amount of milk produced and the perishability of the product resulted in the expansion of the industrialization of goat's milk, worldwide, making possible higher monthly invoices, by adding value to the milk produced [10, 11]. Among the derivatives of goat milk, the various types of cheese arouse greater technological and economic interest. Compared to bovine casein, casein present in goat's milk contains more inorganic calcium and phosphorus, although it is less stable on heating. Consequently, the coagulation will be faster, the yield of the cheese will be lower and the gel formed is more easily absorbed [12].

Cured cheese, obtained by milk coagulation with acid or enzyme, is one of the most popular in the world, presenting high moisture content, white mass and smooth texture. Cheese texture is an essential quality attribute that determine its acceptability. This property influences its appearance, taste, sensory impressions in the mouth and hands [13].

Since during processing, foods are subjected to forces such as stress and shear during the operations of size reduction, pressing, extrusion, heating and cooling, the study of rheology is of fundamental importance in the acceptance of food products because it is associated to the sensorial properties; these characteristics are perceived mainly by the touch [14]. According to Pelegrine & Gasparetto [15], the importance of the knowledge of the rheological behavior of foods is as quality measurements, in projects, evaluation and operation of food processing equipment.

Viscosity is the rheological property that characterizes a liquid food and, for cheese, the rheological property is its texture, considered an important attribute of the food, considering that it affects the process, the storage, the handling and acceptance of the product by the consumer. The texture of the cheese is an essential attribute for its quality, being one of the four factors that determine its acceptability. This property influences its appearance, taste, sensory impressions in the mouth and various hands suitability. By definition, the consistency of the cheese must be such that even when extracted from its container or refrigerator it is able to remain in the solid state [16].

Taking into account the increasing participation of the goat breeding on the agricultural scenario, the objective of this work includes the development of cured cheese, processed from goat milk goat, by acid coagulation. The same product was prepared in a conventional manner, which is, with cow's milk. For the definition of the product with the characteristics desired by the consumers were carried out sensorial acceptance tests, where the evaluated attributes were: flavor, color and texture.

The main objective of this work is to correlate the sensorial parameters of the cheese to the measurement of its texture and water activity, being such measures measured in the texturometer and water meter respectively.

2 Materials and Methodology

Cheeses were prepared using goat and cow's milk. The milk samples were analyzed through physicochemical analysis of pH, method N. 4.7.1 [17], acidity, method N.4.7.2 [17], density at 15 °C, method 16196 [17], fat content [18], protein content, method 380120 [17] and viscosity [19]; these analyses were all carried out in triplicate.

Then the milk (goat or cow) passed in a double tube pasteurizer. Pasteurized milk was adjusted for fat content (3.4% fat), by adding powdered milk and then added calcium chloride in the ratio of 40 mg per 100 L of milk. The cheese was processed according to [20], heated up to 82 °C in a jacketed stainless steel pan and, once this temperature was reached, 0.3% lactic acid in the ratio 1/9 (w/w) was added. The mixture was coagulated (by stirring), and then decanted. It was used cotton cloths to solve and squeeze the rennet, followed by several washes with cold water, until reach pH of 5.2–5.5 range. Cut up the curd into small pieces, in order to obtain grains of 1.5 cm. The curd was slowly stirred (40 min) and, then, heating to 37 °C. The final point of the stirring occurred when the grains were firm. Thereafter, the whole serum was removed and the desorbed mass was then pressed into the tank for 15 min until entire elimination of serum, placed in cylindrical form for 30 min. The pressed cheese went to the salting chamber (brine of 20% salt, at 12 °C, during 24 h), packed and stored in a cold room at 3 °C. Once firm, the serum was removed from the granules and the desorbed mass was crushed and pressed into cylindrical forms for 30 min. Finally, the cheese was packed and stored at 3 °C.

The products were evaluated on pH [17], fat content [18], protein content [17], water activity in a water activity meter (Alpax, 650) and the rheological properties, in a texturometer (Texture Analyser, TA—TX2). These analyzes were performed in triplicate and the results expressed were the average of the three replicates.

The products were also evaluated with respect to sensorial acceptance test with nine point hedonic scale and purchase intent; the samples were served in codified (with three numeral) white plates and sensorial attributes evaluated were flavor, color, appearance and texture [21], by an untrained team of 30 judges (students, emploees and professors of Taubate University). The sample used from 30 testers was based on the references of Meilgaard et al. [21]; these researchers are considered as reference in sensory analysis studies, which recommend, for an acceptance test, a team composed of 30–50 testers.Variance and Tukey Test (with significance level of 5%) were performed.

3 Results and Discussion

Table 1 presents the physical and chemical properties for cow and goat milks, as well for the cheese formulations. For each analysis, the value represents the repetitions average and the standard deviation.

Table 1 Physical and chemical properties of cow and goat milks

	pH	Acidity	Density (g/mL)	Fat content (%)	Protein content (%)	Aw	Moisture content (%)	Peak load (g)	Work (mJ)	Defopeak (mm)
Cow milk	5.34 ± 0.04	0.135 ± 0.00	1.034 ± 0.00	2.63 ± 0.46	3.24 ± 0.13	–	–	–	–	–
Goat milk	5.43 ± 0.03	0.135 ± 0.03	1.023 ± 0.00	2.04 ± 0.13	3.29 ± 0.08	–	–	–	–	–
Sample 1	6.68 ± 0.07	–	–	5.58 ± 1.11	0.55 ± 0.28	0.94 ± 0.00	69.48 ± 1.87	243.5 ± 9.2	21.36 ± 1.3	6.1 ± 0.5
Sample 2	6.98 ± 0.08	–	–	5.42 ± 01.22	0.93 ± 0.16	0.970.00	62.51 ± 0.40	295.5 ± 8.7	21.92 ± 0.8	8.4 ± 0.8

Sample 1—cheese made from cow milk
Sample 2—cheese made from goat milk

The results in Table 1 show that both goat and cow's milks presented characteristic composition of each one, where similar values of fat and protein contents can be observed. These results are in agreement with Barrionuevo et al. [22], and Haenlein [23], who state that the amount of fats and proteins in goat's milk are similar to that of cow's milk, the difference is in the quality (type, configuration) of these nutrients.

Regarding the physical and chemical properties of the cheese, it was observed that when the goat's milk was used, the product had a higher nutritional value, that is, with a higher concentration of nutrients, since the cheese made from goat's milk presented moisture content 11% lower when compared to the product made with cow's milk. Regarding protein content, the use of goat's milk resulted in a 69% higher, compared to the same product made from cow's milk.

Leuthier et al. [24] analyzed the effects of storage temperature on the physical and chemical characteristics of goat cheese, when stored under refrigeration (temperature around 10 °C), detected lower values of humidity and higher values for fat and protein content. According to the author, the difference in the centesimal composition of these products is noted, according to the variety of the cheese. The same happened in the researches of Cavalcante et al. [25], when evaluating the quality of bovine rennet cheese using pasteurized milk and endogenous lactic culture.

In addition, the differences in relation to moisture and fat contents are expected when compared to the different types of cheeses, due to factors related to the composition of the milk used. Milk composition, in turn, may be influenced by several aspects, such as the breed, age, health status and lactation stage of the animal, as well as the environmental conditions and the type of feed [26]. In addition, the process of cutting, forming and pressing time may also influence the ability of the products to retain fat and moisture [27].

Respecting to the rheological parameters Table 2 shows that, for the same work, the product obtained from goat's milk provided a penetration of about 37% higher, which indicates that it is slightly less consistent, when compared with the same product made from cow's milk. These results are in agreement with Haenlein [23], and Cunha, Viotto & Viotto [28].

Table 2 presents the results of the sensorial analysis.

The results presented in Table 2 demonstrate preference for cheese made with goat's milk. The ANOVA results (5% probability) showed no significant difference between the two, only for the texture. In the other attributes, the cheese elaborated

Table 2 Sensorial attributes for the different cheeses

Attribute	Sample 1	Sample 2
Color	7.49 ± 1.38^a	8.14 ± 0.97^b
Appearance	7.68 ± 1.32^a	8.14 ± 0.69^b
Flavor	6.22 ± 2.04^a	7.49 ± 1.31^b
Texture	7.03 ± 1.54^a	7.31 ± 1.49^a

[a]Sample 1—cheese made from cow milk
[b]Sample 2—cheese made from goat milk

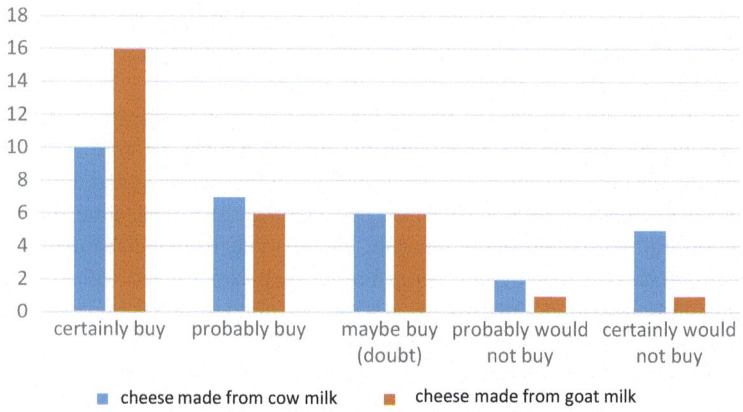

Fig. 1 Purchase intention of the cheese obtained from goat and cow milks

with goat's milk differed from that elaborated with cow's milk, where the former obtained higher averages.

Figure 1 illustrates the purchase intention for the two formulations.

Regarding to purchase intention of the samples, Fig. 1 shows that both products were widely accepted, where more than 50% of the population would certainly buy. The cheese obtained from goat's milk had greater acceptance, where 86% of the population would certainly buy the product and 14% would possibly buy.

4 Conclusions

The results of the nutritional composition of the cheese obtained from the goat milk presented, with protein content slightly above the protein content of the same product elaborated from the bovine milk, besides its excellent sensorial acceptance, demonstrated the potential expansion of the product consumption.

Although the results of the rheological tests attributed less consistency to the cheese made from goat's milk, this difference was not observed in the sensorial tests, reinforcing the viability of the use of goat's milk in the preparation of cured cheese, besides the high acceptance of the product, which presented higher nutritional value, compared to cheese made from cow's milk.

References

1. Gallier, S., Zhu, X.Q., Rutherfurd, S.M., Ye, A., Moughan, P.J., Singh, H.: In vivo digestion of bovine milk fat globules: effect of processing and interfacial structural changes. II. Upper digestive tract digestion. Food Chem. **141**(3), 3215–3223 (2013)
2. Walstra, P., Wouters, J.T.M., Geurts, T.J.: Dairy Science and Technology. CRC Press, New York, NY (2006)
3. Faye, B., Konuspayeva, G.: The sustainability challenge to the dairy sector—the growing importance of non-cattle milk production worldwide. Int. Dairy J. **24**, 50–56 (2012)
4. Selvagi, M., Laudadio, V., Dario, C., Tufarelli, V.: Major proteins in goat milk: an updated overview on genetic variability. Mol. Biol. Rep. **41**, 1035–1048 (2014)
5. Park, T.W., Jeanjulien, C., Siddique, A.: Factors affecting sensory quality of goat milk cheeses: a review. Adv. Dairy Res. **5**(3), 1–9 (2017)
6. Park, Y.W., Juárez, M., Ramos, M., Haenlein, G.F.W.: Physico-chemical characteristics of goat and sheep milk. Small Ruminant Res **68**(1), 88–113 (2007)
7. Santillo, A., Kelly, A.L., Palermo, C., Sevi, A., Albenzio, M.: Role of indigenous enzymes in proteolysis of casein in caprine milk. Int. Dairy J. **19**, 655–660 (2009)
8. Fernandes, M.F., Queiroga, R.C.R.E., Medeiros, A.N., Costa, R.G., Bomfim, M.A.D., Braga, A.A.: Características físico-químicas e perfil lipídico do leite de cabras mestiças Moxotó alimentadas com dietas suplementadas com óleo de semente de algodão ou de girassol. Revista Brasileira de Zootecnia **37**, 703–710 (2008)
9. F.A.O (Food and Agricultural Organization): Homepage. Production of goat milk by continent. http://faostat3.fao.org/browse/Q/QL/E 2016/11/12
10. Rodrigues, A.A., Comeron, E.A., Vilela, D.: Cultivo e utilização da alfafa nos trópicos, Embrapa Pecuária Sudeste (2008)
11. Olalla, M., Ruiz, M.D.L., Navarro, M., Artacho, R., Cabrera, C., Gimenez, R., Rodriguez, C., Mingorance, R.: Nitrogen fractions of Andalusian goat milk compared to similar types of commercial milk. Food Chem. **113**, 835–838 (2009)
12. Cunha, T.M., Castro, F.P., Barreto, P.L.M., Benedet, H.D., Prudêncio, E.S.: Physico-chemical, microbiological and rheological evaluation of dairy beverage and fermented milk added of probiotics. Semina: Ciências Agrárias, **29**(1), 103–116 (2008)
13. Santos, B.N.C., Silva, C.C.C.V., Domingues, J.R., Cortez, M.A.S., Freitas, D.D.G.C., Chiappini, C.C.J., Araújo, K.G.L.: Effect of calcium addition and pH on yield and texture of Minas cured cheese. Brazilian Arch. Vet. Med. Zootechnics **65**(2), 601–609 (2013)
14. Silva, F.C., Pelegrine, D.H.G., Gasparetto, C.A.: Reologia do suco de acerola:efeitos da concentração e da temperatura. Food Sci. Technol. **25**(1), 121–126 (2005)
15. Pelegrine, D.H.G., Gasparetto, C.A., Bezerra, J.R.M.V.: Estudo da viscosidade aparente das polpas de manga (Keitt) e abacaxi (Pérola). Food Sci Tech-Brazil **20**(1), 128–131 (2000)
16. Rosenthal, A.: Food Texture: Measurement and Perception. Aspen Publishing, New York, NY (1999)
17. A.O.A.C.: Official Methods of Analysis. Washington: Sidney Willians (1980)
18. Bligh, E.G., Dyer, W.J.A.: A rapid method of total lipid extraction and purification. Canadian. Journal of of Biochemical Physiol. **7**, 911–917 (1959)
19. Singh, R.P., Heldman, D.R.: Introduction to Food Engineering. Academic Press, Cambridge (2003)
20. Park, T.W., Jeanjulien, C. Siddique, A.: Factors Affecting Sensory Quality of Goat Milk Cheeses: A Review. Adv. Dairy Res. **5**(3) (2007)https://doi.org/10.4172/2329-888x.1000185
21. Meilgaard, M., Civille, G.V., Carr, B.T.: Sensory Evaluation Techniques. CRC Press, New York (2004)
22. Barrionuevo, M., Alferez, M.J.M., Lopez, A.I., Sanz, S.M., Campos, M.S.: Beneficial effect of goat milk on nutritive utilization of iron and copper in malabsorption syndrome. J. Dairy Sci. **85**, 657–664 (2002)
23. Haenlein, G.F.W.: Goat milk in human nutrition. Small Ruminant Res. **51**(2), 155–163 (2004)

24. Leuthier, S.M.F. Efeitos do Armazenamento a 10 ± 2 °C sobre as Características Físicas, Químicas e Condições Higiênico-Sanitárias do Queijo de Leite de Cabra Tipo Coalho Produzido na Fazenda Pendência - Soledade-PB. 1997. 103 f. Dissertação (Mestrado em Ciência e Tecnologia de Alimentos)-Universidade Federal da Paraíba, João Pessoa, (1997)
25. Cavalcante, J.F.M., Andrade, N.J., Furtado, M.M., Ferreira, C.L.L.F., Pinto, C.L.O., Elard, E.: Processamento do queijo coalho regional empregando leite pasteurizado e cultura lática endógena. Ciência e Tecnologia de Alimentos, **27**(1), 205–214, (2007)
26. Alves, F.S.F., Pinheiro, R.R.: A importância do leite de cabra na nutrição humana. Revista Agropecuária Catarinense **16**(1), 25–31 (2003)
27. Naassu, R.T., Araújo, R.S., Guedes, C.G.M., Rocha, R.G.A.: Diagnóstico das condições de processamento e caracterização físico-química de queijos regionais e manteiga no Rio Grande do Norte. Fortaleza, CE. Boletim de pesquisa e desenvolvimento-Embrapa Agroindústria Tropical, 11, **24** (2003). Disponível em: http://www.cnpat.embrapa.br/publica/pub/BolPesq/bd_11.pdf
28. Cunha, C.R., Viotto, W.H., Viotto, L.A.: Use of low concentration factor ultrafiltration retentates in reduced fat "Minas Frescal" cheese manufacture: effect on composition, proteolysis, viscoelastic properties and sensory acceptance. Int. Dairy J. **16**(3), 215–224 (2006)

Viability of the Uses of Effluent for Cultivation of *Chlorella vulgaris* in Order to Integrate the Generation of Biofuels with the Treatment of Water

Daniela Helena Pelegrine Guimarães, Victor Fernandes Marino, Gabriel Costa Blanco, Wallyson Ribeiro dos Santos and Maria Thereza de Moraes Gomes Rosa

Abstract Microalgae constitutes an adequate environmental alternative: besides the accumulation of lipids, they capture CO_2 of the atmosphere and absorb contaminants present in aquatic environments. This work used *Chlorella vulgaris*, cultivated in a pretreated wastewater effluent generated by Lorena School of Engineering. The tests were carried out with effluents collected on different days because of the variation of their concentration along the week. *Chlorella vulgaris* presented good adaptation to the university effluent, reaching a maximum cell density of 1.92×10^7 cells/mL and a dry biomass concentration of 1.74 g/L. It obtained satisfactory concentrations of lipids per gram of biomass (0.095–0.164 mg/g). In addition, it promotes nitrate removal rates between 54 and 78.74%, and COD reduction between 17.73 and 36.13%. The cultivation of the microalgae *Chlorella vulgaris* in effluent proved to be a good alternative to promote the accumulation of lipids and the improvement of water quality.

Keywords Microalgae · *Chlorella vulgaris* · Biofuel · Water treatment · Effluent

1 Introduction

The accelerated growth of population and industrial growing have led to the indiscriminate exploitation of natural resources, without taking into account the negative impacts on the environment. Today most of the world's energy demand is supplied by the use of fossil fuels [1].

On the other hand, the growing concern with environmental issues has been evident in recent years, influencing the decisions of business leaders regarding energy

D. H. P. Guimarães (✉) · V. F. Marino · G. C. Blanco · W. R. dos Santos
Lorena Engineering School (EEL-USP), Lorena, Brazil
e-mail: dhguima@usp.br

M. T. de Moraes Gomes Rosa
Mackenzie Presbyterian University, Campinas, Brazil
e-mail: gomes.mtms@gmail.com

Y. Iano et al. (eds.), *Proceedings of the 4th Brazilian Technology Symposium (BTSym'18)*, Smart Innovation, Systems and Technologies 140,
https://doi.org/10.1007/978-3-030-16053-1_34

sources, creating the need to generate energy from renewable sources, such as the production of biofuels [2, 3].

One method that has been well studied by researchers is the production of lipids using microalgae, the microalgae biomass accumulated lipids that can be extracted and converted into biodiesel via transesterification; besides the energy generation capacity, microalgae have the capacity to absorb and remove contaminants present in water improving its quality [4, 5].

In recent years, energy from microalgae has gained popularity in many parts of the world, such as the United States and several European countries. The encouragement of the use of microalgae as raw material for the production of biofuels is due to the great advances in biosystems engineering, but still has large gaps to be optimized [6, 7]. In this context, a great challenge is the construction of economically viable reactor that allow the cultivation of microalgae on a large scale. The high cost of the construction and operation of photo bioreactors, coupled with the need of large amounts of nutrients, limits the commercial use of it. In order to reduce these costs, several types of effluents have been tested, such as water sources and nutrients for cultivation [8].

The treatment of water is another environmental issue that has been highlighted, because of the development of various human activities where water is used, which is a renewable resource, still present in a great amount on the planet, with a wide variety of applications, easy to obtain and considerably low cost [9].

Despite abundance, water must be used consciously to ensure the preservation of the ecosystems. Water treatment technologies have evolved considerably, and it can be said that any water can be treated and consumed although the costs and risks involved may be extremely high. In this sense, different types of effluents have been tested as sources of water and nutrients for the production of microalgae, which can visibly reduce production costs [10]. The use of effluents as a cultivation form can also improve the energy balance and reduce environmental impacts associated with their disposal in water bodies without previous treatment [11].

Taking into account the depletion of the energy resources available on the planet, trying to minimize the impacts caused by traditional means of energy generation, the present work aims to evaluate the feasibility of integrating the generation of biofuels with the treatment of water through the cultivation of the species *Chlorella vulgaris* effluent from the Engineering School of Lorena (EEL/USP).

2 Materials and Methodology

For the present study, it was used *Chlorella vulgaris* microalgae (clone strain BMAK D1), provided by Aidar &Kutner (Oceanographic Institute) of Sao Paulo University (USP).

2.1 Characterization of the Effluent

EEL/USP Effluents Treatment Station uses a biological reactor with decanter, with capacity to treat all the effluent generated by students and staff (about 3000 people), composed of domestic sewage and small concentrations of metals used in laboratories. The concentration of the produced effluent varies according to the frequency of the students in the university. Therefore, due to these variations, crops were grown on three different dates, as mentioned late.

To evaluate the efficiency of the process, the effluent was characterized, at the entrance and exit of the reactor, using Chemical Oxygen Demand (COD) and Nitrate analyzes, according to the methodologies described in Standard Methods.

The COD analysis was done according to Apha [12], by spectrophotometry (ICP-OES, Optima 8000), in order to verify the amount of organic matter removed from the effluent. Then, the dilutions were made and the calibration curve was prepared. Afterwards, the absorbance was converted to the concentration of dissolved oxygen, consumed per liter.

The nitrate concentration analysis was performed every 3 days to determine the nitrate consumption profile of *Chlorella vulgaris* in the effluent. The analyzes were done in a spectrophotometer (Jenway 7305), at the wavelength of 220 nm which represents the total amount of nitrogen in the sample. However, it was necessary to discard the amount of organic nitrogen produced by the microalgae during its growth. Organic nitrogen was determined by reading the sample at wavelength 275 nm. Finally, the concentration of the organic nitrogen of the total nitrogen was reduced to obtain the amount of inorganic nitrogen absorbed [13].

2.2 Inoculum

The inoculum used was maintained in the Basal Bold Medium (BBM), since these nutrients constitute the structure of the membranes and the cellular environment of organisms, as well as participate in energy exchange processes and regulate the enzymatic activity of microalgae [14].

To sterilize the effluent, the method described by Kawachi and Noel [15] was used to add sodium hypochlorite (0.3–0.5 mL/L). Neutralization was then carried out with sodium thiosulfate (30 mg/L of effluent). The neutralized effluent was inoculated into 5 L photobioreactor.

Fig. 1 Cultivation of Chlorella vulgaris in a medium containing effluent in batch photobioreactor

2.3 Batch Photobioreactor

The microalga cultivation was in a 5 L photobioreactor (in batch) with a constant artificial light source composed of fluorescent lamps, provided with an air compressor for stirring and solubilization of oxygen and carbon dioxide in the effluent (Fig. 1).

Microalgal growth was monitored daily by spectrophotometer reading at 680 nm (UV-Vis, Bel Photonics) and the cell count, under an optical microscope in the Neubauer chamber (TNB-01T).

The effluent generated by EEL/USP shows variations in composition, according to the frequency of students during the semester. Therefore, the cultivation of the microalgae and the characterization of the effluent were carried out on three different dates: Cultivation 1 (end of July, corresponding to school holidays at the university and, consequently, absence of students), Cultivation 2 (beginning of August, beginning of the academic period of the university), Cultivation 3 (mid-August, with high frequency, by students, staff and professors).Cultures were carried out until the number of cells was stabilized.

2.4 Biomass Handling

After each cultivation, biomass was recovered by flocculation, adding aluminum sulphate (1 Eq/L) at a rate of 3 mL/L for decantation of the biomass. The sediment biomass was filtered, dried in the oven at 60 °C and grinded to increase the contact surface area. Then, the biomass was quantified with respect to the lipid content, according to Bligh and Dyer [16].

2.5 Productivity of Biomass and Lipids

After obtain the biomass, the productivity of lipids was calculated by the following Eq. (1):

$$P = \frac{\Delta x}{\Delta t} \tag{1}$$

where:

Δx Variation of lipid or biomass concentration;
Δt Time.

3 Results and Discussion

3.1 Characterization of the Effluent at the Reactor Entrance

The values presented in Table 1 correspond to the parameters analyzed in the treated wastewater of the EEL/USP Effluent Treatment Station (ETS).

For the determination of the oxygen concentration consumed by the Potassium Dichromate in COD, as well as of the nitrate concentration present in the effluent, it was necessary to construct the analytical calibration curves, shown in Figs. 2 and 3.

3.2 Cellular Density (Number of Cells)

The density of *Chlorella vulgaris* cells during cultures in the effluents can be visualized in Fig. 4, which relates the number of cells with time. Each experiment presented

Table 1 Characterization of the effluent at the reactor entrance

Parameter	Cultivation 1	Cultivation 2	Cultivation 3
COQ (mgO_2/L)	251. 54	127.24	88.36
nitrate (mg/L)	23.23	15.74	10.88

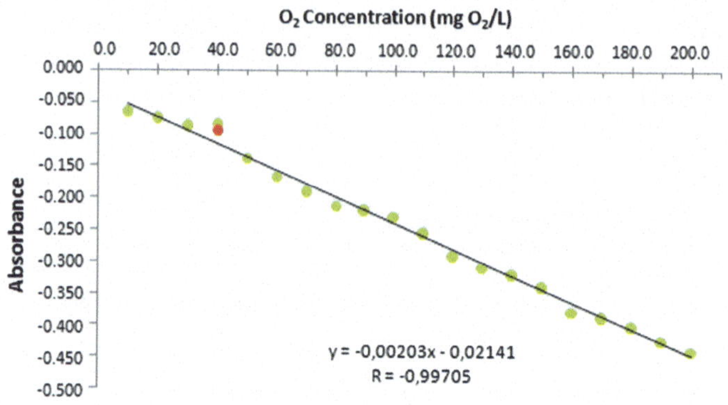

Fig. 2 Analytical curve of DQO

Fig. 3 Analytical curve of nitrate

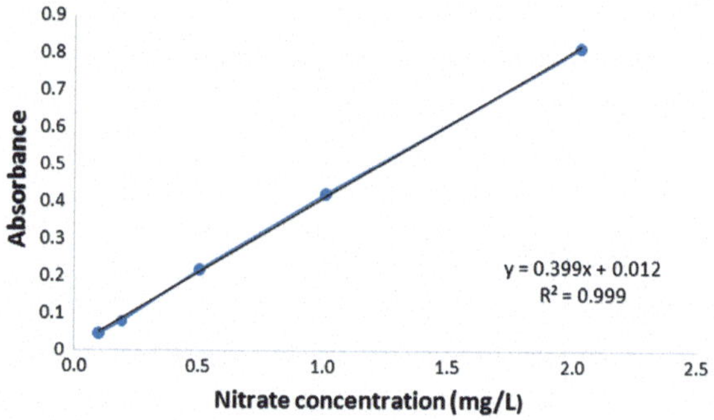

its own culture time, according to the concentration of the effluent (until reaching the stabilization of the number of cells).

The highest growth rate of *Chorella vulgaris* occurred in the effluent corresponding to cultivation 2, reaching 1.915 $\times$ 10^7 cells/mL, while in the effluents used in cultures 1 and 3, the growth occurred until reaching 1.340 $\times$ 10^7 and 1.002 $\times$ 10^7 cells/mL respectively.

Comparing Fig. 4 with Table 1, it is possible to observe that due to the excess of nitrate present in the effluent of cultivation 1, *Chlorella vulgaris* presented difficulties in the process of assimilation of nitrates; consequently, did not present the highest cell density. However, the culture in the effluent of cultivation 3 (with amount of intermediate nitrate) obtained the largest number of cells.

It is also possible to verify that the cell growth in the effluent of the cultivation 3 reached the stabilization of the growth more quickly, due to the smaller amount of nitrate; consequently, nutrient depletion occurred more rapidly.

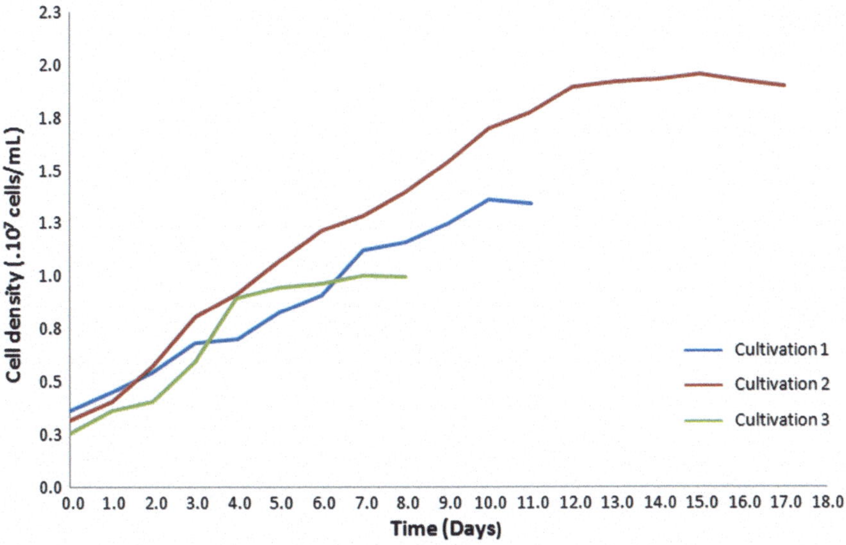

Fig. 4 Growth curve Chlorella vulgaris in wastewaters

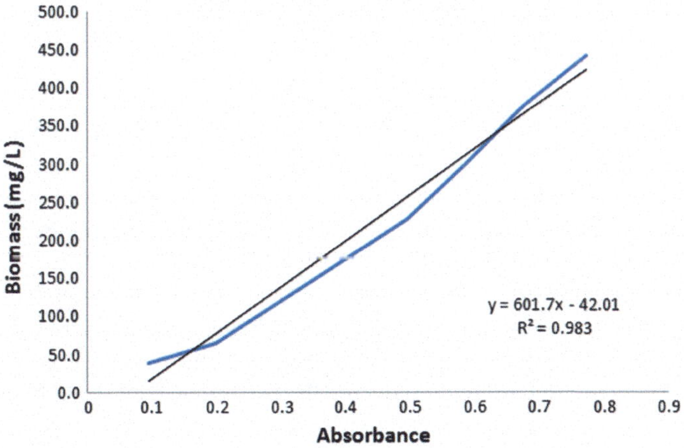

Fig. 5 Analytical curve of biomass growth of the microalga *Chlorella vulgaris*

3.3 Growth Curve (Dry Mass)

The microalgae growth was also evaluated by the concentration of the biomass present in the effluent. Concentrations were determined using the analytical curve given in Fig. 5. The curve was drawn by dilutions of a sample of the culture with known concentration.

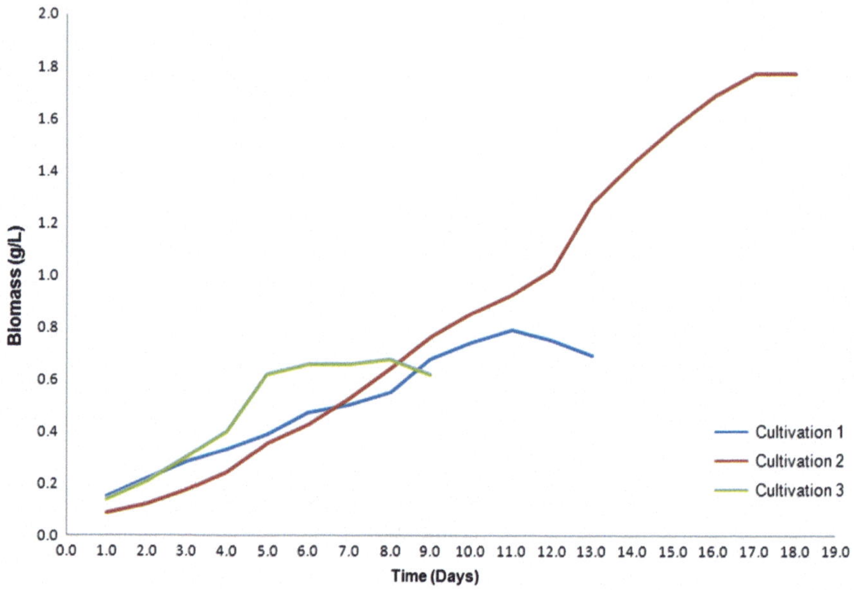

Fig. 6 Concentration of biomass in cultures containing wastewaters

After performing the absorbance readings of the culture, they were converted to biomass concentration (g/L) to construct the graph of its variation with time, according to Fig. 6.

Chlorella vulgaris presented a higher concentration of biomass in culture 2, reaching a concentration of 1.74 g/L, while in cultures 1 and 3 concentrations of 0.68 and 0.62 g/L respectively were obtained.

Comparing Fig. 6 with Table 1, it is possible to observe that although crop 1 had the highest concentration of nitrate, crop 2 had the highest concentration of biomass; according to Melo [17], the excess of nitrate has made the assimilation of nutrients by the cells become more difficult, reducing the performance of the cultivation.

3.4 *Nitrate Consumption*

The nitrate consumption profile of *Chlorella vulgaris* is shown in Fig. 7.

Analyzing the graph of Fig. 7 it can be seen that there were extremely high nitrate removal rates in all growing conditions. According to Shaw et al. [18], nitrate amounts exceeding 0.3 mg/L in bodies of water are sufficient to provide algae proliferation, with that been said, the initial amount present in the effluent corresponding to the culture condition 1 was very high, which may have interfered in the process of nutrient assimilation and development of the cultivation.

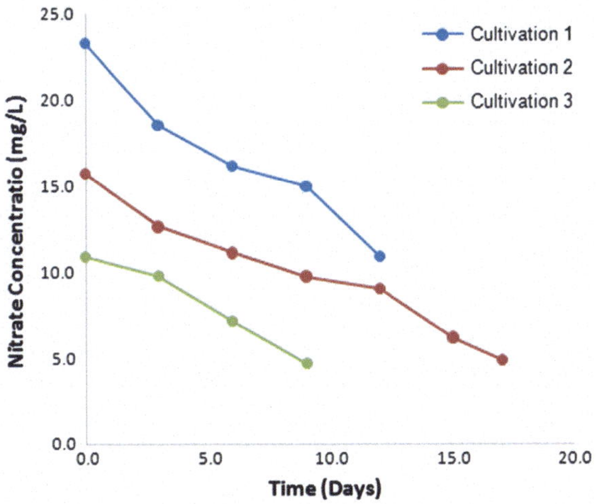

Fig. 7 Nitrate consumption profile

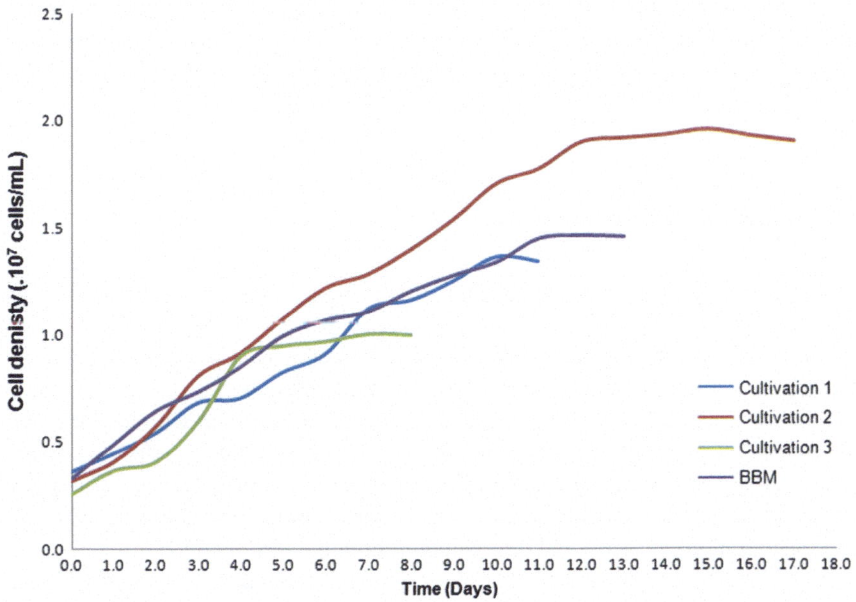

Fig. 8 Comparison of cell density using wastewater with the BBM standard

A comparison of the cell density and biomass production in effluent with the BBM standard

In order to better evaluate the performance of the microalgae in the effluent, a comparison was made between the cell density and the concentration of dry biomass accumulated by the cultivation in effluent and BBM standard medium (Fig. 8, Table 2).

Table 2 Concentration of dry biomass

Parameter	Cultivation 1	Cultivation 2	Cultivation 3	BBM
Biomass (g/L)	0.69	1.74	0.61	0.96

Table 3 Concentration of lipids extracted and ratio lipids per gram of dry biomass

Parameter	Cultivation 1	Cultivation 2	Cultivation 3	BBM
Lipids (mg/g)	0.095	0.124	0.164	0.122
Ratio (%)	9.500	12.400	16.400	12.200

Table 4 Characterization of the effluent at the exit of the reactor

Parameter	Cultivation 1	Cultivation 2	Cultivation 3
COQ (mgO_2/L)	206.92	81.26	64.57
Nitrate (mg/L)	10.69	3.35	2.77

From Table 2 and Fig. 8, the second cultivation showed cell density and biomass concentration higher than the BBM standard; although the cell density and biomass concentration of cultures 1 and 3 were not as high, they were not much lower than the BBM culture.

3.5 *Lipid Concentration*

Table 3 presents the capacity of accumulation of lipids between the different types of culture.

The highest accumulation of lipids was with the effluent corresponding to culture 3, presenting a ratio of lipids per gram of 16.4%. Comparing the results presented in Tables 1 and 3 it is possible to observe that growing medium with low nitrogen concentrations promote a greater accumulation of reserve lipids. On the other hand, in the case of cultures in which the medium presents high concentrations of nitrogen the formation of reserve lipids is low. That is, the accumulation of lipids is related to the concentration of carbon and nitrogen present in the medium [19].

It is also possible to note that *Chlorella vulgaris* presented satisfactory lipid accumulation, compared to standard BBM medium, since the percentage of lipids was only lower in the effluent where the nitrate concentration was very high.

3.6 *Characterization of the Effluent at the Exit of the Reactor*

Table 4 shows the values corresponding to the parameters analyzed in the effluent, after the end of each cultivation.

The first cultivation (at the end of July) showed a removal rate of 17.73% of the COD and 54% of the nitrate concentration of the effluent; the second (at the beginning of August) had a removal rate of 36.13% of the COD and 78.74% of the nitrate concentration of the effluent. Finally, the third cultivation (mid-August) showed a removal rate of 26.92% of the COD and 74.57% of the nitrate concentration of the effluent.

Table 4 shows that the excess of nitrate in the effluent (culture 1) interfered in the assimilation process of the nutrients by the cells, resulting in a lower nitrate removal rate [17], whereas the initially lower concentrations of nitrate growing medium, obtained higher rates of nitrate removal. However, all the crops had an excellent nitrate removal potential from the effluent and a reduction in the concentration of considerable organic matter.

3.7 Productivity of Biomass and Lipids

The productivity results in biomass and lipids are presented in Table 5.

From the data presented in Table 5 it is possible to observe that the cultivation with the highest biomass productivity was made using the effluent with an intermediate nitrate concentration of 15.74 mg/L (cultivation 2), reaching a biomass productivity of 4.0 mg/L.h While the cultures using effluents with concentrations of 23.23 mg/L.h (cultivation 1) and 10.88 mg/L.h (cultivation 3) obtained a biomass productivity of 1.6 and 2.3 mg/L.h, respectively.

In other words, the excess of nitrate in the cultivation 1 made it difficult to assimilate the nutrients by microalgae while the low concentration of nitrate in the cultivation 3 limited the growth due to the exhaustion of the nutrients present in the medium.

Comparing Table 5 with Fig. 6 it is possible to observe that although the biomass concentration of the cultivation using the more concentrated effluent was higher than that using the less concentrated effluent, the biomass yield was lower due to the longer time interval of the cultivation.

Interpreting the data presented in Table 5 it is possible to observe also that the lipid productivity is directly related to the nitrate concentration in the medium. The higher the concentration of nitrate the lower the lipid yield of the cultivation [19]. The Economic analysisof each cultivation is illustrated in Table 6, which presents the percentage of each input oncapital costs.

Table 5 Biomass and lipids productivity in different growing conditions

Parameter	Cultivation 1	Cultivation 2	Cultivation 3
Productivity of biomass ($mg.L^{-1}.h^{-1}$)	1.60	4.00	2.30
Lipid Productivity ($mg.L^{-1}.h^{-1}$)	2.46×10^{-4}	5.29×10^{-4}	5.26×10^{-4}

Table 6 Percentage of the inputs on capital cost in each cultivation and in BBM

Input	Cultivation 1	Cultivation 2	Cultivation 3	BBM
Reactor building (%)	36.90	28.00	43.69	32.98
Inoculum (%)	0.24	0.18	0.28	0.21
Growing médium (%)	0.00	0.00	0.00	1.86
CO_2 delivery(%)	50.60	59.59	42.79	7.34
Harvesting(%)	0.45	0.34	0.54	0.40
Extraction(%)	4.90	3.75	5.85	4.41
Lighting(%)	6.80	0.08	5.87	53.72
Capital cost (U$)	54.53	71.90	46.06	61.02

Table 7 Efficiency and cost of crops

Parameter	Cultivation 1	Cultivation 2	Cultivation 3	BBM
Productivity of biomass ($mg.L^{-1}.h^{-1}$)	1.60	4.00	2.30	2.2
Lipid Productivity ($mg.L^{-1}.h^{-1}$)	2.46×10^{-4}	5.29×10^{-4}	5.26×10^{-4}	3.75×10^{-4}
Nitrate removal (%)	54.00	78.74	74.57	75.22
COQ removal (%)	17.73	36.13	26.92	0.0
Cost (R$)	16.80	22.15	14.19	18.80

Analyzing the values of Table 6, it is possible to observe that the use of effluents as a culture medium is an interesting alternative, since it eliminates the need to prepare synthetic medium and purchase reagents, allowing savings per liter compared to the BBM standard. That is, in large-scale projects and for industrial use, using effluents will bring significant cost reduction, as well as a proportional improvement of water quality, eliminating the need for future investments in effluent treatment. The efficiency and cost of cultivation are present in Table 7.

From Tables 6 and 7 it is possible to compare the main parameters evaluated in this project with the cost of implementing each experiment. It is concluded that the use of effluent as a culture medium actually reduces the cost of the project, since all the cultivation in the wastewater presented lower costs, except for the cultivation 2 where due to its high hydraulic retention time of 17 days the energy demand for lighting and CO_2 release was higher. In addition, the lipid yield in the cultures using effluent was higher than that of the culture using the BBM standard, except for culture 1 due to the presence of a high concentration of nitrate in the effluent.

Therefore, the cultivation of effluents is a good alternative for the optimization of lipid production, providing considerable cost reductions if it is done at larger scales such as in the industrial area.

4 Conclusions

The EEL/USP effluent is a good alternative to cultivate the microalga *Chlorella vulgaris*, reaching cell densities and biomass concentrations of the dried biomass higher or similar using the standard BBM medium. The lipids accumulated by *Chlorella vulgaris* in ETS (EEL/USP) effluent cultures also showed satisfactory lipid concentrations, lower than the BBM standard only in the culture where the effluent had very high nitrate concentrations.

In addition, *Chlorella vulgaris* has been shown to be an interesting option in the treatment of water, presenting high rates of nitrate and organic matter removal, which changes the water quality, causing negative impacts on the environment, such as eutrophication of water bodies. That is, after the cultivation, the effluent returns to the environment with a better quality than the effluent at the entrance.

Finally, the cultivation of *Chlorella vulgaris* in effluents (instead of using the standard BBM medium) eliminates the preparation stage of reagent solutions and consumption, resulting in reduced nutrient purchase expenses. Although the results of the rheological tests attributed less consistency to the cheese made from goat's milk, this difference was not observed in the sensorial tests, reinforcing the viability of the use of goat's milk in the preparation of cured cheese, besides the high acceptance of the product, which presented higher nutritional value, compared to cheese made from cow's milk.

References

1. Baicha, Z., Salar-García, M.J., Ortiz-Martínez, V.M., Hernandez-Fernandez, F.J., de los Rios, A.P., Labjar, N., Elmahi, M.: A critical review on microalgae as an alternative source for bioenergy production: a promising low cost substrate for microbial fuel cells. Fuel Process. Technol. **154**, 104–116 (2016)
2. Wang, X., Feng, Y., Liu, J., Lee, H., Li, C., Li, N., Ren, N.: Sequestration of CO_2 discharged from anode by algal cathode in microbial carbon capture cells (MCCs). Biosens. Bioelectron. **25**(12), 2639–2643 (2010)
3. Schmitz, R., Magro, C.D., Colla, L.M.: Environmental applications of microalgae. Revista Ciatec **4**(1), 48–60 (2012)
4. Vidotti, E.C., Rollemberg, M.C.E.: Algas: da economia nos ambientes aquáticos à bioremediação e à química analítica. Quim. Nova **27**(1), 139–145 (2004)
5. Chu, W.L.: Biotechnological applications of microalgae. IeJSME **6**(1), S24–S37 (2012)
6. Mathimani, T., Uma, L., Prabaharan, D.: Homogeneous acid catalysed transesterification of marine microalga Chlorella sp., Renew. Energy, **81**, 523–533 (2015)
7. Tian, X., Dai, L., Liu, M., Liu, D., Du, W., Wu, H.: Lipase-catalyzed methanolysis of microalgae oil for biodiesel production and PUFAs concentration. Catcom **84**, 44–47 (2016)
8. Lam, M.K., Yusoff, M.I., Uemura, Y., Lim, J.W., Khoo, C.G., Lee, K.T., Ong, H.C.: Cultivation of Chlorella vulgaris using nutrients source from domestic wastewater for biodiesel production: growth condition and kinetic studies. Renew. Energy **103**, 197–207 (2017)
9. Mihelcic, J., Zimmerman, J.B.: Engenharia ambiental: fundamentos, sustentabilidade e projeto. LTC, Rio de Janeiro (2012)

10. He, Y., Wang, Y., Song, X.: High-effective denitrification of low C/N wastewater by combined constructed wetland and biofilm-electrode reactor (CW–BER). Biores. Technol. **203**, 245–251 (2016)
11. Neckel, A., Goellner, E., Bertoldi, T.: The efficiency in the treatment of effluent bullets industry with sludge modification in anaerobic. Reget **20**(1), 427–437 (2016)
12. APHA.: Standard Methods for the Examination of water and wastewater. American Public Health Association, Washington (1998)
13. Nessler, J.A.: Colorimetric determination of ammonia by Nessler reagent. Chem. Zent. **27**, 529–541 (1856)
14. Lourenço, S.O.: Cultivo de microalgas marinhas: princípios e aplicações. Rima, São Carlos (2006)
15. Kawachi, M., Noel, M. H.: Sterilization and sterile technique. In: Andersen, R.A. (ed) Algal Culturing Techniques. Elsevier, Amsterdan, pp. 65–81 (2005)
16. Bligh, E.G., Dyer, W.J.: A rapid method of total lipid extraction and purification. Can. J. Biochem. Physiol. **37**, 911–917 (1959)
17. Melo, T.B.: Remoção de nitrato e fosfato presente em efluente sintético, por meio do cultivo de Chlorella vulgaris, com foco preliminar para produção de biodiesel. Universidade Tecnológica Federal do Paraná, Graduate (2014)
18. Shaw, B., Menchenich, C., Klessig, L.: Understanding Lake Data (G3582). Wisconsin, UWSP (2004)
19. Aquarone, E., Borzani, W., Schmidell, W., Lima, U.A.: Biotecnologia industrial: biotecnologia na produção de alimentos. Blucher, São Paulo (2001)

Modeling a Photovoltaic Panel with the Algorithm Incremental Conductance to Maximize Efficiency using Matlab/Simulink® Software

Domingos Teixeira da Silva Neto, Amélia Moreira dos Santos, Jessica Fernandes Alves, Fernando Ribeiro Arduini and Polyane Alves Santos

Abstract A photovoltaic cell junction forms a photovoltaic panel that has small efficiency, occasioning a maximum power unavailability. The maximum power depends on temperature and irradiation's environmental conditions. To provide maximum power, it is recommended to add an electronic system known as maximum follower. The Maximum Power Tracker, also known as MPPT (Maximum Power Point Tracking), alters the operation of the converter connected to the photovoltaic panel array, thereby allowing maximum power availability regardless of irradiation and temperature values. Some MPPT algorithms are used to allow the optimization of solar energy. As a result of the maximum power efficiency, the MPPT Conductance Incremental algorithm was used. Through the software MatLab/Simulink®, the modeling and simulation of the photovoltaic panel was carried out. The output current and the power of the photovoltaic panel depends on the Irradiation. The Incremental Conductance algorithm was efficient, given that the results obtained with this algorithm compared to the output power values of the photovoltaic panel presented relative errors lower than 10%.

Keywords Panel · Power · MPPT · Incremental conductance

1 Introduction

Global climate change has attracted interest and concern from governments, international organizations and society. The excessive increase in greenhouse gas emissions, caused in large part by the burning of fossil fuels for electricity generation, makes it essential that effective measures be taken to reduce emissions of these pollutants and their effects [1].

D. T. da Silva Neto (✉) · A. M. dos Santos · J. F. Alves
University of Sao Paulo, São Carlos, SP, Brazil
e-mail: domingosneto@usp.br

F. R. Arduini · P. A. Santos
University of Campinas, Campinas, SP, Brazil

Y. Iano et al. (eds.), *Proceedings of the 4th Brazilian Technology Symposium (BTSym'18)*, Smart Innovation, Systems and Technologies 140,
https://doi.org/10.1007/978-3-030-16053-1_35

A new technological concept of energy generation has emerged in recent years. The words renewable energies, sustainable energy and energy efficiency have begun to be pronounced more frequently, although these technologies are not yet well established and they do not already have a guaranteed market because they are expensive [2].

From the first oil crisis, in the 70s, they were considered as possible sources of electric energy, increasing the studies on them, leading to the expansion of their use. Brazil has an intensive energy matrix based on water resources and the energy utilization of sugarcane, and much of the water potential is found in regions that present difficulties in sustainable environmental management practices, such as the Amazon region [1].

The development of the global energy industry should be based on the search for supply security and meeting energy demand in order to take a path towards environmental and economic sustainability. Among the renewable energies, it is mentioned the photovoltaic energy [3].

The solar cell is an electronic component, built by a PN junction, which converts sunlight energy into electricity from the photovoltaic effect, that is, a component that absorbs solar energy and rearranges the energized electrons of a state to an external circuit [4].

The efficiency of the photovoltaic panel indicates the amount of incident radiant power that it converts into electrical energy. After the conversion of solar energy into electrical, electronic processing is necessary through computational models that contemplate the behavior of the panel, considering the atmospheric changes of temperature and irradiation [5].

The reason for having an efficient modeling is due to the need to understand the maximum power that is provided by the photovoltaic panel. According to [3], the implementation of a photovoltaic panel using Newton-Raphson algorithm, is effective in many applications, due to the considerations that the algorithm makes as a function of the temperature and irradiation conditions, panel terminal voltage and the inclusion of the maximum power point of operation [6].

This algorithm system is known as Maximum Power Tracking (MPPT) which corresponds to voltage and cururent values that determine the point of maximum power through instantaneous changes of temperature and irradiation [4].

The most popular MPPT algorithms are: Fixed Cyclic Ratio, Constant Voltage, Disturbance and Observation (P&O), Incremental Conductance (IC), Beta Method, System Oscillation, Ripple Correlation [6].

According to [6] for the maximum power range, the IC algorithm presents relevant efficacy.

The purpose of this article is to implement, through the software Matlab/Simulink®, a photovoltaic panel with the algorithm of MPPT Incremental Conductance and verify the efficiency of this through the results found.

2 Modeling a Photovoltaic Panel

In the modeling of a photovoltaic panel, a traditional photovoltaic cell model is sufficient for most applications as the one used (Fig. 1).

According to [4], the equations that describe a photovoltaic panel are:

$$I{=}I_{ph}{-}I_d[e^{\frac{q(V+IR_S)}{nKT}}-1]-(\frac{V+IR_S}{R_p}) \tag{1}$$

$$I_{ph}{=}[I_{sc}{+}\alpha(T{-}T_r)]\frac{\lambda}{1000} \tag{2}$$

$$I_d{=}I_{rr}(\frac{T}{T_r})^3 e^{[\frac{qE_g}{nK}(\frac{1}{T}-\frac{1}{T_r})]} \tag{3}$$

where: I is the output current of the photovoltaic panel; Iph is the photo current; Id is the cell reverse saturation current; n is the p-n junction quality factor; V is the cell output voltage; q is the value of the electron charge (1.6×10^{-19} C); K is the Boltzmann constant (1.38×10^{-23} J/K); R_s is the series resistance (representing the losses by contact and conduction); R_p is the parallel resistance (representing the losses due to the leakage current); I_{sc} is the short-circuit current; α is the I_{sc} temperature coefficient ; T_r is the reference temperature 298 K or (25 °C); λ is the solar radiation intensity in m/W^2, E_g is the bandwidth energy valued 1.1115 eV according [5].

In order to characterize the photovoltaic panel, some values of the model SPR 305 WHT-D from the manufacturer SUN POWER were taken. The measures were taken under conditions of irradiance of 1000 W/m^2 and temperature of (25 °C). The main characteristics used in this study were: Maximum power, $P_{mp} = 305$ W; Rated voltage, $V_{mp} = 54.7$ V; Nominal current, $I_{mp} = 5.58$ A; Open circuit voltage, $V_{oc} = 64.2$ V; Short circuit current, $I_{sc} = 5.96$ A.

With the catalog values of the SPR 305 WHT-D model from the SUN POWER manufacturer, Eqs. (2) and (3) were solved. The only value to be discovered is I_{rr}. It is known that when $I = 0$, and $V = V_oc$ (open circuit voltage) and making $T = T_r$, we have Eq. (4) through (1).

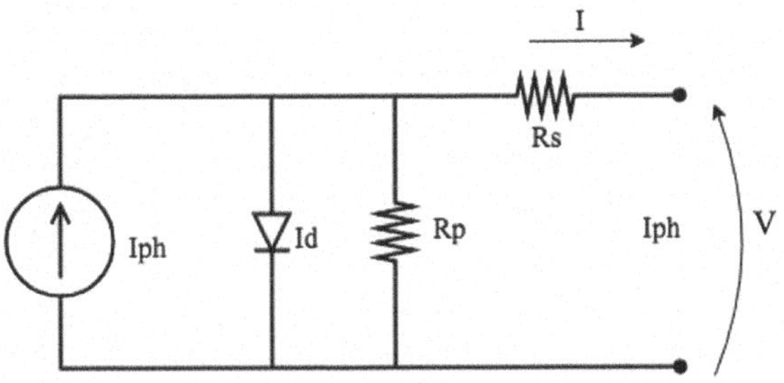

Fig. 1 Equivalent circuit of a photovoltaic cell

$$I_{rr} = \frac{I_{sc} - \frac{V_{oc}}{R_p}}{e^{\frac{qV_{oc}}{nKT_r}} - 1} \tag{4}$$

The used iterative method was the Newton-Raphson by virtue of having recursive calculation and easy implementation, as pointed out by [7]. In mathematical notation, the method is given by:

$$x_{n+1} = x_n - \frac{f(x_n)}{f'(x_n)} \tag{5}$$

where n indicates the nth algorithm iteration and $f(x_n)$ is the f' function derivative in x_n.

MATLAB tools were used to predict the behavior of the C-V and P-V curves represented, respectively, in Figs. 2 and 3 with irradiation scenarios of 1000, 800, 600 W/m^2 and temperature of 25 °C. The simulation shows that the output current and power depend on the irradiation, as stated by [3]. Moreover, the Newton Raphson method proved to be efficient when comparing the curves supplied by the manufacturer SUN POWER with those obtained in the simulations

3 MPPT Algorithm

A photovoltaic system consists of a photovoltaic panel, a boost converter, and an electronic interface to perform the maximum power tracking (MPPT), as shown in Fig. 4.

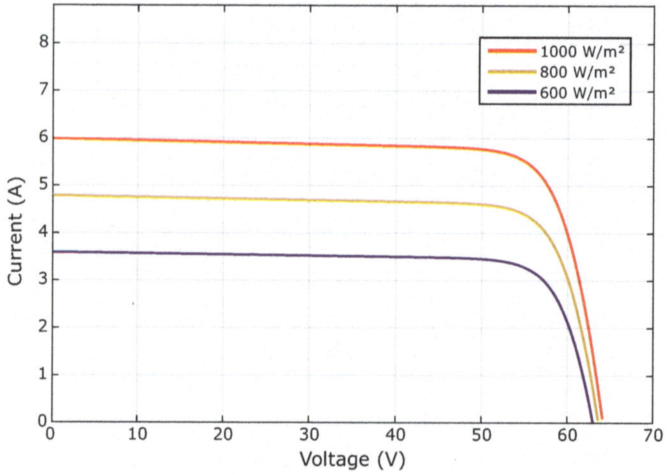

Fig. 2 Differential irradiation curves I-V

The maximum power tracking algorithm is implemented to the system so that the PV panel operates delivering the maximum available power. Thus, the function of the MPPT algorithm is to ensure that regardless of temperature and irradiance variations, the system provides maximum power. This is done by dynamically adjusting the current and voltage variables as well as the duty cycle converter. In this paper, the MPPT Incremental Conductance (IC) algorithm was used, evaluating its efficiency with regard to the maximum power delivery.

The basic flowchart of this approach is presented in Fig. 5 [8]. The IC algorithm uses the derivative technique in the power curve in relation to the voltage. When the maximum power is reached, $dP/dV = 0$. If the derivative is positive, the algorithm needs to advance, otherwise, it should be retreat. According to Eq. (6), the algorithm works as a function of the photovoltaic panel voltage and current [8].

$$dP/dV = d(VI)/dV = IdV/dV + VdI/dV = I + VdI/dV \quad (6)$$

For the IC algorithm MATLAB/Simulink implementation, the work steps of [8] were used. The initial duty cycle rate (D) was 0.5. A DC-DC boost converter was

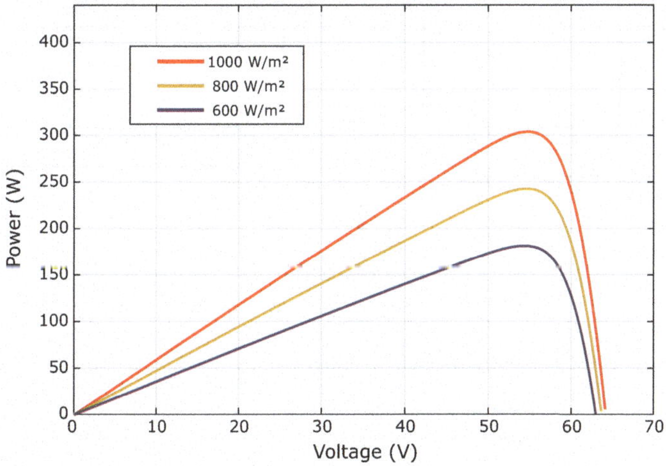

Fig. 3 Differential irradiation curves P-V

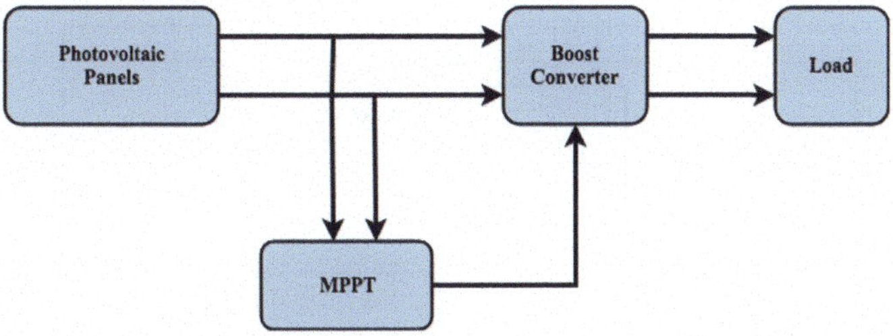

Fig. 4 Photovoltaic system connected to grid

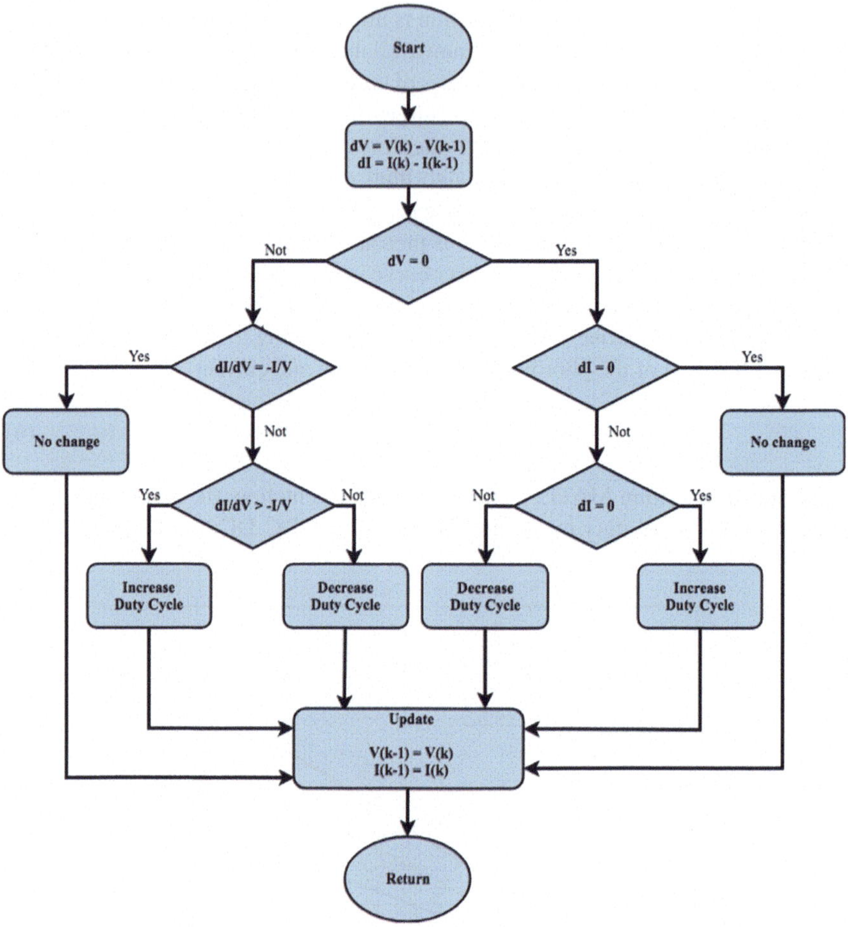

Fig. 5 Incremental conductance algorithm flowchart

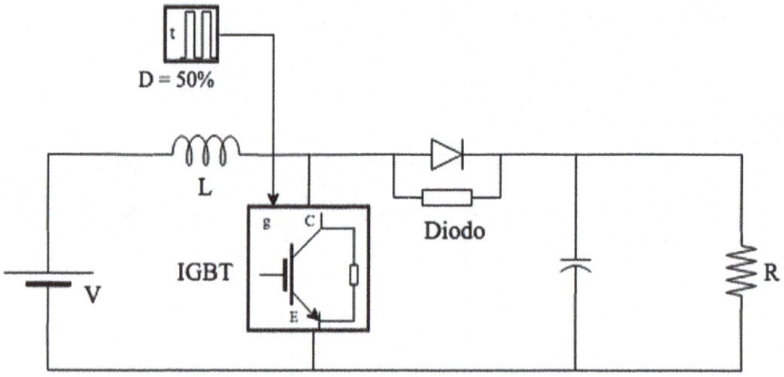

Fig. 6 Boost converter CC-CC

chosen due to its smaller losses when compared to other converters as indicated by [9]. Figure 6, show a boost converter CC-CC.

The parameters of the boost converter were calculated using Eqs. (7–9).

According to [7], when the boost converter operates in continuous mode, there is a relation between the input (V_{in}) and output voltage (V_{out}) of the PV panel, and the duty cycle rate, according to Eq. (7):

$$V_{out} = \frac{V_{in}}{1 - D} \tag{7}$$

The inductor L of the boost converter is given by Eq. (8) as a function of the peak-to-peak current Δi_L, whose value was chosen as 0.762 A.

$$L = \frac{V_s DT}{\Delta i_L} = \frac{V_s D}{\Delta i_L f} \tag{8}$$

The capacitor for the boost converter is found using Eq. (9) and the value of the output ripple was considered to be 0.2%.

$$C = \frac{D}{R(\Delta V / V_{out}) f} \tag{9}$$

With values of $V_{in} = 54.7$ V and $V_{out} = 109.4$ V, and setting a resistor value (R) of 10 Ω and a frequency (f) of 5000 Hz, inductor and capacitor values of $L = 14.35$ mH and $C = 5$ mF were obtained, respectively.

4 Results and Discussion

To measure the efficiency of the MPPT (IC) algorithm the temperature was set at 25 °C and temperatures irradiations ranged at 600, 800 and 1000 W/m^2 for 2.5 s, as can be seen from Fig. 7.

The results for the mentioned irradiations are found in Fig. 8, where it shows that the power increase is proportional to the irradiation variation increase. Table 1 compares the maximum output power values taken from the catalog values of the SPR 305 WHT-D model from the SUN POWER manufacturer of the irradiations mentioned above in the work with those obtained by the simulation.

5 Conclusions

A photovoltaic panel was modeled using the Newtonian Raphson iterative method. The method showed efficiency regarding the representation of the mathematical

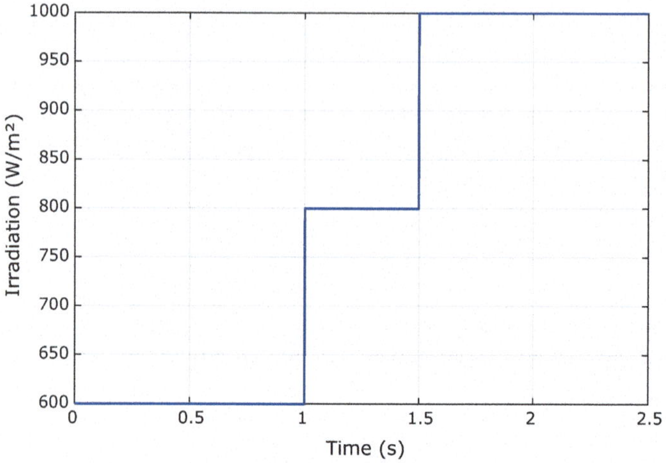

Fig. 7 Irradiance variance

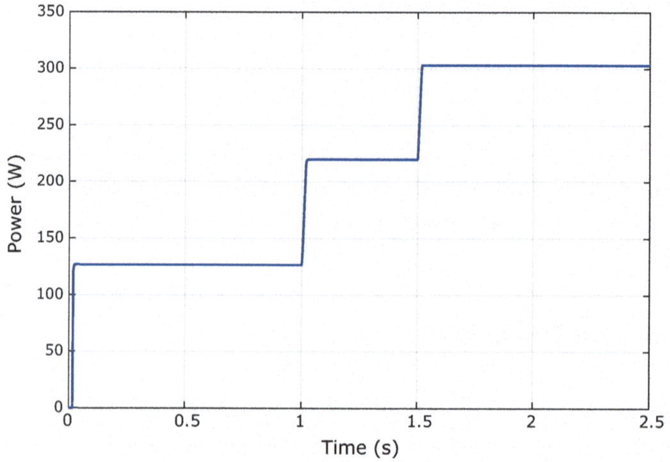

Fig. 8 Power variance

Table 1 Comparison of the output power of the photovoltaic panel and the values obtained by means of the MPPT algorithm Incremental conductance (IC)

Analysis time period of 2.5 s			
Constant temperature of 25 °C			
Irradiation W/m^2	Maximum output power (W)	Maximum output using IC (W)	Relative error %
1000	305.00	303.40	0.52
800	242.90	223.60	7.94
600	140.81	128.80	8.52

model. Subsequently, the IC algorithm was added for the efficiency maximization, by means of computational simulation using Simulink®. The signal resulting from the MPPT algorithm acted at the gate of the boost converter. The output of the photovoltaic panel presents non-linear characteristics due to the inputs that are irradiance and temperature. In the simulation, the temperature was set at 25 °C and the irradiation levels varied at 600, 800 and 1000 W/m^2 a period of 2.5 s. The IC algorithm showed relevant efficacy since the results obtained in simulation compared to the output power values of the SPR 305 WHT photovoltaic panel presented relative errors lower than 10%.

References

1. Tundisi, H.S.F.: Uso de Energias Alternativas para o século XXI. **112**, (2013). https://doi.org/10.8535715908
2. Tolmasquim, M.T.: Energia Renovável: Hidráulica, Biomassa, Eólica. Solar. Oceânica. **452** (2016). https://doi.org/10.978-85-60025-06-0
3. Centro de Referência para Energia Solar e Eólica Sérgio de Salvo Brito. http://www.cresesb.cepel.br
4. Pereira, P.F., Souza, A.C., Santos, I.N.: Modelagem e Avaliação de Desempenho das Principais Técnicas de Controle MPPT empregados na Geração Fotovoltaica, Uberlândia (2016)
5. Brito, M.A.G., Junior, L.G., Sampaio,L.P., Canessin, C.A.: Avaliação das Principais Técnicas para Obtenção de MPPT de Painéis Fotovoltaicos. In: 9th IEEE/IAS International Conference on Industry Applications , pp. 1–6, São Paulo (2010). https://doi.org/10.1109/INDUSCON.2010.5740002
6. Challa, D.T.R.: Implementation of incremental conductance MPPT with direct control method using cuk converter. In: Int. J. Mod. Eng. Res. (IJMER), 4491–4496 (2010). doi: 2249–6645
7. Casaro,M.M., Martins, D.C.: Modelo de arranjo fotovoltaico destinado a análises em eletrônica de potência via simulação
8. Mohamed, S.S., Devaraj, D.:Simulation of incremental conductance MPPT based two phase interleaved boost converter using MATLAB/Simulink. In: 2015 IEEE International Conference on Electrical, Computer and communication Technologies (ICECCT), pp. 1–6. IEEE (2015). https://doi.org/10.1109/ICECCT.2015.7225987.
9. Barbi, I.: Eletrônica de Potência. Universidade Federal de Santa Catarina, Florianópolis (2006)

Stability Analysis of the Interval Systems Based on Linear Matrix Inequalities

Navid Razmjooy, Mehdi Ramezani, Vânia Vieira Estrela, Hermes José Loschi and Douglas Aguiar do Nascimento

Abstract Positive definiteness and Hurwitz stability of the interval systems are discussed. A linear matrix inequality representation is introduced to simplify the analysis of the interval system. First, it is shown that the interval matrix can be stable if it has 2 conditions. Afterward, they converted to linear matrix inequalities for simplifying the conditions solution. A Lyapunov function is introduced to prove the new representation based on linear matrix inequalities.

Keywords Stability analysis · Lyapunov stability · Kharitonov stability · Linear matrix inequality · Interval analysis

1 Introduction

In the recent years, there are different methods which have been introduced based on interval analysis [1–4]. Ordinary methods can not capable of solving the systems with interval uncertainties.

Generally, uncertainties are in the nature of any practical system. These uncertainties can be generated because of different reasons like neglecting some small phenomena, truncation error, neglecting some unknown processes, rounding off errors

N. Razmjooy (✉) · M. Ramezani
Department of Electrical and Control Engineering, Tafresh University, 39518 79611 Tafresh, Iran
e-mail: navid.razmjooy@hotmail.com

V. Vieira Estrela
Departamento de Engenharia de Telecomunicações, Universidade Federal Fluminense, Rio de Janeiro, Brazil

H. J. Loschi · D. A. do Nascimento
Departamento de Comunicações (DECOM), Faculdade de Engenharia Elétrica e de Computação (FEEC), Universidade Estadual de Campinas (UNICAMP), Campinas, Brazil

Y. Iano et al. (eds.), *Proceedings of the 4th Brazilian Technology Symposium (BTSym'18)*, Smart Innovation, Systems and Technologies 140,
https://doi.org/10.1007/978-3-030-16053-1_36

and etc. as it is represented, using the ordinary methods to solve these types of systems make totally wrong answers for the system.

This issue leads researchers to study different methods under uncertainties [5, 6]. There are different types of these methods, like Stochastic methods [7], fuzzy programming [8], and interval arithmetic methods [4].

Among different problems which have been discussed, robust stability for these types of systems has a valuable cost for researches. For system and control engineers, the stability analysis is the most significant step for the identification or controlling the system [6, 9–11]. Therefore, the requirement of an interval based method for analyzing the systems in the presence of interval uncertainties is so important.

Robust stability analysis problem of the interval uncertain systems has been addressed in many papers during the last years. In [12], parameter-dependent Lyapunov functions were employed to supply the robust stability conditions in continuous-time uncertain systems.

There are also some different methods for analyzing the stability of discrete-time interval systems [13].

In [14], an extended version of the described methods for D-stability analysis is proposed. All of the explained methods have a big issue: computational complexity. In this paper, we introduce a new method based on LMI representation for reducing the system complexity and simplifying that.

The rest of the paper is organized as follows: in Sect. 2, the concept of the interval analysis is introduced briefly. Section 3 includes the main part of the paper and presents the definitions of the stability of the interval analysis. The proposed method is also introduced in this section. The paper is finally concluded in Sect. 4.

2 Interval Analysis

As it is represented before, an interval variable $[x]$ can be represented in the way of interval arithmetic. Consider the interval integers $[x]$ and $[y]$ with lower and upper bounds where $[x] = \left\{x|x \in \mathbb{R} \cup \{-\infty, \infty\}, \underline{x} \le x \le \bar{x}\right\}$ and $[y] = \left\{y|y \in \mathbb{R} \cup \{-\infty, \infty\}, \underline{y} \le y \le \bar{y}\right\}$. The general arithmetic operations on these interval integers are as follows:

$$[x] + [y] = [\underline{x} + \underline{y},\ \bar{x} + \bar{y}], \tag{1}$$

$$[x] - [y] := [\underline{x} - \bar{y},\ \bar{x} - \underline{y}]. \tag{2}$$

$$[x] \times [y] = [\min\{\underline{xy}, \bar{x}\underline{y}, \underline{x}\bar{y}, \bar{x}\bar{y}\},\ \max\{\underline{xy}, \bar{x}\underline{y}, \underline{x}\bar{y}, \bar{x}\bar{y}\}], \tag{3}$$

$$[x]/[y] = [x] \times \frac{1}{[y]},$$

$$\frac{1}{[y]} = [\frac{1}{\bar{y}}, \frac{1}{\underline{y}}],\ 0 \notin [\underline{y}, \bar{y}] \tag{4}$$

The centered value (x_c) and the radius (x_r) of the matrix $[x]$ can be also defined as follows:

$$\begin{aligned} x_c &= \frac{\bar{x} + \underline{x}}{2}, \\ x_r &= \frac{\bar{x} - \underline{x}}{2}, \end{aligned} \tag{5}$$

More detailed about interval arithmetic can be found in [20].

3 Analyzing the Stability of the Interval Systems

After introducing the Kharitonov stability analysis, many researchers start to analyze this method on the dynamic systems [15–17]. In this part, a new method for stability analysis of the interval systems has been introduced. The stability analysis in this method is based on the positive definiteness of the system matrix which is explained in the following.

3.1 The Positive Matrix of the Interval Matrix

In general, one can say that an interval matrix of $[A]$ is positive definite, provided that each selected matrix $A \in [A]$ is positive. This case (Positive Determination of the interval Matrices) was studied in [18].

A square matrix (whose symmetry feature is not necessary) is called a positive definite if, $f(A) > 0$; i.e. for each $x^T \neq 0$, $x^T Ax > 0$. Accordingly, an interval matrix is called a positive definite, if it is positive definite for all $A \in [A]$, so it can be said that $\min\{f(A); A \in [A]\} > 0$.

Theorem 1 [19] *Assuming that the interval matrix* $[A] = [A_c - A_r, A_c + A_r]$ *is symmetric and provided that* A_c *is a positive definite matrix, and* $\rho(|A_c^{-1}|A_r) < 1$, *the interval matrix* $[A]$ *will be positive definite.*

Proof Because A_c is positive definite, it is invertible and it will supply the condition of the assertion theory for being definiteness of the interval matrix $[A]$ ($[A]$ is definite if, for all $A \in [A]$, it will be nonsingular); hence, based on [17] the interval matrix $[A]$ is positive definite.

Theorem 2 [20] *The interval matrix* $[A] = [A_c - A_r, A_c + A_r]$ *is positive definite if* $\rho(\Delta) < \lambda_{\min}(A_c)$.

Proof According to the definition [161], $\min\{f(A); A \in [A]\} > \lambda_{\min}(A_c) - \rho(A_r) > 0$, therefore, for each $A \in [A]$, $f(A) > 0$ and therefore, the interval matrix $A \in [A]$ will be positive definite.

3.2 Stability Analysis of the Interval Systems

Definition 1 A square matrix A is Hurwitz stable, if all its singular values are located on the left side of the imaginary axis ($\lambda < 0$). Accordingly, an interval matrix $[A]$ is called stable if each matrix $A \in [A]$ is stable. The problem of the stability of interval matrices in here is based on the nature of the control theory in relation to the behavior of invariant linear systems $\dot{x}(t) = Ax(t)$ under interval disturbances.

Theorem 3 *Let's suppose* $[A] = [A_c - A_r, A_c + A_r]$ *is an interval matrix. In this condition, the following cases are equivalent to each other* [20]:

1. The interval matrix $[A]$ is symmetric and A_c is stable.
2. $\rho(|A_c^{-1}|A_r) < 1$ is positive definite.

Proof Suppose that $[A_0]$ is positive definite. Consider a singular value, λ from the interval matrix $A \in [A]$. According to the Bendixen's theory, $\mathrm{Re}\,\lambda < (\frac{1}{2}(A + A^T))$ where, $[A] = \frac{1}{2}(A + A^T)$ is symmetric and it belongs to $[A]$; so, $[-A] \in [A_0]$.

Therefore, $[-A]$ is positive definite and hence, all of the singular values of $[A]$ will be negative; hence, all values of A in the interval will be stable.

As is clear from the above definition, here we deal with an inequality; therefore, the best idea for analyzing these systems is to employ *Linear Matrix Inequalities.*

3.3 Linear Matrix Inequalities (LMI)

In the following, an LMI representation of the Hurwitz stability will be discussed for interval systems to make the problem simpler. A linear matrix inequality is a symmetric and affine matrix which can be defined as follows:

$$F(x) = F_0 + \sum_{i=1}^{\infty} x_i F_i > 0 \tag{6}$$

where, $F_i = F_i^T,\ F_i \in \mathbb{R}^{n \times n},\ x \in \mathbb{R}^m$.

By the following definition, $F(x) = F(x)^T$:

In general, it can be said that any symmetric and affine matrix inequality which depends on its variable, can be expressed as an LMI. For instance, consider the maximum singular value where $\bar{\sigma}(A(x)) < 1$. This equation can be written as an LMI as follows:

$$\bar{\sigma}(A(x)) \prec 1 \Leftrightarrow A(x)A(x)^T \prec I \Leftrightarrow I - A(x)I^{-1}A(x)^T \succ 0$$
$$\Leftrightarrow \begin{pmatrix} I & A(x) \\ A(x)^T & I \end{pmatrix} \succ 0 \tag{7}$$

Based on Theorem 3, two features should be supplied for the interval stability, which we will redefine, respectively:

(A) The interval matrix $[A]$ is symmetric and A_c is stable:

Consider the Lyapunov function as $V(x) = x^T Px$. Each matrix can be formulated as a summation of symmetric and asymmetric definition as follows:

$$P = \frac{1}{2}(P + P^T) + \frac{1}{2}(P - P^T),$$
$$P_1 = \frac{1}{2}(P + P^T) = P_1^T,$$
$$P_2 = \frac{1}{2}(P - P^T) = -P_2^T. \tag{8}$$

Therefore,

$$P = \frac{1}{2}(P + P^T) + \frac{1}{2}(P - P^T),$$
$$V(x) = x^T Px = x^T P_1 x + x^T P_2 x = \underbrace{\frac{1}{2}x^T(P + P^T)x}_{V_1(x)} + \underbrace{\frac{1}{2}x^T(P - P^T)x}_{V_2(x)},$$
$$V_2(x) = \frac{1}{2}x^T Px - \frac{1}{2}x^T P^T x = \frac{1}{2}x^T P^T x - \frac{1}{2}x^T P^T x = 0 \Leftrightarrow V(x) = V_1(x) \tag{9}$$

i.e. the matrix is absolutely symmetric.

Considering the Lyapunov stability, in this case, the system is stable if its derivative is negative.

$$\dot{V}(x) = \dot{x}Px + x^T P\dot{x} = x^T A_c^T Px + x^T PA_c x, \tag{10}$$

Here, by considering the LMI definition,

$$P \succ 0,$$
$$\begin{bmatrix} -[1\ A_c^T]\begin{bmatrix} 0 & P \\ P & 0 \end{bmatrix}\begin{bmatrix} 1 \\ A_c \end{bmatrix} & 0 \\ 0 & P \end{bmatrix} = \begin{bmatrix} -A_c^T P - PA_c & 0 \\ 0 & P \end{bmatrix} \succ 0. \tag{11}$$

The above definition shows that the matrix A is Metzler.

Definition 2 Matrix A is Metzler if all of its all elements outside the original diameter are non-negative; i.e. $\forall_{i \neq j},\ x_{ij} \geq 0$.

(B) $\rho(|A_c^{-1}|A_r) < 1$ is positive definite:

In general, the spectral radius has different definitions, each of which can be used to provide a definition to LMI. Assuming $A = |A_c| \times A_r$, for instance:

$$\textbf{(D1)}\quad \begin{array}{l} \rho(A) = \max\limits_i \{A \in C^{n \times n} : |\lambda_i|, \lambda_i \in \sigma_i\} \\ \sigma_i := \{Z \in C : \det(Z\, I_n - A) = 0\} \end{array} \tag{12}$$

$$\textbf{(D2)}\quad \begin{array}{l} \rho(A) = \lim\limits_{k \to \infty} \|A^k\|^{\frac{1}{k}} \\ k : Type\ of\ Norm \end{array} \tag{13}$$

$$\textbf{(D3)}\quad \rho(A) = \bar{\sigma}(A(x)) \tag{14}$$

$$\textbf{(D4)}\quad \rho(A) = Trace(A^T(x)A(x)) \tag{15}$$

In this work, (D3) is employed to generate LMI.

Definition 3 Perron-Frobenius theory: By assuming $t \in \mathbb{R}$ and $A \in \mathbb{R}_+^{n \times n}$,

$$\exists\,(t\, I_n - A)^{-1} \geq 0 \Leftrightarrow \rho(A) \prec t \tag{16}$$

Therefore, based on the theory and assuming $t = 1$, $\rho(A) \prec 1 \Leftrightarrow (I_n - A)^{-1} \geq 0$. Since, based on [21], it can be said that the system is stable if the $(A - I)$ is Hurwitz.

Therefore, according to the non-negative hypothesis of the matrix A, the matrix $(A - I)$ is Metzler and it can be stated that the matrix $A \in \mathbb{R}_+^{n \times n}$ is a Schur matrix if and only if there is a diagonal positive definite matrix, $P \succ 0$, that makes the matrix Hurwitz stable. By defining this theory based on LMI,

$$(A - I)^T P + P(A - I) \prec 0 \tag{17}$$

Or in a different way, based on Schur's stability,

$$\begin{array}{l} P \succ 0 \\ \begin{bmatrix} I & (A - I)^T P + P(A - I) \\ I & 0 \end{bmatrix} \succ 0 \end{array} \tag{18}$$

If the definition (D4) has been employed for the stability of LMI,

$$\bar{\sigma}(A(x)) \prec 1 \Leftrightarrow A(x)A(x)^{-1} \prec I \Leftrightarrow I - A(x)A(x)^{-1} \succ 0 \tag{19}$$

i.e. based on Schur's definition,

$$P \succ 0$$
$$\begin{bmatrix} I & A(x) \\ A^T(x) & I \end{bmatrix} \succ 0 \tag{20}$$

Consequently, in general, for the stability of an interval system, there must be both of the stated LMIs should be established; in other words, by combining the two above expressions, there is a stable time interval system, where,

$$P \succ 0,$$
$$\begin{bmatrix} F_1(P) & 0 \\ 0 & F_2(P) \end{bmatrix} \succ 0. \tag{21}$$

Where,

$$F_1(P) = \begin{bmatrix} -A_c^T P - PA_c & 0 \\ 0 & P \end{bmatrix},$$
$$F_2(P) = \begin{bmatrix} (I - A)^T P + P(I - A) & 0 \\ 0 & P \end{bmatrix}. \tag{22}$$

Or,

$$F_1(P) = \begin{bmatrix} -A_c^T P - PA_c & 0 \\ 0 & P \end{bmatrix},$$
$$F_2(P) = \begin{bmatrix} I & (|A_c|A_r) \\ (|A_c|A_r)^T & I \end{bmatrix}. \tag{23}$$

4 Conclusions

A sufficient condition is presented for the Hurwitz stability of the interval systems. LMI representation is utilized for simplifying the stability problem. For more reducing the ordinary properties of Hurwitz stability, a Lyapunov function is introduced into the interval systems. Schur technique, Peron Frobenius and Metzler matrices' definitions are used to identify the LMI representation.

References

1. Petre, E., Tebbani, S., Selişteanu, D.: Robust-adaptive control strategies for a time delay bio-electrochemical process using interval observers. Asian J. Control **17**, 1767–1778 (2015)
2. Khadraoui, S., Rakotondrabe, M., Lutz, P.: Design of a fixed-order RST controller for interval systems: application to the control of piezoelectric actuators. Asian J. Control **15**, 142–154 (2013)
3. Razmjooy, N., Ramezani, M.: Solution of the Hamilton jacobi bellman uncertainties by the interval version of adomian decomposition method. Int. Rob. Auto. J. **4**, 113–117 (2018)
4. Razmjooy, N., Ramezani, M.: Optimal control of two-wheeled self-balancing robot with interval uncertainties using Chebyshev inclusion method. Majlesi J. Electr. Eng. **12**, 13–21 (2018)
5. Wu, J.: Uncertainty analysis and optimization by using the orthogonal polynomials (2015)
6. Khalilpour, M., Valipour, K., Shayeghi, H., Razmjooy, N.: Designing a robust and adaptive PID controller for gas turbine connected to the generator. Res. J. Appl. Sci. Eng. Technol. **5**, 1544–1551 (2013)
7. Touzi, N.: Optimal stochastic control, stochastic target problems, and backward SDE. Bull. Amer. Math. Soc. **54**, 333–339 (2017)
8. Hosseini, H., Tousi, B., Razmjooy, N.: Application of fuzzy subtractive clustering for optimal transient performance of automatic generation control in restructured power system. J. Intell. Fuzzy Syst. **26**, 1155–1166 (2014)
9. Razmjooy, N., Khalilpour, M., Ramezani, M.: A new meta-heuristic optimization algorithm inspired by FIFA World cup competitions: theory and its application in PID designing for AVR system. J. Control Autom. Elect. Syst. **27**, 419–440 (2016)
10. Hosseini, H., Tousi, B., Razmjooy, N., Khalilpour, M.: Design robust controller for automatic generation control in restructured power system by imperialist competitive algorithm. IETE J. Res. **59**, 745–752 (2013)
11. Razmjooy, N., Ramezani, M., Namadchian, A.: A new lqr optimal control for a single-link flexible joint robot manipulator based on grey wolf optimizer. Majlesi J. Electr. Eng. **10**, 53 (2016)
12. Dasgupta, S., Chockalingam, G., Anderson, B., Fe, M.: Lyapunov functions for uncertain systems with applications to the stability of time varying systems. IEEE Trans. Circuits Syst. I: Fundam. Theor. Appl. **41**, 93–106 (1994)
13. de Oliveira, M.C., Bernussou, J., Geromel, J.C.: A new discrete-time robust stability condition. Syst. Control Lett. **37**, 261–265 (1999)
14. Leite, V.J., Peres, P.L.: An improved LMI condition for robust D-stability of uncertain polytopic systems. IEEE Trans. Autom. Control **48**, 500–504 (2003)
15. Swain, S., Khuntia, P.: Kharitonov based robust stability for a flight controller. Int. J. Syst. Signal Control Eng. Appt. **7**, 26–32 (2014)
16. Pastravanu, O., Voicu, M.: Necessary and sufficient conditions for componentwise stability of interval matrix systems. IEEE Trans. Autom. Control **49**, 1016–1021 (2004)
17. Lin, C., Lam, J., Wang, J., Yang, G.-H.: Analysis on robust stability for interval descriptor systems. Syst. Control Lett. **42**, 267–278 (2001)
18. Skalna, I.: Positive definiteness and stability of parametric interval matrices. arXiv preprint arXiv: 1709.00853 (2017)
19. Neumaier, A.: Interval Methods for Systems of Equations, vol. 37. Cambridge university press (1990)
20. Rohn, J.: Positive definiteness and stability of interval matrices. SIAM J. Matrix Anal. Appl. **15**, 175–184 (1994)
21. Twardy, M.: An LMI approach to checking stability of 2D positive systems. Tech. Sci. **55** (2007)

Phase Locked Loop PLL-Based Frequency Synthesizers: A General Overview

R. N. S. Raphael, Agord M. Pinto Jr., Leandro T. Manera and Saulo Finco

Abstract This work summarizes the operating features for type II second-order Phase-Locked Loop (PLL)-based Frequency Synthesizers (FS) from the set of equations of linear dynamic modeling for defining the corresponding block level and system level transfer functions and developing the reference design equations for the Low Pass Filter (LPF) components sizing.

Keywords Frequency synthesizer · Phase-locked loop · Third-order system · Linear dynamic modeling · Low-pass filter components sizing

1 Introduction

Frequency Synthesizers (FS) are widely employed for a diversity of applications involving wireless communication systems: generation of clock signal for driving Analog-to-Digital Converters (ADC), Clock and Data Recovery (CDR) circuits for digital systems, and generation of Local Oscillator (LO) reference signal in Radio Frequency (RF) transceivers.

From the referred applications, **Phase-Locked Loop-based Frequency Synthesizers (PLL-FS)** has been employed as a proper solution for providing a higher performance for critical operating requirements as stability, low noise and tunable local oscillation. In this context, a significant number of works has been proposed in

R. N. S. Raphael (✉) · S. Finco
Center of Information Technology Renato Archer (CTI), Campinas, Brazil
e-mail: raphael.souza@cti.gov.br

S. Finco
e-mail: saulo.finco@cti.gov.br

A. M. Pinto Jr. · L. T. Manera
University of Campinas (Unicamp), Campinas, Brazil
e-mail: agord.junior@gmail.com

L. T. Manera
e-mail: manera@fee.unicamp.br

Y. Iano et al. (eds.), *Proceedings of the 4th Brazilian Technology Symposium (BTSym'18)*, Smart Innovation, Systems and Technologies 140,
https://doi.org/10.1007/978-3-030-16053-1_37

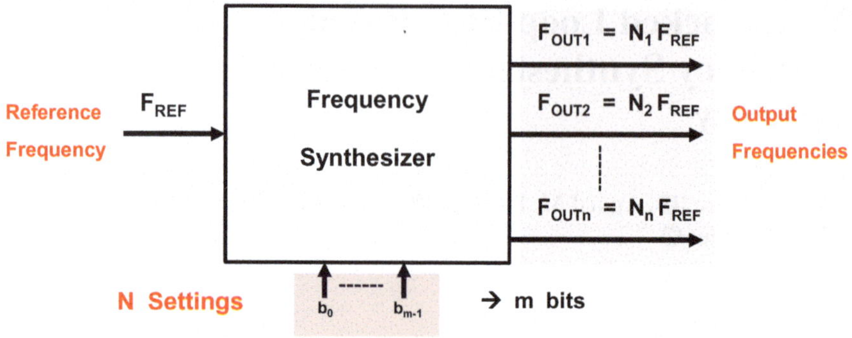

Fig. 1 Frequency synthesizer (general representation): reference (F_{REF}) and outputs (F_{OUT})

the literature considering several strategies for **PLL-FS** design and implementation [1–17].

Thus, considering type II second-order **PLL-FS**, this work presents a detailed set of reference algebraic steps for time and frequency-domain-based linear dynamic modeling from the block level and system level transfer functions to final expressions for LPF components parameters sizing.

Toward this end, Sect. 2 summarizes the operating features for PLL-based systems, considering a general block diagram for defining a correlation between signal nodes and building blocks, Sect. 3 develops the algebraic process for block level and system level dynamic modeling, and finally, Sect. 4 summarizes the final conclusions.

2 General Structure and Operation

Frequency synthesizers (**FS**) can be defined as electronic structures for the generation of a limited set of output frequencies ($\mathbf{F_{OUT1}}$ to $\mathbf{F_{OUTn}}$) with high level of accuracy and precision from a given input reference frequency $\mathbf{F_{REF}}$ [1–6], as represented in the Fig. 1.

In this case, as indicated in the synthesizer bock level representation, considering a general multiplying factor **Ni**, a corresponding output frequency $\mathbf{F_{OUTi}}$ can be expressed as a function the input reference frequency $\mathbf{F_{REF}}$, according to the Eq. (1):

$$F_{OUTi} = Ni\, F_{REF} \tag{1}$$

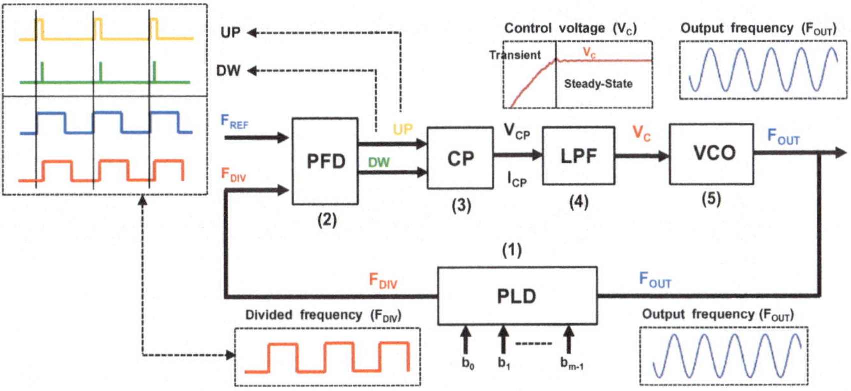

Fig. 2 General PLL block diagram: structure and functional representation

2.1 *Phase-Locked Loop (PLL): General Description*

For effects of frequency generation, **PLL-FS** are composed by a nonlinear feedback system for output frequency $\mathbf{F_{OUT}}$ and phase $\boldsymbol{\phi_{OUT}}$ control. Thus, considering a set of analog and digital building blocks with different operating features, the **PLL** structure-based feedback system performs a comparison in frequency and phase between two input signals (reference frequency $\mathbf{F_{REF}}$ and divided frequency $\mathbf{F_{DIV}}$) for generating an output frequency $\mathbf{F_{OUT}}$ proportional to the phase difference $\boldsymbol{\Delta\Phi} = \boldsymbol{\Phi_{REF}} - \boldsymbol{\Phi_{DIV}}$.

A structural and functional representation is provided from the Fig. 2 considering the general **PLL** block diagram with the set of building blocks and the corresponding waveforms in each system level voltage node for blocks interconnection.

Thus, integer-N PLL-based systems are usually composed by a **Phase-Frequency Detector** (**PFD**), a **Charge Pump** (**CP**), a **Low-Pass Filter** (**LPF**), a **Voltage-Controlled Oscillator** (**VCO**), and a frequency divider or **Programmable Loop Divider** (**PLD**) for allowing a variable output frequency $\mathbf{F_{OUT}}$ generation [3–6], according to the Fig. 2:

(1) **PLD**: from $\mathbf{F_{OUT}}$, performs a frequency division generating a CMOS standard divided frequency $\mathbf{F_{DIV}} = \mathbf{F_{OUT}/N} = \mathbf{F_{VCO}/N}$.
(2) **PFD**: through the phase comparison between two inputs ($\mathbf{F_{REF}}$ and $\mathbf{F_{DIV}}$), generates a sequence of rail-to-rail pulses (**UP** and **DW**) with length proportional to the input phase difference or error signal $\boldsymbol{\Delta\Phi} = \boldsymbol{\Phi_{REF}} - \boldsymbol{\Phi_{DIV}}$.
(3) **CP**: from **UP** and **DW** signals, generates output voltage $\mathbf{V_{CP}}$ and current $\mathbf{I_{CP}}$ for driving the **LPF**.
(4) **LPF**: from the voltage and current provided by **CP**, performs a linear filtering operation, generating a voltage signal $\mathbf{V_C}$ for **VCO** frequency control and output frequency generation $\mathbf{F_{VCO}} = \mathbf{F_{OUT}}$.
(5) **VCO**: from $\mathbf{V_C}$, generates an output phase $\boldsymbol{\phi_{VCO}}$ and output frequency $\mathbf{F_{VCO}}$.

Table 1 PLL building blocks and interfacing parameters: general summary

Block feature	Building blocks				
	PFD	CP	LPF	VCO	PLD
Operation	Analog/Digital	Analog	Analog	Analog	Digital
Linearity	Nonlinear	Nonlinear	Linear	Nonlinear	Nonlinear
Input parameter	Phase difference [$\Delta\phi$]	Average voltage [V]	Voltage/Current [$\mathbf{V_{CP}}/\mathbf{I_{CP}}$]	Control voltage [$\mathbf{V_C}$]	Phase/Frequency [$\boldsymbol{\phi_{OUT}}/\mathbf{F_{OUT}}$]
Output parameter	Average voltage [**V**]	Current [$\mathbf{I_{CP}}$]	Control voltage [$\mathbf{V_C}$]	Phase/Frequency [$\boldsymbol{\phi_{OUT}}/\mathbf{F_{OUT}}$]	Phase/Frequency [$\boldsymbol{\phi_{DIV}}/\mathbf{F_{DIV}}$]

Thus, for allowing a multiple output frequency $\mathbf{F_{OUT}}$ generation in the **PLL** operation, a **PLD** is applied for integer-**N** frequency division, considering a given set of input control bits ($\mathbf{b_0}\ \mathbf{b_1}\ldots\mathbf{b_{m-1}}$) for defining the divide ratio **N**, and a corresponding frequency division range $\mathbf{\Delta N} = \mathbf{N_{MAX}} - \mathbf{N_{MIN}} + \mathbf{1}$, according to the Eq. (2):

$$N = \frac{F_{OUT}}{F_{REF}} \leftrightarrow F_{OUT} = N\, F_{REF} \tag{2}$$

2.2 *Phase-Locked Loop (PLL): General Operation*

From a feedback system-based structure, the **PLL** operation can be divided in two states:

- **Acquisition state** (**transient operation**): the time variant phase difference $\mathbf{\Delta\phi}$ between the fixed $\mathbf{F_{REF}}$ and the variable $\mathbf{F_{DIV}}$ generates a sequence of pulses with a time variant length and, as result, an oscillating LPF output or control voltage $\mathbf{V_C}$ establishes a variable output frequency $\mathbf{F_{OUT}}$.
- **Locked state** (**steady-state operation**): the stabilized phase difference $\mathbf{\Delta\phi}$ between $\mathbf{F_{REF}}$ and $\mathbf{F_{DIV}}$ generates a sequence of pulses with a time constant length and, as result, establishes a constant output frequency $\mathbf{F_{OUT}}$.

Thus, Table 1 summarizes a general categorization for each building block, considering linearity and interfacing parameters [3, 10, 15, 17].

3 Phase-Locked Loop (PLL): Building Blocks

This section presents a structure and operating description for each **PLL** building block focusing on frequency domain-based linear dynamic modeling for defining

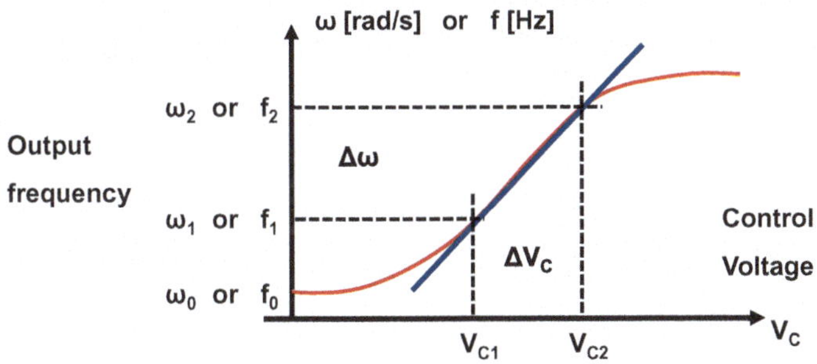

Fig. 3 VCO transfer curve (frequency tuning characteristic): $\omega_{OUT} \times V_C$

performance parameters to be applied for system level design, considering a given requirement of stability condition.

3.1 Voltage-Controlled Oscillator (VCO)

Considering different operating principles, **Voltage-Controlled Oscillators** (**VCO**) operate as active structures for autonomous generation of output signals, from a non-linear correlation between the output frequency $\mathbf{\Delta F_{OUT}}$ range and a corresponding input control voltage range $\mathbf{\Delta V_C}$. From this nonlinear correlation, Fig. 3 illustrates the frequency tuning characteristic curve [5] indicating the corresponding tuning sensitivity $\mathbf{k_{VCO}}$ [Hz/V] or $\mathbf{K_{VCO}}$ [rad/Vs], according to Eq. (3).

$$k_{VCO} = \frac{\Delta f}{\Delta V_C} = \frac{f_2 \quad f_1}{V_{C2} - V_{C1}} \overset{2\pi}{\longleftrightarrow} K_{VCO} = \frac{\Delta\omega}{\Delta V_C} = \frac{\omega_2 - \omega_1}{V_{C2} - V_{C1}} \tag{3}$$

Thus, from the Eq. (3), the ideal relationship between the **VCO** output frequency $\mathbf{f_{VCO}}$ [Hz] or $\boldsymbol{\omega_{VCO}}$ [rad/s] and the input control voltage $\mathbf{V_C}$ is commonly represented through the linear equation in (4), considering the running frequency $\mathbf{f_{IN}}$ ($V_C = 0$), according to the indicated scales ($\boldsymbol{f_{VCO} = f_2, f_{IN} = f_1}$ and $\boldsymbol{\omega_{VCO} = \omega_2, \omega_{IN} = \omega_1}$):

$$f_{VCO} = f_{IN} + k_{VCO} V_C \overset{2\pi}{\longleftrightarrow} \omega_{VCO} = \omega_{IN} + K_{VCO} V_C \tag{4}$$

From the previous definition in the Eqs. (3) and (4), and considering $\mathbf{\Delta V_C = V_C}$, $\mathbf{K_{VCO}}$ can be represented as follows: $\boldsymbol{\omega_{VCO} - \omega_{IN} = \Delta\omega = K_{VCO}.V_C}$. In this case, the application of Laplace Transform results in the set of equations in (5), from time domain to frequency domain, as follows:

$$L[\Delta\omega(t)] = K_{VCO} L[V_C(t)] \overset{L[\ldots]}{\longleftrightarrow} \Delta\omega(s) = K_{VCO} V_C(s) \tag{5}$$

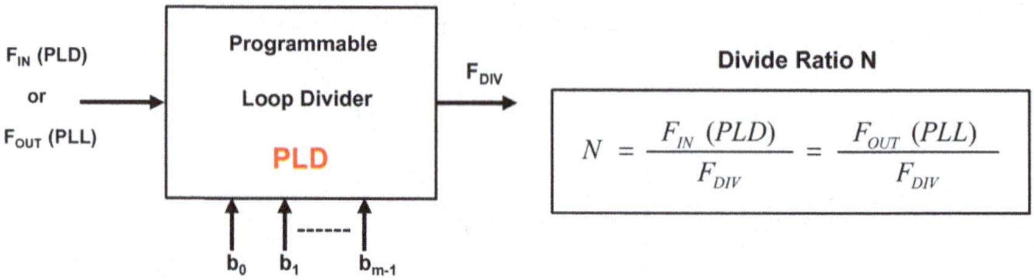

Fig. 4 Programmable loop divider (PLD): general representation (pins diagram)

Since the frequency can be taken as a time derivative of the phase difference ($\boldsymbol{\omega(t)} = \boldsymbol{\partial\Phi(t)/\partial t}$ and $\boldsymbol{\Delta\omega(t)} = \boldsymbol{\partial\Delta\Phi(t)/\partial t}$), the application of Laplace Transform results in the equations in (6):

$$L\left[\frac{\partial\Delta\Phi(t)}{\partial t}\right] = K_{VCO}L[V_C(t)] \overset{L[\ldots]}{\longleftrightarrow} s\,\Delta\Phi(s) = K_{VCO}V_C(s) \tag{6}$$

Finally, from the expression in (6), a resulting frequency domain-based transfer function is obtained for describing the correlation involving the input control voltage $\mathbf{V_C}$ and the corresponding phase variation $\boldsymbol{\Delta\phi}$ in the **VCO** operation, according to the block level transfer function $\mathbf{H_{VCO}(s)}$ in the Eq. (7):

$$H_{VCO}(s) = \frac{\Delta\Phi(s)}{V_C(s)} = \frac{K_{VCO}}{s} \tag{7}$$

3.2 *Programmable Loop Divider (PLD)*

Frequency dividers or **Programmable Loop Dividers** (**PLD**) can be generally represented through the Fig. 4. In this case, the corresponding output frequency or divided frequency $\mathbf{F_{DIV}}$ is generated from the input frequency $\mathbf{F_{IN}}$ (PLL output frequency $\mathbf{F_{OUT}}$), considering a predefined and configurable divide ratio $\mathbf{N = F_{OUT}/F_{DIV}}$, according to a given division range $\mathbf{\Delta N}$. The set of **m** control bits ($\mathbf{b_0}$ to $\mathbf{b_{m-1}}$) indicates a maximum division range ($\mathbf{\Delta N = N_{MAX} - N_{MIN} + 1}$), considering a continuous or non-continuous **N** interval (composed by integer numbers).

The conventional solution presented in the literature [3–6, 16] for programmable frequency division or **pulse swallow frequency divider** involves an asynchronous architecture composed by 3 sequential modules: (1) **Dual Modulus Prescaler** (**DMP**), (2) **Program Counter** (**PC**) and (3) **Swallow Counter** (**SC**), according to the diagram in the Fig. 5.

Considering the block diagram, the input signal with frequency $\mathbf{F_{IN}}$ (PLD) = $\mathbf{F_{OUT}}$ (PLL) at CLK pin in **DMP** module generates a divided output frequency $\mathbf{F_{DIV}} = \mathbf{F_{IN}/N} = \mathbf{F_{OUT}/N}$ in **PC** module, considering a divide ratio **N** defined through the

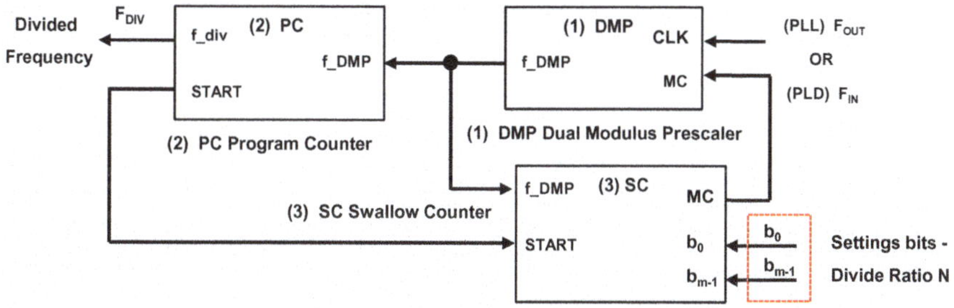

Fig. 5 Programmable loop divider (PLD): reference pulse swallow architecture

settings bits in **SC**. Thus, the operating features for each module can be described as follows:

(1) **Dual Modulus Prescaler** (**DMP**): programmable divider with 2 modes of operation (division by M/(M + 1), according to Mode Control (**MC**) input.
(2) **Program Counter** (**PC**): counter module driven through **f_DMP** signal from **DMP** and designed as fixed binary counter for counting from 0 to P.
(3) **Swallow Counter** (**SC**): second counter driven through **f_DMP** signal from **DMP** and designed as variable binary counter from 0 to S ($S_{MAX} = P - 1$), according to a set of $m = \log_2(S_{MAX})$ control bits for defining the resulting divide ratio **N**.

Thus, the operation of the **PLD** architecture can be described through the following set of 2 steps (modes of operation), from the building features shown in the Fig. 4.

An input frequency (Divider $\mathbf{F_{IN}}$ = PLL $\mathbf{F_{OUT}}$) applied to **DMP** (CLK pin), generates an intermediary output frequency (**f_DMP**), as a clock signal for driving the counters **PC** and **SC**, according to the mode of operation:

- **Mode 1** (**MC = 1** during **SC** counting): the both counters (**PC** and **SC**) operate with a common clock signal **f_DMP** = $\mathbf{F_{IN}}$/(**M** + 1) up to the overflow of **SC**, considering $S \leq P - 1$ or $S_{MAX} = P - 1$. After overflow, **SC** generates **MC** = 0 for starting the second operation mode (mode 2).
- **Mode 2** (**MC = 0** after **SC** overflow state with no counting): **SC** remains without counting after overflow and **PC** operates with a clock signal **f_DMP** = $\mathbf{F_{IN}}$/**M** up to overflow. After this second overflow, the both counters restart the counting process, and **SC** generates **MC** = 1 for defining the first operation mode (mode 1).

Considering the processes in the sequence of operation modes, mode 1 involves **S(M + 1)** counting cycles, and mode 2 involves (**P** − **S**)**M** and, as result, the division ratio **N** = $\mathbf{F_{OUT}/F_{DIV}}$ represents a complete division cycle [3–6] and can be determined according to the model in Eq. (8):

$$N(S) = S(M + 1) + (P - S)M = PM + S \tag{8}$$

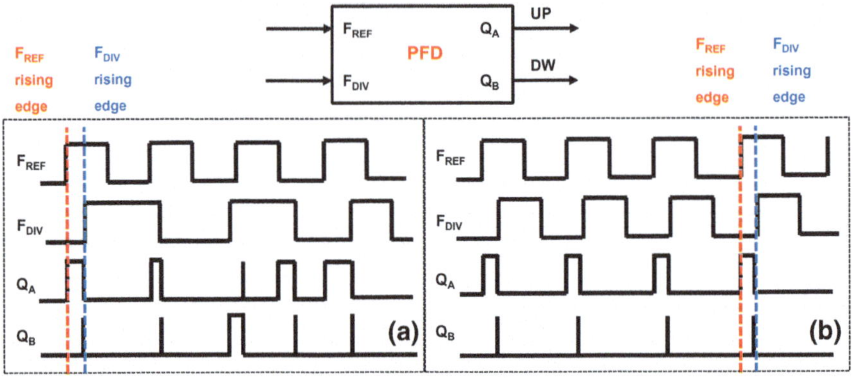

Fig. 6 Phase-frequency detector (PFD): **a** transient and **b** steady-state waveforms

3.3 *Phase-Frequency Detector (PFD)*

The **Phase-Frequency Detector** (**PFD**) is the building block whose operation generates the error signal e(t) to be minimized and controlled through the **PLL** structure-based feedback system.

Thus, from 2 inputs (**V_{REF}(t)** with frequency **F_{REF}** and **V_{DIV}(t)** with frequency **F_{DIV}**) for phase and frequency comparison, **PFD** outputs a sequence of rail-to-rail pulses (signals **UP** and **DW**) whose lengths are linearly proportional to the input phase difference $\Delta\Phi$, according to the Fig. 6.

Thus, from the diagram, **F_{REF}** and **F_{DIV}** indicates the set of inputs for phase comparison, and **UP** (**Q_A** pin) and **DW** (**Q_B** pin) represent the corresponding outputs. The resulting output waveforms are composed by a sequence of pulses indicating the time scale difference between **F_{REF}** rising edge and the **F_{DIV}** signal rising edge, according to the scheme:

- **Signal Q_A**: rises with **F_{REF}** rising edge and falls with **F_{DIV}** rising edge.
- **Signal Q_B**: rises with **F_{DIV}** rising edge and falls with **F_{REF}** rising edge.

Considering the generated sequence of pulses, the time-based profile of the output signals (**Q_A** and **Q_B**) is composed according to the indicated pattern: (1) **transient operation** (pulses with variable length and frequency), and (2) **steady-state operation** (pulses with constant length and frequency).

The described operation pattern is usually implemented by applying the reference **tri-state architecture** [5, 6], according to the Fig. 7, with two possible variants. In this case, Fig. 7a illustrates a first **PFD** variant composed by a NAND2 logic gate (highlighted in grey) and 2 rising edge-triggered flip-flops D with low logic-based asynchronous reset RST (highlighted in blue). Finally, Fig. 7b indicates the second **PFD** variant composed by an AND2 logic gate and 2 flip-flops D with high logic-based asynchronous reset (RSTb in red).

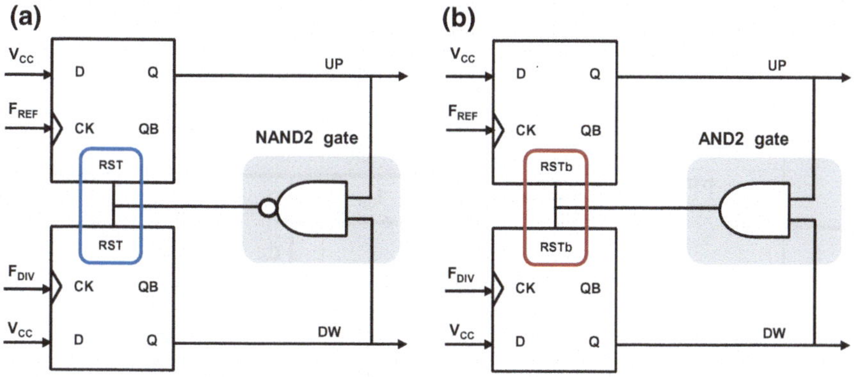

Fig. 7 Phase-frequency detector (PFD): variants (**a**) and (**b**) for tri-state architecture

From the described operating features, a correlation between time-based and frequency-based models can be established and thus, considering the waveforms illustrated in the Fig. 6, the resulting average value for the signal **UP** or $\mathbf{UP_M}$ (DC component from the time variable output) is given through the Eq. (9). In this case, the model includes the power supply $\mathbf{V_{DD}}$, a reference period **T** and the time intervals for high voltage level $\mathbf{t_H}$ and low voltage level $\mathbf{t_L}$:

$$UP_M = V_{DD}\frac{t_H}{t_H + t_L} = V_{DD}\frac{t_H}{T} \tag{9}$$

In the sequence, Eq. (10) indicates an equivalence relationship in different scales (time and angle information) and the resulting expression for $\mathbf{UP_M}$ as a function of the corresponding input phase difference $\mathbf{\Delta\Phi}$:

$$\frac{t_H}{T} = \frac{\Delta\Phi}{2\pi} \rightarrow UP_M = V_{DD}\frac{t_H}{T} = V_{DD}\frac{\Delta\Phi}{2\pi} \tag{10}$$

Finally, by rearranging the variables in (10) and applying the previous definitions, the resulting model is indicated in the Eq. (11):

$$UP_M = \frac{V_{DD}}{2\pi}\Delta\Phi = K_{PFD}(\Phi_{REF} - \Phi_{DIV}) = K_{PFD}\Delta\Phi \tag{11}$$

A frequency domain model can be obtained considering the application of Laplace Transform in the Eq. (11) and hence, the linearized **PFD** transfer function ($\mathbf{K_{PFD}} = \mathbf{V_{DD}/2\pi}$) indicates the proportionality constant between the average value of the **PFD** output voltage $\mathbf{UP_M}$ and the corresponding input phase difference $\mathbf{\Delta\Phi} = (\mathbf{\Phi_2} - \mathbf{\Phi_1}) = (\mathbf{\Phi_{REF}} - \mathbf{\Phi_{DIV}})$, according to (12):

$$UP_M(s) = K_{PFD}\Delta\Phi(s) \rightarrow H_{PFD}(s) = \frac{UP_M(s)}{\Delta\Phi(s)} = K_{PFD} = \frac{V_{DD}}{2\pi} \tag{12}$$

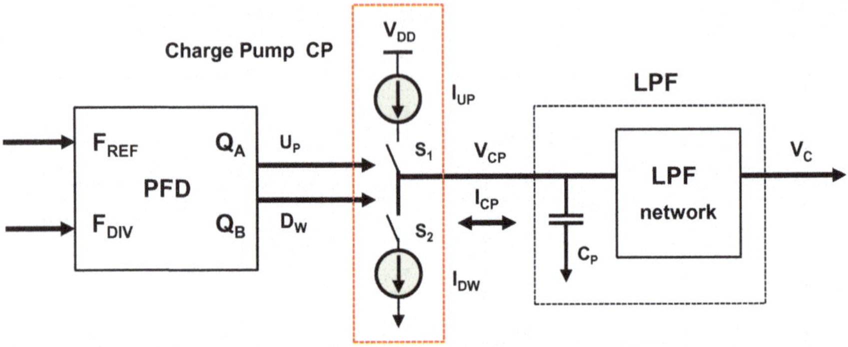

Fig. 8 Charge pump (CP): conceptual representation of the switching operation

According to the linearized **PFD** transfer function in the Eq. (12), the average output voltage is proportional to the phase difference between the 2 inputs. Thus, the corresponding linear model describes the **PFD** as a phase subtractor whose output is proportional to the resulting slope of the straight line or **PFD** gain ($\mathbf{K_{PFD}}$).

3.4 *Charge Pump (CP)*

Considering the operation of a tri-state PFD, a **Charge Pump** (**CP**) can be defined as a three position-based switch controlled by the corresponding PFD states for the generation of output voltage $\mathbf{V_{CP}}$ and current $\mathbf{I_{CP}}$ from an input voltage (PFD outputs—**UP** and **DW** signals) with phase and frequency information.

As reported significantly in the technical literature [3–6], Fig. 8 illustrates a combined representation for the referred **PLL** building blocks: tri-state PFD, a general conceptual **CP** model, and a general Low-Pass Filter (LPF) network. Thus, according to the illustration, the PFD outputs (signals Q_A (**UP**) and Q_B (**DW**)) drive the **CP** switches (S_1 and S_2 respectively) for the control of the current flow (from $\mathbf{I_{UP}}$ and $\mathbf{I_{DW}}$) over the capacitor $\mathbf{C_P}$.

In this case, **CP** operation can be summarized according to a limited set of input possibilities:

- **PFD State 1** ($Q_A = Q_B = 0$): S_1 and S_2 are off and the output $\mathbf{V_{CP}}$ remains in the same state (constant).
- **PFD State 2** (Q_A = high and Q_B = low): S_2 is off, S_1 is on and the current $\mathbf{I_{UP}}$ charges the capacitor $\mathbf{C_P}$.
- **PFD State 3** (Q_A = low and Q_B = high): S_1 is off, S_2 is on and the current $\mathbf{I_{DW}}$ discharges $\mathbf{C_P}$.

In this case, Eq. (13) defines the time for switch activation $\mathbf{t_{ON}}$ (from $\mathbf{I_{UP}}$ or $\mathbf{I_{DW}}$) considering the **PLL locked state** and additionally, indicates the output current $\mathbf{I_{CP}}$

from the differential relationship between the charge and discharge currents ($\mathbf{I_{UP}}$ and $\mathbf{I_{DW}}$), for allowing a simplified expression from the symmetry between the currents.

$$t_{ON} = \frac{\Delta\Phi}{2\pi\, F_{REF}} \rightarrow I_{DIF} = \frac{I_{UP} - I_{DW}}{2\pi}\Delta\Phi = \frac{I_{CP}}{2\pi}\Delta\Phi \tag{13}$$

Hence, the resulting **PFD-CP** linearized transfer function is given by (14):

$$K_{CP} = \frac{I_{DIF}}{\Delta\Phi} = \frac{I_{CP}}{2\pi} \tag{14}$$

The computing of the phase difference $\boldsymbol{\Delta\phi}$ generated from PFD inputs as input parameter for the model in the Eq. (14) implies on the embedded computing of the **PFD** transfer function $\mathbf{K_{PFD}}$ effects through the described **CP** transfer function $\mathbf{K_{CP}}$. In this case, the Eq. (14) could be referred as $\mathbf{K_{PFD}.K_{CP}} = \mathbf{I_{CP}}/2\pi$.

3.5 *Low Pass Filter (LPF)*

As analog module in the **PLL** feedforward path, **LPF** operates as an interface between the **CP** output current $\mathbf{I_{CP}}$ and the control voltage $\mathbf{V_C}$ for VCO output frequency $\mathbf{F_{VCO}} = \mathbf{F_{OUT}}$ generation. Considering active or passive structures for implementation, **LPF** operation can be described under a frequency domain and time domain perspectives. Under the frequency domain perspective, **LPF** provides a proper set of poles and zeros for defining the **PLL** dynamic features, considering transient response parameters and steady-state operation parameters. Finally, under a time domain perspective, **LPF** performs a suppression or reduction of high frequency components for ensuring a proper stability along the time of the resulting control voltage $\mathbf{V_C}$ level.

The switching-based operation in the **CP** does not generate an output signal with proper time stability levels for driving VCO and defining a stable frequency. In this case, a simple output capacitor $\mathbf{C_2}$ can be provided in the **CP** output, as an impedance-based first-order **LPF**, according to Fig. 9a, for allowing a signal leveling with reduced variations. On the other hand, the resulting **PLL** system operates in this condition with 2 poles at origin (VCO and **LPF** poles), compromising the resulting stability margin. The addition of a series resistor $\mathbf{R_2}$ in the capacitive branch, according to Fig. 9b, modify the system dynamics preserving the first-order filter structure by changing the position of one of the poles and, on the other hand, the system operation demonstrates remaining voltage surges in the **LPF** node. In this case, a second capacitor $\mathbf{C_1}$ is applied in parallel with the previous branch ($\mathbf{R_2}$ and $\mathbf{C_2}$), what results in a second-order filter, according to Fig. 9c.

Thus, considering the second-order linear passive topology, the corresponding impedance-based transfer function $\mathbf{H_{LPF}(s)}$ is presented in the Eq. (15), as a function of 3 coefficients: time constants $\boldsymbol{\tau_1}$ and $\boldsymbol{\tau_2}$ and total capacitance $\mathbf{C_T} = \mathbf{C_1} + \mathbf{C_2}$:

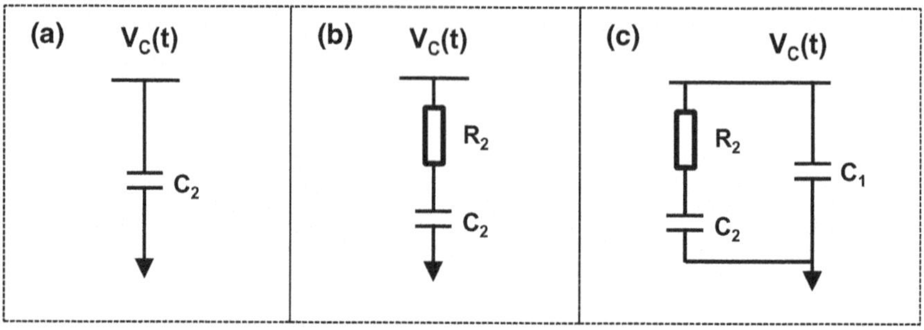

Fig. 9 Low-pass filter (LPF): conceptual evolution for passive implementation

Table 2 PLL: dynamic description by block level transfer function

Block name Notation	PFD/CP $G_{PFD}(s)$ $G_{CP}(s)$	LPF $G_{LPF}(s)$	VCO $G_{VCO}(s)$	PLD $G_{PLD}(s)$
Transfer function	$K_{PFD}K_{CP} = \frac{I_{CP}}{2\pi}$	$\frac{(1+\tau_2 s)}{C_T s(1+\tau_1 s)}$	$\frac{K_{VCO}}{s}$	$\frac{1}{N}$

$$H_{LPF}(s) = \frac{(1+\tau_2 s)}{C_T s(1+\tau_1 s)} \tag{15}$$

In this case, the corresponding expressions for the transfer function coefficients are indicated as a function of the filter components (**C_1**, **C_2**, **R_2**), according to the equations in (16):

$$\tau_2 = R_2 C_2 \leftrightarrow \tau_1 = \frac{R_2 C_1 C_2}{C_T} = \frac{R_2 C_1 C_2}{C_1 + C_2} \leftrightarrow C_T = C_1 + C_2 \tag{16}$$

For modeling effects, considering the block level characterization developed in the previous sections, the dynamic model applied for each **PLL** building block is presented through the Table 2, as a reference for system level design equations development, considering a set of coefficients from the block level operating parameters: **CP** output current **I_{CP}**, **VCO** tuning sensitivity **K_{VCO}**, **PLD** divide ratio **N**, **LPF** time constants **τ_1** and **τ_2** and total capacitance **C_T**.

Thus, in a second modeling stage, a system level open loop transfer function **$H_{OL}(s)$** is represented in the Eq. (17), considering the block level operating parameters summarized in the Table 2:

$$H_{OL}(s) = \frac{I_{CP}}{2\pi}\frac{(1+\tau_2 s)}{C_T s(1+\tau_1 s)}\frac{K_{VCO}}{s}\frac{1}{N} \tag{17}$$

As a result, considering $s = j\omega$ and **$K = (I_{CP}/2\pi) K_{VCO}$**, modulus **$|H_{OL}(\omega)|$** and phase **$\Phi(\omega)$** can be represented through the Eqs. (18) and (19), respectively:

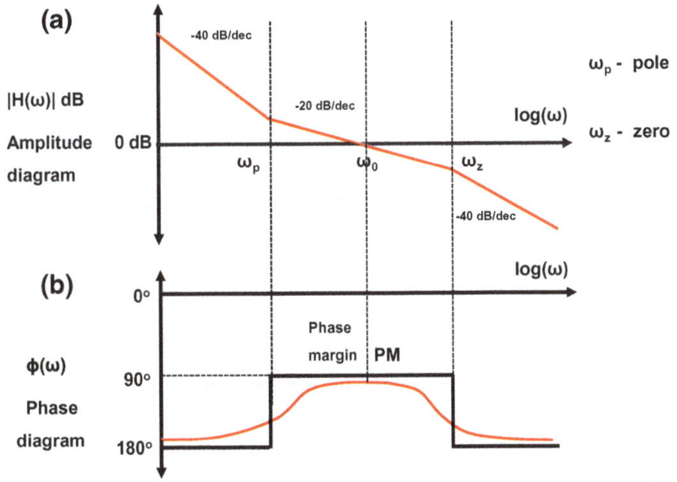

Fig. 10 PLL stability analysis (Bode diagram representation): **a** amplitude and **b** phase

$$|H_{OL}(j\omega)| = \frac{K\sqrt{1+(\tau_2\omega)^2}}{C_T N\,\omega^2\sqrt{1+(\tau_1\omega)^2}} \tag{18}$$

$$\Phi(\omega) = -180 + \arctan(\omega\,\tau_2) - \arctan(\omega\,\tau_1) \tag{19}$$

Thus, the corresponding Bode diagram is illustrated for characterizing the system dynamic features, considering the amplitude variation profile in logarithm scale, according to Fig. 10a and phase, according to Fig. 10b:

From the Eq. (19), a correlation can be established between the bandwidth ω_0 and the time constants ($\boldsymbol{\tau_1}$ and $\boldsymbol{\tau_2}$) or poles and zeros (**p** and **z**), considering the derivative of the phase angle $\Phi(\omega)$ at the frequency $\boldsymbol{\omega = \omega_0}$, according to (20):

$$\omega_0 = \frac{1}{\sqrt{\tau_1\tau_2}} = \sqrt{pz} \tag{20}$$

Thus, Eq. (20) defines a first design correlation involving design parameters ($\boldsymbol{\omega_0}$) and resulting system level dynamic parameters (**p** and **z**). In the sequence of the process, a new format can be established for $\boldsymbol{\Phi(\omega)}$ from the denominator rationalization and, in this case, an alternative expression can be defined resulting in the indicated phase margin (**PM** $= 180 + \boldsymbol{\Phi(\omega_0)}$) model, according to the equations in (21):

$$\Phi(\omega) = -180 + \arctan\left(\frac{\omega\,(\tau_1-\tau_2)}{1+\omega^2\tau_1\tau_2}\right) \rightarrow PM = \arctan\left(\frac{\omega_0\,(\tau_1-\tau_2)}{1+\omega_0^2\tau_1\tau_2}\right) \tag{21}$$

The **PM** equation leads to a second-order polynomial as a function of the variable ($\boldsymbol{\omega_0\tau_1}$), whose solution corresponds to a first **PLL** design equation, allowing a time constant $\boldsymbol{\tau_1}$ sizing and the corresponding pole **p** as a function of **PM** and $\boldsymbol{\omega_0}$, according to the equations in (22):

$$\tau_1 = \frac{\sec(PM) - \tan(PM)}{\omega_0} \rightarrow p = \frac{1}{\tau_1} = \frac{\omega_0}{\sec(PM) - \tan(PM)} \tag{22}$$

In a subsequent process, the second time constant $\boldsymbol{\tau_2}$ and the corresponding zero **z** can be determined from a second design equation, according to (23):

$$\tau_2 = \frac{1}{\omega_0^2\,\tau_1} \rightarrow z = \omega_0^2\,\tau_1 \tag{23}$$

Considering the bandwidth $\boldsymbol{\omega_0}$ (frequency for unitary modulus of the open loop transfer function $|\mathbf{H_{OL}(s)}| = 1$), a third design equation can be defined for the sizing of total capacitance ($\mathbf{C_T = C_1 + C_2}$) as a function of **PLL** building blocks parameters, according to the Eq. (24):

$$C_T = \frac{K}{N}\frac{\sqrt{1 + (\tau_2\,\omega_0)^2}}{\omega_0^2\sqrt{1 + (\tau_1\,\omega_0)^2}} \tag{24}$$

By taking the ratio between the time constants, a fourth design equation can be defined for the sizing of $\mathbf{C_1}$ (first filter component) as a function of $\mathbf{C_T}$, according to the Eq. (25):

$$\frac{\tau_2}{\tau_1} = \frac{C_1 + C_2}{C_1} = \frac{C_T}{C_1} \rightarrow C_1 = C_T\frac{\tau_1}{\tau_2} \tag{25}$$

Additionally, the time constants-based ratio establishes a fifth design equation for capacitor C_2 sizing (second filter component) as a function of C_1 (previously determined through the design sequence), according to (26):

$$\frac{\tau_2}{\tau_1} = \frac{C_1 + C_2}{C_1} = 1 + \frac{C_2}{C_1} \rightarrow C_2 = C_1\left(\frac{\tau_2}{\tau_1} - 1\right) \tag{26}$$

Finally, a sixth and last design equation can be defined for $\mathbf{R_2}$ sizing (third filter component), by applying the time constant $\boldsymbol{\tau_2}$, as indicated in the Eq. (27).

$$\tau_2 = R_2C_2 \rightarrow R_2 = \frac{\tau_2}{C_2} \tag{27}$$

Hence, considering the application of impedance-based second-order passive **LPF** structures, the resulting set of reference design equations for **LPF** components sizing can be summarized in 6 steps, through the sequence in the Table 3.

Table 3 Sequence of design steps for LPF components sizing

Step 1 (τ_1)	Step 2 (τ_2)	Step 3 (C_T)
$\tau_1 = \frac{\sec(PM)-\tan(PM)}{\omega_0}$	$\tau_2 = \frac{1}{\omega_0^2 \tau_1}$	$C_T = \frac{I_{CP}}{2\pi}\frac{K_{VCO}}{N\omega_0^2}\sqrt{\frac{1+(\tau_2\,\omega_0)^2}{1+(\tau_1\,\omega_0)^2}}$
Step 4 (C_1)	Step 5 (C_2)	Step 6 (R_2)
$C_1 = C_T\frac{\tau_1}{\tau_2}$	$C_2 = C_1\left(\frac{\tau_2}{\tau_1} - 1\right)$	$R_2 = \frac{\tau_2}{C_2}$

4 Conclusions

This work presented a summary about the operating features for type II second-order integer-N PLL-based Frequency Synthesizers considering the reference set of linear dynamic equations applied as an approximation for analyzing the nonlinear PLL dynamics. Thus, from frequency domain-based linear analysis, the development of system level design equations for LPF components sizing are demonstrated by applying block level performance parameters and stability parameters.

References

1. Reinhardt, V., Gould, K., McNab, K., Bustamante, M.: A short survey of frequency synthesizer techniques. In: 40th annual frequency control symposium (1986)
2. Farazian, M.: Fast Hopping Frequency Generation in Digital CMOS. Springer Science + Business Media (2013)
3. Sánchez-Sinencio, E., Shu, K.: CMOS PLL Synthesizers Analysis and Design. Springer (2005)
4. Razavi, B.: RF Microelectronics. Prentice-Hall PTR (1998)
5. Razavi, B.: Design of Analog CMOS Integrated Circuits. Tata McGraw-Hill (2002)
6. Plett, C., Rogers, J.W.M.: Radio Frequency Integrated Circuit Design. Artech House (2010)
7. Gardner, F.M.: Charge-pump phase-lock loops. IEEE Trans. Commun. (1980)
8. Hsieh, G.C., Hung, J.C.: Phase-locked loop techniques—a survey. IEEE Trans. Ind. Electron. **43** (1996)
9. Moon, S.T., Valero-López, A.Y., Sánchez-Sinencio, E.: Fully-integrated frequency synthesizers: a tutorial. Int. J. High Speed Electron. Syst. (2011)
10. Best, R.E.: Phase-Locked Loops Design, Simulation and Applications. McGraw-Hill (2007)
11. Singh, M., Ranjan, S.M., Ali, Z.: A study of different oscillator structures. Int. J. Innov. Res. Sci. Eng. Technol. **3** (2014)
12. Singh, S.K., Bhattacharyya, T.K., Dutta, A.: Fully integrated CMOS frequency synthesizer for ZigBee applications. In: 18th International Conference on VLSI Design (2005)
13. Gupta, J., Sangal, A., Verma, H.: High speed CMOS charge pump circuit for PLL applications using 90 nm CMOS technology. Middle-East J. Sci. Res. (2012)
14. Hsieh, G.C., Hung, J.C.: Phase-locked loop techniques—a survey. IEEE Trans. Ind. Electron. 43 (1996)
15. Banerjee, D.: PLL Performance, Simulation, and Design (2006)
16. Dabhi, R.A., Napgara, B.H.: A low power 1 MHz fully programmable frequency divider in 45 nm CMOS technology. IJIRST–Int. J. Innov. Res. Sci. Technol. 1 (2014)
17. Prasad, V., Sharma, C.: A review of phase locked loop. Int. J. Emerg. Technol. Adv. Eng. IJETAE 2 (2012)

Phase-Locked Loop (PLL)-Based Frequency Synthesizer for Digital Systems Driving

R. N. S. Raphael, Agord M. Pinto Jr., Leandro T. Manera and Saulo Finco

Abstract This work describes the implementation and operation features for a Phase-Locked Loop (PLL) architecture-based frequency synthesizer for clock generation and digital systems driving. From a programmable structure, considering an input reference frequency $F_{REF} = 50$ MHz, schematic level simulation results indicate the possibility for generation of 3 distinct output frequencies, according to the transient response limits: T_P (peak time) = 1.9 μs, T_S (settling time) = 2 μs, and M_P (maximum overshoot) <8%. The system was implemented through Cadence Virtuoso Analog Environment (ADE) from UMC CMOS technology (0.18 um), considering power supply $V_{DD} = 1.8$ V.

Keywords Frequency synthesizer · Phase-locked loop · Third-order system · Dynamic modeling

1 Introduction

Frequency Synthesizers (**FS**) are widely employed for a diversity of applications involving wireless communication systems: generation of clock signal for driving Analog-to-Digital Converters (ADC), Clock and Data Recovery circuits (CDR) and digital systems, and generation of Local Oscillator (LO) signal for receiver and transmitter modules, considering the oscillator structure in Radio Frequency (RF) transceivers. From the presence of operating trade-offs and conflicting requirements for design, a proper **FS** implementation represents a critical challenge in fully integrated communication systems [1–9].

R. N. S. Raphael (✉) · S. Finco
Center of Information Technology Renato Archer (CTI), Campinas, Brazil
e-mail: raphael.souza@cti.gov.br

S. Finco
e-mail: manera@fee.unicamp.br

A. M. Pinto Jr. · L. T. Manera
University of Campinas (Unicamp), Campinas, Brazil

Y. Iano et al. (eds.), *Proceedings of the 4th Brazilian Technology Symposium (BTSym'18)*, Smart Innovation, Systems and Technologies 140,
https://doi.org/10.1007/978-3-030-16053-1_38

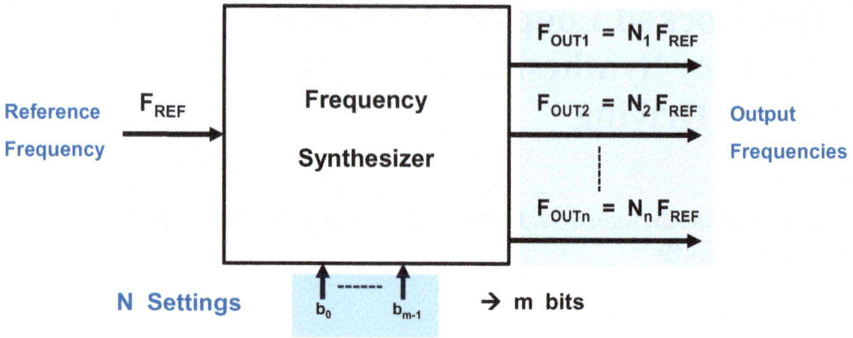

Fig. 1 Frequency synthesizer: general representation for output frequency generation

From the general representation in the Fig. 1, **FS** can be defined as an electronic module for a programmable generation of a given set of output frequencies $\mathbf{F_{OUT}}$, from an input reference frequency $\mathbf{F_{REF}}$, considering a predefined frequency range $\Delta\mathbf{F_{OUT}}$ and a constant or variable step for frequency shifting [8, 9].

Thus, according to the diagram, considering the steady-state operation of **Phase-Locked Loop** (**PLL**) architecture-based **FS**, the resulting output frequency $\mathbf{F_{OUT}}$ can be generally determined from the product between the reference input frequency $\mathbf{F_{REF}}$ and the corresponding divide ratio or frequency division factor $\mathbf{N = F_{OUT}/F_{DIV}}$.

Representing the focus of this work as structure for **FS** implementation, a **PLL**-based architecture is a nonlinear feedback system applied to the frequency and phase control for allowing a programmable output frequency generation. From this operation concept, distinct application or modeling criteria have been proposed through the technical literature considering different block level or system level approaches [3–5].

In this way, this work presents the set of technical features for **FS** implementation considering a type II third-order integer-**N PLL**-based frequency synthesizer and, in this case, the operating features of the system include (1) input reference frequency $\mathbf{F_{REF}} = 50$ MHz, (2) divide ratio $\mathbf{N} = 4$, 8 or 16 and (3) output frequency $\mathbf{F_{OUT}} =$ 200 MHz, 400 MHz and 800 MHz.

From Cadence Virtuoso Analog Design Environment (ADE), the schematic level implementation of the system was performed by applying CMOS UMC 0.18 μm. Thus, Sect. 2 describes the technical data for implementation (system level block diagram and block level building features), Sect. 3 analyzes the system level characterization curves considering 3 steps of settings and simulation, and Sect. 4 presents the final considerations.

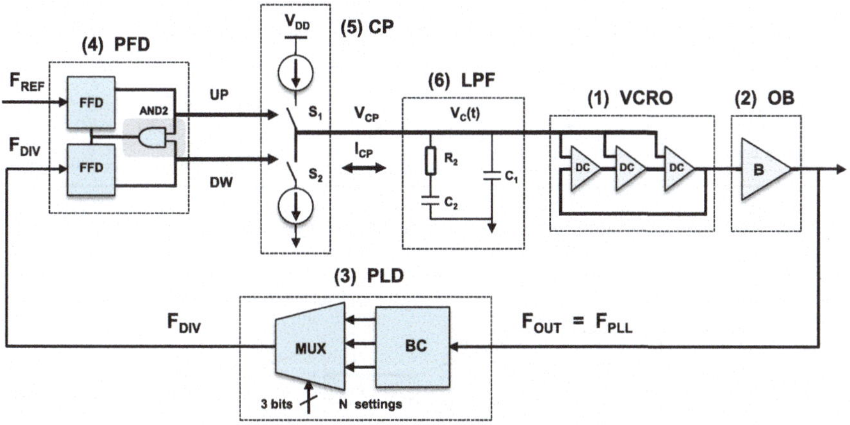

Fig. 2 PLL general block diagram: block level structure representation

2 PLL Block Level Implementation

For direct representation effects, Fig. 2 schematizes the reference **PLL** architecture-based **FS** with the applied structure for each building block, as follows [8, 9]:

(1) **Voltage-Controlled Ring Oscillator** (**VCRO**): analog module for **PLL** output phase $\mathbf{\Phi_{OUT}}$ and frequency $\mathbf{F_{OUT}}$ generation from an input control voltage $\mathbf{V_C}$.
(2) **Output Buffer** (**OB**): analog module for **PLL** output generation (CMOS standard format) for driving external loads and **PLD** (feedback structure).
(3) **Programmable Loop Divider** (**PLD**): digital module for frequency division and divided frequency signal $\mathbf{F_{DIV} = F_{OUT}/N}$ generation considering a given divide ratio or division factor **N**.
(4) **Phase-Frequency Detector** (**PFD**): digital module for **UP** and **DW** signals generation for driving **CP**, considering a phase and frequency comparison from two inputs ($\mathbf{F_{REF}}$ and $\mathbf{F_{DIV}}$).
(5) **Charge Pump** (**CP**): analog module for output current $\mathbf{I_{CP}}$ generation for driving **LPF** from **UP** and **DW** signals (**PFD** outputs).
(6) **Low-Pass Filter** (**LPF**): analog module for a linear filter operation, generating a control voltage $\mathbf{V_C}$ from $\mathbf{I_{CP}}$ for **VCRO** frequency control (output frequency $\mathbf{F_{OUT}}$).

Thus, the next subsections schematize the block level implementation features for each building block.

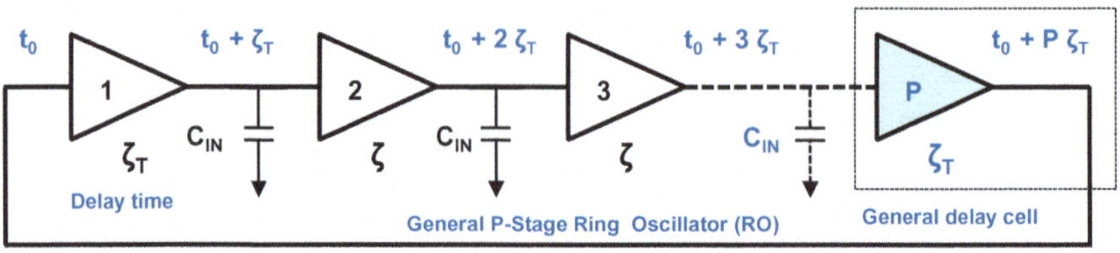

Fig. 3 VCRO structure: general representation for nonlinear analysis

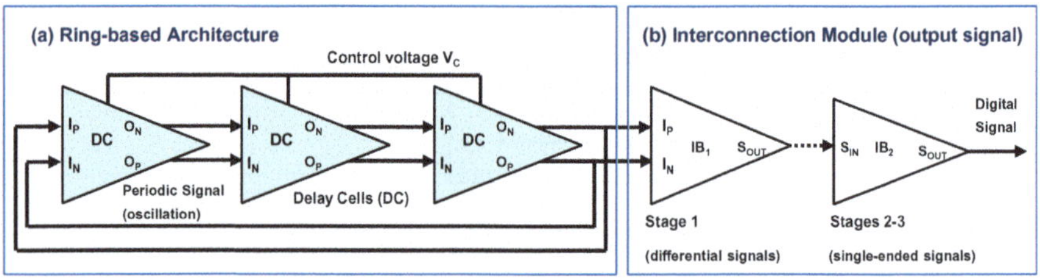

Fig. 4 VCRO block diagram: **a** ring-based structure (RO) and **b** interconnection buffer (IB)

2.1 Voltage-Controlled Ring Oscillator (VCRO)

A **Voltage-Controlled Ring Oscillator** (**VCRO**) was designed for **PLL** output frequency generation, considering area restrictions and the frequency range for application. In a general representation, an oscillating ring-based structure is composed by a set of cascaded stages or delay cells in a closed loop [8–13], as schematized in the Fig. 3.

In this case, for nonlinear analysis effects, considering **P** cascaded equivalent stages with an input capacitance $\mathbf{C_{IN}}$, a rise time $\boldsymbol{\tau_R}$, a fall time $\boldsymbol{\tau_F}$ and a resulting delay time (total delay time) $\boldsymbol{\tau_T = \tau_R + \tau_F}$, the corresponding oscillating frequency $\mathbf{F_{OUT}}$ [10] can be given through the Eq. (1):

$$F_{OUT} = \frac{1}{N(\tau_R + \tau_F)} \overset{\tau_1=\tau_2=\tau_T}{\longleftrightarrow} \frac{1}{2N\tau_T} \tag{1}$$

From these reference concepts, the proposed structure for **VCRO** implementation is composed through 2 operating modules, according to the Fig. 4. Thus, as the oscillating structure of the system, the primary module is composed by 3 self-biased fully differential delay cells for frequency generation through input voltage-based frequency control, as illustrated in the Fig. 4a.

As an interface for signal processing between the frequency generation and the output load, the secondary module is composed by a 3 stages-based **Interconnection Buffer** (**IB**) for (a) insulating of the ring structure (provision of a constant capacitive load), (b) signal conditioning for adequation to the CMOS standard format, and (c) increasing of the driving capability. Thus, the first **IB** stage generates a single-ended

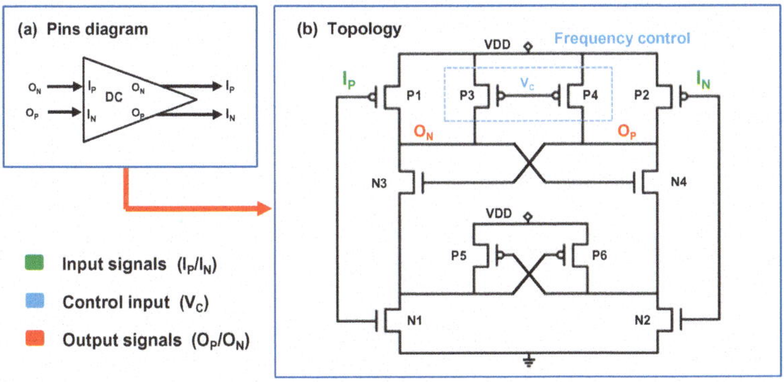

Fig. 5 VCRO (schematic level): **a** delay cell pins diagram and **b** delay cell topology

output from differential inputs and the 2 subsequent stages work in a single-ended operation for driving the feedforward path loads (**PLL** loads and **PLD** in the feedback path), as indicated in the Fig. 4b.

Considering this general scheme for several classes of applications, different transistor level structures and block level architectures can be identified in the literature for **delay cell** topology and **interconnection** pattern implementation [10–14].

According to the schematic level illustration in the Fig. 5b, the **delay cell** proposed in this work considers the application of a fully differential self-biased topology for output frequency generation with voltage-based frequency control. In this case, the transistor level description involves 2 pairs of complementary inputs (P1/N1 and P2/N2), a PMOS-based parallel cross-coupled pair (P5/P6), and a NMOS-based series cross-coupled pair (N3/N4). Additionally, as highlighted in the blue dotted line, a PMOS-based parallel pair (P3/P4) is applied for output frequency definition considering the induction of current and input capacitance $\mathbf{C_{IN}}$ variation with the input control voltage $\mathbf{V_C}$.

From the sequence in the schematic level representation through the Fig. 6b, the first **IB stage** topology applies a double NMOS-based differential pair (N1/N3 and N2/N4) for collecting differential inputs, a parallel PMOS-based cross-coupled pair (P3/P4), and a series PMOS-based pair (P1/P2) for single-ended output generation in the common drain region. Finally, as represented in the Fig. 6c, the remaining stages considers the application of the inverter-based topology for single-ended signal processing and CMOS standard output generation.

2.2 *Phase-Frequency Detector (PFD)*

From the application of customized digital cells, the **Phase-Frequency Detector** (**PFD**) implementation considers the tri-state architecture (2 flip-flops and 1 logic

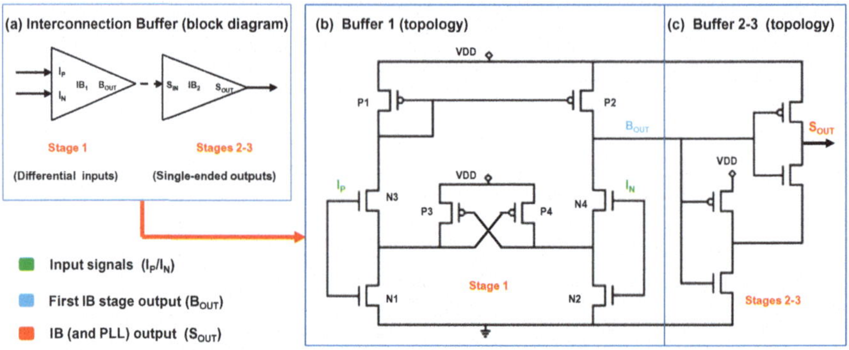

Fig. 6 VCRO (schematic level): **a** IB block diagram, **b** buffer 1 and **c** buffers 2–3

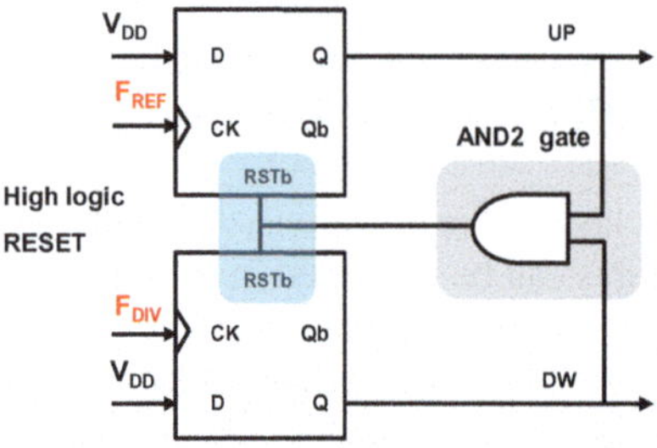

(b) Truth Table

UP	DW	F_{REF}	F_{DIV}	UP (next)	DW (next)
0	0	↑		1	0
0	0		↑	0	1
1	0	↑		1	0
0	1		↑	0	1
1	0		↑	0	0
0	1	↑		0	0

Fig. 7 PFD: **a** Tri-state architecture and **b** corresponding truth table

gate) [8, 9], according to the Fig. 7a, whose operation is summarized according to the truth table in the Fig. 7b.

Thus, the customized digital library involves a clock rising edge sensitive flip-flop D (FFD) with asynchronous reset and differential data processing from single-ended signal input, as represented in the Fig. 8.

In this case, from the master-slave-based structure, the FFD building features include 2 input inverters for complementary signal generation (clock and data), as indicated in the Fig. 8a, 2 C^2MOS structure-based latches for composing the data flow transfer and control mechanism, according to the Fig. 8b, and an additional SRAM (with proprietary topology) memory device-based output unit for data storage and driving capability empowering, according to the Fig. 8c.

2.3 *Programmable Loop Divider (PLD)*

The **Programmable Loop Divider** (**PLD**) operates as an interface in the **PLL** feedback path between the output frequency $\mathbf{F_{OUT}}$ and the input reference frequency

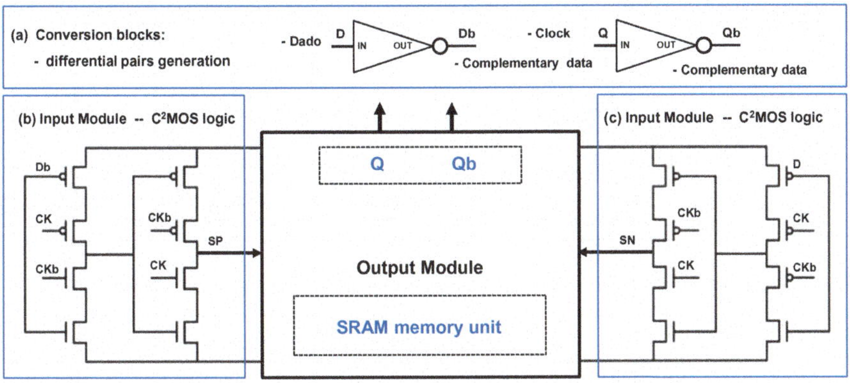

Fig. 8 Customized FFD: **a** input inverters, **b** latches and **c** output SRAM memory

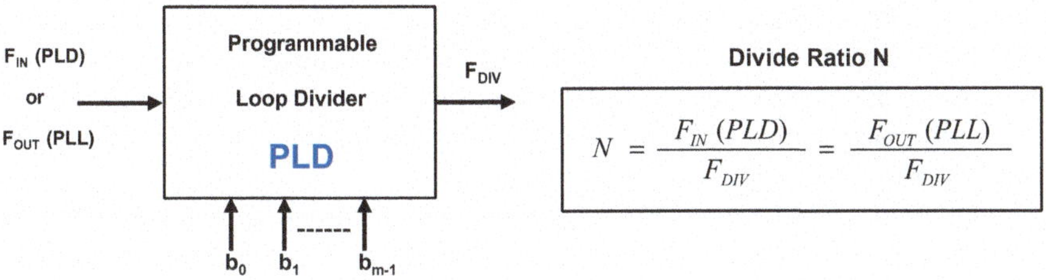

Fig. 9 PLD (programmable loop divider): general pin diagram representation

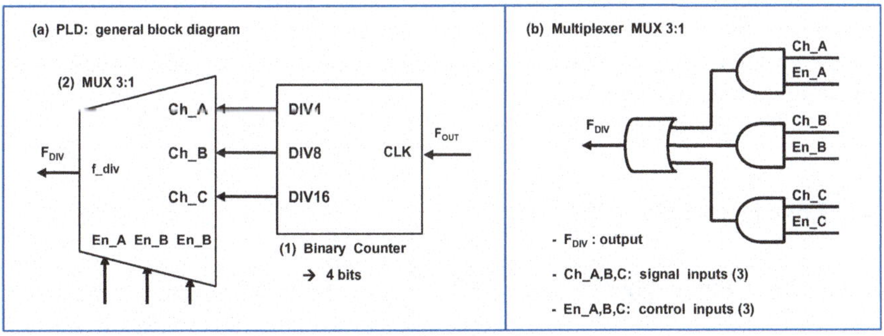

Fig. 10 PLD: **a** general block diagram and **b** Multiplexer MUX3:1

$\mathbf{F_{REF}}$, considering the divided frequency $\mathbf{F_{DIV}}$ generation for a given divide ratio **N** (defined from m setting bits) for phase and frequency comparison effects in the PFD [8, 9], as represented in the Fig. 9.

According to the block diagram in the Fig. 10a, the implementation model adopted in this work involves an integer-**N**-based programmable structure, considering 2 operating modules: (1) Binary Counter and (2) Multiplexer.

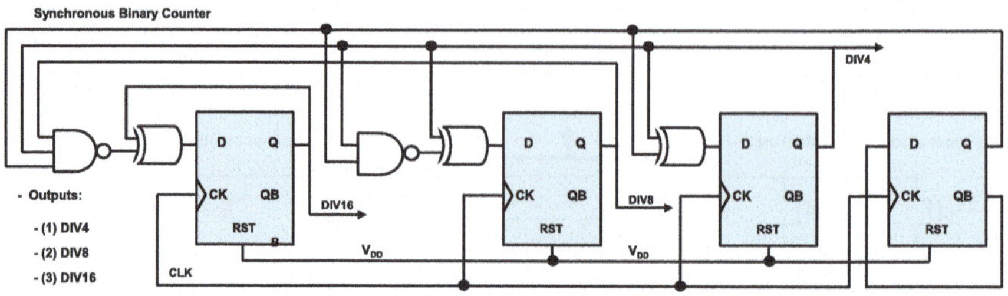

Fig. 11 PLD: synchronous structure with 3 outputs for parallel frequency generation

Implemented with a simplified combinational structure, according to the Fig. 10b, the multiplexer MUX3:1 allows a controlled selection of the frequency generated from the Binary Counter (divided frequency $\mathbf{F_{DIV}}$) through a binary code with 3 mutually exclusive enable bits (En_A, En_B, En_C) for divide ratio settings ($\mathbf{N} =$ 4, 8, 16). Comparatively to MUX architectures with m setting bits and 2^m possible setting states, the applied MUX involves m setting bits and m setting states. In this case, for the same number of setting bits, there is a comparatively reduced number of setting states and logic gates for implementation.

Additionally, implemented in a synchronous architecture by applying the referred FFD structure, according to the Fig. 11, the 4 bits Binary Counter allows the simultaneous or parallel frequency generation from different outputs. Comparatively to counters with asynchronous structure, synchronous counters operate with higher stability or lower sensitivity to phase oscillations (timing jitter) and, on the other hand, demands a wider set of hardware components (combinational logic in the feedback structure), what implies in increased area and power consumption and limitations in the maximum operating frequency.

2.4 Low-Pass Filter (LPF)

The **Low-Pass Filter** (**LPF**) is applied for reducing the high frequency components in the **CP** output and providing a stable voltage level $\mathbf{V_C}$ for **VCRO** frequency control. In this case, according to the Fig. 12, a linear second-order structure with passive components was employed for **LPF** implementation.

Thus, considering the third-order operating features of the resulting **PLL** dynamics, the application of stability parameters is required for system modeling and development of the set of design equations, usually associated with the linear components sizing [4, 5, 8, 9]. In this context, a system level Laplace Transform-based linear analysis generates a set of design equations for the **LPF** components sizing from stability parameters (phase margin PM), operating parameters (bandwidth ω_0) and block level behavioral parameters.

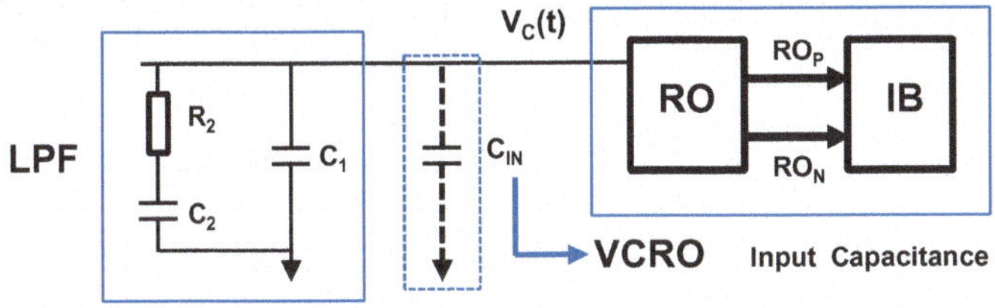

Fig. 12 LPF: linear topology with VCRO input capacitance representation

In this case, from the first modeling step, Eq. (2) indicates a second-order linear transfer function for dynamic representation of the **LPF** topology:

$$H_{LPF}(s) = \frac{(1+\tau_2 s)}{C_T s(1+\tau_1 s)} \tag{2}$$

From the model in Eq. (2), a set of complementary expressions can be derived for the coefficients (time contains τ_1/τ_2 and total capacitance $C_T = C_1 + C_2$) as a function of the filter components (C_1, C_2, R_2), as indicated through the equations in (3):

$$\tau_2 = R_2 C_2 \leftrightarrow \tau_1 = \frac{R_2 C_1 C_2}{C_T} = \frac{R_2 C_1 C_2}{C_1 + C_2} \leftrightarrow C_T = C_1 + C_2 \tag{3}$$

In a second modeling step, PM and ω_0 can be correlated with the system level open loop dynamic properties for the time constants determination, according to the equations in (4):

$$\tau_1 = \frac{\sec(PM) - \tan(PM)}{\omega_0} \rightarrow \tau_2 = \frac{1}{\omega_0^2 \tau_1} \tag{4}$$

In the sequence of the modeling, considering the bandwidth ω_0 as the frequency for unitary modulus of the open loop transfer function ($|\mathbf{H_{OL}}(\boldsymbol{\omega_0})| = 1$), the expression for the total capacitance $C_T = C_1 + C_2$ can be derived, according to the Eq. (5), by applying the previously determined time constants and the set of block level parameters: divide ratio N, output current I_{CP} and VCRO tuning sensitivity K_{VCO}.

$$K = \frac{I_{CP}}{2\pi} K_{VCO} \rightarrow C_T = \frac{K}{N} \frac{\sqrt{1+(\tau_2\omega_0)^2}}{\omega_0^2 \sqrt{1+(\tau_1\omega_0)^2}} \tag{5}$$

In the final sequence of the process, the **LPF** capacitances (C_1, C_2) can be sized from the correlation with C_T and with the previously determined open loop time constants, and the resistance (R_2) can be sized from the second time constant and the corresponding capacitance, according to the equations in (6).

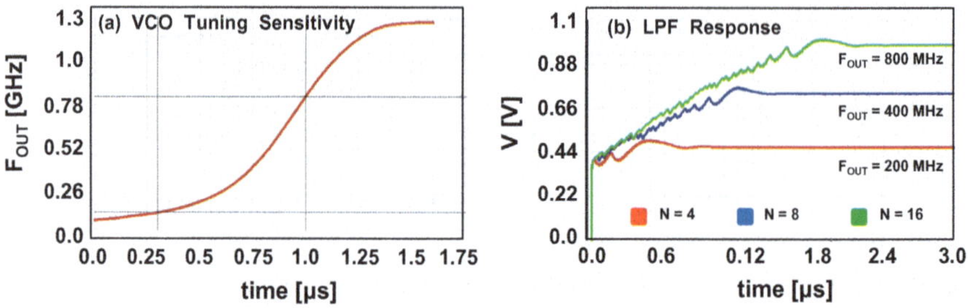

Fig. 13 PLL characterization: **a** VCRO tuning sensitivity, **b** LPF response (N = 4, 8, 16)

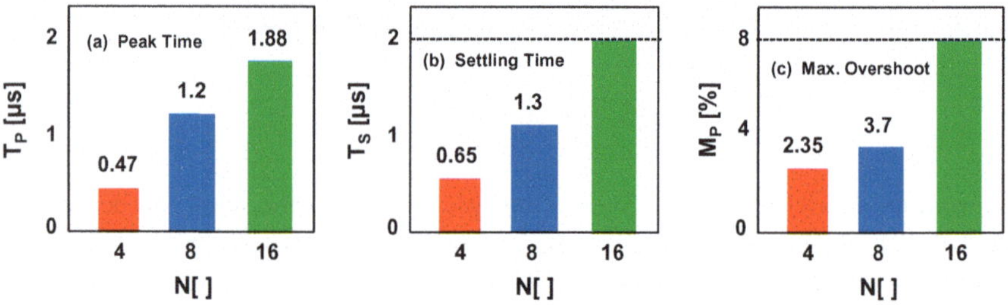

Fig. 14 LPF response characterization (bars chart): **a** T_S, **b** T_P and **c** M_P

$$C_1 = C_T \frac{\tau_1}{\tau_2} \rightarrow C_2 = C_1 \left(\frac{\tau_2}{\tau_1} - 1 \right) \rightarrow R_2 = \frac{\tau_2}{C_2} \tag{6}$$

3 Simulation Results

Considering a block level verification, Fig. 13a illustrates the obtained VCRO characterization curve (nonlinear correlation between the output frequency F_{OUT} and the input control voltage V_C), indicating a frequency range $\Delta F_{OUT} = 1.04$ GHz ($F_{MIN} = 110$ MHz, $F_{MAX} = 1.25$ GHz) and the corresponding input voltage range $\Delta V_C = 1.5$ V. In this case, the region highlighted in the curve indicate $K_{VCO} = 984$ MHz/V.

Additionally, from the system level performance verification, the LPF response (variation profile) is characterized for different divide ratio settings (N = 4, 8, 16) considering $F_{REF} = 50$ MHz, as indicated in the Fig. 13b.

In this case, from LPF variation profile in the Fig. 13b, the set of bar charts in the Fig. 14 summarizes the numerical characterization for 3 transient response-based performance parameters: (a) $\mathbf{T_P}$ (peak time), (b) $\mathbf{T_S}$ (settling time) and (c) $\mathbf{M_P}$ (maximum overshoot).

Thus, the set of data from the bar charts illustrates that each setting state (N = 4, 8 or 16) and the corresponding frequency ($F_{OUT} = 200$, 400 or 800 MHz) impacts over the dynamic features of the system, generating a given transient response and an associated stability condition. Hence, a rising sequence for N variation is associated to a rising trend for $\mathbf{T_P}$ and $\mathbf{T_S}$ and a falling trend for $\mathbf{M_P}$ and, as result, the setting state for the maximum output frequency ($N_{MAX} = 16$, $F_{MAX} = 800$ MHz) corresponds to the maximum time interval for the transient response and, in the other setting extreme, the minimum frequency ($N_{MIN} = 4$, $F_{MIN} = 200$ MHz) corresponds to the fastest transient response.

4 Conclusions

This work presented the technical and implementation characteristics for a third-order PLL-based programmable frequency synthesizer for clock generation and digital systems driving. The programmability for output frequency generation is associated with different input states for divide ratio N definition, and results in the variability of the system dynamics, what implies in the variation and dispersion for the transient a steady-state performance parameters. In this case, even considering the rising and approximately linear variation profile for T_S and T_P, M_P demonstrates a higher level of dispersion around an average line, indicating the nonlinear operating features of the system. Hence, critical variations in the system performance require additional compensation or control structures for minimizing the parameters dispersion and allowing operation with wider stability condition.

References

1. Reinhardt, V., Gould, K., McNab, K., Bustamante, M.: A short survey of frequency synthesizer techniques. In: 40th Annual Frequency Control Symposium (1986)
2. Hanumolu, P.K., Brownlee, M., Mayaram, K., Ku-Moon, U.: Analysis of charge-pump phase-locked loops. IEEE Trans. Circuits Syst. **51**, 1665–1674 (2004)
3. Singh, M.K., Bhattacharyya, T.K., Dutta, A.: Fully integrated CMOS frequency synthesizer for ZigBee applications. In: 18th International Conference on VLSI Design held jointly with 4th International Conference on Embedded Systems Design, pp. 780–783 (2005)
4. Mandal, D., Bhattacharyya, T.K.: 7.95 mW, 2.4 GHz fully-integrated CMOS integer N frequency synthesizer. In: 20th International Conference on VLSI Design held jointly with 6th International Conference on Embedded Systems (VLSID'07) (2007)
5. Mandal, D., Battacharyya, T.K.: Implementation of CMOS low-power integer-N frequency synthesizer for SoC design. J. Comput. **3** (2008)
6. Loveless, T.D., Massengill, L.W., Bhuva, B.L., Holman, W.T., Witulski, A.F., Boulghassoul, Y.: A hardened-by-design technique for RF digital phase-locked loops. IEEE Trans. Nucl. Sci. **53** (2006)
7. Benguang, H., Zhongjie, G., Longsheng, W., Youbao, L.: A single-event-hardened phase-locked loop using the radiation-hardened-by-design technique. J. Semicond. **33** (2012)
8. Sánchez-Sinencio, E., Shu, K.: CMOS PLL Synthesizers Analysis and Design. Springer (2005)
9. Plett, C., Rogers, J.W.M.: Radio Frequency Integrated Circuit Design. Artech House (2010)
10. Mandal, M.K., Sarkar, B.C.: Ring oscillator: characteristics and applications. Indian J. Pure Appl. Phys. **48** (2010)

11. Singh, M., Ranjan, S.M., Ali, Z.: A study of different oscillator structures. Int. J. Innov. Res. Sci. Eng. Technol. **3** (2014)
12. Aranda, M.L., Días, O.G., Álvarez, C.R.B.: A performance comparison of CMOS voltage-controlled ring oscillators for its application to generation and distribution clock networks. Sci. J. Circuits (2013)
13. Demartinos, A.C., Tsimpos, A., Vlassis, S., Souliotis, G.: Delay elements suitable for CMOS ring oscillators. J. Eng. Sci. Technol. Rev. (2016)
14. Loveless, T.D., Massengill, L.W., Bhuva, B.L., Holman, W.T., Casey, M.C., Reed, R.A., Nation, S.A., McMorrow, D., Melinger J.S.: A probabilistic analysis technique applied to a radiation-hardened-by-design voltage-controlled oscillator for mixed-signal phase-locked loops. IEEE Trans. Nucl. Sci. **55** (2008)

Non-intrusive and Intrusive Energy Monitoring Methods Overview and Their Relation with Household Appliances State Sensors Devices

Talita Benedá and Leandro T. Manera

Abstract This work deals with different types of residential energy monitoring methods, comparing their advantages and disadvantages. More specifically, this works focus on the relationship between these methods and the usage of state sensors devices. They are classified as: ILM (*Intrusive Load Monitoring*), where power information is acquired per household appliance and NILM (*Non-intrusive Load Monitoring*), where only one meter is used to get the total energy consumption information. Although, the first method is more effective for regular consumers to help saving energy, both of them provide good energy estimation values. We have shown that when comparing both methods, the state sensor devices can assist to reduce computational efforts for NILM disaggregation algorithms and to infer energy estimate with ILM techniques. Comparing with the commercial devices, the last method can be considered an important and low cost solution, with only 10% error rate.

Keywords NILM and ILM · State sensor devices · Energy efficiency

1 Introduction

The continuous feedback about the energy consumed can lead to significant energy savings. The annual costs reducing can achieve around 12% considering a real-time feedback energy consumption per appliance for end users [1]. This level is reduced when the information is gathered at no-real time or at the building-level. So, the behavior change effectiveness about electric power usage is directly associated with level information methods.

Supported by CnPq and Unicamp.

T. Benedá (✉) · L. T. Manera
University of Campinas (Unicamp), Campinas, Brazil
e-mail: t.beneda@dsif.fee.unicamp.br

L. T. Manera
e-mail: manera@dsif.fee.unicamp.br

Y. Iano et al. (eds.), *Proceedings of the 4th Brazilian Technology Symposium (BTSym'18)*, Smart Innovation, Systems and Technologies 140,
https://doi.org/10.1007/978-3-030-16053-1_39

Hart [2] has suggested two concepts to obtain the amount of energy consumption for residential loads. He introduced the ILM (*Intrusive Load Monitoring*) and NILM (*Non-Intrusive Monitoring*) concepts based on the energy meter intrusiveness in a house, and the electrical load signatures information.

This paper is organized as follows. Section 1.1, provides a brief explanation of NILM and ILM methods and their implementation forms. Section 2 presents some types of load and state sensors devices. Section 3, reports how these sensors devices and these methods can aid saving energy and costs.

1.1 NILM and ILM

The NILM method determines the energy consumption per appliance, it is based on the detailed analyses of current and voltage acquired by one meter installed on the distribution panel. Then, it is possible to obtain the active power per time, (as depicted Fig. 1), as long as the reactive power and power factor.

By using computational techniques, the individual load power can be obtained. This method is called Disaggregation and it is divided in three steps: Event Detection, Recognizing and Consumption Estimation [3].

From the total real power signal acquired, the first part is to determine the rising and falling edges in order to employ the events changes for the next step load operation. It will indicate if the load is turned on or off. According to literature, Hart presents an event detection method that uses arithmetic mean among real power samples. It is the oldest detection method and it is named derivation method.

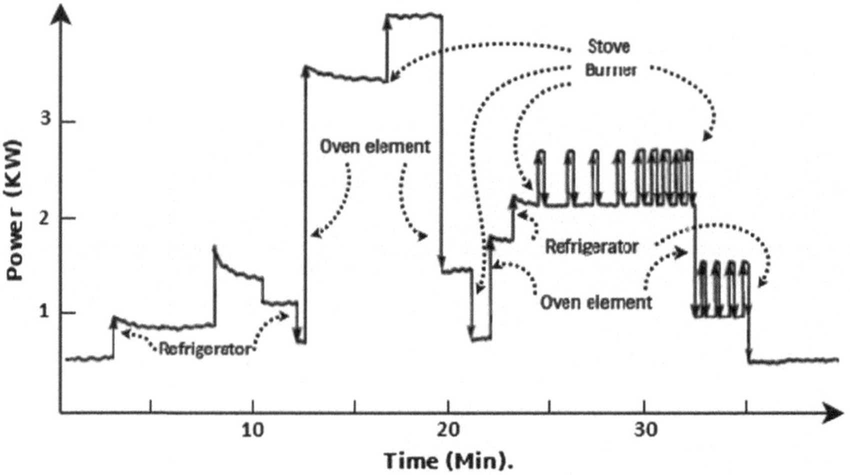

Fig. 1 Real power *versus* time—the aggregated measurement power of the total building [2]

This detection methodology was detailed by Dong [4]. Their approach determines noise limits and the samples below of its are wasted. Monzani [5] has suggested even more analyses for noisy situations by using moving average. New methods, as proposed by Azzini [6] considers two calculus windows to filter transient signals for shifted samples and dispersion measurements.

In addition, Sultanem [7] presents an event method that is appropriated for commercial and industrial environments. Usually, these places deploy electrical machines that presents large starting torque. He proposes an algorithm that identify several changes on real power signal to determine load turning on or off for motors.

The second part uses electrical load signatures to recognize load operation. It can be done by soft computation, supervised and no-supervised approaches [6]. The supervised approach needs a specific data set of the individual load signatures acquired earlier. This step is named as training data set.

The non-supervised approach does not consider the training data set since it uses similar data to determine each appliance. The soft computational methodologies explore tolerances using imprecision and uncertainties results to obtain treatable, robust and low cost solutions [8].

In the last procedure, the load signature (i.e. total energy consumption) is estimated by integrating the Real Power along the time curve, as presented in Eq. 1. Usually, the integration procedure estimation uses squares and trapeziums geometric shapes placed over the load signatures area curves [6].

$$Energy[W.h] = \int_{t_{On}}^{t_{off}} P(t)dt \tag{1}$$

The NILM method implementation needs just one meter. It is a cheaper solution but the computational cost is higher and complex. Moreover, its installation is not feasible for regular consumers.

A second monitoring method is called ILM. This method consists installing power meters between the outlet and appliance, allowing them to obtain appliance-specific energy consumption. Then, the resulting sum of each meter produces the total energy estimation of a house. Commercially, this meter device is named smart plugs.

Although, to install these meter devices is required enter into the user house, turning it in an intrusive method, it has become increasingly easy to install for regular consumers, since it is a plug.

Nowadays, the management systems of NILM and ILM together have wireless communication facilities, allowing them to present the energy consumption results for end users in real time. Commonly, the smart plugs are linked with Apps that provides amount energy estimation per appliance, operation time duration and daily report linked with the current tariff so to estimate real time costs. Some devices can also actives loads to turn on or off. Some commercial brands of smart plugs are Wemo [9], TP-link [10] and Peanut [11].

2 State Sensor Devices

As described above, ILM monitoring method provides an energy estimation per appliance. The metering power is instantaneous and the data acquired provides the energy progress on load operation. By observing the power load behavior is possible to look for similarities to find systematic function shapes called load signature.

The load signatures can be defined as the variation of the power load with time. They are grouping in electrical and non-electrical signatures. The first, voltage and current load parameters vary in a specific according to the type of load. The non-electrical signatures, such as sound, magnetic field and light, are also used to infer appliances operation (see Fig. 2).

Metering electricity per load can be done directly and indirectly. In general, direct power devices need to be installed or coupled into appliance. They are used to acquire electrical signatures.

In the indirectly measurements the meter devices are installed near to appliances and gather no-electric characteristics (light, magnetic field or sound) to deduce the loads operation. They provide no-electric signatures.

Usually for both direct and indirectly measurement the signal presence (or not) can indicate the appliances state. However, associated works about indirect methods combines more than one load characteristic to determines the load operation. They look for redundancy check.

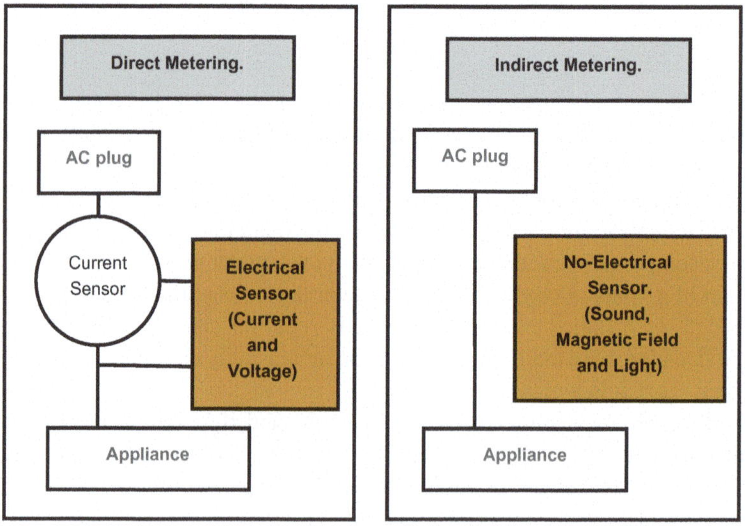

Fig. 2 Diagram of direct and indirect energy intrusive metering per appliance—illustration based on [12]

The device that senses several power changes level could indicate whether the appliances are turning on or off. These state sensor circuits is called state sensor devices (SSD) in this work.

Most of the circuits monitors line or neutral conductor to predict the state of one appliance. Then, the load sensing is based on electrical load signature. To appoint if the load is On/Off, these SSD uses transformer current (TCs) coupled to line to measure the current flowing on conductor. Other possibility is to measure the current flow based on differential voltage, using a shunt resistor in parallel to neutral [13].

Other specific load characteristics can be used to determine the appliances state. A refrigerator, for example, produces noise that indicates the compressor operation. So one can use acoustic signature to determine its state. One also can use RF signal to identify household appliances on/off states [14].

According to Srivastana, the magnetic field can change near to a computer and so it can infer whether the PC is activate or off, as presented in Fig. 3. In Srivastana's work magnetic field characteristics, aid electrical signatures to load identification through a system called *Viridi Scope* [12].

Associated works with no-electric signatures can achieve good energy consumption estimation around 10%. Some of them use the signatures to testify the true values and waste falses values to aid the electrical signatures, as in the *Viridi Scope* project [12], others works estimate energy directly with the same information, as present by *TinyEars* [15].

So, besides so many procedures to identify appliances consumption, one of the best approach per appliance is ILM technique since it is cheaper (depending the devices quantify), easy to install and is able to obtain the real time feedback per load helping end users to save energy to simple mode.

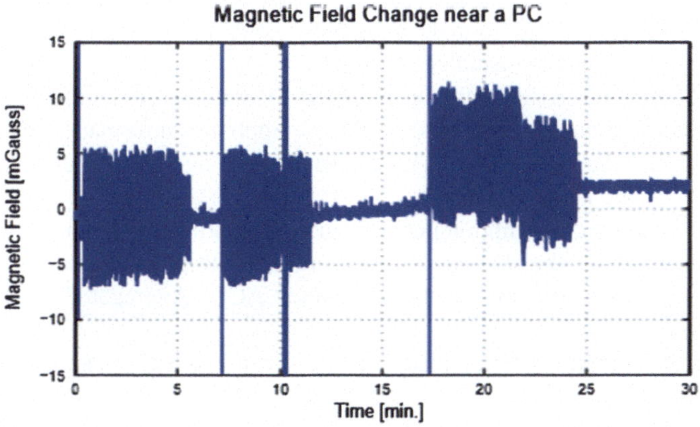

Fig. 3 The magnetic field near a PC. The noisy region indicates that PC is active—Ref. [12]

3 NILM and ILM Assisted by State Sensors Devices

Associated works have been revealed electrical and no-electrical SSDs information can be aid NILM and ILM monitoring methods. These states information can support cheap solutions, optimize and reduce computational costs.

As described above to measure the building energy amount NILM need to carries out Event Detection, Recognize and Consumption Estimation steps. The key of the first and second step is the load identification and activate/deactivate instants, because the total power signal draw acquired by the main meter is aggregated and laborious to select each load.

To diminish NILM computational efforts, these steps can be mitigated due the knowledge about states of appliances and a their identifiers (ID), [2]. It could be indicated how appliance is in operation eliminating the Event Detection.

If a load is detected and associated with an ID, the supervised and clustering non-supervised approaches from Recognize step is unnecessary. According to Azzini [6] the first approach needs labeled events set trained earlier. It could be eliminated due the load previously identified.

The non-supervised approach doesn't require previously training but needs an-others information as power factor (fp), real and reactive power as function time to define an appliance. These information are important to clustering data where the loads will be classified according the similar groups.

Moreover the supervised approach can be eliminated with SSDs and IDs information. Just real power draw signal is considered (fp and reactive power are unnecessary) to Recognize and Consumption Estimation steps.

Considering the whole NILM process, all adjustment made for Event Detection and Recognition steps provide lower storage data and more soft systems. Although it has numerous advantages as smaller computational efforts, the extras low costs sates sensor (per appliance) in addition the main meter could be an expensive solution and uncomfortable to regular consumers.

The ILM method monitoring can be aided by state sensors, supported by the electrical load signatures. The SSDs and IDs per appliance information are gathered on database and they are related with electrical load signature adjusting the time operation indicated by SSDs.

Then, the energy consumption per appliances are estimated and the sum of their energy values provide the amount of total energy consumption estimation of house. According to Azzini [6], this method is named ACE (Aggregated Consumption Estimation).

In addition, Azzini [6] formalized these new forms to gather energy amount by state sensors. Their work has presented ILM− and NILM+ monitoring methods devices. The first is the ACE presented techniques and the second, is SSDs usage supported to electrical load signatures to avoid sames steps NILM the disaggragation method as described above in this work (see Table 1).

It is necessary to take into account that the described ACE measurement method are applied to on/off load types. This loads doesn't work with several switching

Table 1 Brief taxonomic about energy monitoring concepts—according to Azzini [6]

Type of energy monitoring	Device installed per appliance	Device installed on distribution panel
NILM	No	Energy meter
ILM	Energy meter	No
NILM+	State sensors devices (SSDs)	Energy meter
ILM−	State sensors devices (SSDs)	No

states, hence switching states, hence the SSDs associated to signatures constant models can be infer the energy consumption estimation. In addition, most of the common household appliances are of this type.

According to Zeifmam and Roth [3], an house has four loads types: *Permanent Consumer Devices*, devices that remain active 24 h and with constant real and reactive power (as external power supplies), *Finite State Machines (FSM)*, includes devices that switching states on operation cycle (as washing machine), *Continuous Variable Consumer Devices*, that includes devices with variable and aperiodic power draw (dimmer lights) and, *On/Off Appliances* as described before (as toaster, light bulb etc.).

Comparing NILM and ILM methods aided by SSDs, both of them use the load signature with interesting purposes, more especially, the NILM to simplify computational effort and ILM to saving costs, because most of state sensors devices has cheaper project and materials.

Note that, NILM is not an intrusive method but NILM+ needs SSDs to aid detect and infer an load, then it needs to be also intrusiveness. Then, look at the intrusiveness concepts they have to turning inappropriated to classified a energy monitoring methods, but it has been the best term to associated works as yet.

In spite of these advantages, in general they are not an accurate energy estimation methods and also the exactly SSD depends depends on the communication network stability and speed. And in most of the cases, the communication can also be a limiting factor for energy estimation of a house.

4 Conclusion

There are many low cost ways to estimate energy consumption for a house. This work has highlighted the NILM and ILM monitoring concepts to validate them and also to have an overview of many associated works.

The key concept is the meters intrusiveness installation. If the meter device is installed on each outlet, the embraced monitoring method can define ILM (Intrusive Load Monitoring). Usually, this method for each appliance measurement is a feasible installation for end users.

If only one meter is installed on the distribution panel, the metering is called NILM (Non-Intrusive Load Monitoring). The total building energy is aggregated and one uses a computational method to load disaggregation.

As described above, devices have used NILM and ILM methods to reduce the computational data efforts. More especially, Azzini [6] summarized and formalized these concepts in: NILM, NILM+, ILM and ILM−.

Associated works indicates that SSDs have presented another household appliances characteristics beyond electrical signatures to indicate their states, as for example, TinyEars with acoustic signatures.

The SSDs are able to inform loads activation/deactivation patterns using a simple and low cost way, but they are not able to detect the appliances power stages operation, as washing machines for example. Then, SSDs should be employed only in solutions that considers On/Off load's type.

Although the assisted solutions by SSDs have good results in order to saving energy and money, each energy monitoring method has its disadvantages. Therefore, it is important to define how parameters are relevant for an specific usage and their technical installation intrusiveness limitations, so to mitigate computational efforts, to have good estimation values of energy consumption estimation per appliance and a suitable solution for each scenario.

References

1. Karen, E., Kat, A.D., John, A.: Advanced Metering Initiatives and Residential Feedback Programs: A Meta-Review for Household Electricity Saving Opportunities, Tech Rep., vol. E105, pp. 39–42. American Council for Energy Efficient Economy, Washington, DC (2010)
2. Hart, G. W.: Nonintrusive appliance load monitoring. In: Proceeding of the IEEE, Washington, DC, USA, vol, 80, pp. 1870–1891 (1992). https://doi.org/10.1109/5.192069
3. Zeifmam, M., Roth, K.: Nonintrusive appliance load monitoring: review and outlook. IEEE Trans. Consum. Electron. 76–84 (2011). https://doi.org/10.1109/ICCE.2011.5722560
4. Dong, M.: Decomposition Techniques for Power System Load Analysis. University of Alberta, Ph.D (2013)
5. Monzani, R.C.: Monitoramento não Invasivo e Identificação do consumo de energia individual em cargas residenciais. Ph.D.–University of Campinas, Campinas, Brazil (2016)
6. Azzini H.A.D.: Sistema de Monitoramento de Cargas Residenciais Usando Informações On/Off e Modelos de Consumo, Brazil, Ph.D. Unicamp–University of Campinas pp. 19–70 (2017)
7. Leebs, S., Kirtley Jr., J.L., Levan, M.S., Sweeney, J.P.: Development and validation of a transient event detector. AMP J. Technol. **3**, 69–74 (1993)
8. Zadeh, L.A.: Soft computing and fuzzy logic. IEEE Softw. IEEE **11**, 48–56 (1994). https://doi.org/10.1109/52.329401
9. Wemo Switch Smart Plug: http://www.belkin.com/us/p/P-F7C027/
10. TP link plug: http://www.tp-link.com
11. Peanut Plug: http://www.securifi.com/rg/securifi-peanut-plug
12. Kim, Y., Schimid, T., Charbiwala, Z.M., Srivastava, M.B.: ViridiScope: design and implementation of a fine grained power monitoring system for homes. In: ACM. Proceedings of the 11th International Conference on Ubiquitous Computing (2009). https://doi.org/10.1145/1620545.1620582

13. Wu, T., Srivastava, M.: Low-cost appliance state sensing for energy disaggregation. In: ACM. Proceedings of the Fourth ACM Workshop on Embedded Sensing Systems for Energy-Efficiency in Buildings (2012). https://doi.org/10.1145/2422531.2422542
14. Jung, D., Savides, A.: Estimating building consumption breakdowns using on/off state sensing and incremental sub-meter deployment. In: ACM. Proceedings of the 8th ACM Conference on Embedded Networked Sensor Systems (2010). https://doi.org/10.1145/1869983.1870006
15. Taysi, Z.C.; Guvensan, M.A., Melodia, T.: Tinyears: spying on house appliances with audio sensor nodes. In: ACM. Proceedings of the 2nd ACM Workshop on Embedded Sensing Systems for Energy-Efficiency in Building (2010). https://doi.org/10.1145/1878431.1878439

Part II
Emerging Trends in Human, Smart and Sustainable Future of Cities

Organizational Digital Transformation: A Proposed Instrument for Digitalization Management

Laís Santiago da Costa, Luciana Pereira and Alessandra Cristina Santos Akkari

Abstract Nowadays the business environment is very dynamic and has been influenced by the development of new technologies that affect from the factory floor to the business model. This is a phenomenon that entails the digital transformation of several spheres of the organization, such as leadership, strategy, operations, culture, governance, among others. This work aimed to develop and validate (content and semantics) an instrument to evaluate the maturity degree of organizational digital transformation. The process consisted of four main phases beginning with the (i) establishment of the instrument; followed by (ii) initial validation and restructuring; (iii) then a second evaluation, with representatives of the academic and entrepreneurial spheres; (iv) and ending with layout and settings for digital research platform. It was observed that the evaluators presented a high level of agreement in each one of the questions, also considered relevant for the subject studied (>58%). However, they disagreed on which question relates to each organizational dimension. Therefore, new adjustments were made in order to leave the writing of the questions clearer and more specific in relation to their respective organizational dimension. The evaluation process resulted in a digitalization management instrument with 16 multiple-choice questions, two at each of the eight dimensions: strategy, leadership, products, operations, culture, people, governance and technology. The developed instrument can be applied by any organizational profile and the simplified data processing allows the manager to identify the digitalization maturity degree of the company.

Keywords Digital transformation · Digitalization management · Digital organization

L. S. da Costa · L. Pereira
Federal University of ABC, Santo André, Brazil
e-mail: lais.costa@ufabc.edu.br

A. C. S. Akkari (✉)
Mackenzie Presbyterian University, Campinas, Brazil
e-mail: alessandra.akkari@mackenzie.br

Y. Iano et al. (eds.), *Proceedings of the 4th Brazilian Technology Symposium (BTSym'18)*, Smart Innovation, Systems and Technologies 140,
https://doi.org/10.1007/978-3-030-16053-1_40

1 Introduction

The interaction of consumers with tools, platforms and the digital market changes the business performance scenarios in their entrepreneurial systems, which is a phenomenon that stimulates an organizational digital transformation. About these changes, some variables lead to this movement, for example the market expansion that goes beyond physical barriers, due to constant digital evolution [1, 2]. Other aspects refer to changes in the profile of the costumer, which reaches high levels of interaction and exchange of experiences, as well as the high-digitalized performance of the entire value chain, facilitating communication among stakeholders [3, 4].

In fact, business scenarios are changing with internet, communication technologies, connectivity and others devices that combine the most varied functions by applying these resources at the company processes [5]. This new market scenario affects all parts of the organization, which in turn must evaluate the procedural, structural and functional changes, even go further and shift the business model adopted. The company has to check what digital technologies become alternatives for them and its consequences for the organization, considering specific strategies for each phase of digital transformation [6].

The digital companies present different business models, in which the products, distribution or workplace can have virtual configuration. Then, the more digital the company is, the more international it is likely to become [7]. Therefore, business that trade in the value chain using the digitalization as a base, have some returns in terms of efficiency, lean stock, better customer relationships and sales, market introduction and financial results [8].

It is therefore important to understand how companies can take advantage of the opportunity to innovate, differentiate and grow, managing the digital transformation changes. The purpose of this work was to present the development and validation (content and semantics) of an instrument to evaluate the degree of maturity of organizational digital transformation, considering the applicability for any organization profile.

2 Background

Nowadays the business environment is very dynamic and has been influenced by the development of new technologies that affect from the factory floor to the business model, emphasizing IoT, big data, intelligent robotics, industrial automation, analytics, social media, cloud storage, 3D printing devices, among others [4].

The digitization of information enables the knowledge to be processed universally and stored as data. These data are then available in different execution forms and are disposable for processing, storage or filters, among other applications and utilities. This cycle can happen innumerous times without wear, presenting high speed and low costs. Through the internet, the cycle enters the global environment, reducing multiple barriers. Therefore, digitalization involves the deployment or increase of the use of digital technologies by companies, organizations or any other component of an entrepreneurial ecosystem, affecting its internal scenarios and the external relations spheres [9].

The progressive development of digital technologies influences companies to transform themselves and to reflect on the organizational digital transformation management and innovation development [10]. The organization tries to discover which technological innovations can be attributed in its structures aiming at differentiation and growth in the market. Also, new practices are developed to improve organizational processes at low cost and effectively to respond appropriately to new technologies and customer demands. At that moment the company has the challenge of applying the necessary changes in the business model to obtain the full digital transformation of the organization [2].

That is when, the organization must select which strategy is appropriate to lead to digital transformation. There is no simplified method for this, however it can emerge a participatory process model that engages business leaders in four steps: to identify points of view and interests; analyze agreements and disagreements; discuss the consequences and create a replenishment strategy [11].

In addition to strategy and process management, is highly important that the leader of the change management can lead the transition of knowledge. For that a significant collaboration is required to understanding the new technologies, so the leadership can surpass the role of facilitator inside the process. With the insertion of digital technologies into personal and professional spheres, business leaders require new sets of cognitive and emotional skills to deal with the dimensions of digitalization [12].

Other relevant aspect to the occurrence of the digital transformation of a business is the relationship between the various organizational dimensions. It is reported that the application of technology, data integration, digital leadership, human resources management and processes influence the digitalization cycle. These are converted into the resources necessary for the organization to implement the digital transformation strategy, among them external collaboration; preparation of leadership for the dynamism of the digital environment; and the formation of flexible and agile operations. It is also important to present a consumer service unit that is digital, to analyze the necessary changes in the value proposition and facilitate the adaptation of products or services through digital devices [13].

3 Research Approach

The development of organizational digital transformation instrument based on a framework of Azhari et al. [14] which was developed by Neuland consultancy in cooperation with the Digital Business Research Center in Germany. The instrument developed foresees a data collection to evaluate the degree of digitalization maturity of companies (*unconscious, conceptual, defined, integrated, and transformed*)

Table 1 Maturity in digital transformation of the organization

Score	Percentage	Degree of maturity
Up to 35 points	Up to 20%	Unconscious
Between 36 and 70 points	Between 21 and 40%	Conceptual
Between 71 and 105 points	Between 41 and 60%	Defined
Between 106 and 140 points	Between 61 and 80%	Integrated
Above 141 points	Between 81 and 100%	Transformed

Source Authors (2018)

according to organizational dimensions (*strategy, leadership, products, operations, culture, people, governance and technology*) to ensure that digital transformation occurs in the entire organization [14].

The study began with the translation of the material, once the model was published in German. Then, it was formulated multiple choice questions (closed-ended) using the phrase completion scale which presents a continuous score from 0 to 10, in which "0" corresponds to an association with the missing attribute, while "10" relates to the maximum intensity of attribute presence. Thus, the ordinals 0, 1 and 2 are related to the *very low existence* of the attribute, whereas, the numbers between 3 and 7 appear *moderate existence* of the attribute. Finally, the values 8, 9 and 10 correspond to a *very high existence* of the attribute [15].

The instrument allows the data processing to be performed by the company itself through the execution of a simple average. Thus, according to the response interval and the maximum score, it is possible to categorize the maturity level of the organizational digital transformation, according to Table 1.

The instrument validation stages considered the content, application time, structure and semantics. A first phase occurred with university professors (3) and the second step went with a different team, composed by entrepreneurs (4). An analysis roadmap was developed in Microsoft Excel to guide specialists.

After the identification of the perception about the instrument, the consonance among the evaluators was analyzed by numerical indicators represented by the Inter-rater Agreement (IRA) and the Content Validation Index (CVI) [16]. IRA was used for the purpose of estimating the agreement of the results of expert evaluations, considering the number of participants agreeing to the original item by total number of participants. CVI was calculated by the proportion of the specialists that evaluated the item with relevance in grade 3 or 4.

4 Results and Discussion

The process presented to develop and validate the digitalization management instrument consisted of four main phases beginning with the establishment of the instru-

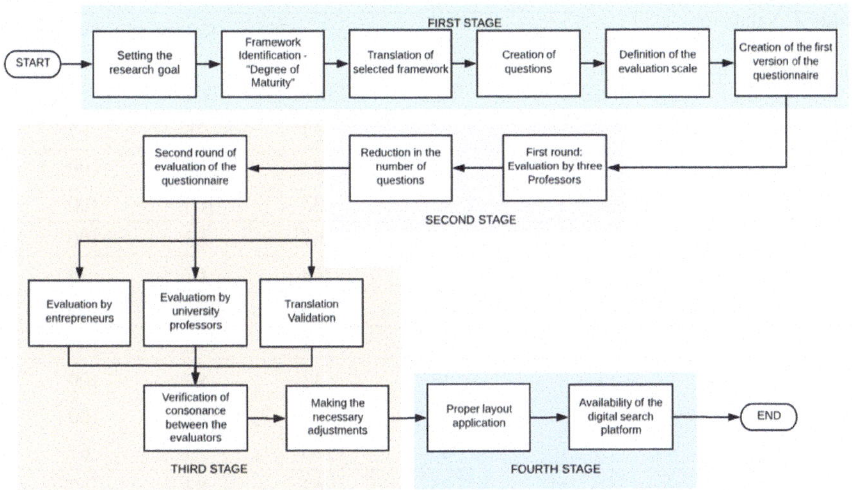

Fig. 1 Validation of the digitalization management instrument

ment (I), followed by initial validation and restructuring (II), a second evaluation (III) and ending with layout and settings for digital research platform (IV) (Fig. 1).

The first stage was composed of six phases, starting by assigning objectives to the research and phenomena studied. Then, there is the identification of the framework that can answer the question-problem and response to the objectives. At this point, the framework of Degree of Maturity in Digital Transformation [14] was selected. This framework was published in German, so it was necessary to translate the material, which initially occurred through the Microsoft translation system in the Office 365.

Then, the formulation of 24 closed-ended questions for data collection occurred. However, in order to make the questionnaire applicable to different types of segments and organizational sizes, it would be necessary to have a scale that would be easy to understand and meet different profiles, reduce response time and possible doubts.

To meet these requirements, the "Phrase Completion" scale was chosen. It was observed that the response procedure used in the instrument made the choice of the respondent easier, assigning a numerical scale with 11 options, since there is usually a natural familiarity of the individuals with that mention in other environments, such as in organizational evaluations. Another motivating factor for choosing this scale is that it represents a logical structure and better probabilistic results, promoting an increase in the reliability of the questionnaire. Following the process of Fig. 1, the first version of the questionnaire was finished, with the representation of eight organizational dimensions and three questions each.

Then, the second stage was started, characterized by the evaluation of the instrument by three professors in digital area, whose responsibility was to analyze the content of the questionnaire versus the objectives defined for the study. After this evaluation, it was indicated that the digitalization management instrument had a high

Table 2 Validation indicators of the instrument for digitalization management

Question	Organizational dimension		Relevance	IVC (%)
	%	IRA (%)	%	
1	25	67	75	67
2	13	33	100	100
3	13	33	100	100
4	25	67	83	100
5	25	68	67	33
6	13	30	100	100
7	13	33	100	100
8	13	33	83	100
9	13	33	92	100
10	40	100	75	67
11	25	67	83	67
12	25	67	83	67
13	13	33	75	67
14	13	33	92	100
15	30	67	83	67
16	25	67	58	33
17	25	30	83	67

Source Authors (2018)

number of questions which would cause a very high average response time and possibly decrease the interest of application by the organizational managers. Therefore, the instrument was adjusted according to the observations of the specialists.

The third stage started with a new evaluation process, which was made by another two university professors and two entrepreneurs, so the process was performed by seven different people. It was considered four levels of relevance of the content covered in the instrument, the number "1" was categorized as *non-relevant or non-representative item* and the "4" as *relevant or representative item*. After identifying the perception of the participants, the consonance among the evaluators was verified by numerical indicators (Table 2).

It was observed that the assessors presented a high level of agreement on each of the issues, also judged relevant to the subject studied (>58%). However, they disagreed on which question relates to each organizational dimension. This aspect is important for respondents to be able to understand and classify maturity in digitalization of their company. Therefore, new adjustments were made in order to leave the writing of the questions clearer and more specific in relation to their respective organizational dimension.

Also, the evaluation criterion of the instrument resulted in the exclusion of question number 17, which the team was about the implementation of new digital technologies.

Since the technology dimension already presented other two questions, the question 17 would only lengthen the questionnaire, increasing the average response time. It was also necessary to improve other two questions 4 and 7, fixing the texts available in the scale, because they had no words that quantified gradually the degree of existence of that criterion in the company.

Another point of adjustment was the change in the introductory text of the instrument, so that the respondent understands better how to make use of the management tool. This was also an observation made by all participants in the validation process, which also showed their interest in knowing how the instrument could be applied or be useful in the digital transformation management.

The evaluation process resulted in a digitalization management instrument with 16 questions, two at each of the eight dimensions: strategy, leadership, products, operations, culture, people, governance and technology.

5 Conclusion

The instrument was developed and validated, considering the content and the semantics. It was verified that digitalization management instrument meets the identification of the dimensions of the digital transformation as indicated in the literature, because it presents questions to recognize the main organizational areas of the business.

It is also effective to visualize the interaction between the dimensions encompassing digital transformation and the need to consider aspects of strategy, leadership, products, operations, culture, people, governance and technology. Therefore, the instrument directs the spheres that need to be managed in the process of digitization, collaborating to reach the status of the digital company.

The developed instrument can be applied by any organizational profile and the simplified data processing allows the manager to identify the digitalization maturity degree of the company. Besides that, when analyzing the score obtained in each organizational dimension, the manager can observe the weaknesses of the business regarding the digital transformation. This finding allows directing efforts and optimizing the digitization process in order to gain market competitiveness.

References

1. Gimpel, G., Westerman, G.: Shaping the Future: Seven Enduring Principles for Fast Changing Industries. MIT Center for Digital Business (2012)
2. Berman, S.J.: Digital transformation: opportunities to create new business models. Strateg. Leadersh. **40**(2), 16–24 (2012)
3. Hafezieh, N., Akhavan, P., Eshraghian, F.: Exploration of process and competitive factors of entrepreneurship in digital space: a multiple case study in Iran. Educ. Bus. Soc. Contemp. Middle East. Issues **4**(4), 267–279 (2011)

4. Nambisan, S.: Digital entrepreneurship: toward a digital technology perspective of entrepreneurship. Entrepreneurship Theor. Pract. **41**(6), 1029–1055 (2017)
5. Gottschalk, P.: Research propositions for knowledge management systems supporting electronic business. Int. J. Innov. Learn. **3**(6), 593–606 (2006)
6. Matt, C., Hess, T., Benlian, A.: Digital transformation strategies. Bus. Inf. Syst. Eng. **57**(5), 339–343 (2015)
7. Hair, N., Wetsch, L.R., Hull, C.E., Perotti, V., Hung, Y.T.C.: Market orientation in digital entrepreneurship: advantages and challenges in a Web 2.0 networked world. Int. J. Innov. Technol. Manage. **9**(6), 1250045 (2012)
8. Oliveira, T., Martins, M.F.: Understanding e-business adoption across industries in European countries. Ind. Manage. Data Syst. **110**(9), 1337–1354 (2010)
9. Organisation for Economic Co-Operation and Development: Data-Driven Innovation: Big Data for Growth and Well-Being. OECD Publishing (2015)
10. Nambisan, S., Lyytinen, K., Majchrzak, A., Song, M.: Digital innovation management: reinventing innovation management research in a digital world. Mis Q. **41**(1) (2017)
11. Hansen, A.M., Kraemmergaard, P., Mathiassen, L.: Rapid adaptation in digital transformation: a participatory process for engaging is and business leaders. MIS Q. Executive **10**(4) (2011)
12. Sahyaja, C., Rao, Dr. K.S.Sekhara.: New leadership in the digital era: a conceptual study on emotional dimensions in relation with intellectual dimensions. Int. J. Civil Eng. Technol. **9**(1), 738–747 (2018)
13. Nadeem, A., Abedin, B., Cerpa, N., Chew, E.: Digital transformation & digital business strategy in electronic commerce: the role of organizational capabilities. J. Theor. Appl. Electron. Commer. Res. **13**(2), i–viii (2018)
14. Azhari, P., et al.: Digital Transformation Report. Neuland GmbH & Co. KG, Köln (2014)
15. Hodge, D.R., Gillespie, D.: Phrase completions: an alternative to Likert scales. Soc. Work Res. **27**(1), 45 (2003)
16. Serra, M.M.P., Ayres, L.F., Santos, M.A., Ribeiro, G.S., Pansonato, V.: Validation and adaptation Scale Denilson Culture (DOCS) for large brazilian companies. Espacios **37**(35), 3 (2016)

Data Science Supporting Smart City Management: A Predictive Analysis Perspective

Giovanni Moura de Holanda, Cristina Y. K. Obata Adorni and Jorge Moreira de Souza

Abstract This paper aims at presenting an R&D view on how Data Science may be inter-related with smart city management, especially in terms of supporting predictive analyses. Trends on this fast-growing scenario are pointed out as well as some experiences and applications that the authors' institution has built up or may come to develop.

Keywords Data science · Smart city · Predictive analysis

1 Introduction

Technologies and solutions for smart cities generate large amount of data, requiring special structure for dealing it. Digital technologies become present in various sectors of a city, in fields of application such as mobility, transport, energy, security, water supply, business and ambient intelligence [1–3]. The underlying technologies, such as internet of things (IoT), sensor networks, image-based services, geographic information systems and real time applications, among others, respond by a dizzying growth of data and contribute to leverage the complexity of digital world to an unpreceded level [4, 5].

Application features that contribute to print a smart atmosphere in an urban conglomerate pose specific problems for engineers, statisticians, strategists and public administrators [1, 6]. Beyond the challenge of dealing with a flood of data, the new technologies and applications demand intelligent management support in order to explore their full potential functionalities.

G. M. de Holanda (✉) · C. Y. K. Obata Adorni · J. M. de Souza
FITec—Technological Innovations, 79, Aguaçu St, Campinas, SP 13098-321, Brazil
e-mail: gholanda@fitec.org.br

C. Y. K. Obata Adorni
e-mail: cadorni@fitec.org.br

J. M. de Souza
e-mail: jmdsouza@fitec.org.br

Y. Iano et al. (eds.), *Proceedings of the 4th Brazilian Technology Symposium (BTSym'18)*, Smart Innovation, Systems and Technologies 140,
https://doi.org/10.1007/978-3-030-16053-1_41

In smart city initiatives, these issues are being addressed by means of big data techniques (cf. [4, 7–9]). The large volume of data that began to emerge with this evolving scenario has opened new fields of study and allowed new opportunities for industry (products and services), society and public administration. Depending on the context, the complexity becomes even greater, involving a multitude of applications with different levels of criticality and requirements, and with data being generated by diverse sources.

Increased complexity due to the combination of data collected from many sources may hinder the applicability of methods and techniques to classify them, in such a way that by overcoming this first analytical obstacle can be a big step towards making smart use of big data [10]. Data Science (DS) has demonstrated potential to overcome the new challenges and reveal alternatives to enable the opportunities arising with the expanding data mass.

Data mining and artificial intelligence techniques have allowed machines to learn what to do as well as analyses able to predict and prevent problems before they happen. The results of this kind of analysis have been applied, for example, in traffic prediction [8] and predictive maintenance [11]. Operational practices that anticipate failures and enable to schedule maintenance with minimal impacts on service delivery have met increasingly pressing needs, both in terms of reducing energy waste and operating costs, as well as increasing the quality of services and the effectiveness of public policies.

In this context, this paper present an R&D view on how Data Science may be interrelated with smart city management, especially in terms of supporting predictive analyses. Some principles and analytical methods are shortly introduced in order to provide a conceptual background and facilitate the visualization of predictive efforts. Trends on this fast-growing scenario are pointed out as well as some experiences and applications that the authors' institution has built up or may come to develop.

2 A Brief on Smart City Concepts and Data Science Principles

In order to clarify some underlying concepts and terminology that scientists, engineers and implementers have been using to put data science into practice, this section shows two descriptive items. The first discusses about the label "smart city" and the second is related to the classification and short description of some methods that make data science a discipline adequate for prediction analyses and able to support the management of complex systems.

2.1 When a City Is Considered Smart?

The idea of "smart city" has been widely used in recent years, initially by suppliers of information systems and digital sensors, which can be integrated into urban services and infrastructure. Since then, the label "smart city" is often used interchangeably with other expressions, such as digital city, virtual city. Furthermore, it is associated to a number of definitions in the literature and most of them is centered on the technological dimension, specifically in terms of networked infrastructure. However, for Caragliu et al. [12], the focus on the internet as the most significant smart city identifier is insufficient.

Besides the industry, many institutions have adopted this label and, with the maturation of the discussions and experiences, the "smart city" concept has been extended and incorporated other dimensions. The European Commission, for example, has aligned the use of new technologies with the objective of sustainable development of cities, encouraging partnerships to promote progress in areas such as energy production, transport and ICT, and thus improve services and contribute to reducing natural resource consumption [13].

Even though it is widely disseminated, the concept of smart city is still not clearly defined. Many efforts to define or describe it are observed in the literature (e.g., [3, 12–16]) and many of them highlight the role that digital technologies play in the process of attributing intelligence to cities and urban conglomerates.

Among the technological characteristics that have contributed to designating cities as smart, Harrison and Donnley [15] emphasize the widespread use of sensor networks and digital control systems in the operation of urban infrastructure, and the development of computational power and new algorithms that allow analyzing the flow of information and providing guidance to improve operational performance. These authors also draw attention to the need of establishing a theoretical foundation in order to guide the actions in smart cities, which are conducted by players in various urban segments.

From a practical point of view, the presence of digital technologies supporting services and applications in cities has been verified both in *ad hoc* implementations—focused on a specific segment of infrastructure—and in the level of service integration, acting at a more municipal level. Irrespective of model cases, where a city experiences an integrated solution of intelligent services, coordinated by the municipality and covering various sectors of urban infrastructure (e.g., Stockholm [17], Barcelona [17] and Singapore [18]), some cities start with more localized applications, and often the initiative comes from the very sectors that feel the need to improve their services and operational processes.

In this sense, there is a growing demand for more efficient, effective and sustainable solutions [6], based on intelligence technologies that allow increasing prediction and operational management capabilities—a condition often observed in utilities and Industry 4.0, for instance. It is based on this demand that we highlight the potential of data science applications and guide the analytical contextualization presented as follows.

2.2 Some Analytical Methods in Data Science

Data Science is a nascent field of science, requiring a multidisciplinary approach and keeping a close relation with data driven technologies [19]. The term has been adopted in several contexts although not in an established way. Regarding its main objective, we share the vision presented by Hayahsi et al., i.e., this science aims to reveal the hidden structure of complex phenomena with data from a different point of view, which implies multidimensional, dynamic and flexible ways of thinking [20].

Data Science has been established as a discipline derived from the fusion of fields of knowledge as Artificial Intelligence (AI), mathematics, statistics and, naturally, computer science. Most of these disciplines are part of the daily lives of many analysts and researchers, but what does AI really mean?

The definition of AI is field of controversy: for example, sometimes considered as computational intelligence, sometimes the adjective "artificial" is questioned under the argumentation that it may connote "simulated" leading to other misunderstandings [21].

Several disciplines have contributed to background and evolving AI: philosophy, mathematics, economics, neuroscience, psychology, control theory, computer science, and linguistics (see [22]). It is interdisciplinary by nature and its origins goes back to the beginnings of human civilization.

A more recent historical trajectory can be shortly traversed from Charles Babbage, in the 19th century, with his idea of "analytical machine", passing through some researchers in the 30s and 40s, until assuming a more current format with the influence of the computational theory of Alan Turing in the early 50s (cf. [22, 23]). In his article entitled "Computing Machinery and Intelligence", Turing [24] inquiries about the possibility of machines thinking and discusses the idea of digital computers.

This field of knowledge has grown tremendously since then, with the development of many methods and techniques applied in a number of areas. Its rapid diffusion and wide reach have also led to some conceptual inaccuracies—for example, there is a certain ambiguity in the use of terms like AI and computational intelligence (CI), on the one hand, and machine learning and data mining on other. Obviously, such a polysemic trend does not compromise the effectiveness and potential of this analytical body, but it hampers the initiation of new AI developers and does not favor the consolidation of a common terminological basis between the various disciplinary fields.

In the midst of the rich terminology and several techniques that have been characterizing the discipline, Rutkowski [23] points out that the term "computational intelligence", which is currently used in English literature, is understood as computational means to perform numerical calculations aimed at solving AI problems. Among the techniques that allow this approach are, to name a few: neural networks, fuzzy logic, evolutionary algorithms and probabilistic methods [23].

In an evolutionary line, it is possible to instantiate in three stages the escalating of AI techniques. Such partitioning has only the intention of making clearer the conceptualization and terminology underlying the Data Science. According to the

Table 1 Methods and techniques for predictive modeling

Artificial intelligence (AI)			
Decision tree		Numeric calculation for AI	Predictive modeling ↓
Statistical tools			
Probabilistic methods			
Neural networks		Computational intelligence supporting AI	
Fuzzy systems			
Computational linguistics			
Evolutionary algorithms			
Machine learning	Clustering	Big data and computational intelligence for AI	
	Random forest		
	Markov models		
	Deep learning		

Synoptic table based on Refs. [22, 23, 25]

historical perspective developed by Rutkowski [23], classical probabilistic methods and induction methods, such as the decision tree, formed the initial basis for calculating AI problems. With neural networks and fuzzy logic, a new step was achieved. And more recently, data mining and machine learning have promoted a further analytical and technological leap, making machines more "smart" and predictive models more accurate. Table 1 presents a particular summary of this evolutionary process, based on the view of some academics and adapted to our analytical practices.

An in-depth description of the entire set of methods and techniques listed in Table 1 is beyond the scope of this paper, however it is possible to bring the focus to the latter step and introduce some information on data mining and machine learning. Recalling that the objective is only to contextualize such techniques regarding predictive analysis.

Big data is a phenomenon driven mainly by the increase in computational capacity observed in the last two decades, as well as in data storage capacity. Moreover, the very characteristics of new technologies and applications (e.g., IoT, smart phones, and a plethora of sensors—from commercial goods up to vehicles) have contributed to shape big data era, as usually named in the literature.

In general, Big Data is characterized by 4 Vs: Volume, Variety, Velocity and Veracity (some authors have included more Vs into the list). Data mining is therefore an enormous challenge to be faced. In a first view, Roiger and Geatz [25] relate data mining to a process in which computer learning techniques are applied to obtain knowledge from large amount of data.

Machine Learning assumes an important role concerning such a challenge. For Jordan and Mitchell [26], machine learning is not only a matter of computers able to learn from data, but it is also related to laws that applies to learning systems in general. For these authors, machine learning is in the core of artificial intelligence and data science.

Neural networks, clustering, decision trees, random forests (as an ensemble method based on several decision trees) algorithms have been properly applied for training and learning, elevating the computational intelligence to a higher floor. Even before machine learning development, Markov modeling has been supported innumerous approaches in reliability prediction, significantly since 70s (cf. [27]), and more recently it has also been used as machine learning technique for pattern recognition.

With big data opportunities, some data representations result in high level of abstraction, demanding automated mechanisms to be extracted. Deep Learning algorithms arise as an effective method to deal with such a need [28–30].

In short, all these techniques and approaches may be used in models and algorithms with the purpose of carrying out predictive analysis. The output generate by such models may be either numeric or categorical, and the nature of data that analysts have available does condition whether the model is adequate for predicting or simply for classifying the inputs.

Many applications developed to attend urban needs may benefit from prediction and control analysis. When combined, the aggregate effect can help not only the operational management but also policy makers and managers in charge of smart cities initiatives.

3 Predictive Applications to Urban Issues

Monitoring indicators, data integration and predictive analyses may be used for controlling operations in strategic sectors of urban infrastructure. Figure 1 suggests how an R&D cycle may run to help monitoring and control functions in such sectors, developing, for example, applications to ambient intelligence, transportation and mobility, industry, energy and water utilities. Anticipating failures and avoiding service interruptions, reducing operational costs, improving service quality and supporting decisions, are benefits that big data, computational intelligence and predictive analysis may provide.

Such a scenario becomes more definite as computational intelligence techniques advance and the demand for urban solutions and sustainable development grows continuously. In this section, we describe generically how data science is being applied to some sectors and point out R&D alternatives that can be adopted in Brazilian scenario.

3.1 A Fast-Growing Scenario

New demands and emerging technologies have shaped an ever-changing scenario that meets the predictive potential of artificial intelligence and data science. Sectors of urban infrastructure may make use of intelligent algorithms able to provide predictive

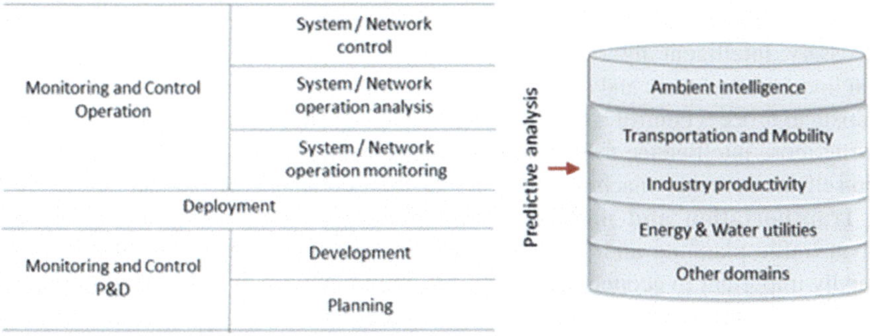

Fig. 1 R&D supporting predictive analysis for urban issues

information and thus improve their operational performance. Sectors such as utilities, transportation and mobility, industry and ambient intelligent systems are beneficiaries of this potential, by using methods and techniques as described below.

Ambient intelligence. Smart algorithms acting on consumer electronic devices has arisen the need for computational intelligence also in this field of application. An intelligent environment has mechanisms to adapt itself to people live in it. In such an environment people lives surrounded by networked devices that provide information and digital media in accordance to individuals' demand, and even anticipating their needs.

This area introduces new challenges to R&D field, mainly by bringing together qualitative aspects and computational intelligence [31, 32]. The models and algorithms in this field of research have to consider the involvement of the user, learning from the ambient and adapt it to user's preferences.

Energy and water utilities. In power generation, transmission and distribution, including renewable sources, machine learning and big data analytics are used to monitor and analyze plant availability, weather and historical series in order to predict future outcomes. Data may be in various format—image, audio, video and sensors, for example—and after being extracted are integrated to mining and analysis process. In general, analytics and artificial intelligence applies to prediction of outage and power demand [33], fault detection and classification [34], corrective and preventive actions [35], and asset management (see [36]), to mention a few field of applications.

In the operation of a water supply system, the inherent complexity of its network imposes obstacles to obtaining real-time measures and even significant data in order to monitor and control service quality. In this context, Artificial Intelligence may be used to estimate the distribution network health in order to more quickly detect anomalies. In contributions to this field, Izquierdo et al. [37], for example, use neural network and Gamboa-Medina and Reis [38] use a genetic algorithm for leak detection.

Industrial productivity. The transformations that have driven the so-called 4th Generation Industrial Revolution (or Industry 4.0) are based mainly on IoT and cyber-physical systems. As more automation, data exchange and embedded intelligence are integrated in industrial production, predictive technologies are further necessary. Intelligent algorithms have been used to predict degradation in production line, reduce losses and autonomously optimize production and product quality. Statistical process control is now reinforced with data mining and machine learning, giving more intelligence to production management, at a time in which self-aware and self-maintenance machines arise in industrial big data environment (see [39]).

Transportation and mobility. Predictive maintenance in transportation, the urban traffic planning and prediction of traffic conditions represent other aspects equally important to economics and quality of life in modern cities. Transportation and mobility are crucial to a healthy functioning of a city. Again, computational intelligence is applicable to improve operation in a sector with such challenges. By way of illustration, Li et al. [11] apply statistical methods and machine learning to historical and real-time data to predict conditions of failure in rail networks.

Other domains. Education, public health and government may benefit from big data and artificial intelligence as well. This data can be used for analysis and extracting useful trends and using them to promote enhanced education by means of learning analytics that can proportionate, for example, student's performance prediction and attrition risk detection [40]. Big data approaches can contribute to health sector by providing insights into the causes of diseases, as well as by enhancing prediction and prevention of epidemics [41]. Yet, the processing and integrated analysis of large amount of data are helpful for evaluating public policies (see, e.g., [42]).

Obviously, the scenario is dynamic and other sectors not mentioned above may make use of computational intelligence. Furthermore, intelligent algorithms can be part of integrated solutions of monitoring and control, congregating two or more urban infrastructures and services, for example, in public policy evaluation or in the logistic domain.

3.2 Some R&D Experiences and Potential Applications: An Own Vision

Several R&D projects applying Data Science in predictive analysis have been conducted by the institution of the authors of this paper. The expertise thus built provides a vision of how to meet specific needs of sectors such as those presented and discussed heretofore. The contextualization of the solution to attain such needs, which often gain particular contours, not only gives originality to the projects, but also makes it possible to introduce innovations of processes and products in the organization demanding R&D activities.

Specific characteristics of the context in which these organizations operate—such as climatic, topological and socioeconomic conditions—pose development challenges whose solutions require the exercise of adaptation and flexibility. Frequently, the technological challenge requires a multidisciplinary composition of experiences, resulting in partnerships with two or more research institutes, in addition to the orga-

Table 2 Some R&D experiences applying data science

R&D Experience	CI technique	Institutions
Predictive system of fault occurrence in electrical networks [43]	Data mining and Laplace trend chart	Celg D, FITec, Unicamp
Fuzzy system for defect point identification in electric power distribution systems [44]	Fuzzy Logic	Celg D, FITec, PUC-Goiás
Remote monitoring of transformers in self-healing systems for substations and electrical distribution networks [45]	Data mining	Energisa, FITec
Optimization methodology for the allocation of automated keys in the Celg electric distribution network [46]	Genetic algorithms	Celg D, FITec
Fault monitoring using data mining and statistical analysis [47]	Data mining, Control chart and OLAP[1]	Eletrosul, FITec
Alarm management using statistical analysis and data mining [48]	Data mining, and OLAP	CPFL, FITec
Method for correlating alarms of electric power supply and telecom networks [49]	Fuzzy logic	AES Eletropaulo, FITec
Analysis of financial default [50]	Data mining and clustering (K-means)	AES Eletropaulo, FITec
Reference model for smart grid applied to smart cities deployment [51]	Decision tree [52], fuzzy inference [53]	Elektro, FITec, USP-SC, PUC-Rio, UNESP

OLAP On-line Analytical Processing

nization requesting the project itself. In this way, the sum of acquired knowledge allows drawing an own view on the potential of application of the Data Science to meet urban needs of a market with the endogenous characteristics of Brazil. To illustrate the vision thus consolidated, some themes of projects already executed are presented in Table 2, as well as a list of solutions being formatted for possible application in this scenario of increasing demands is shown in Table 3.

Based on the experience built and on the prospection of sectors demanding solutions to support their operational management, Table 3 lists some Data Science (DS) applications already formatted as R&D project.

The short description presented in Table 3 is only an enumeration of computational intelligence (CI) methods and techniques that have been identified as potentially adequate to address the technological challenge of the proposed theme. This is to

Table 3 Potential DS applications focusing Brazilian R&D scenario

Project's theme	CI technique
Stochastic models of reliability and availability of wind generators: prediction of unscheduled stops and optimization of preventive maintenance	Laplace trend chart, data mining and clustering
Machine Learning for real-time monitoring of the transformer performance in the electrical distribution network	Deep learning and clustering
Methods and techniques of AI for mapping of vegetation near electrical distribution network, in order to plan and guide pruning, according to ecological principles and operational safety	Deep learning, data mining and clustering
Determination of the optimal number of transformer meters in an electric power network, for more precise location of faults and greater efficiency of maintenance	Data mining, clustering and genetic algorithms
Detection of "hidden disturbances" in electric power network, providing information to avoid disturbances that lead to the burning of equipment or service interruption	Deep learning, clustering and recursive neural network
Predictive algorithms for monitoring assets in electric power networks	Clustering and recursive neural network
Analysis of large volumes of data in real time to optimize the attendance of emergencies in electric power networks	Machine learning with clustering and recursive neural network
Intelligent systems to support operation and decision making in municipal water supply networks	Machine learning with neural networks
IoT based solutions for smart environments, collecting data from distributed sensors and providing support to smart city services and applications	Pre-analysis: measurement, collection, data integration and dashboard

illustrate artificial intelligence approaches for predictive purposes of problems faced by Brazilian companies. The methodological approaches were designed with basis on the experience constructed and according to the taxonomy discussed at the beginning of this paper (which is summarized in Table 1).

4 Discussion and Final Remarks

The fast-growing generation of data has posed some challenges to scientists, engineers, managers and policy makers. At the same time that digitalization and computational features embedded in new equipment are reality in various sector of information society and open unprecedented opportunities to improve quality and efficiency of processes, products and services, many stakeholders do not know exactly what to do with the emergent data flood. Data mining, artificial intelligence, computational intelligence—or whatever we use to name the possibility of learning from data and supporting decisions—have surfaced as pillars to sustain a data science.

This new science fits like a glove to solve typical issues of problems that lead cities to be considered as intelligent. Urban services and infrastructures can directly benefit from intelligent algorithms and all analytical potential that data science may provide to elevate operational management to an unparalleled level. As observed in the literature and market indications, this is a worldwide trend.

In Brazilian scenario, there is a clear demand from sectors aiming at deploying predictive management, in order to improve maintenance and operational procedures as well as quality of service or product. Solutions based on computation intelligence can be developed to support specific segments of infrastructure or help managing several urban services. In the latter case, the solution can behave as a general monitoring and control tool for a smart city.

Nevertheless, alongside the benefits data science may bring to management tasks, new ethical issues arise from data science application. Privacy and personal security are among the main aspects to be faced and preserved in this fast-growing scenario. Big data and computational intelligence applied to fields like ambient intelligence, advertising and fraud detection are pervasive in general. Public policies and ethical codes are fundamental to balance potential business benefits and ethical principles.

In this sense, how Big Data, Analytics and Computational Intelligence may deliver business and social value in a scenario demanding solutions to very particular needs? How to meet all stakeholders' interests and respect individual rights? The answer is not obvious and must be consciously sought in order to accompany the techno-scientific progress, foster sustainable development and assure societal and economics gains.

In Brazilian scenario, there is a clear demand from sectors aiming at deploying predictive resources, in order to improve maintenance and operational procedures as well as quality of service or product.

References

1. Komninos, N.: The age of Intelligent Cities—Smart Environments and Innovation-for-all Strategies. Routledge, New York (2015)
2. Lim, C., Kim, K.-J., Maglio, P.P.: Smart cities with big data: reference models, challenges, and considerations. Cities (2018). https://doi.org/10.1016/j.cities.2018.04.011
3. Novotný, R., Kuchta, R., Kadlec, J.: Smart city concept, applications and services. J. Telecommun. Syst. Manage. **3**(2) (2014)
4. Moreno, M.V., Terroso-Sáenz, F., González-Vidal, A., Valdés-Vela, M., Skarmeta, A.F., Zamora, M.A., Chang, V.: Applicability of big data techniques to smart cities deployments. IEEE Trans. Ind. Inf. **13**(2), 800–809 (2017)

5. Rabari, C., Storper, M.: The digital skin of cities: urban theory and research in the age of the sensored and metered city, ubiquitous computing and big data. Cambridge J. Reg. Econ. Soc. **8**, 27–42 (2015)
6. Guedes, A.L., Alvarenga, J.C., Goulart, M.S., Rodriguez, M.V., Soares, C.A.: Smart cities: the main drivers for increasing the intelligence of cities. Sustainability **10** (2018)
7. Batty, M., Axhausen, K.W., Giannotti, F., Pozdnoukhov, A., Bazzani, A., Wachowicz, M., Ouzounis, G., Portugali, Y.: Smart cities of the future. Eur. Phys. J. Spec. Top. **214**, 481–518 (2012)
8. Hashem, I., Chang, V., Anuar, N., Adewole, k, Yaqoob, I., Gani, A., Ahmed, E., Chiroma, H.: The role of big data in smart city. Int. J. Inf. Man. **36**(5), 748–758 (2016)
9. Nuaimi, E., Neyadi, H., Nader Mohamed, N., Jaroodi, J.: Applications of big data to smart cities. J. Internet Serv. Appl. **6**(25) (2015)
10. Bengio, Y.: Deep learning of representations: looking forward. In: Dediu, A., Martín-Vide, C., Mitkov, R., Truthe, B. (eds.) Statistical Language and Speech Processing—SLSP 2013. Lecture Notes in Computer Science, vol. 7978, pp. 1–37. Springer, Berlin (2013)
11. Li, H., Parikh, D., He, Q., Qian, B., Li, Z., Fang, D., Hampapur, A.: Improving rail network velocity: a machine learning approach to predictive maintenance. Transp. Res. C **45**, 17–26 (2014)
12. Caragliu, A., Del Bo, C., Nijkamp, P.: Smart cities in Europe. In: 3rd Central European Conference in Regional Science, pp. 45–59 (2009)
13. European Commission: Communication from the commission. Smart Cities and Communities—European Innovation Partnership. Brussels (2012)
14. European Union: Cities of tomorrow - Challenges, visions, ways forward. Brussels (2011)
15. Harrison, C., Donnelly, I.A.: A theory of smart cities. In: 55th Annual Meeting of the International Society for the Systems Sciences (2011)
16. Arafah, Y., Winarso, H.: Redefining smart city concept with resilience approach. IOP Conf. Ser.: Earth Environ. Sci. **70** (2017)
17. Angelidou, M.: Four European smart city strategies. Int. J. Soc. Sci. Stud. **4**(4), 18–30 (2016)
18. Lee, S.K., Kwon, H.R., Cho, H., Kim, J., Lee, D.: International Vase Studies of Smart Cities—Singapore, Republic of Singapore. IDB Inter-American Development Bank (2016)
19. Demchenko, Y., et al.: EDISON data science framework: a foundation for building data science profession for research and industry. In: 2016 IEEE International Conference on Cloud Computing Technology and Science (CloudCom), pp. 620–626 (2016)
20. Hayashi, C.: What is data science? Fundamental concepts and a heuristic example. In: Hayashi, C., Yajima, K., Bock, H.H., Ohsumi, N., Tanaka, Y., Baba, Y. (eds.) Data Science, Classification, and Related Methods. Studies in Classification, Data Analysis, and Knowledge Organization. Tokyo: Springer (1998)
21. Poole, D., Mackworth, A., Goebel, R.: Computational Intelligence: A Logical Approach. Oxford University Press, New York (1998)
22. Russell, S. J., Norvig, P.: Artificial Intelligence—A Modern Approach. Prentice-Hall (2010)
23. Rutkowski, l.: Computational Intelligence—Methods and techniques. Springer, Berlin (2008)
24. Turing, A.M.: Computing Machinery and Intelligence. Mind **49**, 433–460 (1950)
25. Roiger, R., Geatz, M.W.: Data Mining—A Tutorial-Based Primer. Addison Wesley, Boston (2003)
26. Jordan, M.I., Mitchell, T.M.: Machine learning: trends, perspectives, and prospects. Science **349**(6245), 255–260 (2015)
27. Arnold, T.F.: The concept of coverage and its effect on the reliability model of repairable system. IEEE Trans. Comput. **22**(3), 251–254 (1973)
28. Witten, I.H., Frank, E., Hall, M.A., Pal, C.J.: Data Mining: Practical Machine Learning Tools and Techniques, 4th edn. Elsevier, Cambridge, USA (2017)
29. Najafabadi, M., et al.: Deep learning applications and challenges in big data analytics. J. Big Data **2**(1) (2015)
30. Lee, K.-S., Lee, S.-R., Kim, Y., Lee, C.-G.: Deep learning–based real-time query processing for wireless sensor network. Int. J. Distrib. Sens. Netw. **13**(5) (2017)

31. Verhaegh, W., Aarts, E., Korst, J. (eds.): Algorithms in Ambient Intelligence. Springer Science & Business Media (2004)
32. Sharma, A., Kumar, A., Bhardawaj, A.: A review of ambient intelligence system: bringing intelligence to environments. Int. J. Inf. Comput. Technol. **4**(9), 879–884 (2014)
33. Jiang, H., Wang, K., Wang, Y., Gao, M., Zhang, Y.: Energy big data: a survey. IEEE Access **4**, 3844–3861 (2016)
34. Jamil, M., Singh, R., Sharma, S.K.: Fault identification in electrical power distribution system using combined discrete wavelet transform and fuzzy logic. J. Electr. Syst. Inf. Technol. **2**, 257–267 (2015)
35. Maia, A.T., Morais, J.M., Pires, Y.P., Rocha, A.B., Martins, D.: Data mining techniques for shutdowns prediction of electric power systems. In: Proceedings of XI Brazilian Symposium on Information System, pp. 155–162 (2015)
36. Momin, A.A.S., Kolekar, S.: Mapping of the assets and utilities: a vision for the development of smart cities in India. Int. J.Appl. Eng. Res. **12**(24), 15378–15383 (2017)
37. Izquierdo, J., López, P., Martínez, F., Pérez, R.: Fault detection in water supply systems using hybrid (theory and data-driven) modelling. Math. Comput. Modell. **46**, 341–350 (2007)
38. Gamboa-Medina, M.M., Reis, L.F.R.: Sampling design for leak detection in water distribution networks. Proc. Eng. **186**, 460–469 (2017)
39. Lee, J., Kao, H.-A., Yang, S.: Service innovation and smart analytics for Industry 4.0 and big data environment. Proc. CIRP **16**, 3–8 (2014)
40. Sin, K., Muthu, L.: Application of big data in education data mining and learning analytics—A literature review. ICTACT J. Soft Comput. **5**(4) (2015)
41. Khoury, M.J., Ioannidis, J.: Big data meets public health. Science **346**(6213), 1054–1055 (2014)
42. Alves, A.M., Holanda, G.M., Silva Junior, D.C.: A web platform for evaluating public policies in smart city initiatives. In: BTSym 2016 Proceedings, vol 1 (2016)
43. Adorni, C.Y.K.O., Passos, L.F.N., Machado, B.B., Murari, C.A.F., Junior, M.A.M.: Modelo de um sistema preditivo de ocorrência de falta. In: Proceedings of XXIV Seminário Nacional de Produção e Transmissão de Energia Elétrica – SNPTEE (2017)
44. Teles, L.H., Carvalho, W., Souza, J.M., Adorni, C.Y.K.O., Sousa, M.A.: Sistema Fuzzy para Identificação de Pontos de Defeito em Sistemas de Distribuição. In: Proceedings of IX Congresso de Inovação Tecnológica em Energia Elétrica – CITENEL (2017)
45. Alves, T.A., Santos, T.T., Filho, E.P., Souza Jr, J., Avelino, M.D., Dias, R.G., Medeiros, A.A., Mafra Jr., J.J., Souza, J.M., Segundo, M.A., Lacerda, R.N.. Experiência da Energisa Paraíba na Implantação de Sistema Self-healing Auxiliado por Monitoramento Remoto de Transformadores. In: Proceedings of IX Cong. Inov. Tec. Ener. Elét. – CITENEL (2017)
46. Moraes, Jr, G.M., Oliveira Jr, J., Silva, S.R., Souza, J.M., Adorni, C.Y.K.O., Cavelucci, C.: Metodologia de otimização para alocação de chaves automatizadas na rede CELG. In: Proc. VI Congresso de Inovação Tecnológica em Energia Elétrica – CITENEL (2011)
47. Scheunemann, L., Silva Neto, E.A., Maciel, A.C., Souza, J.M., Sanches, M.: Utilização de sistemas inteligentes para o processamento de alarmes – Mineração de dados usando OLAP. Technical Report, FITec, Campinas (2007)
48. Menezes, R.S., Curtarelli, S.R., Adorni, C.Y.K.O, Souza, J.M.: Sistema de processamento de alarmes para uso no centro de operação do sistema da CPFL. In: Proceedings of XX Seminário Nacional de Produção e Transmissão de Energia Elétrica, Recife, Brazil (2009)
49. Rizzi, N., Pierozzi, A., Adorni, C.Y.K.O., Barbosa, W., Souza, J.M.: Correlação entre os alarmes do SCADA e da rede de Telecom para identificação de faltas. Technical Report, FITec, Campinas (2015)
50. Coccetrone, R., Albino, S., Adorni, C.Y.K.O., Souza, J.M., Pierozzi, A., Alves, J.: Análise da inadimplência em função do envelhecimento da fatura (Aging). Technical Report, FITec, Campinas (2013)
51. Silva, A., Denis, I.F.E.D.: Modelo de Referência para Implantação de Redes Elétricas Inteligentes (Smart Grid). Relatório Analítico Referente à Finalização de Projeto de P&D (PD 0385 0062/2013), Elektro/FITec (2018)

52. Araújo, M.A.: Methodology based on dispersed voltage measures and decision trees for fault location in modern distribution systems. Ph.D. Thesis, USP—São Carlos, Brazil (2017)
53. Batista, O.U.: Intelligent system based on orthogonal decomposition technique and fuzzy inference for high impedance location fault in distribution systems with distributed generation. Ph.D. Thesis, USP—São Carlos, Brazil (2016)

Hydrothermal Dispatch Considering Natural Gas Restrictions

Jean Rodrigo dos Santos, Guilherme V. Vargas and Juan Carlos Galvis Manso

Abstract Brazil is a country whose electrical energy generation comes mainly from hydroelectric and thermoelectric power plants. In order to optimize the water resource and minimize the cost of electricity generation, the National System Operator—ONS solves an optimization problem known as hydrothermal dispatch. In this way, the hydrothermal dispatch problem was solved using the stochastic dual dynamic programming, taking into account constraints imposed by hydroelectric plants and also restrictions imposed by thermoelectric plants, such as transport and generation of natural gas. For the implementation of the dispatch, we used real operation data of the Brazilian National Integrated System of the year 2015, referring to three subsystems: North, Northeast and Southeast/Midwest. The results of the simulations are presented for the different operating situations and explain how the restrictions imposed by natural gas influence the operational cost and the modė of operation of the electric power generating units.

Keywords Hydroelectric dispatch · Natural gas network · Electrical planning · Dynamic programming · Energy operation

1 Introduction

Electrical energy has characteristics that make it different from all other energy sources, due to the fact that it has very small storage possibilities, occurring indirectly through reserves of other energy sources, such as water, diesel, coal, natural gas, etc.

J. R. dos Santos (✉)
Federal University of Itajubá (UNIFEI), Itajubá, Brazil
e-mail: jeanrdsantos@hotmail.com

G. V. Vargas · J. C. G. Manso
Federal University of Ouro Preto (UFOP), Ouro Preto, Brazil
e-mail: guivargas93@yahoo.com.br

J. C. G. Manso
e-mail: juancgalvis@ufop.edu.br

Y. Iano et al. (eds.), *Proceedings of the 4th Brazilian Technology Symposium (BTSym'18)*, Smart Innovation, Systems and Technologies 140,
https://doi.org/10.1007/978-3-030-16053-1_42

Great amount of electricity generation in Brazil comes from hydroelectric plants, 61% of the total, and from thermoelectric plants, 26% [1], existing a specific production cost for electric energy generation for each one. Therefore, it is necessary to exist planning actions on the resources generation [2].

In order to optimize the resources that serve as raw material for hydroelectric and thermoelectric plants, it is necessary to solve the hydrothermal generation dispatch problem, determining the amount of energy to be generated [3]. However, the resolution of this problem is somewhat complex, since the amount of water available in the reservoirs of hydroelectric plants in future periods is unknown, making the problem of nature stochastic. In the thermoelectric plants, for example, there are restrictions on the transport and availability of natural gas.

The objective of this work is to perform the hydrothermal dispatch with the insertion of the natural gas restrictions, being possible to provide a greater representation of the real situation of the system, and verify the influence of these restrictions on the cost of operation.

2 Methodology

2.1 Dispatch Model

Given the high dependence of hydropower on power generation, planning actions face a difficulty with respect to operation [4], since the rain factor is a stochastic variable, that is, it is not 100% sure how much and when it will rain. Consequently, the natural flow of the rivers that feed the hydroelectric dams are affected. Thus, it creates a great doubt on when to use water for generation and when to stock it.

When defining how to use water, it is necessary to plan the use of thermoelectric plants, since they are the largest source of complementary generation. In this sense, an important factor to take into consideration is the availability of fuel, in order to supply the requested demand.

2.2 Objective Function

The hydrothermal flow model used in this work is described in [5], with some adaptations. This model aims to minimize the marginal cost of operation, given by the cost of generating thermoelectric plants, since the fact that for the generation in the hydroelectric plants water is used and it has no cost of use, as opposed to gas, besides the cost related to the Deficit, which is a way to supply the demand, if the generating units fail, and this cost is adapted in the model. Thus, the dispatch model aims to solve the following linear programming problem, where J is the number of thermoelectric power plants in the system and T is the number of stages that the problem deals with:

$$z = \min \sum_{t=1}^{T} \left(\sum_{j=1}^{j} (c_{j,t} g_{ter_{j,t}} + C_{Def_t} Def_t) \right) \tag{1}$$

In (1), the cost of immediate operation $c_{j,t}$ represents the thermal generation cost required to meet the load in stage t. Part of this load is served by the hydroelectric plants, and the remainder is supplied taking into account the operational decisions $g_{ter\,j,t}$, by the thermoelectric plants of the system. Finally, there is the Deficit (Def_t), which aims to compensate for a probable lack on the part of the generation units, which has a cost (C_{Deft}) at each stage.

2.3 Generation-Demand Balance Sheet Equation

The generation-demand balance equation aims to establish a relation and limits for the generation and Deficit according to the demand required in the analyzed period.

$$\sum_{j=1}^{J} g_{ter_{j,t}} + \sum_{i=1}^{I} p_i u_{i,t} + Def_t = D_t \tag{2}$$

Thus, in (2), the sum of all thermoelectric generation (g_{ter}) together with hydroelectric generation (pu) and Deficit (Def) must be equal to the demand (D) required in the analysis.

2.4 Hydraulic Balance and Hydropower Equation

The relation on how the volume of water in the hydroelectric plants varies is described in a function of several variables.

$$v_{i,t+1} - \sum_{m \in M} \left[u_{m,t} + s_{m,t} \right] = v_{i,t} + a_{i,t} - u_{i,t} - s_{i,t} \tag{3}$$

Thus, in (3) the volume present in the reservoir of a hydroelectric plant i at the end of a stage $t + 1$, ($v_{i\,t+1}$), must be equal to the volume at the beginning of stage t ($v_{i,t}$) to the volume of water received, denominated affluence ($a_{i,t}$), plus the turbine flow and the volume of the turbines immediately upstream ($u_{m,t+Sm,\,t}$), discounting the turbine volume ($u_{i,t}$) and pouring ($s_{i,t}$) at plant i.

Where: mϵM: set of plants immediately upstream of plant i.

The electric power generated by a hydroelectric plant is proportional to the gross fall (H_b) of the dam and to the turbine flow ($u_{i,t}$).

$$Pel_{i,t} = \rho g k H_b u_{i,t} \tag{4}$$

where: ρ = density of water in kg/m^3; g = acceleration of gravity in m/s^2; k = turbine-generator set efficiency and $u_{i,t}$ = turbine flow in plant i in stage t, in m^3/s.

2.5 Thermoelectric Generation Equation

The natural gas consumption in a given thermoelectric plant is described in (5).

$$\delta_{i,j} = \frac{\varphi_j}{HR} g_{t,j}^{E} \tag{5}$$

where: $\delta_{i,j}$ = Gas consumed by thermal j, in stage t, in Mm^3/month; φ_j = Thermal efficiency of production of plant j, in KW/J; HR = rate of heating of natural gas in KJ/m^3 and $g_{t\,j}^{E}$ = Generation of electric energy by thermal j, in stage t, in MW/month.

Since it is known that in a thermoelectric plant, the generation is related to the gas consumed by it, then it is generally defined according to [5], that the gas that is injected into the thermoelectric power plant j, ($g_{j,t}^{G}$) plus all gas transported from the gas generating plant m to the thermoelectric plant j, through the gas pipeline s, ($f_{s,t,m,j}$), must be equal to the consumed demand of the thermal plant j, ($\delta_{j,t}$).

$$\sum_{j \in G_m} g_{j,t}^{G} + \sum_{j \in \phi_m} f_{s,t,m,j} = \sum_{j \in \Gamma_m} \delta_{j,t} \tag{6}$$

where: $g_{j,t}^{G}$ = Gas injected at plant j, at stage t; $f_{s,t,m,j}$ = Gas transported by pipeline s, from gas generating plant m to thermal plant j, in stage t; and $\delta_{j,t}$ = Demand for gas consumed at plant j, at stage t.

3 Test Scenarios

Employing the Dual-Stochastic Dynamic Programming (PDDE), the dispatch model was implemented in a computational environment. The data utilized by the simulations were obtained from the official website [6] and from the official websites of the other hydroelectric and thermoelectric administrators, as follows:

Data from the hydroelectric plants are: Influence values of all the plants in the year 2015; Installed power in each hydroelectric plant; Maximum and minimum volume of hydroelectric reservoirs; Maximum and minimum turbine of each hydroelectric plant; Maximum amount of each hydroelectric plant; Gross fall of each hydroelectric plant and Schematic of the positioning of each hydroelectric plant in the river basins.

The illustration of how the hydroelectric plants of the North subsystem are arranged, in the study period, is shown in Fig. 1. It is possible to observe which plants operate in cascade with others and also which operate in isolated rivers, containing also the information about the type of each reservoir.

The data referring to thermoelectric, gas pipelines and natural gas generating units are: Power installed in each thermoelectric plant; Cost of generation of each thermoelectric plant; Gas consumption in each thermoelectric plant; Production capacity in each gas generating unit; transport capacity in each pipeline and Schematic of connection of the pipeline system.

Regarding the thermoelectric plants that use natural gas in the region, it is possible to observe in Fig. 2 how they are distributed. In the same figure, it is also possible to observe the representation of the pipelines that interconnect the natural gas and thermoelectric processing units.

The detailed representation of all the subsystems and all the data used in the problem, with respect to hydroelectric plants, the data relationship of thermoelectric, gas and gas pipelines, and how the method of configuring the interconnection of hydroelectric power plants in rivers and of the thermoelectric plants with the pipelines can be seen in [7].

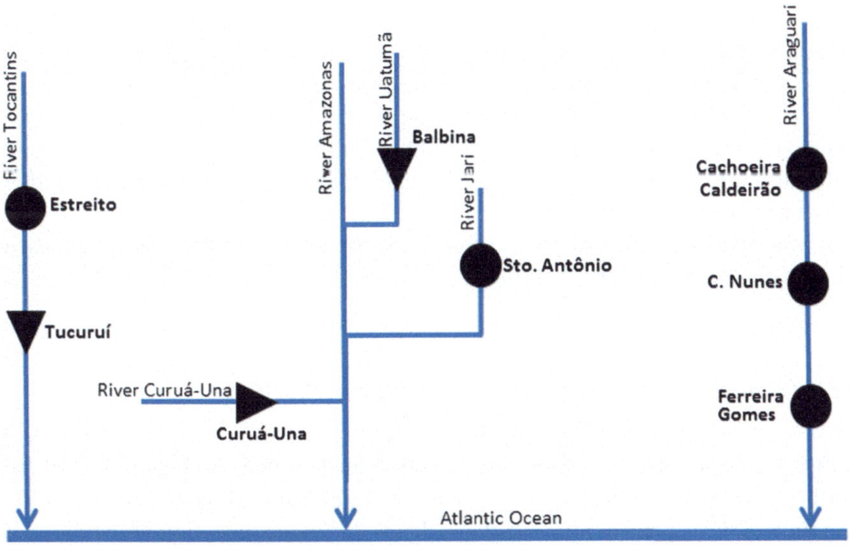

Fig. 1 North region hydroelectric plants system

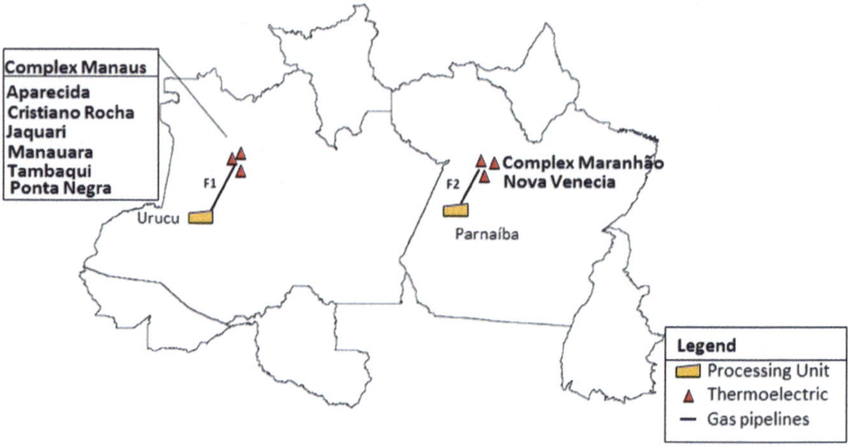

Fig. 2 North region thermoelectric and pipelines system

4 Results and Discussion

The subsystems were then simulated in order to obtain an analysis on the insertion of gas restrictions, evaluating how much energy was generated in the thermoelectric plants and how the operation cost was varied. For all scenarios presented, the following considerations were made: Test period—1 year (2015) discretized in monthly stages; Initial reservoirs volumes such as those in January 2015, according to [6]; Subsystems analyzed—North, Northeast and Southeast/Midwest; Demand for subsystems equal to those recorded in 2015 (Table 1); Inflow scenarios of 10% in relation to the inflow data recorded in the year 2015 and Values of production, transportation and consumption of natural gas.

It was verified according to [6], that in 2015, the average monthly demand in each subsystem varied according to Table 1.

4.1 Generation in Thermoelectric Power Plants

In Fig. 3, it is possible to observe the operation mode of the Tambaqui plant of the North region. In this, when adding + 10% in the affluence, it is not necessary the operation of the same from the 6th stage, this is due to the reduction of the demand after this stage and the increase of the affluence in the hydroelectric plants, in case of no increase in the affluence, Tambaqui operates at its maximum constant capacity, and there is also no need to operate the Deficit.

By including the natural gas restriction, it is possible to observe how much it limits the generation in the thermoelectric plants, Fig. 3. In the illustration, the generation capacity of the Tambaqui plant is conditioned to the availability of gas; therefore, it

Table 1 Monthly medium demands (MW) of each subsystem [6]

Stages	Northeast	North	Southeast/Midwest
1	9757	7543	22,008
2	9251	8391	20,620
3	8987	9279	21,320
4	7929	10,160	21,205
5	8143	9812	20,781
6	8304	7910	19,797
7	8024	5652	19,319
8	8924	5968	18,872
9	9095	6919	19,934
10	9159	5184	22,461
11	8765	4508	23,241
12	8774	4353	22,266

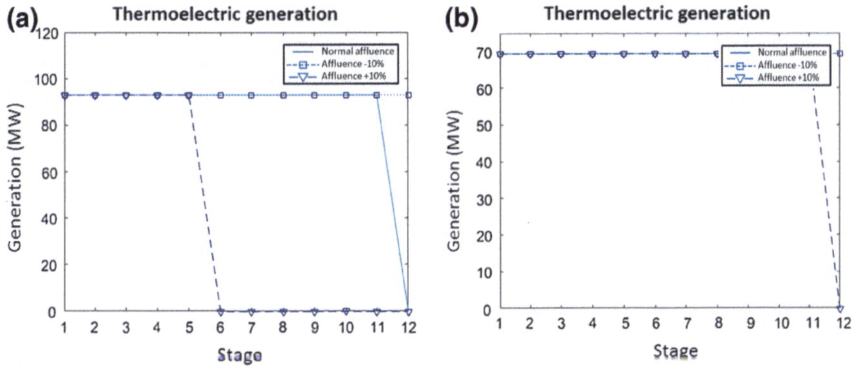

Fig. 3 Generation at the Tambaqui thermal power plant: **a** Without gas restriction and **b** with gas restriction

always operates at a capacity less than the maximum and constant, except the case in the last stage, which with an optimistic affluence, does not operate.

4.2 *Cost of Operation*

The values used in relation to the operating cost of thermoelectric plants were taken from the PMO (Monthly Operating Program) provided by [6]. In Fig. 4, it is possible to observe the variation of this cost along the stages; it is perceived that the initial cost is high and tends to decrease. This fact is due to the fact that the initial volume of the reservoirs were low in the year evaluated. Thus, with the passing of the stages, thermoelectric plants that are not fueled by natural gas and have a higher cost of

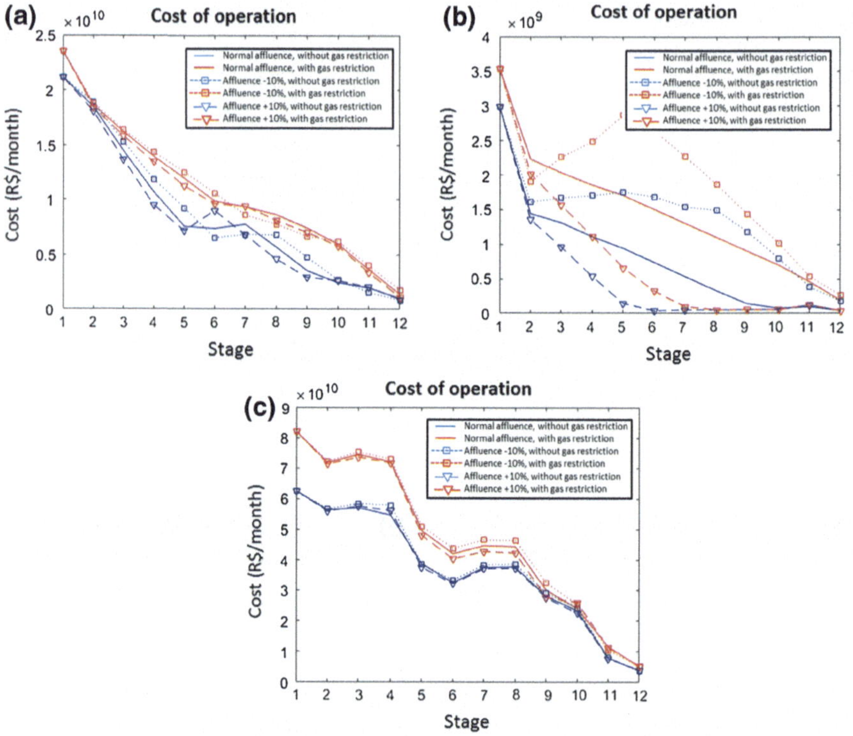

Fig. 4 Cost of operation in each subsystem. **a** Northeast, **b** North and **c** Southeast/Midwest

operation, together with the Deficit, stop operating, reducing the cost at the end of the analyzed period.

Another fact to be observed in Fig. 4 is that the gas restriction makes the operation more expensive, since gas plants are limited in generation and other types of plants are used, which have a higher cost of operation. The affluence is also an important factor to consider, since when it is increased, the cost has reduced, and in the case of the North region, by decreasing the affluence the cost has increased considerably.

A comparison on how much the gas restriction influences the cost, with respect to annual values, has for the Northeast region, considering only the normal affluence, are shown in Table 2.

Table 2 Annual operation cost of the northeast region for normal affluence

	R$/year
Unrestricted cost of gas	9,757,102.365.225.517,60
Cost with gas restriction	130.819.612.492,80
Difference	28.454.386.975,20

5 Conclusion

The execution of hydrothermal dispatch planning is of extreme importance in order to have a continuous supply and quality of electrical energy. Managing how the system's electric power plants should operate avoids major problems such as power failure and minimizes the cost of operation. By including the restrictions imposed by natural gas, there is a closer response to the real case, since generation in thermoelectric plants is conditioned to the transportation and generation of natural gas.

The results obtained showed that the natural gas restriction influences the operating cost of the system, since the fact that the cost tended to rise when the restriction was considered; in some cases the values were considerably high if compared to the cases that not applied the restrictions. It was also noticed that the plants that have the natural gas as fuel, always operated steadily throughout the stages when the gas restriction was inserted, while the plants employing other fuels varied the mode of operation, and the generation by gas thermoelectric was limited by the availability of natural gas.

References

1. ANEEL, A.N.d.E.E.: Matriz de energia elétrica. 2017. Disponível em: www2.aneel.gov.br/aplicações/capacidadedebrasil/OperacaoCapacidadeBrasil.cfm. Acesso em:07.10.17. Author, F., Author, S.: Title of a proceedings paper. In: Editor, F., Editor, S. (eds.) Conference 2016, LNCS, vol. 9999, pp. 1–13. Springer, Heidelberg (2016)
2. ENACOM, A.L.: Tratamento de incertezas climáticas e modelos computacionais de geração hidrelétrica. 2015. Disponivel em: https://pt.slideshare.net/waycarbon/adriano-by-the-way. Acesso em: 06.02.2018
3. Finardi E.C.: Planejamento da Operação de sistemas hidrotérmicos utilizando computação de alto. Tese (Doutorado)- Universidade Federal de Santa Catarina, Departamento de Engenharia Elétrica. Florianópolis (1999)
4. ANEEL, A. N. d. E. E. (ed.): Relatório ANEEL 10 anos.Brasilia: Agência Nacional de Energia Elétrica (2008)
5. Toledo, F., et al.: Energy cost distortion due to ignoring natural gas network limitations in the scheduling of hydrothermal power systems. Proc. IEEE **65**(5), 633–641 (2015)
6. ONS, O.N.d.S.: Histórico de Operacao. 2015. Disponível em http:/nos.org.br/Paginas/resultados-da-operacao/histórico-da-operacao/geração energia.aspx. Acesso em: 12.11.17
7. SANTOS, J.R.: Despacho hidrotérmico considerando restrições do gás natural. Monografia (Monografia em Engenharia Elétrica)-UFOP. João Monelvade (2018)

Human Capital Measurement: A Bibliometric Survey

Neusa Maria Bastos Fernandes dos Santos, Danilo Nunes, Fernanda Cardoso Romão Freitas, Igor Polezi Munhoz and Adriana Caldana

Abstract Human capital is recognized as a pillar of the creation of intellectual capital and is fundamental for the productivity and functioning of organizations. From a contemporary perspective, human capital ceases to be considered only as cost and is seen as an investment, once it is possible to measure it. The objective of this article was to measure some existing approaches to capital human measurement, using as a methodological tool a bibliometric research in the last 11 years of the papers submitted and approved at EnAnpad, which is an event organized by the National Association of PostGraduate Studies and Research in Administration (ANPAD). The bibliometric research was elaborated by quantitative documentary analysis. The results showed, among others, a great diversity of areas in relation to how this theme is treated. This leads to reflections on how human capital has been gaining space for discussions in various organizational environments. There is a sign that from this subject, before, certainly, a discussion almost restricts the area of human resources, now goes beyond the walls and departmental currencies.

Keywords Intangible asset · Human capital · Bibliometry

1 Introduction

In recent years, there has been a significant increase in the interest of companies in the measurement of Human Capital, Intellectual Capital and all that is called Intangible Assets. Many companies have joined and made available to their investors managerial

N. M. B. F. dos Santos · D. Nunes · F. C. R. Freitas
Pontifical Catholic University of Sao Paulo (PUC-SP), Sao Paulo, Brazil
e-mail: admneusa@pucsp.br

I. P. Munhoz (✉)
Federal Institute of Education, Science and Technology of São Paulo (IFSP), Sao Paulo, Brazil
e-mail: igor.munhoz@ifsp.edu.br

A. Caldana
University of Sao Paulo (USP), Ribeirao Preto, Brazil

Y. Iano et al. (eds.), *Proceedings of the 4th Brazilian Technology Symposium (BTSym'18)*, Smart Innovation, Systems and Technologies 140,
https://doi.org/10.1007/978-3-030-16053-1_43

reports such as integrated report, annual sustainability report, social report and others. Regardless of the obligation, the companies began to show interest in these statements due to the demands of the stakeholders and to recognize the added value to the brand through the intangible assets.

In many companies, intellectual capital represents a significant portion of its value. However, despite its importance, it is still not recognized in the financial statements, since its conception does not meet the definition of assets according to accounting standards [1].

According to Mayo [2] the way in which companies are evaluated has been undergoing progressive changes since 1990, when it was assigned a greater value to intangible assets, which are knowledge, skills and the brand itself. According to the author, these assets are also called intellectual capital. Human Capital, object of study of this research, are the people who construct this value.

The term Human Capital was coined by Schultz in 1962 and refers to the knowledge and competence of an individual. After Schultz, other authors such as Edvinsson, Sveiby and Davenport discussed the topic that has gained increasing relevance in this debate on the importance of intangible assets [2]. In this context, human capital is recognized as a pillar of the creation of intellectual capital and its main characteristic is that it can disappear with the exit of employees, and is thus a constant source of strategic renewal [3].

In the period of 2015, 2016 and 2017, the analysis of the studies available in the Spell database and the publications of the EnAnpad (Meeting of the National Association of PostGraduate Studies and Research in Administration—ANPAD Brazil) and Semead (Business Administration Seminars of FEA USP—School of Economics, Business Administration and Account/University of Sao Paulo) pointed out several articles on the human capital theory in order to explain the problems related to education and socioeconomic development and to the understanding of the responsibility of Human Capital in this development process. It was possible to find in the current literature the approach Human Capital *versus* Organizational Performance, but few studies were found with focus on the measurement approach of Human Capital. It was also found that there are considerable difficulties in converting the intangible assets of the company into a projection of future cash flows that are capable of determining the value of a company based on its intangible assets and its human capital [4].

The big question is how to measure intangible assets. Unlike the tangible assets that can be measured by consolidated accounting instruments, it was necessary to develop new means to measure intangible assets. Thus, this article raises the following question: *what studies exist to measure Human Capital and how has this issue been addressed in the main Brazilian congresses of business administration over the last 11 years?*

To answer this question, the next section will make a brief theoretical background on the concepts of human capital, intellectual capital and intangible assets, then will present existing studies to measure human capital. Finally, will be presented a bibliometric analysis with the purpose of identifying the academic production of the last 11 years on the subject and to verify how this discussion has evolved in the field of the science of the administration of companies in the last years in one of the main national congresses.

2 Theoretical Background

2.1 Human Capital as an Intangible Asset

According to The Human Capital Reporter 2016, World Economic Forum report, Human Capital is critical to the productivity and functioning of the political, social and civic institutions of a society. Understanding their capability is valuable to a wide variety of stakeholders. The Report also highlights that today's economies are increasingly knowledge-driven and technology-driven in a way that makes it impossible to state how work on tomorrow's market will work [5].

The Disclosure on Human Capital began with Professors Schultz and Becker, Nobel Prize winners. These seminal studies emphasized the importance of the accumulation of Human Capital through knowledge and skills. Schultz [6] argues that human productive capacity outweighs all other forms of wealth and reinforces the understanding that the development of some countries comes mainly because of the country's investment in people. The specialized Human Capital is seen as an investment made by the organization, which should have a return, although there was some difficulty in calculating it [7].

From then on new researchers emerge with the purpose of developing models and new ways of measuring human capital. A non-financial measurement method is based on the fact that people are the real and only agents of the business. All other tangible assets, as well as everything that corresponds to the intangible, are the results of human action and depend on people for their existence, so one can note the importance of this human capital [8].

Tangible assets are all those found in a company's balance sheet, such as money, buildings, and machinery in general. In contrast, intangibles are those that can not be measured in the traditional accounting model, such as people and their knowledge [9].

In the definition of Bontis [9], Human Capital is the level of knowledge of each employee and is considered the main factor of the creation of the intellectual capital of a company, being this source of innovation and strategic renewal. Thus it is possible to say that people lend their human capital to the organizations, they make available all their intellectual capacity, knowledge and commitment.

Thus, it is important to mention that the intangible asset is accompanied by risks due to its vulnerability, since the company may lose part of it when employees leave. Another five main risks that may be highlighted are: lack of fundamental skills in a company's staff; regulatory issues; succession planning issues; possible differences between current talent resources and business objectives and, ultimately, the lack of essential skills in the workforce [10].

Deloitt, a leading business consulting firm in 2017, has launched the Deloitte Global Rewriting the Rules for the Digital Age 2017 report in which it demonstrates ten new trends for Human Capital and highlights the changes in the world of work. In response to these changes, business leaders will no longer be able to operate on the basis of the old paradigms, since they must adopt new ways of thinking about their companies, their human capital and even their role in global society [11].

Therefore, it is necessary to overcome the old paradigm that the measurement of the assets of an organization can only be done in concrete and tangible objects. The challenge is to rewrite the rules and find ways to make Human Capital, so far intangible, tangible.

2.2 Studies on Measuring Human Capital

Human Capital at Sysco Corporation

Cascio and Boudreau [10] make an important contribution to the creation of value from Human Capital. Syscom Corporation is a world leader in food marketing and distribution and has created a chain of logical connections to help leaders understand how the company generates value from its Human Capital. This chain, according to the authors, was structured and sustained in four stages: effective management practices; employee satisfaction; customer satisfaction, and the latter, the fruit of this sequence, profitability and long-term growth. The guiding focus of this methodology is that processes, technologies and structures can have much similarity in the labor market, but the skilled, committed and fully engaged workforce is typical of each organization, so it is what can differentiate it.

The model has been termed as *five stars* being: ensuring that leaders provide guidance and support; strengthen front-line supervisors; pay performance; address the quality of life of employees and promote the inclusion of employees by engaging them and taking advantage of diversity. Construction starts from an organizational climate survey in which all members of the company, in a process of evaluation and self-assessment, assign values to each of the five principles.

Cascio and Boudrean [10] elucidate several actions that arise as a consequence of the application of the model proposed by the company, emphasizing the use of the Balanced Scorecard as an instrument to leverage the best organizational practices and finalize the study emphasizing how the practices and management processes that human capital indicators influence financial results.

The Human Capital at Universal Weather

Universal Weather and Aviation is a privately held company based in Houston, Texas that provides products and services to the general aviation industry. Cascio and Boudreau [10] portray the experience of this company based on three pillars: alignment, capacity building and engagement. It is the so-called ACE model that emerged from a problematic scenario that seems to be a tonic of several companies, such as the absence of HR policies and processes, including the issue of remuneration, lack of definition of succession issues, non-aligned and inadequate evaluation processes, among others.

According to the authors, these pillars cover: in alignment, the workforce of employees is linked directly to the objectives and strategies of the company; in the training, it is sought to measure the strength of the talent in the HR systems implanted and designed to add strategic value to the company's business; in engagement, the workforce of employees is measured at the minimum of what was offered to the customer, or even reflect the company as a great place to work. For talent risk management, competitive remuneration, career advancement, leadership coaching, etc. were developed [10].

The indicators for follow-up of the actions were structured taking into account HR supervision and preparation of annual reports. In less than five years the company could already see how positive the adoption of this model of human capital approach was.

Intangible Assets Monitor Approach
This approach is an intangible asset monitor that reconciles the measurement of human capital through skills. To measure intellectual capital, this approach provides a framework of indicators from three different perspectives: external structure, internal structure, and individual competencies. Indicators are suggestions that need to be adapted to the reality of each company. The idea is to select one or more indicators by sub-heading in the table and display them in a coherent format [8].

Sveiby [8] explains that the external structure relates to relations with customers and suppliers, trademarks, trademarks and the reputation or image of the company. The internal structure includes patents, concepts, models and administrative and computer systems. So individual competencies involve the ability of professionals to act in diverse situations to create tangible and intangible assets. It also emphasizes that people are the most valuable elements in an organization.

3 Methodology

The method of bibliometry sought to evaluate, in a predetermined space of time and event, what is being produced related to the study proposal of this article. It is understood that scientific production promotes the establishment and monitoring of teaching and research policies, since it signals potentialities and paths taken by researchers.

The present study considered the definition of bibliometrics as the study of the quantitative aspects of the production, dissemination and use of registered information [12]. In his study of bibliometrics, Araújo [13] draws the three dimensions of this method, which are called Lotka's Law, which objectively mentions -sure the productivity of authors; the Bradford Law, which aims to measure the dispersion of journals, and Zipf's Law, which aims to measure the frequency of certain words or themes.

The methodology closest to this study is the Zipf Law, which consists in measuring the frequency of the appearance of words in several texts, generating an ordered list of terms of a particular discipline or subject.

In this bibliometric study, the option was quantitative documentary analysis based on the last 11 years of the papers submitted and approved at EnAnpad. Thus, the analysis considered the year 2007 (from which the event began to gain increasing importance in a national and international context) until the last year of registration (2017).

Barros and Lehfeld [14] emphasize that in quantitative research a statistical set of data is used, and a descriptive analysis of the data collected and ordered according to the order of interest of the researchers' studies is performed.

The technique used was the analysis of content that, that allows the identification of the data collected in the research for framing from the indicators generated by the knowledge and perception of the messages [15].

4 Results and Discussion

The EnAnpad is traditionally divided into sections (in 2017, there were 11 sections) according to the theme of the works. Thus, the total amount of papers in the period of the survey (2007–2017) considering the section of *people management* was approximately 600 articles. So, it was first researched which articles are related to the two themes of this research, Human Capital and Intangible Assets, as observed in Table 1. According to Table 1, it was observed that about 70% of the articles refer to Human Capital its measurement, demonstrating the importance of Human Capital and its debate by the academy as a great asset capable of adding value to an organization.

In terms of percentage, these two themes may be insignificant in the subject of people management, but an increase of articles related to human capital in the last 5 years has been observed.

It was also investigated which institutions of origin of the authors of the articles (Table 2).

It was observed that there is no concentration in one or a few institutions, nor does it present specific regional trends. Then all the works listed in the bibliographical references of the productions were tabulated and stratified, considering the number

Table 1 Number of articles submitted, period from 2007 to 2017

Focus theme	Number of articles
Human capital	18
Intangible assets	08
Total	26

Source Developed by authors (2018)

Table 2 Institutions of origin of the researchers

Institutions	Human capital	Intangible assets
FGV/SP (Brazil)	1	0
MACKENZIE (Brazil)	2	0
UEM (Brazil)	1	0
UFAL (Brazil)	1	0
UFC (Brazil)	2	0
UFS (Brazil)	2	0
UFU (Brazil)	0	3
UNIGRANRIO (Brazil)	4	0
UNIOESTE (Brazil)	3	2
UNISINOS (Brazil)	4	0
USP (Brazil)	2	0
Not disclosed	31	16
Total	53	21

Source Developed by authors (2018)

of times the authors appear in the 26 articles surveyed, not being considered probable repetitions in the same article (Table 3).

According to Table 3, it was found that another 39 researchers were cited in the articles obtained in EnAnpad, in a universe of about 920 citations. Therefore, it is observed that there is a network of knowledge being built by authors who are trying to measure human capital through a tangible approach. There is no polarization which signals an even greater possibility of productions that contribute to the theme proposed in this study.

5 Conclusion

From the literature review it was possible to identify the main approaches to human capital measurement and to verify that, although there are specific tools for this intellectual capital niche in the composition of the intangible value, this measurement is always accompanied by the measurement of intellectual capital as a whole.

After the bibliometric analysis of the 26 articles found in the EnAnpad of the last eleven years, eighteen refer to Human Capital while eight of them have a focus on Intangible Assets in general. This evidence reinforces the idea already mentioned throughout this work of increasing appreciation and recognition of Human Capital as a great asset capable of adding value to an organization.

As limitations of this work, we highlight the fact that bibliometric research was carried out in the annals of only one congress. We suggest future research aimed at broadening the field of research of publications.

Table 3 Researchers cited in the articles

Researcher	Amount	Researcher	Amount
Stewart, Thomas A.	10	Kayo, Eduardo Kazuo	4
Sveiby, Karl Erik	9	Malhotra, Naresh	4
Edvinsson, Leif; Malone, M. S.	8	Perez, M. Monteiro; Famá, Rubens	4
Becker, Gary S.	7	Brooking, Annie	3
Schultz, Theodore W.	7	Dutra, J. S.	3
Schumpeter, Joseph Alois	7	Hendriksen, E. S.; Van B., Michael F.	3
Bontis, Nick	6	Hitt, Michael A.	3
Hair, Joseph R.	5	Iudícibus, Sérgio De	3
Lev, Baruch	5	Norton, David P.	3
Bardin, Lourence	4	Mcclelland, D. C.	3
Barney, Jay B.	4	Mincer, Jacob	3
Gil, Antonio Carlos	4	Porter, Michael E.	3
Kaplan, Robert S.	4	Prahalad, C. K.	3

Source Developed by authors (2018)

Finally, this article has the concern of showing to the academic community and the organizational corporate context the need to always review the practices focused on Human Capital.

Human resource strategies and policies permeate the environmental, institutional, behavioral and human forces at the time of the adoption of their practices, and Human Capital is always the protagonist of the proposed actions, since they reside in the group formed by the competencies and abilities of the people.

By integrating organizational objectives with Human Resources policies and practices, people are treated as a source of competitive advantage, with proactive stance by all those involved. It is hoped, therefore, that the valuation of Human Capital will be increasingly effective and grow so that all can reap those benefits; people as personal goals and organizations as organizational goals.

References

1. Guthrie, J., Petty, R., Ricceri, F.: The voluntary reporting of intellectual capital: comparing evidence from Hong Kong and Australia. J. Intell. Capital **7**, 254–271 (2006)

2. Mayo, A.: O Valor Humano da empresa: valorização das pessoas como ativos. Prentice Hall, Sao Paulo (2003)
3. Bontis, N.: Managing organisational knowledge by diagnosing intellectual capital: framing and advancing the state of the field. Int. J. Technol. Manage. **18**, 433–462 (1999)
4. Dalbem, M.C., Bastian-Pinto, C.L., Andrade, A.M.: O valor financeiro do capital humano e o desafio de retê-lo nas empresas. Braz. Bus. Rev. **11**(1), 51–72 (2014)
5. World Economic Forum: The Human Capital Report 2016. Switzerland (2017)
6. Schultz, T.W.: Investment in human capital. The Am. Econ. Rev. 1–17 (1961)
7. Becker, G.S.: Investment in human capital: a theoretical analysis. The J. Polit. Econ. 9–49 (1962)
8. Sveiby, K.: The-intangible-assets-monitor. J. Hum. Resour. Costing Acc. **02**(1), 73–97 (1997)
9. Bontis, N., Dragonetti, N.C., Jacobsen, K., Roos, G.: The knowledge toolbox: a review of the tools available to measure and manage intangible resources. Eur. Manag. J. **17**(4), 391–402 (1999)
10. Cascio, W.F., Boudreau, J.W.: Gestão Estratégica de Recursos Humanos: uma introdução. Saraiva, São Paulo (2014)
11. Deloitte: Global Human Capital—Trends 2017. Deloitte, London (2017)
12. Vanti, N.A.P.: Da bibliometria à webometria: uma exploração conceitual dos mecanismos utilizados para medir o registro da informação e a difusão do conhecimento. Revista Ciência da Informação **31**(2), 152–162 (2002)
13. Araújo, C.A.: Bibliometria: evolução histórica e questões atuais. Em Questão **12**(1), 11–32 (2016)
14. Barros, A. de J.P., Lehfeld, N.A.de S.: Projeto de Pesquisa: propostas metodológicas. Vozes, Petrópolis (1990)
15. Bardin, L.: Análise de conteúdo. Edições, Lisboa (2011)

Project Knowledge Transfer: A Study with Brazilian Project Managers

Neusa Maria Bastos Fernandes dos Santos, Nelson José Rosamilha, Alessandra Cristina Santos Akkari, Fernando de Almeida and Renan Riedel

Abstract In a rapidly changing business environment, companies are continually striving to increase productivity and maintain competitiveness. Projects become the delivery mechanism for the products and services of these organizations, which use more and more knowledge workers who benefit from accumulated knowledge and learn from other projects. Thus, the transfer of knowledge, in the area of strategic management, has an important role in the integration of knowledge to optimize the delivery of products and services. This study aimed to identify the best practices of knowledge transfer and evaluate if they are applicable in project management. For this, a theoretical survey was carried out identifying the knowledge transfer practices and, later, a instrument was developed, which was applied to the Brazilian project management community. The data treatment involved the application of test of Equal Variances and One Way ANOVA to prioritize results. There were 365 responses and the dimensions Culture, Leadership and Communication were identified as the most relevant for the transfer of knowledge in project management, being critical success factors among the members of the project team.

Keywords Knowledge transfer · Project management · Knowledge management

1 Introduction

Currently, the business environment changes rapidly where knowledge and knowledge management plays a critical role in increasing productivity and maintaining the competitiveness of companies. Organizations that continually increase their knowledge will be better prepared to face the uncertainties of the organizational environ-

N. M. B. F. dos Santos · N. J. Rosamilha · F. de Almeida · R. Riedel
Pontifical Catholic University of Sao Paulo (PUC-SP), Sao Paulo, Brazil
e-mail: admneusa@pucsp.br

A. C. S. Akkari (✉)
Mackenzie Presbyterian University, Campinas, Brazil
e-mail: alessandra.akkari@mackenzie.br

Y. Iano et al. (eds.), *Proceedings of the 4th Brazilian Technology Symposium (BTSym'18)*, Smart Innovation, Systems and Technologies 140,
https://doi.org/10.1007/978-3-030-16053-1_44

ment, market dynamics, economic cycles, technological escalations and social needs, and will remain competitive and sustainable [1].

In this scenario, knowledge is the most important resource in the organization and the most enduring competitive advantage being considered a strategic resource that, when properly managed, allows organizations to create value through the exploitation of production [2].

Given the relevance of knowledge sharing to maintain a competitive market, organizations strive to create conditions for employees to share their knowledge and the ability to share this knowledge is positively associated with organizational performance [3].

According to the Project Management Institute (PMI), more effective knowledge transfer organizations improve project results by 35%. In addition, knowledge transfer is a learning tool that can strengthen team performance, emphasizing collaboration, innovation. The challenge, then, is to identify knowledge transfer practices that can impart introspections and valuable insights from one project to another, regardless of where the professionals reside [4].

Thus, the aim of this study was to identify the best knowledge transfer practices and then verify the applicability of these practices in the project knowledge transfer among Brazilian project managers.

2 Background

Projects are becoming more complex, requiring the integration of diffuse partners, often separated physically and from different cultural backgrounds. Accuracy, punctuality, and effective communication become increasingly important. Organizations increasingly use teams of knowledge workers to carry out projects and believe that these teams can benefit from accumulated knowledge and learn from other projects [5].

Due to the volatility of the projects, an informational problem occurs: they are transient configurations, mostly using distributed resources, partially or totally virtualized in relation to the interaction and intensive use of information flows with large amounts of content to be managed [1].

Thus, in a project team, knowledge sharing is very important in order to reduce project costs; increase team performance; increase the chances of delivery of the project; and increase the team's ability to develop new knowledge within the team itself, being essential to the success of the project [6].

Zhao et al. [7], in an empirical study examining the factors that influence the transfer of knowledge in information technology projects in China, reported that the problem occurs in the dissolution of the members in the conclusion of a project, generating the fragmentation of knowledge and loss of learning based on experience.

In addition, project teams face pressures of strict deadlines to complete their activities and deliver them, making them heavily focused on tasks, so as not to consider the importance of knowledge transfer, leading to lack of learning, lack of integration of critical knowledge and reduced skills training for complex jobs [5].

Knowledge management and knowledge transfer in the project management literature is a field of research that will receive great attention in the coming years. Once the project is finished, the knowledge created must be transferred to the entire organization for the purpose of organizational learning and building a common knowledge base among the projects [1].

Goyette et al. [8], in a case study, concludes that 50% of projects are delivered with budget overflow and 60% of projects are delivered late. The authors discuss the need for a formal knowledge transfer mechanism, since the knowledge produced by the project team is essential to serve as a knowledge base for future projects.

Jafari Navimipuor and Charband [6] concluded that knowledge transfer practices continue to be challenged as to the effective capacity to achieve effective results. The biggest challenge is to get employees to participate, access and reuse knowledge captured, and the main goal of using knowledge management in organizations is to bring awareness of the individual and collective knowledge of the company and to shape it [9, 10]

3 Methodology

It is an exploratory and descriptive work, so that the research strategy consisted of the following steps: bibliographic research; mapping the best practices of knowledge transfer; elaboration of a questionnaire with questions closed through Survey Mokey; pre-test application; adjustments in the questionnaire; application of the questionnaire; and data analysis.

Through the bibliographic research, made by the Web of Science, the best knowledge transfer practices were identified. Thus, the following keywords were taken into account in this research: *Project management, Knowledge exchange, Transfer of knowledge, Knowledge transfer, Exchange of knowledge, Project management, Knowledge dissemination and Dissemination of knowledge*.

The research instrument was divided into three parts: (I) socio-demographic information; (II) professional experience; (III) and best practices. In the first part of the questionnaire, data were obtained on gender, age, schooling, company branch of work, among others. In the second part, it was evaluated the time that the professional worked with project management, as well as the presence of certification in the area. In the third part, the practices identified in this study were presented to the respondents using a 5-point Likert scale (5. *Strongly agree*, 4. *Agree*, 3. *Undecided*, 2. *Disagree*; 1. *Strongly disagree*), in order for the respondent to indicate negative and positive attitudes about each of the practices.

The tool was applied to the Brazilian project management community through social media networks and the tool used was Survey Monkey. A total of 575 responses were collected from Project Managers in May and June 2017, and data analysis was performed using the Minitab® version program.

The results were grouped in six dimensions (groups) and, in order to verify if there was any type of trend in the sample, the Normality of Behavior was verified through the statistical tools Summary Report and I-MR Control Chart. Subsequently, hypothesis tests were developed, based on the *p value*, in order to obtain prioritization of the results.

- *Test for Equal Variances*: in order to compare the standard deviation, checking if at least one group has a different variation from the others.

$$H_o : \sigma_{dimension1} = \sigma_{dimension2} = \sigma_{dimension3} = \sigma_{dimension4} = \sigma_{dimension5} = \sigma_{dimension6} : p\text{ - value} > 0.05$$
$$H_1 : At\, least\, ONE\, dimension\, has\, Different\, Standard\, Deviation : p\text{ - value} < 0.05$$

- *One Way ANOVA*: to compare the average behavior between groups and, to verify if at least one group is less important than the others.

$$H_o : \mu_{dimension1} = \mu_{dimension2} = \mu_{dimension3} = \mu_{dimension4} = \mu_{dimension5} = \mu_{dimension6} =: p\text{ - value} > 0.05$$
$$H_1 : At\, least\, ONE\, dimension\, has\, Different\, Mean : p\text{ - value} < 0.05$$

4 Results and Discussion

During the verification of the instrument through behavior analysis, incomplete and blank answers were identified; after the cleaning, the final number of 365 questionnaires answered.

In the characterization of the sample, 73 and 21% of the respondents were in the service and industry sectors, respectively, and a male prevalence (74.3%) was also observed in the projects area. About 28% of participants had at least 10 years of professional experience in the field, while 55% of respondents claimed to have project management certification.

4.1 Diagnosis of the Database and Data Transformation

The individual analysis of each question would make the study subjective because the answers were on the Likert scale. Therefore, to facilitate the analysis of the data and to carry out the necessary correlations and comparisons, the questions of the questionnaires were grouped in dimensions: Communication, Technique, Leadership, Tool, Procedure and Culture. Table 1 presents the mapping of best practices, results of the bibliographic research, according to each dimension.

Table 1 Best practices grouped in dimensions

Dimension	Best practice	Dimension	Best practice
Technique	Use of multifunctional teams	Leadership	Awards and recognitions
Technique	Use of experts in technical matters	Leadership	Valuing the sense of trust between team members
Technique	Formal training	Leadership	Team autonomy
Technique	Manuals and procedures	Leadership	Time available to share knowledge
Technique	Informal documents	Leadership	Objectives and vision of the project
Technique	Documents and guides	Communication	Dialogue within social networks
Technique	Use of offices	Communication	Adoption of the role of facilitator
Technique	Project office adoption	Communication	Frequent communication
Technique	Narrative techniques	Procedure	Outbound interviews
Technique	Communities of practice	Culture	Organizational culture
Technique	Shared teams	Tool	Strategy for corporate information management
Technique	Lessons learned	Tool	Use of collaborative technologies
Leadership	Leadership support, support and motivation		

Source Developed by authors (2018)

Based on the participants' responses, the relation between the total score of the group according to respondents vs. maximum score of the dimension according to the number of questions, generating a result in percentage, in order to transform the data into index (continuous data). From this, it was possible to develop the analysis and prioritization of the groups.

4.2 Analysis and Prioritization of Dimensions

The normality of the behavior of the sample was verified, concluding that there was no type of trend or abnormal behavior of the sample according to the six dimensions. Thus, to analyze which groups had the greatest impact to transfer knowledge in project management, statistical tests were applied:

According to the hypothesis tests for each of the dimensions and by means of the comparison of the results, it was possible to make the prioritization of the dimensions, according to Table 2. It was observed that the dimension Procedure had a larger variation and the average relatively smaller than the other dimensions, demonstrating that this group was chosen by the respondents as the least important for the transfer of knowledge in project management.

The Technical dimension had a relatively lower mean than the other four dimensions, indicating this group as the second least important, according to the respondents. In fact, Table 3 presents the knowledge transfer practices that the project community pointed out as those with lower average scale (<3.5), being associated with the Technique and Procedure dimensions.

The Culture dimension presented a mean and standard deviation relatively higher than all the dimensions, being pointed out as the most important dimension for the transfer of knowledge in project management. Mueller [3], in investigating the cultural background of knowledge sharing among project teams, concluded that knowledge management initiatives will only succeed if they are aligned with the company's cultural factors. Thus, the employees inserted in this culture of knowledge see the sharing of knowledge as a natural activity. Wiewiora et al. [11] also mention in their research that the feeling of trust among team members facilitates and improves the project knowledge transfer although the authors have recommended in this study that the relationship between culture and trust needs further investigation.

Statistically, the Leadership and Communication group are equal in relation to the mean and standard deviation, representing the most important dimensions after

Table 2 Dimensions prioritization order

	Dimensions	Mean value (%)	Standard deviation (%)
1	Culture	90.9	11.4
2	Communication	86.7	8.9
3	Leadership	86.5	9.9
4	Tool	84.8	9.7
5	Technique	72.4	7.9
6	Procedure	63.5	23.1

Source Developed by authors (2018)

Table 3 Practices that presented the lowest averages according to the respondents

Question	Average	Practice	Dimension
Informal exchange of unstructured documents among project team members	3.10	Informal tips, tricks and documents	Technique
Formal exit interview of the employee of the company to be used by the employee who will occupy the same position	3.06	Outbound interviews	Procedure
Rotation of job roles among team members	3.50	Shared support team	Technique

Source Developed by authors (2018)

Table 4 Practices that presented the highest averages according to the respondents

Question	Avarage	Pratice	Dimension
Consistent and credible communication increases the effectiveness of knowledge transfer among project team members	4.64	Frequent communication	Communication
Leadership values, encourages and motivates the team to share knowledge	4.55	Leadership support, support and motivation	Leadership
The purpose and the scenario of the project are shared with the team	4.41	Purpose and vision of the project	Leadership

Source Developed by authors (2018)

Culture. Table 4 presents the knowledge transfer practices that the project community pointed out as those with the highest average scale (>4.4). It was observed that all these practices are associated with the dimensions Leadership and Communication.

In fact, leadership and communication are skills that the project manager must demonstrate in the execution of the project or the activities belonging to it. Thus, the personal effectiveness of the project manager encompasses: attitude, fundamental characteristics of personality and leadership. These skills are needed to guide the project team during execution, to achieve project objectives, and to balance their constraints [11]. In addition, the highest average score (4.64) for the practice of knowledge transfer was also pointed out in Park and Lee [12] study of software development projects, as a critical success factor among members of the project team, corroborating literature results. Donate and Sánchez de Pablo [9] add that the leader should be the example to be followed where he encourages learning by challenging workers and stimulating them intellectually and promoting an error-tolerant learning culture.

It is observed, therefore, that knowledge transfer practices and research on the subject continue to be challenged as to the effective capacity to achieve effective results [11], presenting different bottlenecks in organizations. Therefore, the different dimensions obtained in the present study complement the view of Goyette et al. [8] and point out a way to operationalize the transfer of knowledge as a subcomponent of knowledge management. Thus, for the transfer of knowledge, formal and informal mechanisms (in Culture, Communication, Leadership, Tool, Technique and Procedure perspectives) are needed that integrate, interpret and share anchored knowledge in individuals or groups of individuals.

It is then valid to identify not only the practices internal to the organization but also external transfer of knowledge that can transmit insights and valuable knowledge from one project to another, so that the execution of knowledge transfer is successful [13].

5 Conclusion

It was obtained 375 responses from Brazilian project managers regarding the feasibility and importance of applying knowledge transfer practices in project management. The present work identified 25 good practices of knowledge transfer, which were categorized in the dimensions of Culture, Communication, Leadership, Tool, Technique and Procedure.

According to Brazilian project managers, the importance of factors linked to organizational culture, leadership and communication as essential for effective transfer of knowledge in project management was verified. Differently, Technical and Procedural dimensions represented the lower averages, suggesting the less importance of technical and procedural requirements for the transfer of knowledge in the area. These results can guide organizational policies that favor the project knowledge transfer, optimizing the performance of companies.

References

1. Almeida, M.V., Soares, A.L.: Knowledge sharing in project-based organizations: overcoming the informational limbo. Int. J. Inf. Manage. **34**, 770–779 (2014)
2. Bradshaw, A., Pulakanam, V., Cragg, P.: Knowledge sharing in IT consultant and SME interactions. Australas. J. Inf. Syst. **19**, S197–S217 (2015)
3. Mueller, J.A.: specific knowledge culture: cultural antecedents for knowledge sharing between project teams. Eur. Manag. J. **32**(2), 190–202 (2014)
4. PMI. Pulse of the Profession® 2015: Capturando o Valor do Gerenciamento de Projetos. Annual Pulse of the Profession® Report. PMI (2015)
5. Zhao, D., Zuo, M., Deng, X.: Examining the factors influencing cross-project knowledge transfer: an empirical study of IT services firms in China. Int. J. Project Manage. **33**(2), 325–340 (2015)
6. Jafari Navimipour, N., Charband, Y.: Knowledge sharing mechanisms and techniques in project teams: literature review, classification, and current trends. Comput. Hum. Behav. **62**, 730–742 (2016)
7. Zhang, X., Pablos, P.O., Zhou, Z.: Effect of knowledge sharing visibility on incentive-based relationship in electronic knowledge management systems: An empirical investigation. Comput. Hum. Behav. **29**(2), 307–313 (2013)
8. Goyette, S., Cassiv, L., Courchesne, M., Elia, E.: The ERP post- implementation stage: a knowledge transfer challenge. Int. J. Inf. Syst. Project Manag. **3**(2), 5–19 (2015)
9. Donate, M.J., Sánchez de Pablo, J.D.: The role of knowledge-oriented leadership in knowledge management practices and innovation. J. Bus. Res. **68**(2), 360–370 (2015)
10. Duffield, S.M., Whitty, S.J.: Application of the systemic lessons learned knowledge model for organisational learning through projects. Int. J. Project Manage. **34**(7), 1280–1293 (2016)
11. Wiewiora, A., Murphy, G., Trigunarsyah, B., Brown, K.: Interactions between organizational culture, trustworthiness, and mechanisms for inter-project knowledge sharing. Project Manag. J. **45**(2), 48–65 (2014)
12. Park, J.G., Lee, J.: Knowledge sharing in information systems development projects: explicating the role of dependence and trust. Int. J. Project Manage. **32**(1), 153–165 (2014)
13. Vicente-Oliva, S., Martínez-Sánchez, Á., Berges-Muro, L.: Research and development project management best practices and absorptive capacity: empirical evidence from Spanish firms. Int. J. Project Manage. **33**(8), 1704–1716 (2015)

The Use of Lyophilization as a Technological Platform in the Preservation of a Vaccine of the National Immunization Program

Sérgio Luiz de Lima Assumpção and Celso de Farias Crespo

Abstract Facing the importance of knowing the correlation between the controllable variables of a lyophilization and their impact on quality parameters established by regulatory agencies of health products the correlation between the freezing rate and the speed of sublimation achieved during the lyophilization of an immunobiological product. For this, performed in which the lyophilization parameters have been set, the temperature and pressure cycles were identical, with the exception of the two freezing rates which in the first experiment (batch A) was −0.22 °C/min and second experiment (batch B) was −1.77 °C/min. For these experiments, the mass reduction velocity was measure during the sublimation step in the lyophilization process. 1. The effects of freezing stability of the proteins vary significantly. 2. The immunobiological product evaluated was sensitive to the two freezing rates applied. With the control of parameters of the lyophilization process, in particular nucleation, is reduce the drying time and consequently the process. 3. The freezing rate of − 0.22 °C/min can be used as a freezing rate appropriate to the lyophilization process of the immunobiological product studied since this condition optimized the mass reduction time during the sublimation stage and above all preserved the requirements of the final product quality. 4. With this, it becomes possible to rationalize the costs of inputs and energy related to the production of this immunobiological, as well as conscious maintenance of natural and ecological resources available for this purpose.

Keywords Lyophilization · Preservation · Nucleation · Vaccines

1 Introduction

Lyophilization is a preservation technique that removes water from a solution at very low levels in which biological activities or chemical reactions are inhibited. In a

S. L. de Lima Assumpção (✉) · C. de Farias Crespo
Bio-Manguinhos/Fiocruz, Rio de Janeiro RJ 21040-900, Brazil
e-mail: sergio.luiz@bio.fiocruz.br

Y. Iano et al. (eds.), *Proceedings of the 4th Brazilian Technology Symposium (BTSym'18)*, Smart Innovation, Systems and Technologies 140,
https://doi.org/10.1007/978-3-030-16053-1_45

simple way, we can dismantle the lyophilization process in three basic steps: freezing, sublimation (primary drying) and desorption (secondary drying). Lyophilization is a preservation technique that stabilizes the product for long periods. Several areas of the food industry as well as different pharmaceutical industries use this technology to increase the stability of their products.

The lyophilization process only be performed when the vapor pressure of the ice and the system temperature are below the triple point. In order to obtain the maximum lyophilization speed, it is essential that the heat supply and sublimation conditions are favored under the high vacuum conditions considered [1].

With the control of physical and chemical parameters, it is possible to inhibit or foster the nucleation step and thus the growth of any system. The freezing velocity influences both the location where the ice crystal is formed and can be in intracellular or extracellular regions as a function of solutes concentration, as well as in the size and quantity of ice crystals [2].

For this work, it was propose to study the correlation between the effect of nucleation evaluated with two different freezing rates and the respective gains in the sublimation rate during the lyophilization of an immunobiological product used in public health.

2 Objective

2.1 General Objective

Check the sensitivity of the formulation of a specific immunobiological product at two different freezing rates and evaluate the correlation between these rates and the rate of sublimation achieved during the lyophilization process for a possible optimization of the mass reduction time during the sublimation step.

Evaluate the final product quality requirements: residual moisture, potency, thermostability, appearance and solubility for batches of a specific immunobiological product at two different freezing rates.

Rationalize the costs of inputs and energy related to the production of this specific immunobiological by developing the conscious use of natural and ecological resources available for this purpose.

3 Methodology

3.1 Temporal Evolution of Mass Reduction

In the evaluation of the mass reduction during the sublimation of the formulation, a systematized methodology of weighing of samples was use. For this, a set of

50 vials-samples was prepare initially with the identification of the vials. The vials were weigh empty, and these values were record in an experiment control worksheet. Then these vials-samples were fill, re-weighed and had these values recorded in the worksheet. Thus, the initial mass of the product that was fill in each vial was obtain by subtracting the values of the mass of the vial filled from the mass of the empty vial.

3.2 Residual Moisture Analyzes by Coulometric Titration

Preceding the analysis, the moisture present in the vessel containing the Karl Fischer reagent was reduce, so that it reached values approximately 5%. Then, with the macerated cake, the mass of the product that was dispense in the titration cell of the equipment was determine, with the use of Sartorius analytical balance, until, after a time interval, it presented in its screen the percentage value of the residual moisture present in the sample.

3.3 Determination of Potency and Thermostability

Six wells of the dish were inoculate 0.1 mL of each dilution, splitted from the highest to the lowest dilution. 0.1 mL of medium 199 was inoculate into all wells into the cell control dish. Dishes were incubate for 1 h in a 37 °C in 5% CO_2 oven for adsorption of the virus. The dishes were shake every 20 min so that the inoculum covered the entire surface of the wells. After the incubation period, it was aspirate to the inoculum of each well and 1.0 mL of the Overlay, prepared at the time of use, was distribute, consisting of: (90 mL 3% Carboxymethylcellulose (CMC), 10 mL Eagle Medium 199 Earle's [10×], 10 mL FBS, 5.0 mL 4.4% sodium bicarbonate, 0.1 mL Garamycin and 0.1 mL Fungizone). Again, the dishes were incubate in a 38 °C incubator.

3.4 Review by Visual Inspection and Solubility Test

The visual inspection review was perform to evaluate the appearance of the lyophilized product of the two batch, A and B after their respective processing. The inspection was perform by two operators certified internally and with visual acuity test in day. The vials were shake and subjected to rolling movements on the display stand, thus verifying the final volume of the product, the appearance of the cake as well as its structure and coloration.

Water for injections (WFI) was use as the diluent for the test. The 2.5 mL volume was fill slowly into the vials containing the cake with the aid of a needle and syringe for ten seconds. The vials was shake gently without allowing foaming to obtain a uniform suspension. The time was time until complete reconstitution of the product.

4 Results and Discussion

The study demonstrated through the results of the behavior of the average temperature of the products, the time evolution of the mass reduction of residual moisture analysis, potency tests, thermostability, and solubility parameters quality remains preserved in function of the use of two different freezing rates. For each processing, was measure the time course of mass reduction during the sublimation where they observe two curves adjusted on the processing conditions as shown in Fig. 1.

The value of the average set of registered bottles on the worksheet to control the experience generated a worksheet in Origin, plotted Software 8.6 the experimental points and makes the best curve adjustment according to the mathematical function. It has been show in Tables 1 and 2.

Residual moisture values for batches approved of immunobiological products are typically below 3%. The results of the assessment of the percentage of residual moisture of the batches A and B after their respective process are present in Table 3 and in Fig. 2. The evaluation of Student's t statistic test applied to the average of the residual moisture of the two batches, presented the value $\cong 0.8735 < 2.101$ (limit

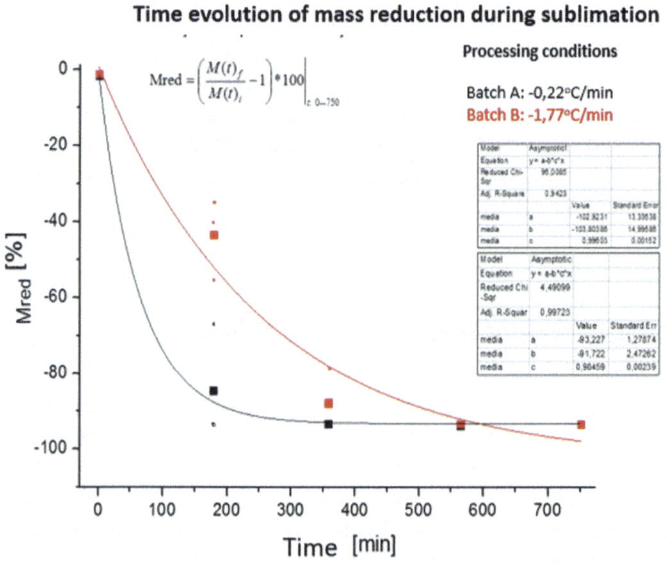

Fig. 1 Time evolution of mass reduction during sublimation

Table 1 Course of mass reduction during the sublimation

Vial (n)	Time (min)	Empty vial (g)	Filled vial (g)	Lyophilized vial (g)	Initial mass (g)	Final mass (g)	Mred (%)	Average Mred (%)
1	0	5.962	6.468	6.460	0.506	0.498	−1.58103	−1.5044
2	0	5.874	6.378	6.370	0.504	0.496	−1.58730	
3	0	5.942	6.425	6.418	0.483	0.476	−1.44928	
4	0	5.989	6.489	6.482	0.500	0.493	−1.40000	
1	178	5.951	6.449	6.116	0.498	0.165	−66.86747	−84.4812
2	178	5.911	6.420	5.946	0.509	0.035	−93.12377	
3	178	6.024	6.528	6.057	0.504	0.033	−93.45238	
4	178	6.002	6.499	6.035	0.497	0.033	−93.36016	
1	358	6.028	6.533	6.061	0.505	0.033	−93.46535	−93.2096
2	358	5.872	6.373	5.907	0.501	0.035	−93.01397	
3	358	6.015	6.513	6.049	0.498	0.034	−93.17269	
4	358	6.063	6.562	6.097	0.499	0.034	−93.18637	
1	564	5.925	6.431	5.958	0.506	0.033	−93.47826	−93.5073
2	564	5.938	6.444	5.971	0.506	0.033	−93.47826	
3	564	5.943	6.417	5.974	0.474	0.031	−93.45992	
4	564	5.944	6.445	5.976	0.501	0.032	−93.61277	
1	750	6.039	6.543	6.072	0.504	0.033	−93.45238	−93.2265
2	750	5.948	6.453	5.980	0.505	0.032	−93.66337	
3	750	6.056	6.549	6.088	0.493	0.032	−93.50913	
4	750	5.979	6.480	6.011	0.501	0.032	−93.61277	

Batch A

value table t). Thus, to the level of significance of 5%, the average residual moisture of two batches do not have significant fluctuation.

The potency test to determine the concentration of viral particles contained in the final product after the lyophilization process. The results of the tests of potency and thermostability of the batches A and B after their respective processes are present in Table 4. The result was express in 10 Log PFU/mL. Considered values between 10 and 60 PFU, the criterion for approval of tests is when this present value ≥ the Log10 PFU/mL 106 and disapproval when present value less than 106 Log10 PFU/mL. The evaluation of Student's t statistic test applied to the averages of the two batches presented the value ≅ 0.0453 < 2.101 (limit value table t). Therefore, conclude at a significance level of 5%, that average the potency of two batches do not have significant fluctuation. The evaluation of Student's t statistic test applied to averages of thermostability of two batches presented the value ≅ 1.9685 < 2.101 (limit value table t). Thus, to the level of significance of 5%, that average the two batches not have significant fluctuation.

Table 2 Course of mass reduction during the sublimation

Vial (n)	Time (min)	Empty vial (g)	Filled vial (g)	Lyophilized vial (g)	Initial mass (g)	Final mass (g)	Mred (%)	Average Mred (%)
1	0	6.070	6.573	6.562	0.503	0.492	−2.18688	−1.06031
2	0	6.000	6.509	6.507	0.509	0.507	−0.39293	
3	0	5.985	6.458	6.452	0.473	0.467	−1.26850	
4	0	5.996	6.505	6.503	0.509	0.507	−0.39293	
1	178	5.975	6.481	6.279	0.506	0.304	−39.92095	−43.2646
2	178	6.071	6.575	6.297	0.504	0.226	−55.15873	
3	178	6.056	6.563	6.387	0.507	0.331	−34.7140	
1	358	5.960	6.471	5.997	0.511	0.037	−92.7593	−87.6435
2	358	5.957	6.467	5.995	0.510	0.038	−92.54902	
3	358	6.014	6.501	6.119	0.487	0.105	−78.43943	
4	358	5.969	6.470	6.035	0.501	0.066	−86.82635	
1	564	5.931	6.440	5.966	0.509	0.035	−93.12377	−93.1828
2	564	5.975	6.480	6.009	0.505	0.034	−93.26733	
3	564	5.897	6.405	5.932	0.508	0.035	−93.11024	
4	564	5.978	6.495	6.013	0.517	0.035	−93.23017	
1	750	5.999	6.503	6.033	0.504	0.034	−93.25397	−93.2311
2	750	6.013	6.515	6.047	0.502	0.034	−93.22709	
3	750	5.958	6.466	5.993	0.508	0.035	−93.11024	
4	750	6.015	6.525	6.049	0.510	0.034	−93.33333	

Batch B

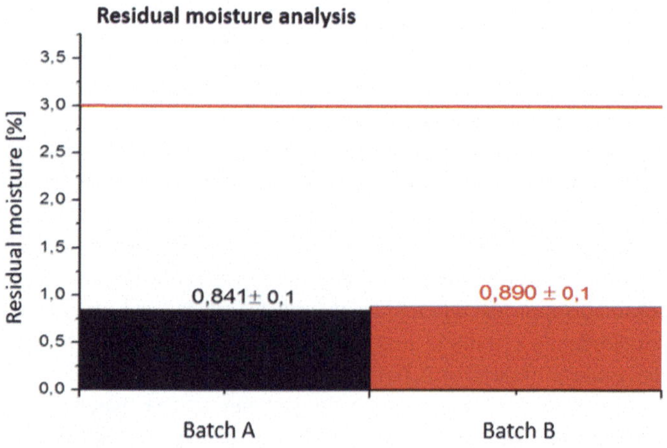

Fig. 2 Residual moisture analysis

Table 3 Residual moisture analyzes

Sample	Batch A		Batch B	
	Mass (g)	Residual moisture (%)	Mass (g)	Residual moisture (%)
01	0.082	0.628	0.089	0.869
02	0.084	0.735	0.087	0.862
03	0.084	0.845	0.088	0.908
04	0.084	0.742	0.089	1.148
05	0.084	0.987	0.087	1.039
06	0.087	0.954	0.087	0.771
07	0.084	0.943	0.088	0.869
08	0.087	0.802	0.086	0.842
09	0.085	0.766	0.087	0.878
10	0.083	1.009	0.089	0.718
m		0.841		0.890
s^2		0.01625344		0.01521644
s		0.127		0.123

Table 4 Determination of potency and thermostability

Sample	Dish	Dilution	Rectification	Title PFU/mL	Title average
Θ_A de 2 a 8 °C	58/54/35	3.0	0.7	5.39	5.345
Θ_A de 2 a 8 °C	36/46/38	3.0	0.7	5.30	
Θ_B de 2 a 8 °C	49/37/33	3.0	0.7	5.30	5.305
Θ_B de 2 a 8 °C	42/47/33	3.0	0.7	5.31	
Θ_A a 37 °C	17/17/23	3.0	0.7	4.98	5.005
Θ_A a 37 °C	20/23/22	3.0	0.7	5.03	
Θ_B a 37 °C	22/28/19	3.0	0.7	5.06	5.055
Θ_B a 37 °C	26/22/19	3.0	0.7	5.05	
Bulk	28/33/40	3.0	0.7	5.83	

The results of the review by visual inspection of batches A and B after their respective productions are present in Fig. 3. The review by visual inspection of the lyophilized product held of two batches presented the final volume as the initial volume of the product, the aspect of the healthy cake and loose from the bottom of the bottle, homogeneous structure and white. Thus, it was conclude that the two batches do not have significant difference in relation to the appearance of the final product.

The test results of solubility of batches A and B after their respective productions are present in Fig. 4. The images were record on 2, 4, 6, 10, 20, 30 and 120 s and are in the order of their respective times. The solubility of lyophilized product testing of the two batches presented a obtaining a uniform suspension without formation

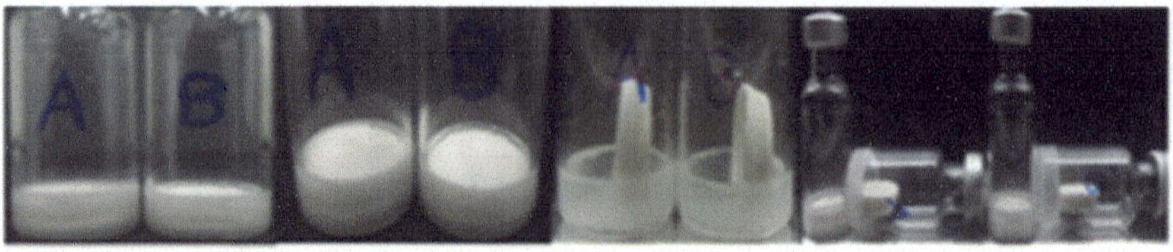

Fig. 3 Review by visual inspection

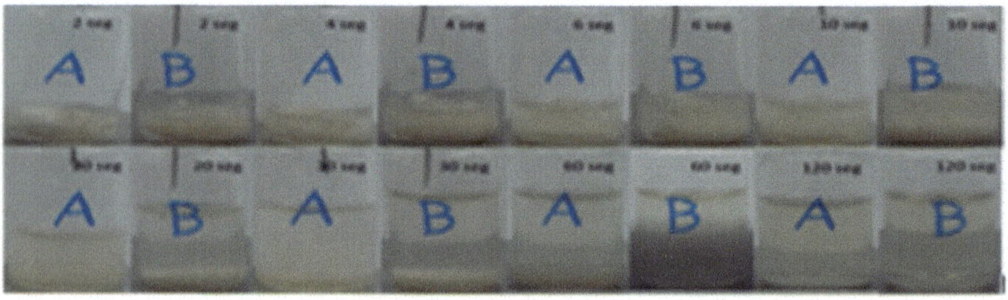

Fig. 4 Solubility test

of lumps or filaments. The complete reconstitution of the product lasted about two minutes. Thus, it was conclude that the two batches have no significant difference regarding the solubility of the final product.

5 Conclusion

The immunobiological product evaluated proved sensitive in front of two freezing rates applied, which opened doors to the optimization of the time required to complete the phase of sublimation in lyophilization cycle.

According to [2], when evaluated the influence of freezing rate on physical-chemical and structural behavior during lyophilization of native bovine albumin concluded that the slow freezing favored the elimination of structural water of this protein and thus reducing the drying time for lyophilization. Drying time of measured immunobiological product has been reduced nearly in half to reach stable levels of mass reduction using the freezing rate of −0.22 °C/min.

The study demonstrated through the analysis of residual moisture, potency tests, thermostability, and solubility that the final product quality requirements established by regulatory agencies remained preserved in the function use of two different freezing rates.

The use of −0.22 freezing rate °C/min proved favorited, since the rate of sublimation for this freezing condition performed more accelerated. Therefore, the rationalizing costs with inputs and energy related to the production of this specific immunobiological can be implement in the optimization of processes.

References

1. Jennings, T.A.: Lyophilization, Introduction and Basic Principles. 1st. ed. Intherpharm Press, Englewood, CO (1999)
2. Tattini, Jr., V., Parra, D.F., Pitombo, R.N.M.: Influência da taxa de congelamento no comportamento físico-químico e estrutural durante a liofilização da albumina bovina. Braz. J. Pharm. Sci. **42**(1) (2006)

Adaptation of the Raise Borer Elaboration Method to a Short Ore Pass by Evaluating Its Efficiency

Andree Huaynate, Juan Jara and Carlos Raymundo

Abstract The recent climate change has forced people to live in extreme conditions, either excessive heat or cold, implying that they must adapt to survive in these situations. However, there are people who, because of their geographical condition and lack of resources, lack the means and tools to combat these climate changes. The context of this study is provided in a rural town located in the Arequipa region (Peru), whose inhabitants have to fight against frosts of up to −20 °C in an area without electricity. A viable solution to this problem is found through the design and implementation of a heating system using geothermal and photovoltaic energy, which are resources found in the area, according to a report of the Ministry of Energy and Mines. This study analyzes and researches the geographical and meteorological conditions of the region, for validating, through theory and simulations, whether the proposed system can supply the thermal energy required to maintain the indoor temperature at a minimum of 15 °C under extreme conditions. The system is designed after analyzing the best technological options and techniques currently available in the context studied for its ultimate financing and establishing guidelines and indicators for monitoring results.

Keywords Renewable energy · Geothermal · Photovoltaic · Frost · Heat exchanger · Outdoor · Efficiency

A. Huaynate (✉) · J. Jara
Escuela de Ingeniería de Gestión Minera, Universidad Peruana de Ciencas Aplicadas (UPC), Lima, Peru
e-mail: U201212625@upc.edu.pe

J. Jara
e-mail: juan.jara@upc.pe

C. Raymundo
Dirección de Investigación, Universidad Peruana de Ciencas Aplicadas (UPC), Lima, Peru
e-mail: Carlos.raymundo@upc.edu.pe

Y. Iano et al. (eds.), *Proceedings of the 4th Brazilian Technology Symposium (BTSym'18)*, Smart Innovation, Systems and Technologies 140,
https://doi.org/10.1007/978-3-030-16053-1_46

1 Introduction

For many years in Peru, shafts have been implemented in short-length sections; all mining companies with underground works need to excavate shafts for constructing ore passes or sub-shafts to access the mineral to be extracted. Mining companies invest large sums of money in the elaboration and implementation of shafts because shafts are of great importance in the extraction process of minerals. In general, the time required to excavate a shaft is an issue, because traditional methods that use Stoper drills only advance 1.2 m, which delay the time of implementation and elaboration. That time is accepted within the mining and design plans. However, an adapted method is proposed, which can increase the effectiveness of the method by 100% and reduce the time required to implement and elaborate shafts by 80%. Furthermore, methods of support and self-support of conventional shafts are not functional because of the shape and the geomechanical characteristics of shafts. Therefore, bolts adjusted by wooden wedges are implemented, which generate a compression between boxes or side walls of a shaft for permanent support. Nevertheless, it is expensive and lacks instantaneous effectiveness.

The implementation of the Raise Borer (RB) method is a very effective way to build a shaft in a circular way with effective self-support. However, indirect results on the adaptation of the RB method would be elimination of the use of bolts and wedges, which have a limited shelf life. To avoid that, the support method is improved with a 3-inch wet Shotchere method in order to provide permanent support.

The direct benefit and purpose of the research are reduction of time of implementation of a shaft, which will provide considerable improvement in the mining plan and reduction of unit costs for the construction of shafts. In this article, the development of the proposal and the validation and adaptation of the method will be presented in order to arrive at the results and the validation of the method compared with the results of the traditional method. For the following article, we used the results of a theoretical validation research conducted previously with similar adaptations in order to perform experimental validation and verify the data in the field.

The article consists of the following parts: first, the state of the art will present the details of previous related investigations that will support our results. The second part, contribution, will detail a new alternative method to provide a solution to the problem previously raised; the third, validation, will present the implementation of the proposed contribution and demonstrate the efficiency of the proposed contribution.

2 State of the Art

For the elaboration of this article to depict greater academic consistency, we present previous investigation of different authors and reviews. Natural ventilation has been used for many years in the exploitation of minerals by underground methods. In the middle of the 19th century, the use of mechanical fans began in the various galleries of mines. These fans were an exclusively centrifugal type with large diameters and reduced velocities, moving by means of windmill hydraulic wheels [1] and [2]. Under this premise, ventilation was a point of great importance in mining facilities. With the deepening of the mines, as the author stated, the centrifugal fans that currently work with air extractors to expel toxic gases have gained interest. At present, fans have been improved substantially according to the increase in the production and depth of mines. The necessity of airflows has grown enormously, surpassing the 1000 m^3/s of circulating air [3].

Airflow is an important factor in ventilation because the balance of gases has to compensate for the air that enters with the one that leaves. In this case, the outgoing air contains polluting gases that are highly harmful to workers' health. The flow of air that enters has to be efficient and clean so that the regulation proposed by the Peruvian law, is complied, which states that a flow of 20 m per minute is the optimal flow for an adequate temperature in mining facilities [4].

In contrast, some mines currently do not have good air intake and exhaust through fans, resulting in low efficiency and low possibility of maintaining good and necessary environmental conditions for the protection of workers. The general reason for this may be an inadequate distribution of air inside the mine or a regulation of wrong circuits, because the geometry of the shaft directly impacts the airflow [5].

The Raise Borer (RB) method is evaluated to achieve the required efficiency and reduction in costs because it allows a continuous advance in the elaboration of shafts. The RB method was developed to meet the demands of the mining industry but has also given numerous applications in the construction of tunnels or infrastructure projects for ventilation, even in very hard rock formations or in the opening of deep wells [4].

The machine used with the RB method is one of the most important factors in determining the efficiency and economy of the excavation since daily advance rates are directly related to these factors. Therefore, a daily control of unit costs and economic comparison with other methods is necessary.

The implementation of the RB machine incurs a high cost at the beginning. However, the long-term benefits are enormous so it is recommended that the relevant investment be made. On the other hand, it is expected that the return on investment will be approximately in 3 years; the machine would devaluate 50%. However, the shelf life or use can exceed 10 years depending on the maintenance and drilling used [6].

The efficiency of shaft elaboration using this method as well as the economic advantages is evaluated in all cases. However, the most important aspect would be a new system that considerably reduces the concentration of gases. As mentioned previously by the author [7], an adequate flow of air inside the mine means more optimal working conditions and reduction of the costs of mechanical ventilation.

In the present work, the objective is to optimize the natural ventilation of deep mines by implementing the RB system such that the costs of processing and ventilation will be reduced with the improvement in the quality of the tunnel and the continuity of the hole.

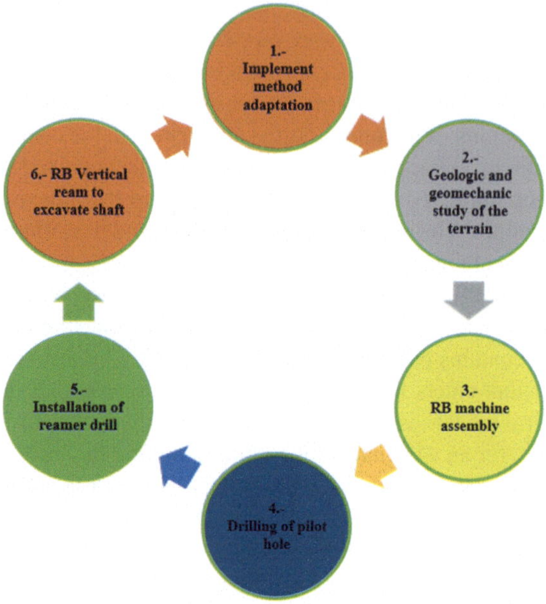

Fig. 1 Suggested methodology

3 Contribution

The following procedure was followed for the method proposed, as described in Fig. 1. It details the process and the implementation of the system, developing it as a work methodology that is based on clear experimental data with quantitative variables presented in tables for comparison.

3.1 RB Method Adaptation Implementation

The implementation of the adaptation of the RB method is new because past methods have not been done this; reference is made to the adaptation of the method and in a specific type of condition because, as it is generally known, not all mining units have the same design and the same dimensions. Therefore, during this implementation, an adaptation of the RB method will be made for the new galleries of the company in which the design and validation of the experiment will be performed.

Proper elaboration on the adaptation of the method will be key to the investigation because it will be possible to evaluate the efficiency of the elaboration of the RB method.

To validate the implementation, a list of documents will be used to compile the data and the most appropriate way to use the RB machine for its implementation, as shown in Table 1.

Table 1 Necessary hours to install equipment

Hours	
Reamer installation	Work time
8	8
3.45	3.45

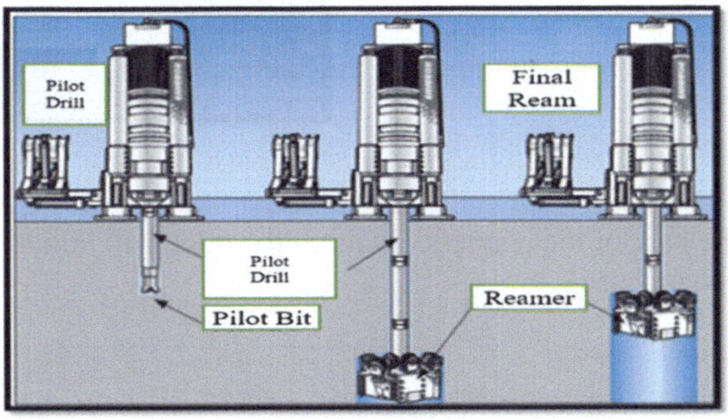

Fig. 2 Perforation of the pilot hole (Atlas Copco. Product 30 catalogue—raise boring equipment", Secoroc Rock Drilling Tools Pag. 56)

For this adaptation, the size of the gallery and its access was controlled, since the entrance of the gallery was reduced to 2.7 m high, 3 m wide, and 4 m deep, resulting in a new configuration. There was a separation of 1.2 m between the three elements and a new dimension of hydraulic hoses was applied. Table 2 presents the average installation time. In the same way, control of the installation time of the equipment has to be recorded, since it will help to know the number of hours worked.

3.2 *Drilling of the Pilot Hole*

The importance of this stage is that the principle of the RB method is to communicate two sub-levels through a pilot hole of diameter 2–5 inches. The sub-levels have access cameras to excavate the shaft. This drilling starts from the need to create a guide so that the reamer rises vertically for the shaft to be elaborated in a constant manner.

The pilot hole is made in the following way, as shown in Fig. 2.

3.3 *Installation of the Reamer Drill*

Once the pilot hole is properly communicated, it will be free and will have to extend to the floor of the gallery in which the reamer drill will be placed.

The process of disassembling the drill bit of the pilot drill must be done anti-clockwise by removing the main thread and releasing the coupling for the reamer drill. The reamer drill couples with a splined shaft with 8 teeth, which is fastened to

Fig. 3 Reamer drill (Palmieri reamer drill)

the base of the reamer column or the pilot drill initially developed. A reamer drill is shown in Fig. 3.

3.4 RB Vertical Reaming Excavating the Shaft

With the start of vertical reaming, the shaft will be excavated with an efficiency of 3–5 m per 8-h shift. The 8-h guards are considered according to the standard work schedule. In the same way, only 2 shifts per day of work will be considered so that there is time for change of staff, resting of the machine, and checking of the basic functions, such as hydraulic pressure, reaming, and advance.

4 Validity

For the validation phase, the results obtained were presented first, and later, a matrix was generated using the scenarios that were generated. They were then compared with the results obtained for analyzing in a quantitative way and to validate the next stage of discussions to reach the final results.

The validation of the variables and the studies in planning as the elaboration of the design in AutoCAD. Regarding experimental validation, it was implemented in a mining company that already has the equipment and was interested in doing the study for the improvement and determining the viability of this equipment.

The pilot hole was drilled once the machine was installed so as to implement the first stage of the shaft. A summary chart of the implementation times of the machine is published once it was moved, and its service connections have been already installed. The above chart shows that the installation time is 11.45 effective hours of work. Once the equipment is installed and with corresponding verifications, it is possible to start drilling the pilot hole to communicate the two galleries or sub-levels. The table in which the drilling times are determined is attached. The drilling advance varied between 4 and 6 m for each effective shift of 8–9 h of work, which indicates that the initial drilling showed a better than expected estimate by applying the same reaming and speed of reaming. It was tested in the shaft. Previously, there was a minor touch and advance due to the requirements of the manual. However, the new configuration of the implementation with a hydraulic motor that provided greater pressure allowed the continuous advance, as presented in the following table. Perforation remained between 80 and 90 KN in some cases, increasing due to the hardness of the rock and the geology of the area.

It can be observed that the reaming is more oscillating than the perforation of the pilot hole. This is due to the hardness of the terrain and the faults or discontinuities present in the terrain. This type of clay material that was found in some areas caused a delay of the reamer since it had to lower the reamer drill so that it could raise with a greater torque. That is controlled by the operator of the equipment, who has to have the basic knowledge of the geology of the land and a previous qualification so as not to affect the equipment. Results show that no material blockage was presented during reaming. That benefitted from the study. Blockages can be present in longer lengths and in situations where the dip angle of the shaft is less than 45°. An AutoCAD graph was created to determine the angle and length of the shaft between the galleries. Comparative table between the flow rate of air between shaft VR150 excavated by the RB method and the VR80 excavated by the Alimak method shows that the RB system has a higher flow rate, as shown in Table 4, in which all the calculations were made by taking into account all the variables that were previously established.

It was possible to determine that between the drilling of the pilot hole and the reaming, a working time of 43 days resulted necessarily to excavate a shaft of 182 m. That is optimal for the elaboration of a shaft which, in comparison with other methods, would have taken longer. This analysis will be done on the following pages.

5 Conclusion

To conclude the present article, the elaboration time required when the proposed RB method is used is less than that of a conventional method. In addition, the advance distances are more effective and the speed of reaming and perforation is related to the type of rock. In this study case it can be determined that, as there were no failures nor blockages, the advance was related to the type of rock. Therefore, the following conclusions are presented:

- The airflow is higher because variables such as the friction coefficient were reduced by 20% and the airflow became more continuous during the investigation. The use of mechanical fans was rejected due to the existence of logistical problems in the mining unit.
- The advance was 178.4 m, which allowed to leave a bridge for the uninstallation of the equipment, and later, to use the blast to release that face from the shaft. The total time used for excavating the shaft was 465.03 effective hours of work of the

machine, which showed us that the advance was of 25 days taking into account the 2 shifts per day.

- Drilling remained between 80 and 90 KN, in some cases, increasing due to the hardness of the rock and the geology of the area.
- This shows that those values can be taken as constants in a terrain of rock type 3A or 2A and 3B.
- The adaptation method can still be improved by reducing the size of the equipment with more efficient ones, such as the hydraulic pump. A smaller pump with a hydraulic pressure could reduce the gallery and allow for the implementation of the equipment in places of short accessibility. By contrast, it is necessary to improve the drilling force because, in practice, it is a determining factor in advance. It has to be evaluated to determine it according to the geology and geotechnics of the rock mass.

References

1. Zhou, L., Yuan, L., Thomas, R., Iannacchione, A.: Determination of velocity correction factors for real-time air velocity monitoring in underground mines. Int. J. Coal Sci. Technol. **4**(4), 322–332. Clerk Maxwell, J.: A Treatise on Electricity and Magnetism, 3rd ed., vol. 2, pp. 68–73. Oxford: Clarendon, 1892
2. Bustamante-Rúa, M.O., Daza-Aragón, A.J., Bustamante-Baena, P.: Simulation software VENTSIM™ the influence of implementation of work abandoned sealings ventilation of an underground coal mine. Boletín de Ciencias de la Tierra **43**, 5–13 (2018)
3. Wengui, L., Xianping, X., Houzao, X., Chao, C., Xiaoyu, Y., Meiwei, S., … Xingyu, T.: Reliability calculation of mine ventilation network. Proc. Eng. **84**, 752–757 (2014)
4. Shuyu, Z.Y.M.Z.K.: Determination of weight value by AHP. J. North. Jiaotong Univ. **5**, 025 (1999)
5. Cong, R.G., Wei, YM.: Experimental comparison of the impact of auction format on carbon allowance market. Renew. Sustain. Energy Rev. **16**(6), 4148–4156 (2012)
6. El-Nagdy, K.A., Shoaib, A.M.: Alternate solutions for mine ventilation network to keep a pre-assigned fixed quantity in a working place. Int. J. Coal Sci. Technol. **2**(4), 269–278 (2015)
7. Maleki, M.R.: Rock Joint Rate (RJR); a new method for performance prediction of tunnel boring machines (TBMs) in hard rocks. Tunn. Undergr. Space Technol. **73**, 261–286 (2018). LNCS Homepage, http://www.springer.com/lncs. Last accessed 21 Nov 2016
8. Shaterpour-Mamaghani, A., Bilgin, N.: Some contributions to the estimation of performance and operational parameters of raise borers–a case study in Kure Copper Mine, Turkey. Tunn. Undergr. Space Technol. (2016)

Application of PMBOK to Improve the Deadline of Projects in SMEs Engineering Consultancies

Maricielo Hinostroza, Pedro Chavez, Victor Nuñez and Carlos Raymundo

Abstract In recent years, project management in consulting firms has grown due to greater demands for goods and services. However, this growth has meant that due to inefficiencies in the management, delays in project delivery times are generated. In addition, the same situation occurs for performance of personnel's competences, as well as in defining the scope of the project. Therefore, this article proposes a model for the application of Project Management Body of Knowledge (PMBOK) based on knowledge management, which consists of the following areas: stakeholder management, risk management, cost management, time management, human resources management, quality management, integration management, and scope management. To show the ease of use of this model, a case study was conducted in an engineering consulting firm. The results of the application of the model were successful: the delay in project delivery times was reduced by 70% and the risks that may exist during execution were reduced by 67%. In addition, the company highlighted the success of the application of knowledge management because it improved the interaction and skills of staff by facilitating the implementation of the model.

Keywords Project management · PMBOK · SMEs · Scope management · Knowledge management

M. Hinostroza (✉) · P. Chavez · V. Nuñez
Escuela de Ingeniería de Gestión Empresarial, Universidad Peruana de Ciencas Aplicadas (UPC), Lima, Peru
e-mail: U201314465@upc.edu.pe

P. Chavez
e-mail: pedro.chavez@upc.pe

V. Nuñez
e-mail: victor.nunez@upc.pe

C. Raymundo
Dirección de Investigación, Universidad Peruana de Ciencas Aplicadas (UPC), Lima, Peru
e-mail: Carlos.raymundo@upc.edu.pe

Y. Iano et al. (eds.), *Proceedings of the 4th Brazilian Technology Symposium (BTSym'18)*, Smart Innovation, Systems and Technologies 140,
https://doi.org/10.1007/978-3-030-16053-1_47

1 Introduction

Project management is defined as "the application of knowledge, skills, tools, and techniques to satisfy one or several requirements" [1]. According to Chaos Report [2], in 2015, 50,000 projects, both small and large, were globally analyzed, of which 6% were very large and more than 62% of successful projects were small. In addition, 29% were successful and 19% failed due to various factors affecting their development.

Several researches show how the guide of Project Management Body of Knowledge (PMBOK) generates benefits in different organizations conducting construction projects [3] or in different areas; however, medium and small Latin American (MYPES) engineering consultancies denote a lower percentage of application [4]. This is because of various factors that are creating problems in project development activities, such as the lack of coordination, low budget, limited resources, and the lack of trained professionals or inadequate planning [5].

This research proposes the implementation of PMBOK in MYPES so companies can improve on project management, the lack of procedures, and avoid delays in delivery. Therefore, the article consists of five sections: Section 1 covers a brief introduction; Section 2 contains a bibliographic review of the model implementing knowledge in the areas of PMBOK in MYPES; Section 3 displays the proposed model; Section 4 explains the validated model in an engineering consulting MYPE to expand drinking water and sewer system projects and the results thus obtained; and Section 5, concludes the paper.

2 State of the Art

In the literature, there are several methods used to apply the PMBOK for reducing the delay in project delivery [6]. In [1], PMBOK mathematical methods are used to determine the degree of risk that may exist and thus corrective measures to consider the impact of real performance on the project are adopted. The model in [7] relates to best practices and managing the scope of the project through the structure of the scope, defining the scope, continuously evaluating, balancing execution, and evaluating the project. In addition, through change management [8], one can include the scope of the project, size, and duration, as well as the expectations of the organization on the speed of its implementation. In contrast, there is knowledge management [9], which provides better results with an innovative organization that achieves selective integration of business and activities based on intellectual capital. This allows processes to improve their technical performance. In model [10], open innovation is considered a way to improve the acquisition of knowledge from external sources. In [11], it was used to overcome critical factors: opportunity recognition, business commitment, credibility, and sustainability, all of which relate to the internal knowledge of the company [12].

Table 1 Benchmarking of the proposed model and existing models

Authors criteria	[1]	[13]	[3]	[4]	Proposed model
Organizational leadership			x		x
Expert opinion	x		x		x
Statement of project scope		x			x
Implementing risk response plan	x				x
Validating scope					x

The implementation of the PMBOK Guide in small projects can generate both benefits and limitations. The main benefits according to [8] are better team performance and project management, in addition to establishing better solutions. PMBOK also helps improve customer satisfaction and determines time, quality, and cost for the benefit of the organization. [13] agrees with [8] that the PMBOK collaborates with any project, small or large, to develop best practices in the organization. In [6], there are some limitations such as a lack of understanding in all areas required knowledge about how to obtain best practices from lessons learned.

3 Contribution

Compared to other models, this proposal includes managing the project's scope as well as the knowledge to improve implementation of the PMBOK. Managing the scope is important when companies need to validate and measure the scope, and knowledge in this area is required for the project to be successful [8]. Besides improving business knowledge, acquiring value and enabling an organization to share intellectual capital, both explicit and implicit as well as people, experiences, and diverse perspectives of the same problem for promoting collaboration among workers is required. This is needed to achieve change in culture and to strengthen the identity of the company and staff.

According to Table 1, it can be seen that the models of other authors did not consider organizational leadership as part of their research. In addition, they did not take into account scope validation because most of them did not consider project scope.

3.1 Vision. General

The proposed model (Fig. 1) is based on the PMBOK guide, five pillars that together represent an improvement process (stakeholder management, risk management, cost management, quality management, and integration management), a pillar that sup-

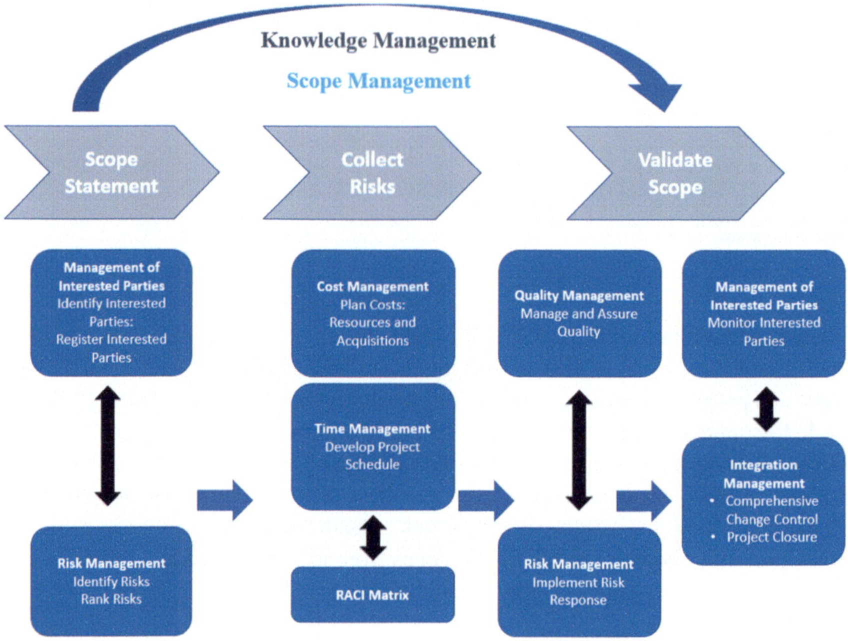

Fig. 1 Model proposed PMBOK

ports the whole model transversally is scope management, which helps to compile the requirements to establish a scope plan for managing the scope of the project.

The contribution considered in the model is Knowledge Management as a reinforcement of the Project's Scope Management because the workers involved in this area must have the necessary skills to resolve conflicts effectively, achieve objectives, project restrictions, and the time scheduled for each activity. Therefore, it is necessary that prior to implementation of the knowledge areas, the staff brings together the organizational leadership and the required competencies.

The model in (Fig. 1) consists of four stages: the first stage involves implementing knowledge management, which requires an organizational leader for meetings held with the project team. In the second stage, scope management will be implemented, along with the development of a scope statement, the compilation of requirements, the management of stakeholders, and risks. The third stage consists of cost and time management, where cost planning will be conducted and the project schedule as well as human resources matrix will be developed. In the fourth stage, there is quality management; quality will be managed and a risk response plan will be implemented. Finally, stakeholder and integration management will be performed by monitoring interested parties and comprehensive change control.

4 Result Analysis and Discussion

The model was implemented in the project "Expansion of drinking water and sewer system in the center of Machu Picchu" by the company Consultores Paccha SAC. This was conducted in January 2018 with a 60-day standstill that ended in August 2018, having duration in accordance with the 150-day contract. The results obtained from each component of the PMBOK guide, as well as the knowledge management are shown below:

4.1 *Knowledge Management*

Organizational leadership, open innovation, and training were used. This resulted in an improvement in staff satisfaction to 100%, which was suitable for the company (Table 2).

4.2 *Risk Management*

Identifying risks and the risk response plans caused a decrease in risks for this project as compared to those conducted between 2015 and 2017 (Table 3).

4.3 *Quality Management*

Better quality control and continuous supervision from before the start of the project until the finalized project was delivered resulted in a reduction in resource costs to be used for the project procedures by S/. 148,890.00 soles (Table 4), which provided better performance and productivity.

Table 2 Indicators obtained from knowledge management

Indicators	Objective	Initial situation	Implementation	Variation
		Projects from 2015–2017	2018 project	
Staff satisfaction (%)	100	75	100	25
Category				
Critical		Less than 40%		
At risk		40–70%		
Suitable		Greater than 70%		

Table 3 Indicators obtained from risk management

Indicators	Objective	Initial situation	Implementation	Variation (%)
		Projects from 2015–2017	2018 project	
Internal risks	0	7	3	57.14
External risks	1	9	3	66.67
Risks that cause delays	1	12	4	66.67

Table 4 Indicators obtained from quality management

Indicators	Objective	Initial situation	Implementation	Variation (%)
		Projects from 2015–2017	2018 project	
Procedures performed by contract time	10	32	13	59.38
Resource costs for procedures executed	S/. 70,000	S/. 235,000	S/. 86,110	63.40
Risks that caused delays	1 month	3 months	1 month	66.67

Table 5 Indicators obtained from integration management

Indicators	Objective	Initial situation	Implementation	Variation (%)
		Projects from 2015–2017	2018 project	
Cost of the changes	S/. 40,000	S/. 133,183	S/. 45,920	52.38
Total changes	5	15	6	60.00

4.4 Integration Management

More frequent monitoring and control of the project resulted in a decrease in requests for a change of project because good compilation of requirements was put together, and validation of the scope and control of the risks prevented changes that could have existed. Additionally, approximately S/. 87,263.00 soles were saved due to changes that existed during execution (Table 5).

Table 6 Indicators obtained from scope management

Indicators	Objective	Initial situation	Implementation	Variation (%)
		Projects from 2015–2017	2018 project	
Time to put scope into practice	10 days	30 days	15 days	50.00
Cost of validating the scope	9000	3650	9650	66.67

Table 7 Indicators obtained from time and cost management

Indicator	≥90%	<90%
CPI	2	10
	17%	83%
SPI	8	4
	67%	33%

4.5 *Scope Management*

Knowledge was provided to the team regarding project objectives, making its management possible during the execution (Table 6).

4.6 *Time Management and Costs*

The activity schedule and cost planning was carried out, calculating the planned value and cost variation of the project (Table 7).

5 Conclusion

From this research study, it can be concluded that the project was good at time management, since 83% met the dates set during the project's schedule. However, the project performed average in cost management, since 67% of the projected costs were improved; although, it was not possible to exceed the projected costs. Finally, the project was successful regarding scope management, since it delivered 100% of the project. Likewise, implementing knowledge management is beneficial for the company since these professionals can be considered for future projects as they demonstrate the necessary skills.

References

1. Muriana, C., Vizzini, G.: Project risk management: A deterministic quantitative technique for assessment and mitigation. Int. J. Proj. Manage **35**(3), 320–340 (2017)
2. Laboratorio, T.I.: Informe del Caos 2015 (2016)
3. Basu, R.: Managing quality in projects: an empirical study. Int. J. Project Manage. **32**(1), 178–187 (2014)
4. Demirkesen, S., Ozorhon, B.: Impact of integration management on construction project management performance. Int. J. Proj. Manage. **35**(8), 1639–1654 (2017)
5. Doloi, H., Sawhney, A., Iyer, K.C., Rentala, S.: Analysing factors affecting delays in Indian construction projects. Int. J. Proj. Manage. **30**(4), 479–489 (2012)
6. Arditi, D., Nayak, S., Damci, A.: Effect of organizational culture on delay in construction. Int. J. Proj. Manage. **35**(2), 136–147 (2017)
7. Abrantes, R., Figueiredo, J.: Feature based process framework to manage scope in dynamic NPD portfolios. Int. J. Proj. Manage. **32**(5), 874–884 (2014)
8. Lines, B., Sullivan, K., Smithwick, J., Mischung, J.: Overcoming resistance to change in engineering and construction: change management factors for owner organizations. Int. J. Proj. Manage. **33**(5), 1170–1179 (2015)
9. Cerchione, R., Esposito, E.: Using knowledge management systems: a taxonomy of SME strategies. Int. J. Inf. Manage. **37**(1), 1551–1562 (2017)
10. Jasimuddin, S., Naqshbandi, M.: Knowledge-oriented leadership and open innovation: role of knowledge management capability in France-based multinationals. Int. Bus. Rev. **27**(3), 701–713 (2017)
11. Dalmarco, G., Maehler, A., Trevisan, M., Mortari, J.: The use of knowledge management practices by Brazilian startup companies. J. Bus. Innov. **14**(3), 226–234 (2017)
12. Hwang, Y., Lin, H., Shin, D.: Knowledge system commitment and knowledge sharing intention: the role of personal information management motivation. Int. J. Inf. Manage. **39**, 220–227 (2018)
13. Sanchez, O., Terlizzi, M., de Oliveira, H.: Cost and time Project management success factors for information systems development projects. Int. J. Proj. Manage. **35**(8), 1608–1626 (2017)

Improvement Proposal for the Logistics Process of Importing SMEs in Peru Through Lean, Inventories, and Change Management

Izon Riquero, Christian Hilario, Pedro Chavez and Carlos Raymundo

Abstract The logistics processes of companies seek an adequate supply of merchandize, streamline processes, reduce costs, and distribution of their products to the customer in an efficient manner and allow them to develop corrective planning in order to be competitive in the market. However, these processes are less efficient in importing companies of the SME sector. Therefore, this article proposes the application of a change management model to improve the logistics process that generates a real transformation. This model is structured in three phases: first, pre-Implementation, which refers to participatory leadership and resistance to change; second, implementation of logistics management; and third, sustainability of the model over time. The model was validated in a Peruvian SME importer of household goods, whose satisfactory results implied an increase in available storage capacity, reduction of logistics costs, satisfaction of change, and organizational culture.

Keywords Logistics management · Change management · Knowledge management · SMEs

I. Riquero (✉) · C. Hilario · P. Chavez
IEscuela de Ingeniería de Gestión Empresarial, Universidad Peruana de Ciencas Aplicadas (UPC), Lima, Peru
e-mail: u201310633@upc.edu.pe

C. Hilario
e-mail: u201311467@upc.edu.pe

P. Chavez
e-mail: pedro.chavez@upc.pe

C. Raymundo
Universidad Peruana de Ciencas Aplicadas (UPC), Lima, Peru
e-mail: Carlos.raymundo@upc.edu.pe

Y. Iano et al. (eds.), *Proceedings of the 4th Brazilian Technology Symposium (BTSym'18)*, Smart Innovation, Systems and Technologies 140,
https://doi.org/10.1007/978-3-030-16053-1_48

1 Introduction

An important element of economic transformation in any growing country is the strengthening of SMEs because they create employment, boost competitiveness, and spur entrepreneurship and innovation [1]. Among the Organisation for Economic Co-operation and Development (OECD) member countries, SMEs account for 60–70% of employment [2].

This sector is becoming an attractive alternative for the Peruvian market, considering trade brings together the highest number of SMEs [3]. According to the Peruvian Ministry of Production, SMEs account for 4.4% of all companies across the country, and, from that percentage, the enterprises engaged in imports (40%) collected USD 6.685 billion from January to February 2018. The good performance of imports during the period was mainly fueled by purchases from China (24%) [4].

Although imports from Peruvian SMEs reflect an upward trend, several issues hinder competitiveness. According to [5], SMEs struggle internally to boost competitiveness because of administrative issues (43%), legal issues (16%), failure to comply with requirements (15%), deficient release to the market (11%), input and production errors (8%), and financing (7%).

In 2014, Peruvian logistics performed as expected for an upper-middle income economy. However, due to its logistics performance index, Peru is far behind leading countries such as Chile, Panama, and Mexico [6]. Streamlined logistics and proper management are key in ensuring SME successful competition [7]. Within this context, other authors proposed models that focused on the characteristics of SMEs; however, the internal information flow remains unheeded, preventing proper implementation [8].

Therefore, the purpose of this article is to develop proper supply planning, increase storage capacity, cut costs, build awareness, and gain the understanding required to become more competitive. The article is divided into the following section: Sect. 1 is the introduction; Sect. 2 contains a review of existing articles; Sect. 3 provides a description and examination of the proposed model; Sect. 4 includes the results obtained from validating the model by an importing company; and Sect. 5 concludes and summarizes the paper.

2 State of the Art

The research conducted in [9] discloses a logistics management methodology aimed at enhancing SME supply, storage, distribution, cost planning, and customer service, but this procedure is not comprehensive. The model proposed by [10] identifies a first cycle related to supply, a second cycle concerning storage, and a third cycle related to the sales plan; however, the model falls flat upon identifying the indicators to be met by a SME. The model designed by [11] relies on the SCOR model and is targeted to SMEs; however, this research was based on a logistics best practices manual. These models highlight the importance of logistics management, but fail to describe how organizations should transmit knowledge and sustain the model [12]. Consequently, the relationship between each logistics process is weak, leading to disintegration and posing a challenge for SMEs [8].

The purpose of lean production is to streamline logistics by minimizing waste in the chain value, cutting costs, and focusing on the customer in order to deliver top quality services [13]. Although the companies have implemented lean tools to drive their business results, they still encounter several implementation barriers [14]. Some of these challenges are poor personnel training and experience, lack of knowledge, and cultural resistance to change [15]. In [16], the authors propose a staircase road map, which identifies four main topics: the type of lean being adopted, how lean is used in SMEs, the impact of lean implementation and the critical success factors.

Change management arises from the need to perform some type of change within the organization; it may exert a positive or negative effect in personnel performance. In [14] it is asserted that managing employee commitment and resistance is essential to embrace change. Besides, the research performed in [17] implies that acceptance of change converts unease into positive energy to boost performance. In prior studies on knowledge management, the model proposed in [18] states that leadership is the cornerstone of high trust for achieving successful knowledge management and organizational performance. Moreover, the model described in [19] reveals that the characteristics of the culture should be borne in mind to plan and make decisions. Thus, in [20], the author proposes integrating existing knowledge management models based on the characteristics of culture.

3 Contribution

A management model aimed at streamlining the logistics process of importing SMEs should differ from that adopted by large enterprises due to the technological resources, structure, language, and culture in which they operate [21]. In this context, other authors presented models focused on the characteristics of SMEs and logistics; however, the implementation flow remains neglected. Table 1 includes a review of the most relevant articles for the purposes of our proposal.

For the proposal included herein, [6] is used as a basis because it combines elements from Lewin, Kotter, and Prosci change models.

This proposal consists of developing a change management model that considers the logistics tools, personnel awareness, and knowledge sharing that allows improving the competitiveness across the sector. The chart below includes the proposed change management model.

The method is implemented in three stages: (1) pre-implementation is related to leadership development and resistance to change; (2) implementation is the development of change management and the skills required for the logistics processes, as well as change monitoring; and (3) post-implementation deals with the sustainability of the organizational culture satisfaction (Fig. 1).

Table 1 I. Article review

	Change management					Knowledge management			Logistics management				Lean
	Organizational performance	Capacity	Resistance to change and commitment	Organizational culture	Efficient communication	Leadership	Sharing and transferring	Technology infrastructure	Planning	Costs	Production	Storage	Best practices
C. Gonzales A. Velásquez Instituto Pyme									x	x	x	x	
J. Calderón L. Francisco									x	x			x
L. Zhang J. Bhamu R. Mason		x		x		x			x	x			x
E. Profka M. Karambelkar A. Abdula R. Busse		x		x	x	x	x		x				
S. Abdullah A. Tsai A. Velásquez I. Hasan					x		x	x					
T. Ramadas A. Raza M. Grupta						x		x	x				

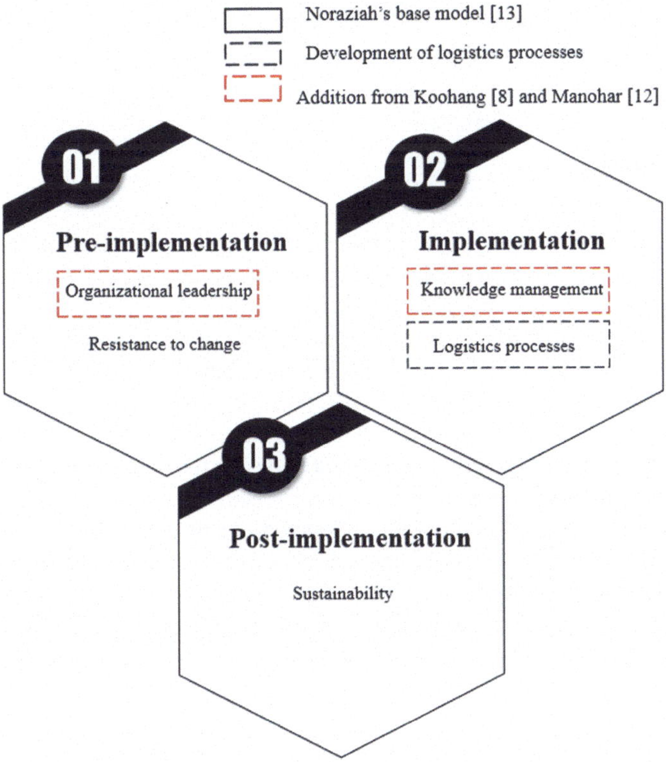

Fig. 1 Change management model proposed

4 Validity

This proposal was developed in a Peruvian SME that was engaged in the import of household items from China to Peru. The research surveyed 10 workers, including the General Manager, in May 2018. Although the company has just started implementing change management, knowledge, lean, and logistics processes, the outcomes have been favorable. Table 2 compares the Baseline and Endpoint regarding the change management model proposed for 2018.

Upon implementing the model proposed, group work was encouraged and awareness was raised on the importance of a proper logistics process. The 5s methodology reduced the occupied capacity in the pilot warehouse; therefore, an appropriate layout was designed based on inventory turnover that decreased dispatch time despite the great demand.

The costs incurred due to the orders requested changed within the same period; hence, a purchase order plan was proposed for the rest of the year.

Finally, organizational knowledge and culture satisfaction also increased. This was driven by the scheduling of meetings, transferring knowledge acquired about the ongoing campaign and the relevant information about current inventory, among other organizational considerations.

5 Conclusion

It can be determined that leadership, diagonal communication, organizational culture assessment, employee commitment, and change acceptance drive organizational performance.

The development of the logistics process is closely associated with the human factor, which, with proper leadership and training, could lead to expected results.

The pre-implementation stage favors the creation of an organization and leaders for proper organizational development. At the implementation stage, change management and monitoring ensure training in order to design the proper tools and methods for a logistics process.

The model suggested promotes the development of soft and hard skills. In addition, the implementation of this model is considered successful since it was adopted and improved using the criteria detailed under Baseline in Table 2.

Unlike other studies, this model incorporates components that crush resistance to change and promote the engagement of all the parties involved in implementing new tools.

Table 2 Comparison between the baseline and endpoint regarding the change management model

	Baseline	Endpoint
Personnel comprising the operations team	10 people	10 people
Meetings and/or training sessions in 2018	1 made	10 made
	2 scheduled	17 scheduled
	(Holidays)	
% training sessions delivered	0.00	0.59
Supplier assessment	No	In process
Total cost of orders requested (Jan 2018–May 2018)	PEN 628,000	PEN 523,000
Storage capacity	0.08	0.21
Average supply time to store	54 min	38 min
2018 annual budget for purchase orders	PEN 1,292,000	PEN 765,000
Organizational culture satisfaction	0.76	0.93
Acquired/transferred knowledge satisfaction	0.83	0.91

References

1. Orlandi: Problemas y Soluciones en la Gestión Logística y Almacenes en PYMES (2014)
2. OCDE: Small and Medium-sized Enterprises: Local Strength, Global Reach (2000)
3. Arbulú, J.: PYME: La Socia Mayorista del Perú. Escuela de Dirección de la Universidad de Piura (2007)
4. ComexPerú, Resultados de Importaciones Enero–Febrero (2018)
5. Palomo, M. A.: Los Procesos de gestión y la problemática de las Pymes (2015)
6. Koohang, A., Paliszkiewicz, J., Goluchowski, J.: The impact of leadership on trust, knowledge management, and organizational performance: a research model. Ind. Manage. Data Syst. **117**(3), 521–537 (2017)
7. Díaz, H.B., García, R.G., Porcell. N.: Las Pymes: costos en la cadena de abastecimiento. Revista Escuela de Administración de Negocios, mayo-agosto, 5–22 (2008)
8. Secretaría de Economía, Políticas públicas para el desarrollo de la logística y la gestión de la cadena de suministro en México. (2011)
9. Velásquez, A.: Modelo de gestión de operaciones para Pymes innovadoras. Escuela de Administración de Negocios, enero-abril, 66–87 (2003)
10. Instituto PYME: Logística y competitividad de las Pyme (2007)
11. Calderón, J.L., Francisco, E.L.: Análisis del modelo SCOR para la gestión de la cadena de suministro. Ponencia, IX Congreso de Ingeniería de Organización, Gijón, España (2005)
12. Mundial, B.: Análisis Integral de Logística en Perú, Parte 1: Resultados y Lineaminetos Estratégicos (2016)
13. Bellido, Y., Rosa, A.L., Torres, C., Quispe, G., Raymundo, C.: Modelo de Optimización de Desperdicios Basado en Lean Manufacturlng para incrementar la productividad en Micro y Pequeñas Empresas del Rubro Textil. CICIC 2018—Octava Conferencia Iberoamericana de Complejidad, lnformatica y Cíbemetica, Memorias, **1**, 148–153 (2018)
14. Appelbaum, S.H., Profka, E., Depta, A.M., Petrynski, B.: Impact of business model change on organizational sucess. Ind. Commercial Training (2018)
15. Bhamu, J., Sangwan, K.S.: Lean manufacturing: literature review and research issues. Int. J. Oper. Prod. Manage. **34**(7), 876–940 (2014)
16. Gupta, S., Kumar Jain, S.: An application of 5S concept to organize the workplace at a scientific instruments manufacturing company. Int. J. Lean Six Sigma **6**(1), 73–88 (2015)
17. Busse. R., Doganer, U.: The role of compliance for organisational change–qualitative evidence from German SMEs. J. Organ. Change Manage. **17** (2017)
18. Al Saifi, S.A.: Positioning organisational culture in knowledge management research. J. Knowl. Manage. **19**(2), 164–189 (2015)
19. Tsai, A.: An empirical model of four processes for sharing organisational knowledge. Online Inf. Rev. **38**(2), 305–320 (2014)
20. Velásquez, A.: Modelo de gestión de Operaciones para PYMES innovadoras. Escuela de Adminitración de Negocios, Enero-Abril, 66–87 (2003)
21. Cano. P., Orue, F., Martínez, J., Mayett. Y, López, G.: Modelo de Gestión Logística para pequeñas y medianas empresas en México. Contaduría y Administración **60**, 0186–1042, enero-marzo, 181–203 (2015)
22. Manohar Singh, R., Gupta, M.: Knowledge management in teams: empirical integration and development of a scale. J. Knowl. Manage. **18**(4), 777–794 (2014)
23. Che Pa, N., Md Jasin, N.: A prevention model for the failure of hospital information systems in malaysian government hospitals. J. Telecommun. Electron. Comput. Eng. **10**(1–10), 67–73 (2018)

Circulino: An IoT Solution Applied in the University Transport Service

R. A. Barbosa, R. P. Sousa, F. A. Oliveira, H. C. Oliveira, P. D. G. Luz and Leandro T. Manera

Abstract Smart bus monitoring systems are used to provide real-time information about vehicles such as location, arrivals, average speed, journey time, and so on. These systems add values to the transportation service and bring benefits to both users and managers. This paper presents Circulino, an IoT solution for monitoring a University campus public transportation system. This solution was developed within the University's Smart Campus project and includes collecting geolocation data from the public transportation vehicles and generating useful information for the university community. The solution developed provides information that allows service managers to automatically monitor in real time the bus schedule accomplishment. In addition, the IoT Circulino solution allows the university community to monitor in real time the bus arrival at a specific bus stop, and other information that aid in planning the use of the transportation system. The solution implements the smart mobility concept and can be used in smart cities as well.

Keywords Vehicular monitoring · GNSS/GPS · SIM808 · Arduino · Smart campus · Smart cities · Smart mobility

---

R. A. Barbosa (✉) · R. P. Sousa · F. A. Oliveira
University of Campinas, Campus City Hall, Campinas, Brazil
e-mail: ricardob@unicamp.br

R. P. Sousa
e-mail: rafaelps@unicamp.br

F. A. Oliveira
e-mail: falveso@unicamp.br

H. C. Oliveira
School of Civil Engineering, University of Campinas, Campinas, Brazil
e-mail: oliveira@fec.unicamp.br

P. D. G. Luz · L. T. Manera
School of Electrical and Computer Engineering, University of Campinas, Campinas, Brazil
e-mail: garcez@utfpr.edu.br

L. T. Manera
e-mail: manera@dsif.fee.unicamp.br

Y. Iano et al. (eds.), *Proceedings of the 4th Brazilian Technology Symposium (BTSym'18)*, Smart Innovation, Systems and Technologies 140,
https://doi.org/10.1007/978-3-030-16053-1_49

1 Introduction

Intelligent Vehicle Monitoring Systems (IVMS) are used to improve the delivery and management of transportation services by using different types of information, such as the geolocation. These systems work with the smart mobility concept making transportation more connected, efficient and flexible. Smart mobility is therefore a multifaceted topic, involving all the smart city paradigms and generating a set of heterogeneous benefits for all the smart city stakeholders [1]. The IVMS add values to transportation services and aid managers to maintain customer satisfaction levels, optimize vehicle use, and control costs.

IVMS are made up of electronic devices for data transmission and computerized vehicle management and monitoring systems. The electronic devices are based on GNSS (Global Navigation Satellite System), popularly known as GPS, for geo-positioning data and GSM/GPRS (Global System for Mobile/General Packet Radio Services), radio or Wifi for data transmission. The monitoring systems have instruments that allow a data analysis and the verification of the behavior related to the vehicle movement. These systems have an interface based on map visualization and real-time monitoring of vehicles location.

IVMSs can be used in different areas and platforms, such as in cargo transport, security, fire trucks, ambulances, etc. In the area of passenger transportation, these systems have been used to provide information on schedules and information in real time about the current location and the expected time for the bus arrival. These information help the transportation system users to plan their own routes and time. In addition, this possibility brings comfort and security to users while they wait the bus arrival since they can follow their approximation in real time. IVMS have been a useful tool for identifying travel patterns, traffic jams, delays and advances on arrival of boarding stops, etc.

The Internet of Things (IoT) is a new technological paradigm conceived as a global network of machines and devices capable of interacting with one another [2]. According to Minerva et al. [3], IoT is a network that connects "Things" exclusively identifiable on the internet. This paper presents the IoT Circulino solution for intelligent monitoring of an internal passenger transportation system of the University of Campinas—Unicamp. The content of this work is structured as follows: Section 2 presents the related works. The Circulino solution is described in Sect. 3. Section 4 highlights the study case followed by the results obtained (Sect. 5).

2 Related Works

In this section we present some works related to the use of bus monitoring devices as well as researches regarding the usage of microcontrollers and IoT modules.

According to Miranda et al. [4], the creation of a bus localization system using Internet of Things, nowadays, can be done at affordable cost. According to the authors the use of this kind of system can help increase the public transportation quality. The device proposed by the authors is assembled with a RaspBerry Pi microcomputer, Ublox Neo GPS and the Telit HSPDA module for data transmission. Fleischer et al. [5] describes the need to monitor the real-time location of intercity buses in order to increase passenger safety. The authors proposed a device that includes a panic button to be used in case of robbery and for accident alerts. The device proposed by the authors uses a Olimex AVR-GSM card with ATMega32 microcontroller and a PMB-648 GPS module.

For Elmedany et al. [6], it is possible to develop real-time low cost GPS/GPRS tracking devices as these technologies become more accessible and functional. The device proposed by the authors uses a PIC microcontroller and a GM862 GPS/GPRS module. Ibrahim and Audah [7] present a real-time bus monitoring system using the Arduino Uno board and a Adafruit GPS card. The authors send latitude and longitude data to the Google Earth software and use the ManyCam software and Google+ Hangouts as a way to broadcast live video of bus routes.

Kassé et al. [8] developed an IoT solution to deal with problems of delays in the public transport system in Dakar, Senegal. The authors have developed a system that makes use of GPS data, Web service and SMS messages. The user sends an SMS message as the bus number and the address and receives in response, also via SMS, the number of stops remaining for the arrival and the distance between the bus and the bus stop. The system proposed by the authors use the Arduino Uno board and shield GPS and GSM. As the user interface is based on the use of SMS the solution does not allow the real-time monitoring of the position of the bus.

The IoT solution proposed by Saad et al. [9] shows the real-time location of the buses, the number of seats available and the time of arrival at the bus stop. The solution makes use of Espresso Lite V2.0 based on Arduino and ESP8266 and a GPS module. Unlike the solution proposed in this paper, the authors' solution requires a Wi-Fi network infrastructure with high availability in addition to the requirement of authentication by the devices. The solution has no monitoring and management system. Eken and Sayar [10] present a system to follow the position of the bus through Google Maps and receive alerts by SMS and e-mail regarding the estimated time of arrival of the bus to the bus stop. The system proposed by the authors uses Machine Learning algorithms to estimate the arrival time of buses. The solution proposed by the authors depends on the availability of the SMS service to the end users.

Jisha et al. [11] present a bus monitoring solution that allows you to visualize the position and route of the bus as well as monitor students' entry and exit using RFID technology. The system alerts parents when the vehicle changes course or when a student enters or exits the bus. The solution is composed of RFID, GPS and GSM/GPRS modules and by an Arduino Mega board, besides making use of machine learning algorithms to predict the arrival of buses. The Arduino Mega board used by the authors is a plate with unfavorable dimensions for commercial application.

The works cited in this section present solutions focused on the visualization and monitoring of the buses in real time. In a different way, the IoT solution presented in this paper deals with the visualization and availability of the data in real time, as well as making use of instruments for data analysis to support managers in decision making, as a post-processing step, and in some cases in real time.

3 IoT Solution for a Campus Public Transportation System

Unicamp has the free internal transportation service to facilitate the movement of the university community within the perimeter of Barão Geraldo campus. This transportation service is known as internal circular and consists of three routes being served by 5 buses in enclosed routes (start and begin at the same bus stop) during the entire period of campus activities.

Each bus is monitored by a tracking device which sends data about its location for the system control. Once processed, the data is made available to the transport service users through a mobile application and a website (in real time). The geolocation data reported by each device is also visualized by a computerized management system that allows the executed routes to be controlled and monitored.

3.1 IoT Device

The IoT device is an electronic equipment designed to send data about geolocation. It collects data such as geographic coordinates (latitude, longitude and altitude) and average speed, as well as adding them to a database for future processing and availability. It is composed by GSM and GPS antennas for mobile and satellite navigation usage, respectively, as well as indicative LEDs and route switch buttons. Figure 1 shows the developed IoT device.

The IoT device was built using low cost electronic components and microcontrollers and it was customized to meet existing demand. It includes an Arduino Pro Mini microcontroller [12] based on the ATmega328P and a SIM808 shield [13] with the features of reading GPS signals and sending data via mobile network (GSM/GPRS).

An interface card has been developed to interconnect the Arduino Pro Mini and the shield SIM808. This board has a button for the bus route selection (which is indicated by four LEDs) and one device status LED. The selection button allows the driver to enter the route that will be made by the bus. The LEDs show the current route that the bus is performing. The status LED indicates the device's operation by color: blue for GPS reading, green for GSM/GPRS transmission and red for communication failure. The code was done by using C programming language. The communication between Arduino board and the SIM808 shield is done via serial by AT commands. The Arduino IDE was used to compile and write the code on the microcontroller.

Fig. 1 IoT Device

3.2 IoT Solution Architecture

The solution proposed in this work consists of tracking devices installed on the buses that send geolocation data to the servers for storage and processing. Figure 2 shows the layers and components that are part of the IoT Circulino solution, as well as the interaction among them.

The "Data Acquisition" layer acts in the capturing and sending the buses' geolocation data. From this layer the IoT devices send the data to the "Data Storage" layer. Devices pick up and send data every 5 s via GSM/GPRS to the virtual gateway. The virtual gateway validates the source of the data to prevent attacks and the use of corrupted or incorrect data. During the receiving data step, the virtual gateway performs a security check to ensure that only valid data is transmitted to the database server and to the IoT platform. Table 1 describes the data sent by the devices.

The "Data Storage" layer stores and make available the collected data by the IoT devices. In this layer the data is processed to obtain useful information such as average speed per segment, average value of the mobile signal quality, GPS signal quality, start time of the route, and others. The PostgreSQL is used as database server. In the IoT platform [14] the equipment installed in the vehicles become virtual devices. All data sent by IoT devices are in the cloud and can be accessed over the internet using the Rest API on the IoT platform. This API facilitates data access and analysis, as well as providing security and standardization of data collect. The data available

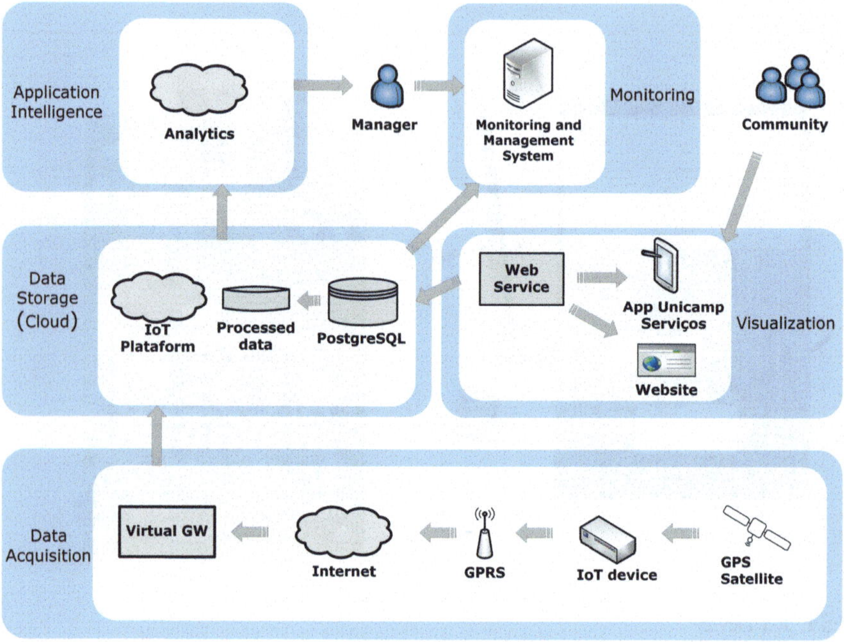

Fig. 2 Circulino architecture

Table 1 Data sent by device

Field	Type	Range	Length (max)
Device ID	int	[1, n]	2
Collect date	char	yyyy: [1980,2039] MM: [1, 12] dd: [1,31] hh: [0,23] mm: [0,59] ss.sss: [0.000,60.999]	18
Latitude	float	[−90.000000,90.000000]	10
Longitude	float	[−180.000000,180.000000]	11
Altitude	float		8
Satellites Used	int	[0, 99]	2
Route Number	int	[1, 4]	1
Temperature (Celsius)	float	[−55, 150]	6
GSM Signal quality	int	[2, 30]	2
PDOP	float	[0,99.9]	4

in the IoT platform can be analyzed using data mining techniques, deep learning, statistical tools, etc.

The "Visualization" layer is intended to present information to the users and to the transport service managers. This layer has a Web Service responsible for making the data accessible to the viewing channels. The Web Service also makes the data available to the Zabbix monitoring system [15] for verifying the operation of the IoT devices. Device information can be obtained by the university community and managers through a mobile application and a website. The visualization layer has a feature to estimate the current position of the bus in case of a momentary communication failure. This feature uses the saved data history to predict the current location and simulate bus movement until communication is reestablished. The simulation can be monitored by the communication status displayed in three colors: yellow showing the position in real time, purple showing estimated position due to momentary loss of signal and black showing prolonged signal loss. Java (Android), Swift (IoS), PHP (website), JSON, HTML + CSS, Javascript and AJAX were used in the "Visualization" layer.

In the "Monitoring" layer, actions are carried out to monitor and supervise the schedules compliance. This layer is composed of a automated service management system that has a module for monitoring and managing the travels performed by the buses. This module allows managers to verify compliances and non-compliances over arrival times at bus stops, in addition to highlighting behaviors or particularities of the travels made. This module processes all the data received considering the comparison between the expected and current schedules and generates new summarized data referring to the bus pass in each bus stop for passengers boarding.

The "Application Intelligence" layer gives to managers an environment for analyzing information for a specific purpose. This layer allows the use of data analysis tools such as data mining techniques for pattern discovery, statistical tools for analyzing indicators or even the use of dashboards to track results and targets. This layer provides relevant information to managers, helping them to make strategic decisions such as changes in schedules and optimization in bus routes. The information extracted from this layer can be used to analyze indicators of the campus public transportation usage.

4 Case Study

The university transportation system provides 4 different routes within the perimeter of the campus, being three routes for the daytime period and one route for the night period. For daytime routes the service starts at 6:30 a.m. and ends at 7:00 p.m., while the night line monitoring starts at 6:00 p.m. and ends at 11 p.m. The service is provided by two companies and it does not work on Saturdays, Sundays and holidays, due to the low passenger request.

The data related to the real-time location of the buses is shared with to the university community through a mobile application called Unicamp Serviços [16] and a website [17]. The data collected by the devices is available to the service managers through the Web application called SGF, acronym for the Sistema de Gerenciamento de Fretados (or Management of Bus Chartered System). By using the SGF, the managers not only check the data acquired by the IoT devices, but also supervise the timing of the buses arrival at the bus stops. At the beginning of the solution implementation, meetings were held with bus drivers, inspectors and transport service managers to present the IoT solution.

Information regarding compliance with bus schedules was extracted from the "Application Intelligence" layer. For this case study, it was considered scheduling compliance when the bus arrives at the bus stop with maximum discrepancy of 3 min (for being late or early when compared to the schedule time). Non-compliance is considered when the bus arrives at the bus stop with an discrepancy more than 3–10 min (early or late) when compared to the scheduled time. Equations 1 and 2 mathematically expresses compliance and non-compliance on schedule.

$$Compliance = 0 \leq |(arrival_{time} - scheduled_{time})| \leq 3 \quad (1)$$

$$Non\text{-}compliance = 3 < |(arrival_{time} - scheduled_{time})| \leq 10 \quad (2)$$

To use these formulations the geolocation data were used from the time that the buses passed by each bus stop. Some coordinates collected by the IoT devices may represent a location distant from the bus stop, due to sending intervals and lack of network signal. For this study, data where the bus is in a distance greater than 30 m from the bus stop were not considered. The removal of this information avoids any influence on the results since such location data may not truly represent an arrival to a bus stop.

5 Results

After 8 months of operation implementation and monitoring using the IoT Circulino solution, it was possible to verify the number of compliances and non-compliances regarding the schedule observance of the transportation system. Table 2 presents the absolute values for compliance and non-compliances identified by the IoT Circulino solution during this period. In August a greater number of conformities were observed when compared with previous months. The greater number of compliance data for the August is justified due to the increase of data collected during that month. The same data collection criterion was used in all the months surveyed.

Figure 3 shows the monthly rate of non-compliance with schedules identified by the IoT Circulino solution. The chart shows a growth in the non-compliance rate between the months of January and April. In the holiday period, which comprises the months of June and July, lower rates of non-compliance are noted. In this period there is a reduction of academic activities which causes a decrease in the number

Table 2 Compliance numbers

Months	Compliance	Non-compliance
January	7822 (82%)	1704 (18%)
February	7034 (79%)	1903 (21%)
March	5567 (75%)	1798 (25%)
April	1845 (69%)	822 (31%)
May	2669 (81%)	567 (19%)
June	4170 (89%)	503 (11%)
July	4429 (84%)	814 (16%)
August	26,437 (75%)	8757 (25%)

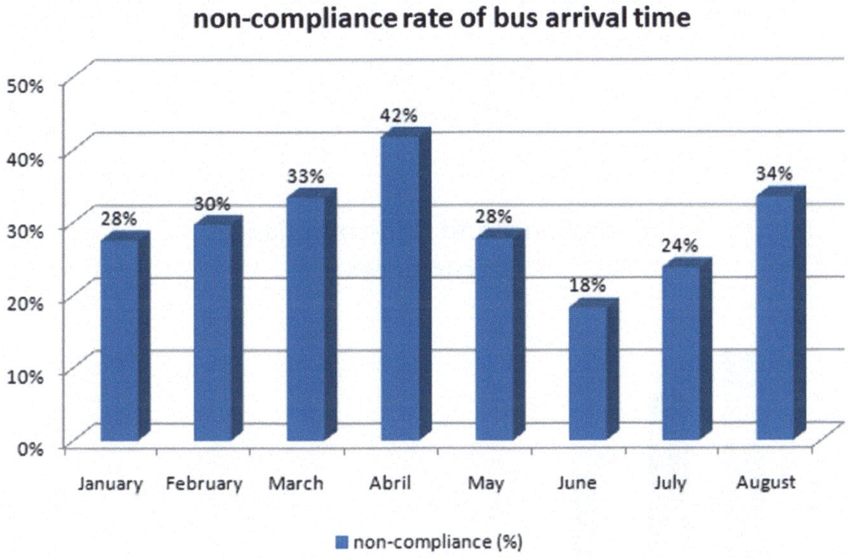

Fig. 3 Monthly non-compliance rate

of people that circulate in the university and consequently less flow of vehicles and passengers. The information about the time of arrival at the bus stop is important for the managers of the transport services for quality evaluation. With a longer period of information it will be possible for managers to create an action plan to decrease these rates.

Figure 4 shows the number of occurrences of non-compliances in the time-by-hour compliance.

Figure 4 shows a greater number of non-compliance occurrences in the hours between midday and 1 p.m. This concentration of non-compliances occurs because it is a period in which there is an increase in the traffic of vehicles due to exits for lunch. The same phenomenon is observed at 9 a.m. when vehicles arrive at the

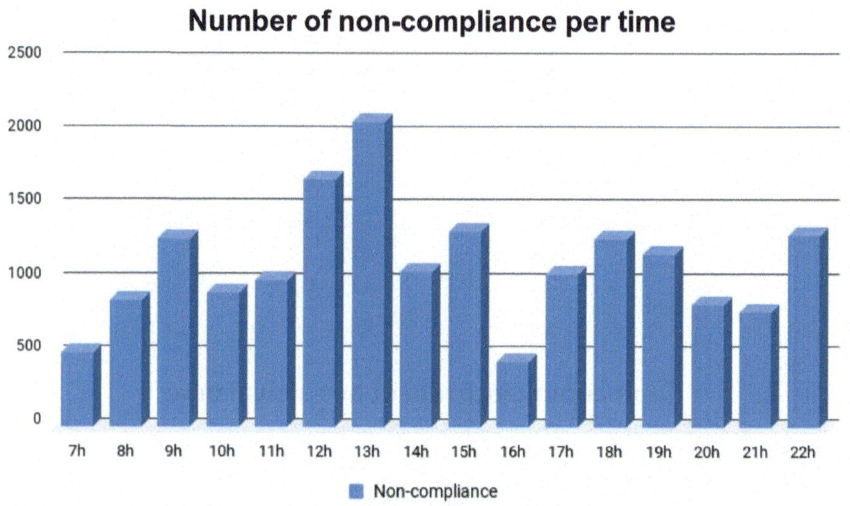

Fig. 4 Number of non-compliance per period of time

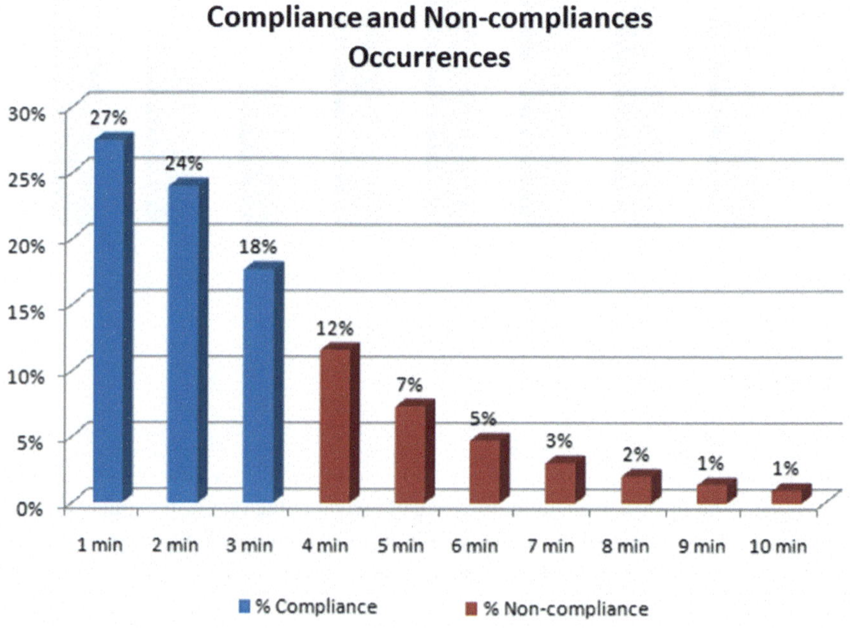

Fig. 5 Rates of compliances and non-compliances occurrences

University and at 6, 7, and 10 p.m., when there are vehicles leaving due to the end of activities at the university.

Figure 5 shows the rates of compliance and non-compliance occurrences between 1 and 10 min collected in 8 months. The columns in the blue color represent the compliance and the red columns non-compliance. Figure 5 shows a concentration of occurrences of compliance between 1 and 2 min (27 and 24%), which indicates that more than half of the occurrences happen around 2 min. It is also noted that non-compliance rates are decreasing as minutes increase. There is a non-compliance concentration in the range of 4–6 min which corresponds to 24% of occurrences.

6 Conclusion and Discussion

This work presented the IoT Circulino solution for monitoring the Unicamp buses responsible for internal passenger transportation. After 8 months of evaluation, the IoT Circulino solution was presented as a feasible and accessible tool since it provides useful information for the users of the transport system and for the managers of these services. IVMS that use geolocation technologies, automated management systems, together with the Internet of Things concept can be considered a great tool and it can help improving the quality of public transportation.

The IoT Circulino solution provided benefits to the university community. It is believed that now the transportation service provided is better planned due to the visualization and quantitative indexes regarding the bus arrivals time related to the schedule. With the information in real time the users save time of waiting in the bus stops, since there is no need to go to bus stop in advance.

It was found that the data provided by the solution support the managers in strategic decision making. The IoT Circulino solution enables managers to identify missed schedules, optimize the timesheet and identify delays and advances. The use of a automated management and monitoring system allowed managers to automatically check the number of travels actually performed by contracted bus companies. This verification facilitated the inspection task and reduced the handwork by transportation inspectors.

The information bases generated by bus geolocation solutions have been shown to be sources for academic and business interest. Considering their great volume, and because they are a source of exploration and generation of knowledge, such information bases are attractive for researchers and companies that work in the areas of statistical sciences, cartography, data mining, artificial intelligence and others.

It is intended in the future to add new features to the IoT device. One of them is the use of SMS communication channel which will allow remotely to make devices settings update. It is also intended to create mechanisms for automatic identification of the bus route. At first, the computerized monitoring system will analyze the data sent by the devices and compare them with unique coordinates of each route. It is also intended to create a repository with the collected data to make it available to researchers and interested companies.

Acknowledgements This work has been supported by Unicamp Smart Campus project partners.

References

1. Benevolo, C., Dameri, R.P., D'Auria, B.: Smart mobility in smart city: action taxonomy, ICT intensity and public benefits. Torre, T. at al. (eds.) Empowering Organizations, Lecture Notes in Information Systems and Organizations, vol. 11, pp 13–28 (2016)
2. Lee, I., Lee, K.: The internet of things (IoT): applications, investments, and challenges for enterprises. Bus. Horiz. **58**(4), 431–440 (2015)
3. Minerva, R., Biru, A., Rotondi, D.: Towards a definition of the internet of things (iot). IEEE Internet Initiative **1**, 1–86 (2015)
4. Miranda, W.M., de Mendonça, R.T.R., da Silva, A.A., de Lima Curvello, A.M., de Souza, F.L.D.S., da Silva, H.J.: BusMe: automatic bus localization system and route registration. Procedia Comput. Sci. 1098–1103 (2017)
5. Fleischer, P.B., Nelson, A.Y., Sowah, R.A., Bremang, A.: Design and development of GPS/GSM based vehicle tracking and alert system for commercial inter-city buses. In: IEEE 4th International Conference on Adaptive Science & Technology (ICAST), pp. 1–6, 2012, Kumasi (2012)
6. Elmedany, W., Al-Omary, A., Al-Hakim, R., Nusaif, M.: Low cost real-time tracking system prototype using GM862 cellular quad band module. J. Eng. **12** (2014)
7. Ibrahim, M.Y.M., Audah, L.: Real-time bus location monitoring using Arduino. In: AIP Conference Proceedings, p. 020016 (2017)
8. Kassé, B., Diallo, M., Gueye, B.: Leveraging GPS and SMS-based bus tracking architecture for an efficient transportation. In: Innovation and Interdisciplinary Solutions for Underserved Areas, Lecture Notes of the Institute for Computer Sciences, Social Informatics and Telecommunications Engineering, vol. 204, pp. 115–124. Springer, Cham (2018)
9. Saad, S.A., Hisham, A.B., Ishak, M.H.I., Fauzi, M.H.M., Baharudin, M.A., Idris, N.H.: Real-time on-campus public transportation monitoring system. In: 2018 IEEE 14th International Colloquium on Signal Processing & Its Applications (CSPA), Batu Feringghi, pp. 215–220 (2018)
10. Eken, S., Sayar, A.: A smart bus tracking system based on location-aware services and QR codes. In: 2014 IEEE International Symposium on Innovations in Intelligent Systems and Applications (INISTA) Proceedings, pp. 299–303, Alberobello (2014)
11. Jisha, R.C., Jyothindranath, A., Kumary, L.S.: Iot based school bus tracking and arrival time prediction. In: 2017 International Conference on Advances in Computing, Communications and Informatics (ICACCI), pp. 509–514. Udupi (2017)
12. ARDUINO PRO MINI [Online]. Available: https://www.arduino.cc/en/Main/ArduinoBoardProMini. Accessed 20 Sep 2017
13. SIM808 [Online]. Available: http://simcomm2m.com/En/module/detail.aspx?id=137. Accessed 27 Sep 2017]
14. KONKER PLATFORM [Online]. Available: http://www.konkerlabs.com. Accessed 17 Nov 2017
15. ZABBIX [Online]. Available https://www.zabbix.com/. Accessed 08 Sep 2018
16. UNICAMP SERVIÇOS [Online]. Available: https://www.ccuec.unicamp.br/ccuec/servicos/unicamp-servicos. Accessed 25 Sep 2018
17. CIRCULINO WEBSITE [Online]. Available: https://www.prefeitura.unicamp.br/servicos/diretoria-de-servicos-de-transporte/mapa-circulares. Accessed 25 Sep 2018

Optimized Ventilation Model to Improve Operations in Polymetallic Mines in Peru

Vladimir Flores, Luis Arauzo, Juan Jara and Carlos Raymundo

Abstract Currently, deficient ventilation systems are a frequently observed problem in mining. Such deficient ventilation systems lead to the recirculation of stale air or air with toxic gases. This translates into an increase in the costs of ventilation and electrical consumption for the mining organizations as conventional and practical solutions simply include options such as buying a greater number of fans. Moreover, this problem also exposes the mine workers to an unsafe work environment with unfavorable conditions that could end in minor, incapacitating, or fatal accidents. Therefore, this research seeks to optimize ventilation systems by introducing the Ventsim software tool to develop efficient coverage, addressing, flow, circuit or network characterization, and air balancing in conjunction with avoiding the stagnation of toxic gases in underground work. The primary result of the proposed model's application in this study was the reduction in ventilation costs related to electricity consumption.

Keywords Underground mining · Ventilation system · VENTSIM software · Optimization

V. Flores (✉) · L. Arauzo · J. Jara
Escuela de Ingeniería de Gestión Minera, Universidad Peruana de Ciencas Aplicadas (UPC), Lima, Peru
e-mail: u201319424@upc.edu.pe

L. Arauzo
e-mail: pccilara@upc.edu.pe

J. Jara
e-mail: juan.jara@upc.pe

C. Raymundo
Dirección de Investigación, Universidad Peruana de Ciencas Aplicadas (UPC), Lima, Peru
e-mail: Carlos.raymundo@upc.edu.pe

Y. Iano et al. (eds.), *Proceedings of the 4th Brazilian Technology Symposium (BTSym'18)*, Smart Innovation, Systems and Technologies 140,
https://doi.org/10.1007/978-3-030-16053-1_50

1 Introduction

Currently, the mining sector has become one of the most important industries in Peru. However, the Peruvian Ministry of Energy and Mines [1] reported 13 fatal accidents due to gassing and poisoning in underground work environments from 2000 to 2018, which is equivalent to 5% of the total number of accidents in the said period. Therefore, mining organizations currently consider that, for the development of operations activities, minimum working conditions must be guaranteed within their work. Although mining operations include water pumping, hydraulic filling, compressed air, electric power, and ventilation as part of the general services provided to mining sites, several mining companies focus their efforts on guaranteeing adequate underground ventilation. According to Supreme Decree 024-2016-EM [2], the main purpose of mine ventilation is to supply the amount of clean air required by the number of workers, the equipment used, wood consumption, high temperatures, and dilution of gases and leaks, to guarantee favorable heat-environmental conditions for the workers, and, consequently, favorable conditions for ore extraction from the mine to the surface. To this end, as the depth at which underground work is conducted deepens every day, mining companies require new or better ventilation systems that may comply with the application criteria of the Peruvian mining regulations.

This paper proposes a ventilation model optimized through the Ventsim tool to prevent stale air recirculation and reduce mining ventilation costs. This research study is divided into six Sections: Sect. 2 contains the state of the art regarding the research topic; Sect. 3 covers the development of the proposed model; the validation of the results is presented in Sect. 4. Finally, Sects. 5 and 6 present the discussion and conclusion respectively.

2 State of the Art

Mechanized ventilation in underground mining means using ducts and auxiliary fans to transport air flows from the surface to the work inside the mine, for which fresh air supply and stale air evacuation circuits are also used [3]. Because this ventilation is forced, it works differently from natural ventilation. There are three types of forced ventilation: vacuum, in which polluted air is sucked from the front through the duct owing to the depression created by fans located at both end points [4]; pressure, which is characterized by the fact that air enters the front of the bottom through the pipe, is driven by fans, and displaces the mass of stale air into the main air stream through the gallery [5]; and the third type is called the "balanced" type, in which both of the abovementioned types are used [6].

In recent years, the first mechanical fans, known as centrifugal fans, were built. These fans offer high static pressure and medium flow and can work at high speeds with their efficiency ranging from 60 to 80% [7]. Next, the first axial flow fans were developed, which are the most commonly used fans today. These fans offer higher air flow as their efficiency ranges between 70 and 80%, and they are capable of working at the highest possible speeds. Noticing the most outstanding differences between the two types of mechanized fans, axial fans produce high noise levels; they are also versatile and cheap. In contrast, centrifuge fans produce less noise and are rigid but much more expensive.

Information technologies serve as support to elaborate and develop an adequate ventilation system plan. Among them, South African software leads the ventilation system improvement market owing to its characteristics. In the studies conducted by Suvar et al. [5], it is evident that using Visual Advanced Ventsim provides solid three-dimensional (3D)-graphics or precise ventilation system layouts for the detailed analysis of the given mining network. Furthermore, this software is an extremely useful tool for technicians, and it fundamentally demonstrates that any type of ventilation network may be modeled, simulated and solved, regardless of its complexity.

Therefore, according to the studies performed by Jing and Cheng [8] in the Majiagou coal mine using the Ventsim software, with air strength being used as the study parameter, the consumption of electric power was reduced almost in half in all ventilation branches. In addition, the authors argue that Ventsim may not only be used for ventilation network calculations, simulations, and air flow dynamics but may also be used to help in short and long term ventilation system planning. Finally, the study performed by Chambergo [3] shows a way to reduce energy costs in the ventilation system, emphasizing that this may directly or negatively affect mine production.

3 Contribution

Ventilation designs are based on the Hardy Cross theory, which is a method of successive approximations that determines the flow running through each pipeline and its direction. This theory is based on Kirchhoff's laws for the conservation of energy and charges in electrical circuits. In this manner, the model proposed herein supports the analysis of the characteristics and behavior of existing ventilation systems [9]. Furthermore, based on the review of previous research studies wherein scenarios are presented in which the use of software in industrial companies has been productive, software implementation in mining planning areas may solve different issues that the mining sector currently faces. Therefore, this study proposes a ventilation design optimization model for an underground mining company.

The methodology focuses on the key exploitation process, which covers mine ventilation. The model presents three components: mine analysis, tests, and computational simulation together with benefits and indicators (Fig. 1). In this manner, the management of ventilation systems in underground mines is comprehensive as it combines advanced computational simulation tools with theoretical and experimental analysis.

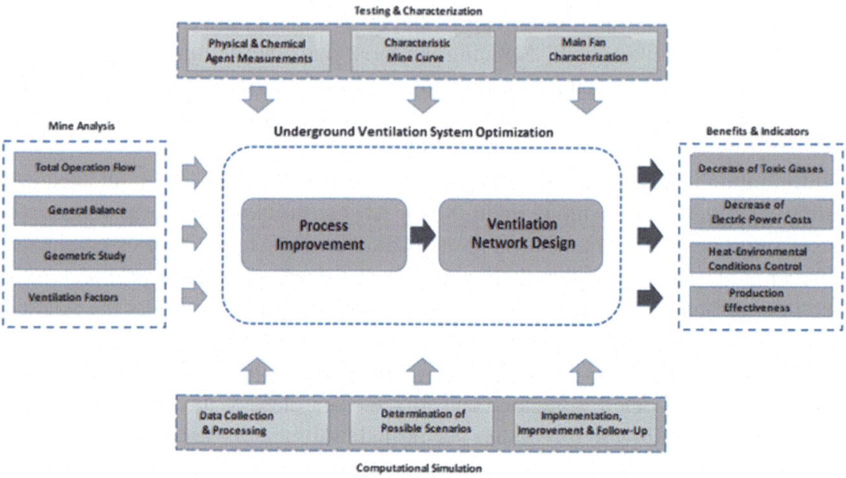

Fig. 1 Flow chart for the optimization of an underground ventilation system

4 Validity

Compañia Minera Condestable (CMC) S.A. owns and operates its Mina Condestable and Mina Raúl underground operations near the town of Mala, Department of Lima, Peru. This organization commercializes copper concentrates as its main product, with concentrations of gold and silver.

4.1 Compañia Minera Condestable (CMC) Mine Analysis

Total Operation Flow CMC employs a total of 408 workers within the mine and the ratio per worker, according to the regulations, is 3/m^3 of air. Thus, the work crew air requirements (QTr) are 1224 m^3/min. Further, CMC has 94 pieces of equipment within the mine, which total 10,755 HP. Therefore, the equipment air requirements (Qeq) are 32,264 m^3/min.

As CMC does not use wood in its operations processes, in this particular case, a zero-flow due to wood consumption (QMa) was considered. With respect to the flow required owing to work temperatures (Qte), a minimum speed of 6.7 m/min was found in the mine (Table 1). The average work area was 16 m^2 and the number of levels in which the temperature exceeds 23 °C was five. The levels that presented high temperatures were NV.-350, NV.-460, NV.-490, NV.-520, and NV.-550. Therefore, the flow required by high temperatures is 536 m^3/min.

The flow rate required to dilute the gases produced by blasting (Qexp) is approximately 7265.25 m^3/min (Table 2).

Table 1 Air requirements due to high temperatures

QTe	536	m^3/min
Vm	6.7	m/min
A	16	m^2
N	5	Levels

Table 2 Air Requirement due to Consumption of Explosives

Method	Explosive	Area (m^2)	Air speed (m/s)	No. of levels
D.S_024_2016_EM	ANFO	15	25	17
	Emulsion	15	20	2
Novitsky	ANFO and emulsion			

Table 3 Flow required owing to leaks

QFu	5103.6	m^3/min
Qtr	1224	m^3/min
QTe	536	m^3/min
Qma	0	m^3/min
Qeq	32,264	m^3/min
Qexpl	6975	m^3/min

Table 4 Total air flow required for operations

QFu	39,127.6	m^3/min
Qtr	1224	m^3/min
QTe	536	m^3/min
Qma	0	m^3/min
Qeq	32,264	m^3/min
QFu	5103.6	m^3/min

As per Table 3, the air flow required owing to leaks (QFu) is equivalent to 5103.6 m^3/min.

Then, the air flow required for operations (Table 4) is 39,127.6 m^3/min.

Finally, based on the previous analysis, the total air coverage in the mine is 92.07%, which means that there is a deficit of 7.93% or 3105 m^3/min.

General Balance On the one hand, the entry of air into the mine, as recorded by the stations, was 36,023 m^3/min, which is equivalent to 1,265,488 cfm. On the other hand, the total stale air extracted to the surface is 36,674 m^3/min, which is equivalent to 1,294,945 cfm. Therefore, the balance between air income and air outflow was 102%.

Geometric Study CMC has two Administrative Economic Units: Mina Condestable and Mina Raúl. The work dimensions are, mostly, 4 × 4 m. These works are classified into GL-Galleries, XC-Cruisers, RP-Ramps, PQ-Pits, and CH-Chimneys.

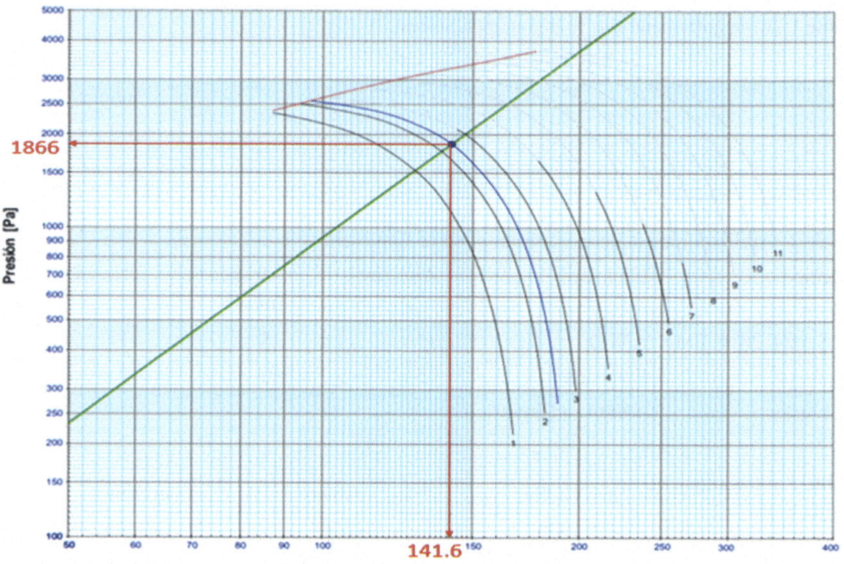

Fig. 2 Compañia Minera Condestable (CMC) characteristic curve

The rock mass presents an RMR classification of 61–80, which means that it is a "good rock". This mine has a depth of approximately 580 m. The exploitation methods used are long-hole drilling, room and pillars, and shrinkage.

4.2 Tests and Characterization

The Condestable Mining Company has a strength of 0.093 Ns^2/m^8 owing to the loss of pressure (Fig. 2).

4.3 Fan Characterization

CMC has one main fan of 300 thousand cfm, four secondary fans of 120 thousand cfm each, and 43 auxiliary fans. Their costs are $5542, $11,816, and $34,823 per month, respectively. This results in a total cost of $52,181/month and, therefore $626,172/year. The power of the main fan is 350 Kw, its pressure is 1865.92 Pa, and its air flow is 141.6 m^3/s.

4.4 Measurement of Physical and Chemical Agents

For the diagnosis of physical and chemical agents, the deepest mine levels were emphasized upon, which are NV.-350, NV.-460, NV.-490, NV.-520, and NV.-550 at the Raúl Economic Unit of the Condestable Mining Company. On the one hand, within the physical agents, the temperature varies from 22.5 to 33.8 °C, the relative humidity presents a range of 68.9–89.6%, and work speed ranges between 19.3 and 34 m/min. On the other hand, within the chemical agents, oxygen (O_2) ranges from 19.8 to 20.8%, carbon monoxide (CO) varies from 3 to 38 ppm, carbon dioxide (CO_2) is within the maximum permissible limits, and nitrous gases are in a range of 0.4–4 ppm.

Computational Simulation. The AutoCAD files were imported to achieve a 3D mine design with flow directions. Then, a simulation of the current CMC system was prepared based on the imported data.

For the simulation, the following parameters were considered:

- Level Reference Surface Elevation: 105 m above sea level
- Strength: 0.093 Ns^2/m^8
- Air Density: 1.2 kg/m^3
- Wet Bulb Temperature: 14 °C
- Dry Bulb Temperature: 19 °C
- Barometric Pressure at Surface: 601 hPa
- Method: Total Pressure
- Air Flow: Compressible
- Energy Costs: 0.022 US$/Kw-h.

In this scenario, the execution of a 45-m conventional chimney, with an inclination of 45°, connecting the RB-4167 chimney with the XC-4357 cruiser was considered. This chimney has a diameter of 8.20 ft, a perimeter of 25.76 ft, an area of 52.81 ft^2 (long term). The purpose of this chimney is to reduce shock losses in the elbow and increase stale air flow to the surface.

Therefore, with this chimney, the new air flow was 521.03 m^3/min, which means that this flow has increased by 28.67% with respect to the flow of 404.93 m^3/min recorded before the simulation of the CH-2507 project. This is also due to the fact that work strength was reduced from $7.5 \times 10e^{-9}$ lb * min^2/ft^4 to $1.7 \times 10e^{-9}$ lb * min^2/ft^4. Similarly, the loss of pressure was reduced from 0.78 to 0.16 "H_2O. The cost of this conventional chimney is $18,000 US dollars, considering a linear advance of $400 per meter.

CMC has one main fan of 8495.05 thousand m^3/min, four secondary fans of 3398.02 thousand m^3/min each, and forty-three auxiliary fans. Their costs are $5542, $11,816, and $34,823 per month, respectively. This results in a total cost of $52,181/month and, therefore $626,172/year. With the execution of the new CH-2507 chimney, main ventilation costs were reduced to $2431/month.

Therefore, VENTSIM predicts that the annual fan operation costs for Option 1 will be approximately US $588,840.

5 Conclusions

The use of the Ventsim software tool optimized the ventilation system of the Condestable S.A. mine, considering the use of fans and energy costs, and presenting a clean air deficiency of 7.93% with respect to the income flow. To improve ventilation conditions in the area, a conventional 45-m-long and 2.4-m-diameter chimney must be developed, intercepting levels NV.-520 and NV.-430, with an approximate cost of US $18,000. These costs are beneficial to the mine as its life span is approximately 9 years. The current energy consumption due to ventilation is 3294.336 Kw, representing 10.62% in main, 22.64% in secondary, and 66.74% in auxiliary ventilation. The monthly cost is US $52,181, and the annual cost is US $626,172. The ventilation ratio of the Condestable S.A. mine is 0.076 S/./Kw-h, with an energy cost of 0.022 US $/Kw-h. The mine requires a total air flow of 39,089 m^3/min, and the air intake to the internal mine is 36,023 m^3/min, the air outlet is 36,674 m^3/min, with a coverage of 102%.

The methodology used for this research is applicable and recommendable for mining companies where the long-hole drilling and room and pillars exploitation methods are applied. In addition, this technique may be improved because the design does not consider the heat of the rock mass found in the mine.

References

1. Ministerio de Energía y Minas (MEM): Estadísticas de Accidentes Mortales en el sector minero (2016)
2. Decreto supremo 024-2016-EM: Reglamento de Seguridad y Salud Ocupacional en Minería (2016)
3. Chambergo, G.: Propuesta de un sistema de ventilación, aplicando tecnologías de información y manejo de escenarios técnico económico en la unidad productiva san Cristóbal, de minera Bateas SAC (2013)
4. Soncco, C.: Diseño de un nuevo sistema de ventilación para la dilución de gases nocivos en la unidad minera Untuca - Puno (2016)
5. Suvar, M., Cioclea, D., Gherghe, V., Pasculesco, I.: Software avanzado para la resolución de redes de ventilación minera (2012)
6. Rodriguez, V.: Diseño del sistema de control para la ventilación de una mina subterránea usando un controlador AC800M (2010)
7. Peruana de uso minero ecológica y técnica: Ventilación de Minas Subterráneas (2006)
8. Jing, S. Cheng, Z.: Estudio para la Optimización del Sistema de Ventilación de una Mina de Carbón basado en Ventsim (2016)
9. Nikolaev, A., Miftakhov, T., Nikolaeva, E.: Modeling of Energy-saving System of Conditioning Mine Air for Shallow Underground Mines (2017)

Six Sigma Model Optimized for Reducing Downtime in an Open-Pit Mine

Josemaria Gargate, Sian A. Fung, Juan Jara and Carlos Raymundo

Abstract Currently, in open-pit mining operations, the highest cost among all processes is that of transporting trucks, whether in fuel, roads, tires, or spare parts, among other factors. Therefore, this article proposes the use of the Six Sigma methodology of continuous improvement as a quality management tool to reduce the downtime of the truck fleet to obtain better productivity in operations. The results of the investigation in the case study allowed to visualize that with the election of an optimal fleet appropriate for the conditions given in a mining project, the values of productivity and efficiency improve considerably. This is reflected in a better use of the machinery and in the reduction of unproductive times.

Keywords Haulage · Dead times · Six sigma · Open-pit mine

1 Introduction

Medium and small open-pit mines operations processes such as drilling, blasting, loading, hauling and unloading must be efficient to render mining activities as profitable, with the aim of reducing downtime and increasing the amount of material moved.

J. Gargate (✉) · S. A. Fung · J. Jara
Escuela de Ingeniería de Gestión Minera, Universidad Peruana de Ciencas Aplicadas (UPC), Lima, Peru
e-mail: u201320884@upc.edu.pe

S. A. Fung
e-mail: u201315802@upc.edu.pe

J. Jara
e-mail: juan.jara@upc.pe

C. Raymundo
Dirección de Investigación, Universidad Peruana de Ciencas Aplicadas (UPC), Lima, Peru
e-mail: Carlos.raymundo@upc.edu.pe

Y. Iano et al. (eds.), *Proceedings of the 4th Brazilian Technology Symposium (BTSym'18)*, Smart Innovation, Systems and Technologies 140,
https://doi.org/10.1007/978-3-030-16053-1_51

In Peru, medium and small mining accounts for 35% of the total number of existing mines, between formal and informal open-pit mines. One of the important activities of the productive process is loading and haulage. Currently, the highest costs in open-pit mining operations are the costs represented by these processes. In [1], it is referenced that these costs account for 45–65% of the total mining cost, over which they have direct impact. However, in many open-pit mines the use of giant trucks does not exceed 55–75% due to factors such as: incorrect selection of haulage equipment, crusher delays, improper fleet distribution, roads in bad condition or dead times. Here, the match factor presented as the main solution is the use of bypass conveyor belts to accelerate and increase production. Also, truck management software such as Dispatch or Mining Star is used. This software monitors the loading and hauling processes in real time through a screen room where decisions are made regarding each truck and each bulldozer.

The Six Sigma method is a Quality Management tool used for process improvement in combination with statistical tool. This process is known as DMAIC (Define, Measure, Analyze, Improve, Control) and its purpose is improving the performance level of a process through successful decisions to allow the organization to better understand the needs of its customers.

For this reason, this research study seeks to detect critical points in the transportation of ore in an open-pit mine, which are likely to generate significant production losses. This article consists of an Introduction in Section I, the Review of Existing Literature in Section II, and the Proposed Model Description in Section III. Finally, Validation Results from an Importing Company are discussed in Section IV and Conclusions are presented in Section V.

2 State of the Art

Most Open-Pit Mining Units employ between 100 and 150 trucks approximately to move hundreds of thousands of tons per day. However, when entering and performing the spotting process, long queues are generated, causing a congestion of trucks in the crushing zone. Therefore, to reduce the long waiting times, which are often caused by system overload, for trucks in the loading or dumping areas, in [2], the authors seek to implement a discrete event simulation model in a mine to help calculate the ideal number of trucks required to make the haulage cycle more efficient.

On the other hand, in [3], they seek to determine the optimal number of trucks in each pit in order to increase the operational use percentage, increase clearing production and infer achievements in terms of reducing environmental impact through simulations using GPSS/H. In addition, the research performed [4] is linked to the previous one because the authors seek to improve and streamline the open-pit ore transportation system by implementing a stochastic simulation model to help solve truck and bulldozer selection and sizing issues, and finding the right number of pieces of equipment that the mining unit should have to maximize production while reducing the operational costs which are often generated by idle equipment due to long queues in the loading and crusher areas. Finally, [5] states that the increase in the number of dump trucks is not able to keep up with the high demand for productivity, which is why the best operational fleet management solution must be found for open-pit mines.

3 Six Sigma Methodology

To better understand Six Sigma methodology and how it behaves in an Open-Pit Mining Unit, Levano [6] developed the phases that comprise the methodology and mentions that there are no previous works in this field where Six Sigma is applied to increase productivity. The methodology includes the following (Fig. 1).

3.1 Define

Fieldwork is performed, where the problem must be defined through planning which considers client expectations and needs. Likewise, the Dispatch or Mine Star large-scale mining management software is used to collect times and formulas for the truck processes, such as:

Theoretical Cycle Time = Loading Time + Loaded Travel Time + Unloading Time + Unloaded Travel Time
Practical Cycle Time = Waiting Time at Bank + Spotting Time + Balancing Time + Loading Time + Loaded Travel Time + Waiting Time at Pad + Unloading Time + Unloaded Travel Time.

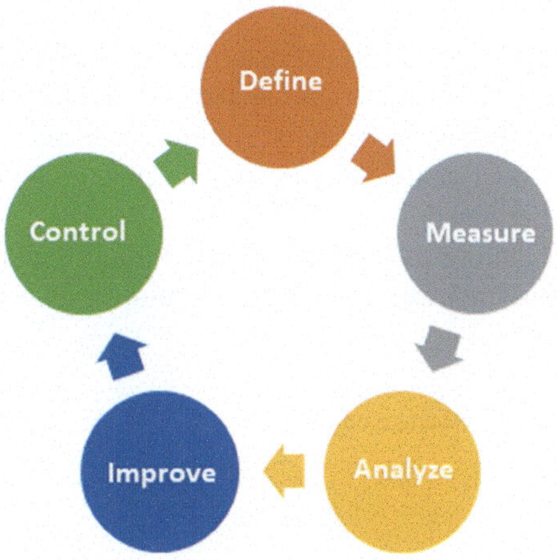

Fig. 1 Six sigma methodology

3.2 Measure

This phase consists of better understanding current trouble operations. That is, making general open-pit mining process charts and detailing hauling activities for further study and measurement.

3.3 Analyze

The aim is to find out the root causes that affect the good performance of the operational process and the error rate generated in order to later propose an improvement using the cause-effect diagram. To find the loading times, loaded travel times to the Pad, unloading times and unloaded travel time to the bank, the MS EXCEL worksheet will be used for the following formulas:

- Loading Time
- No. of Shovelfuls
- Loading and Unloading time.

Likewise, average times are managed to assess the performance of said activity in terms of ore transportation according to operating conditions (Table 1).

Once all the information requested above is obtained, statistical tools, such as IBM SPSS, will be used to create Histograms and Dispersion Charts, allowing an analysis of the different times and determining upper and lower limits in an attempt to identify the cause with greater incidence in dead truck times.

3.4 Improve

This stage consists of determining the Cause-Effect relationship and identifying the possible characteristics that can be improved within the process, as well as proposing solutions to mitigate or eliminate the causes and thus meet client expectations and needs. Finally, the operational range of the process parameters or input variables is determined.

During the collection of the times demanded by dump trucks, situations which presented excess or lack of dump trucks were observed. For this reason, in this

Table 1 Average unloading time

Operation conditions	Average time (min)
Favorable	0.5–0.8
Average	1.0–1.4
Unfavorable	1.6–2.3

Table 2 MF interpretation

Match factor	Complies	Interpretation
MF = 1	$P_{Bulldozer} = N * P_{truck}$	Perfect match
MF < 1	$P_{Bulldozer} > N * P_{truck}$	Insufficient number of trucks
MF > 1	$P_{Bulldozer} < N * P_{truck}$	Excess number of trucks

Improve phase, the Match Factor (MF) will be applied to improve dump truck allocation to the loading equipment in order to reduce dead times and bottlenecks during ore transportation (Table 2).

3.5 Control

Finally, after the Improve stage is implemented to optimize the haulage cycle, information will be collected once more through the Dispatch software in order to perform a comparative analysis of times using Office tools such as MS Excel.

4 Results Analysis and Discussion

The case study was defined as the mine located in the department of Puno, province of Lampa, district of Ocuviri, 20 km from the town center of Parina and 90 km NW from the city of Juliaca. At an altitude between 4500 and 5100 m above sea level. Where the following results were obtained.

4.1 Define

Theoretical Cycle and Practical Cycle times were taken to determine the number of unproductive hours, minutes or seconds. Only 4 times were considered for the Theoretical Cycle because they are the fixed haulage variables. On the other hand, for the Field cycle, the variables which depend on dump truck, bulldozer or front loader operators were taken into account. These variables are not often considered when calculating dump truck round trips from and to the bank.

4.2 *Measure*

Fleet times for 23 trucks were compiled for completing a cycle from the different banks to the Pad. It should be mentioned that a total of 3928 ore transportation trips were recorded to the Jessica Pad. Trip data include truck number, bulldozer or loader type, distance, density and time of arrival at the Bank.

4.3 *Analyze*

The result was that the most relevant times are those generated by the truck loaded with ore on the tracks. The sums of the four times (loading times, loaded travel time towards the Pad, unloading times and unloaded travel time to the bank) formed the Total Theoretical Cycle that the trucks take to complete a round trip loaded with gold ore.

$$\textbf{Total Cycle Time+VL103} = \mathbf{00:04:26 + 00:18:27} \\ \mathbf{+00:01:41 + 00:12:06 = 00:36:40}$$

However, when analyzing the arrival times at the banks of each truck per day, it was found that the actual cycle time is longer than estimated. That is, there are dead times in the process of transferring ore to the Leach Pad.

4.4 *Improve*

In the previous step, it was possible to verify that there are currently idle times during the transfer of ore from the different exploitation banks to the Leach Pad.

Assessment of Volvo Trucks (a&b sac) with Front Loader 992. To determine the production per Volvo truck (t/h), the following parameters must be considered (Table 3).

The coordination between the Loading and Transportation teams is extremely important, with truck number and sizes being the basic factors to be determined in any haulage optimization process in an open-pit mine. Therefore, the number of Volvo trucks was determined in relation to Front Loader 992.

$$\mathrm{N} = (\mathrm{Tc} + \mathrm{Tt})/\mathrm{Tc} = (1.6 + 22.1)/1.6 = 14.81\ \mathrm{trucks}$$
$$\mathrm{Front\,Loader} - 992\,\mathrm{Production} = 14.81 * 80 = 1184.8\,\mathrm{t/h}$$

One way to verify that the number of trucks required in correct for the model and characteristics of the loading equipment would be as follows:

Table 3 Parameters

Parameters	Volvo trucks
Outbound speed (Km/h)	18.23 km/h
Inbound speed (Km/h)	27.79 km/h
Total loading cycle	1.6 min
Total transportation cycle	22.1 min
Efficiency factor	0.92
Tons per cycle (t/cycle)	32 t/cycle
Cycles per hour (cycles/h)	2.50 cycles/h
Production per truck (t/h)	80 t/h

$$\mathrm{N = P\_(Front\,Loader)/P\_Truck} = 1184.8/80 = 14.81\ \text{trucks}$$

It should be noted that if the 14.81 trucks are used to determine the Match Factor, the results will be MF = 1, as shown below:

$$\mathrm{MF} = (14.81 * 80)/1184.8 = 1$$

However, we have to work with the total number of trucks, with two possibilities: Round off the number of trucks to the highest next integer, where 15 trucks would be required. In this case, MF > 1, which means that the system is overloaded, causing the trucks to wait to be loaded by the front loader.

$$\mathrm{MF} = (15 * 80)/1184.8 = 1.013$$

To verify that the 15-truck ore haulage system is overloaded, the truck waiting time was calculated for each of the ore transfer cycles.

$$\mathrm{E\_t} = (\mathrm{N} - 1) * \mathrm{T\_c} - \mathrm{T\_t}$$
$$\mathrm{E\_t} = (15 - 1) * 1.6 - 22.1 = 0.3\,\text{min}$$

However, when estimating the total time of the truck to complete a cycle from the bank to the Leach Pad, the calculated waiting time was not included and instead it was added to the total time:

$$\mathrm{T\_(truck - Volvo)} = \mathrm{T\_c} + \mathrm{T\_t} + \mathrm{E\_t} = 1.6 + 22.1 + 0.3 = 24\,\text{min}$$

These minutes added to truck time will directly affect the production they make in 60 min, since the number of trips per hour estimated above is not correct. For this reason, to determine the production and number of correct cycles, the truck waiting time (E_t) was then considered:

$$\text{N}^{\circ}\,\text{Cycles per Hour} = (60 * 0.92)/(1.6 + 22.1 + 0.3) = 2.3$$
$$\text{Production per Truck} = 2.3 * 32 = 73.6\,\text{t/h}$$

In conclusion, when considering the waiting time, production decreases by 8% with respect to previous estimates. On the other hand, if truck waiting time had not been taken into account, the calculation would be off by 8% for the haulage equipment. If the new production value is replaced per each truck in the MF equation, the following results are obtained:

$$\text{MF} = (15 * 73.6)/1104 = 1$$

4.5 Control

It was evidenced that with the results obtained, dead times decreased in a great percentage. In addition, production in some cases increased and in others decreased, with respect to the current production.

Figure 2 displays a control performed for Frontal Loader 992 using an optimal number of 15 trucks based on simulated haulage process times. The loading time is 1.6 min and the ore transfer time is 22.1 min and the dead time is 0.3 min for trucks waiting to be loaded.

Now, Fig. 3 displays a control performed for Frontal Loader 992 using an optimal number of 14 trucks based on simulated haulage process times. The loading time is 1.6 min and the ore transfer time is 22.1 min and the dead time is 1.3 min for the loader waiting for trucks. Here it is observed that no dead time is generated by the trucks.

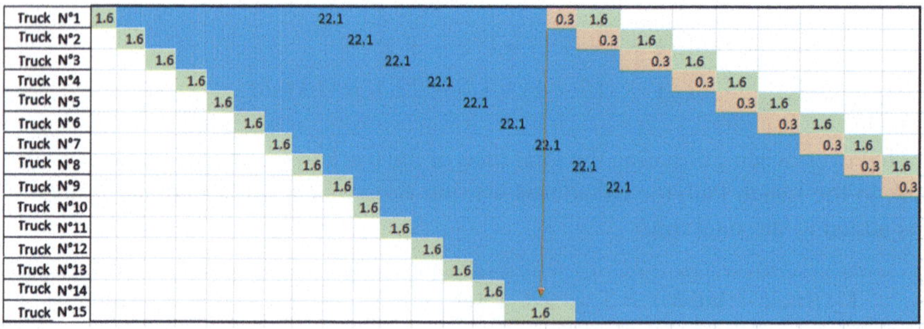

Fig. 2 Front loader 992 control—15 trucks

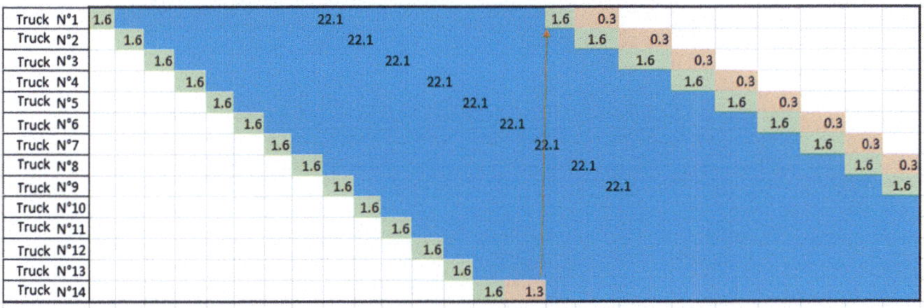

Fig. 3 Front loader 992 control—14 trucks

5 Conclusions

The results of the research study evidence that with the election of an optimal fleet based on the given mining project conditions, the productivity and efficiency values may improve considerably. This is reflected in a better use of machinery and in the reduction of non-productive times.

Therefore, the Six Sigma methodology must be used as a valuable tool for achieving the vision and strategy of the company, as well as for continuous process improvement, but, above all, to improve company competitiveness in these times of constant economic changes at worldwide level.

References

1. Saldaña, A.: Productividad en el ciclo de carguío y acarreo en el tajo chaquicocha bajo clima severo – minera Yanacocha. Tesis para optar el título profesional de ingeniero de minas, universidad nacional de ingeniería, Perú (2013)
2. Kostyuk, D., Thomas, S., Malkin, P., Stone, P.: Is truck queuing productive? In: Proceedings Orebody Modelling and Strategic Mine Planning Symposium, pp 267–274 (2014)
3. Tarshizi, E., Sturgul, J., Ibarra, V., Taylor, D.: Simulation and animation model to boost mining efficiency and enviro-friendly in multi-pit operations. Min. Sci. Technol. **25**, 671–674 (2015)
4. Dindarloo, S., Frimpong, S., Ozanloo, M.: A stochastic simulation framework for truck and shovel selection and sizing in open pit mines. J. South Afr. Inst. Min. Metall. **115**, 209–219 (2015)
5. Saadatmand, A., Sattarvand, J.: Simulation based investigation of different fleet management paradigms in open pit mines-a case study of sungun copper mine. Arch. Min. Sci. **60**, 195–208 (2015)
6. Levano, L.: Incremento de productividad mina mediante herramientas six sigma. Tesis para optar el título profesional de ingeniero de minas, universidad nacional de ingeniería, Perú (2012)

Improvement of Chemical Processes for the Analysis of Mineral Concentrates Using Lean Six Sigma

Alejandro Arone, Miguel Pariona, Ángel Hurtado, Victor Chichizola and José C. Alvarez

Abstract In a chemical laboratory of analysis of concentrates of minerals there are delays in the delivery of results report to the clients, failing to meet the deadlines; so this research aims to solve the problem through the 5 Lean Six Sigma steps that consists in: Defining, Measuring, Analyzing, Improving and Controlling (DMAIC). The findings suggested that the focus on the data collection, by the technical analyst, of the following values: the molar mass in the EDTA solution and the volumetric expense of the chemical reaction of the sample would have the greatest impact on the decrease of the retests.

Keywords Lean six sigma · Continuous improvement · Design of experiments · Process capacity · Mineral concentrate tests

1 Introduction

In This research is carried out in the laboratory M3Q53 that offers the service of mineral analysis through commercial chemical tests of non-ferrous mineral concentrates that are classified according to [1]. In these tests the percentages of Pb, Cu, Zn, Fe and Mn are determined in a sample of mineral concentrate. This chemical analysis

A. Arone (✉) · M. Pariona · Á. Hurtado · V. Chichizola · J. C. Alvarez
Escuela de Ingeniería Industrial, Universidad Peruana de Ciencas Aplicadas (UPC), Lima, Perú
e-mail: u201525485@upc.edu.pe

M. Pariona
e-mail: u201524773@upc.edu.pe

Á. Hurtado
e-mail: pcinahur@upc.edu.pe

V. Chichizola
e-mail: pcinahur@upc.edu.pe

J. C. Alvarez
e-mail: pciijalv@upc.edu.pe

Y. Iano et al. (eds.), *Proceedings of the 4th Brazilian Technology Symposium (BTSym'18)*, Smart Innovation, Systems and Technologies 140,
https://doi.org/10.1007/978-3-030-16053-1_52

laboratory has been registering delays in delivering reports, exceeding five days in most cases, causing customer dissatisfaction. One of the main causes of the delay in the delivery of results reports is due to the retesting of samples that have a monthly economic impact higher than 2000 US dollars on the tests of copper (Cu), lead (Pb), zinc (Zn), manganese (Mn), and iron (Fe). A retest occurs when the difference of results between analyst A and B differ by 0.20%, so the number of retests is greater than 5% of the number analyzed since January 2018.

There is a previous work [2] about the improvement of an ARD chemical laboratory using the Lean Six Sigma methodology, that applied served to address different problems from shortening distance traveled using the Spaghetti diagram tool, reducing activities that do not generate value and reducing costs thanks to a good supply logistics plan. There is another research work [3], which addresses the same problem with a delay of 17 working days and explains the improvement of the processes by applying lean manufacturing tools. Inal [4] analyze the adoption of lean six sigma methodologies to identify and solve problems in a clinical laboratory.

Really the lean methodologies and its integration with others production methodologies, as suggest authors as [5] and [6], are an opportunity for its use in small and medium enterprises as chemical laboratories.

2 Method

The research method started with the collection of primary data following this steps: (i) Determination of the frequency and object of customer complaints; (ii) Determination of the amount of analysis of retests by the type of analysis was recorded in an Excel format; (iii) Determination of the various causes of delays in the delivery of results; (iv) A questionnaire was given to laboratory collaborators to capture other factors that might cause the delay in the delivery of reports; (v) A SIPOC diagram was used to know the input and output elements of the laboratory processes; (vi) The characterization of processes and product quality was carried out to determinate critical factors in each laboratory process.

After that, it is proposal the implementation of 5 Lean Six Sigma steps that consists in: Defining, Measuring, Analyzing, Improving and Controlling (DMAIC). So at the improve phase an experimental research design was carried out, in which certain variables were subjected to experimental conditions, to know the effects this would cause. For this, it was necessary to carry out the following operations:

First, with the collected data, it was used statistics to find out the following points: (i) The existence of normal or abnormal distribution; (ii) The variability of the results
Second, a process capacity report was made with the data collected, with the help of the MINITAB program, to know if the process was capable and statistically competent.
Third, an experimental DOE design was carried out to know the critical factors that affect the dispersion of results.

Finally the technical solution proposal is simulated with the Bizagi software and the results obtained are compared with the preliminary results before the intervention.

3 Results

3.1 Results Before

A general diagnosis was performed of the operative processes involved in the laboratory, to determinate the causes that could originate the delay in delivery of final reports to the client, that in the last 12 months, the trend in the delivery of results reports to the client has mostly been greater than five days after receiving the client's sample. So, the percentages of retests are above 5% between the months of January to May 2018.

3.2 Innovative Proposal

The Lean Six Sigma methodology that follows the DMAIC phases is mentioned below:

1st phase: Define

The processes involved are analyzed for the analysis of technical tests in the laboratory. It begins with the development of a project charter, the definition of a high-level map using the SIPOC diagram, then the process parameters and product characteristics are identified.

2nd phase: Measure

The measurement allows us to obtain real data on the technical characteristics that are closely related to the client's requirements; the decisions taken will be based on real and accurate information.

According the visual flow mapping (VSM), the takt time of 262 s per analysis is determinated; this is the maximum time limit for reporting to customers and meeting their demands.

For the Cu-Fe-Mn-Pb and Zn analyses has the following tolerance. For example, for the range of 13–30%, the maximum difference accepted is 0.2%, as appear in Table 1.

It is considered a retest when the maximum tolerance limit is exceeded. Next, we compare results between two analysts where the lack of precision is presented. Zinc results differ significantly. In the 0–40% range, the tests of analyst B to the greater than A. Between 40–60% the test of analyst A is greater than B.

Table 1 Permissible range and limits

Element	Range	Maximum limit
	%	± (%)
Cu-Pb-Zn-Mn-Fe	13–30	0.20
Cu-Pb-Zn-Mn-Fe	30–50	0.25
Cu-Pb-Zn-Mn-Fe	50–68	0.30

Table 2 Process capacity in copper, lead and zinc pattern

	Cooper	Lead	Zinc
Six Sigma	3.10	2.07	3.16
Cpk	0.43	0.36	0.43
Ppm	55,129	282,836	48,305

The sigma classification is equal to 6 which is the optimum since only 3.4 defects per million results will be obtained when the capacity of the process (Cpk) is 2. With this classification, our process will be evaluated. So, the Cpk is $0.43 < 1.33$.

The zinc pattern analysis process shows high variability with a PPM of 48 305.17 defective results per million results. Under probabilities, the capacity analysis of the 3 chemical elements is between 48,305 and 282,836 defects per million opportunities. (Table 2).

3rd phase: Analyze

In this phase, the potential causes are identified and validated through statistical methods using hypothesis tests.

The Zn results do not have a normal distribution because of $p < 0.05$. The comparison of couples with equal medians is evaluated. The Normal Distribution test is performed on the results of the other elements: Pb, Cu, Fe and Mn, having the same results that they don't present a normal distribution and that must be evaluated by comparing couples with equal medians.

The laboratory has two senior analysts and one practitioner, who has one year of experience in this company and performs the analysis of elements of Cu, Pb, Zn, Mn, and Fe. From this, the following analysis is performed. The analyst's zinc tests have different averages for the same test pattern. This leads to retests due to the variability that it presents. The variable Zn presents greater difference.

The Analysis of Variance test (ANOVA), tests the hypothesis that the means of two or more populations are equal. The ANOVA evaluates the importance of one or more factors by comparing the means of the response variable in different levels of factors. The ANOVA test of a factor is applied to show if there is a significant difference between the testing methods of different analysts. The graph shows that there is no apparent difference because the means are equal to the value of 63.15% concentration in lead patterns.

4th phase: Improve

In this phase, the actions or proposals for improvement that contribute to mitigate or eliminate the root causes that generate problems and defects in the process are defined.

Design of experiments (DOE)

The factors that influence the final results of the analysts are the BD set that is related to the weight factors of the sample and the molar mass of standardization of the EDTA (ethylenediaminetetraacetic acid) solution.

Improvement plans

For the weight of the sample, the variable of pattern dryness is considered. The procedure would be as follows, the standard is weighed under normal conditions presenting humidity, and then it's placed in the desiccator for a period of 60 min and then allowed to warm to 60 min. Finally, the sample is weighed and if the value is not less than 0.1% this weight is taken as the initial weight of the sample and it is followed by the other activities.

For the determination of the molar mass of the standardization of a chemical element, the acquisition of a potentiometric automatic titrator was considered to determine the concentration of a chemical element with greater precision and accuracy when it reached its point of equilibrium of the chemical reaction, previously a digital burette was used and the color change of the reaction was visualized, which could generate variation in the volume of expenditure recorded.

A final improvement plan is the simple average forecast of reagent consumption, to show the logistics area the important reagents that are required month by month.

Reduction waiting times in the VSM

Through a mathematical analysis, considering the cycle times of each type of analysis, we proceed to the application of Little's Law known as "WIP Cap" method, to put a cap on the amount Work In Process (WIP) and in this way reduce the waiting time through improvement actions.

Modal Analysis Failure Effect (AMFE)

A modal failure and effect analysis is developed to identify and prevent potential failure modes. The resulting value known as the acronym RPN (Risk Priority number) determines the variables with important critical risk values.

5th phase: Control

This phase consists of designing control mechanisms and documentation of the processes or procedures modified in order to maintain the sigma level achieved.

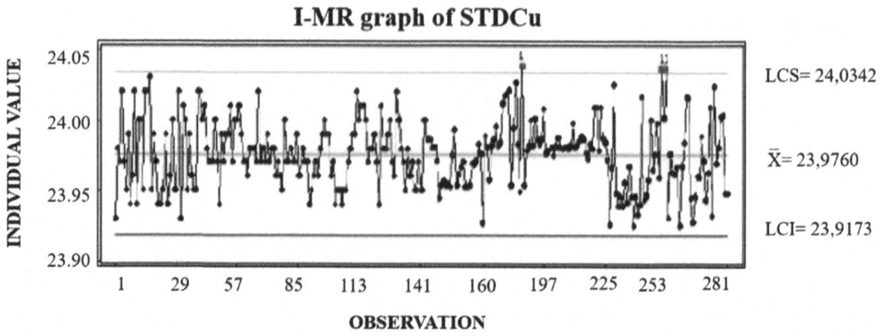

Fig. 1 I-MR graph for copper pattern

Statistical control graphs

– Graph I-MR, to monitor the mean and the variation of the process when it has continuous data that are individual observations which can't be grouped. This control chart (Fig. 1) is used to monitor the stability of the process over time so that it can identify and correct instabilities in a process.

In the previous Fig. 1, the I-MR control chart is shown for the copper pattern results, according to the characteristics of the pattern, this should not exceed the optimum value of 23.975 by ±0.05.

Standard operating procedures (SOP)

The documents contain step-by-step written instructions that laboratory personnel must follow meticulously when performing a procedure.

3.3 Improvement Results with Partial Implementation of the Proposal

The partial implementation is taking place in the volumetric area for Cu, Pb and Zn analysis. So, since May 2018, the number of re-essays from Zn has decreased from 13% to less than 5% as is shown in Fig. 2.

Improvement results with simulation

With the use of the Bizagi software, two scenarios are simulated, in the first scenario the current laboratory re-tests are simulated, and in the second scenario with the acquisition of an automatic titrator and a chiller.

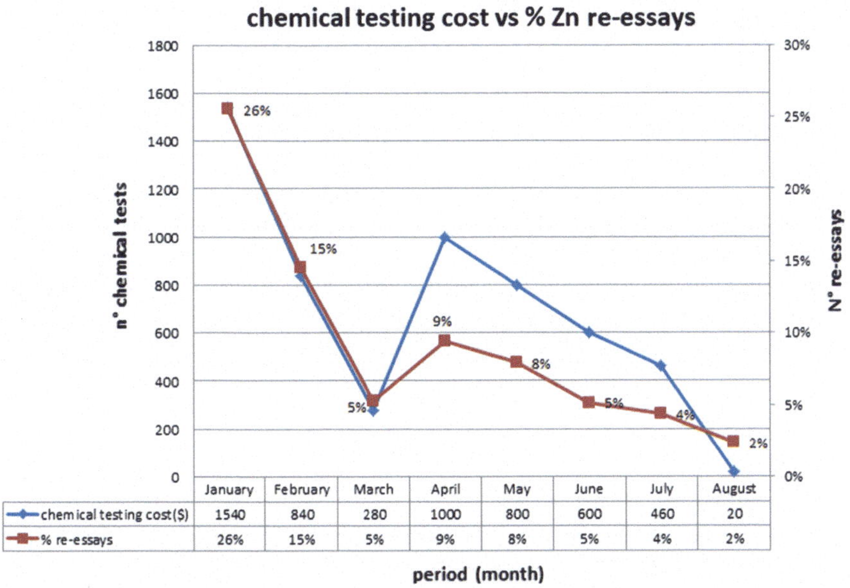

Fig. 2 Evaluation of Zinc re-essays

4 Analysis of Results

The retests have decreased notably since May from 10 to 5% or less. This is due to the requirement in control of the dryness pattern and to the talk with the personnel involved in the retests about the costs associated with this reprocessing.

Wet track patterns do not yet exceed sigma 3, since the value Cpk < 1. To improve this process, it is necessary to have greater control over the processes and reduce the participation of the human factor in order to reduce errors. In the simulation with bizagi a test was carried out for 40 samples.

In scenario 1 the retests are simulated with probability greater than 10%, which results in 27 analyzes that passed quality control and 13 that were retested. These 13 retesayos would have a sale price of loss of 325 dollars. In scenario 2 retests are simulated with probability less than or equal to 5% with which 35 analyzes that passed the quality control and 5 retested trials are obtained. These 5 retested would have a sale price of loss of 125 dollars.

5 Conclusions

The DOE allows the demonstration of optimization models to control the X (independent variable) factors that affect the Y objective (dependent variable). We have worked with a model 2 raised to 5.

To eliminate or reduce the error variables in the technical test, the acquisition of an automatic titrator is proposed; additionally to controlling the moisture level of the pattern used for each test.

To evaluate the accuracy and the certainty of the capacity analysis, a control chart is used, this way the quality of the results is assure. If retests are reduced by 5%, then at least 200 dollars are saved for every 40 trials.

References

1. INACAL, Participación en ensayos de aptitud/Comparaciones Inter laboratorios. Consultado el 24 de agosto del 2018. http://www.inacal.gob.pe/inacal/files/13D.pdf
2. Tran, T.: Implementation of lean six sigma principles in an analytical research and development chemistry laboratory at a medical device company
3. Ruiz, E., Mayorga, M.: Herramientas de manufactura esbelta aplicadas a una propuesta de mejora en un laboratorio químico de análisis de minerales de una empresa comercializadora. Pontificia Universidad Católica del Perú –PUCP. Lima (2013)
4. Inal, T., Goruroglu, O., kibar, F., Cetiner, S.: Lean six sigma methodologies improve clinical laboratory efficiency and reduces turnaround times. J. Clin. Lab. Anal. **32**(1) (2018)
5. Salah, S., Rahim, A., Carretero, J.A.: The integration of six sigma and lean management. Int J. Lean Six Sigma **1**(3), 249–274 (2010). https://doi.org/10.1108/20401461011075035
6. Jiménez, H., Amaya, C.L.: Lean Six Sigma en pequeñas y medianas empresas: un enfoque metodológico. Ingeniare: Revista Chilena de Ingenieria, **22**(2), 263–277 (2014). ISSN 0718-3305

Energy Disaggregation via Data Mining

Pierre Dantas, Waldir Sabino and Maryana Batalha

Abstract Electrical energy consumption of a residence is usually monitored by a meter installed at its entrance and it is composed by the sum of consumptions of installed devices. Energy disaggregation estimates the consumption of each device at each instant of time. This paper presents two main contributions. First, we address disaggregation using data mining techniques by clustering methods, that is, k-Means and Expectation and Maximization (EM). We demonstrate that we can obtain superior disaggregation accuracy from more complex methods. The second contribution, we elaborate clusters (dictionaries) considering that the states of operation of the devices and the signal of total consumption are dependent instances. We use Reference Energy Disaggregation Data Set (REDD), Waikato Environment for Knowledge Analysis (WEKA) and MATLAB.

Keywords Energy disaggregation · Data mining · Dictionaries · Representativeness · WEKA

1 Introduction

Non-intrusive load monitoring (NILM) consists of inferring the individual consumption of equipment that compose a total energy consumption, knowing only the latter [1]. In this approach it's need only a meter installed at the entrance of the installation and this can lead to energy savings. It is also important for smart grids and Internet of Things (IoT). Transitions of energy consumption signals in the time domain

P. Dantas (✉) · W. Sabino
Federal University of Amazonas (UFAM), Manaus, AM, Brazil
e-mail: pierre.dantas@gmail.com

W. Sabino
e-mail: waldirsabino@gmail.com

M. Batalha
Manaus Lutheran University Center (ULBRA), Manaus, AM, Brazil
e-mail: maryanabatalha@yahoo.com

Y. Iano et al. (eds.), *Proceedings of the 4th Brazilian Technology Symposium (BTSym'18)*, Smart Innovation, Systems and Technologies 140,
https://doi.org/10.1007/978-3-030-16053-1_53

were analyzed [1] as well as in the frequency domain analysis [2], PCA, [3], hidden models of Markov [4], neural networks [5] and machine learning algorithms [6]. In this work, we will approach disaggregation using data mining techniques by two clustering methods, k-Means and Expectation and Maximization (EM), demonstrating that we can obtain equivalent disaggregation accuracy from complex methods. We will use Reference Energy Disaggregation Data Set public database (REDD) [7] in which we will apply pre-processing and data transformation procedures using Waikato Environment for Knowledge Analysis software (WEKA) [8]. In a last stage, we use MATLAB.

2 Data Mining Algorithms

Data mining is a process of analysis of large data sets that seeks to identify patterns, associations, groupings, noise, among others, generating new data sets. Statistical classification and clustering algorithms are used.

The k-Means method partitions a set of N observations into k groups where each observation belongs to the group with the nearest mean. Initially, the centroids of the groups are assigned randomly and, iteratively, they are recalculated as the midpoint of the instances of each group. This is done until the assignments do not change.

The Expectation and Maximization (EM) method uses statistics to iteratively find maximum likelihood estimators of the parameters of a probability distribution. First, we calculate the expected value of the logarithm of the likelihood and, then, we find its maximum. These steps are repeated until reaching a convergence.

3 Data Processing

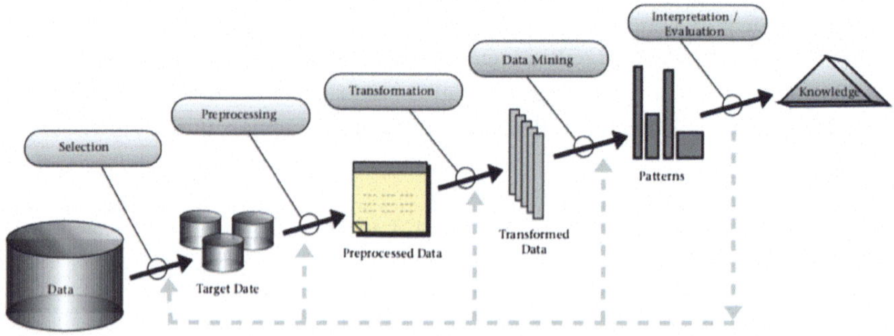

Fig. 1 An overview of the steps that compose the KDD process

Data processing consists of performing operations and transformation procedures on a database. Raw databases often have missing data, noise, redundancies, inconsistencies, among others. We seek better representations of the data. For this, we can use an approach called Knowledge Discovery in Database (KDD) [9], illustrated in Fig. 1.

We initially have a raw database, from which we select a subgroup of data. Then, we apply pre-processing in order to "filter" the data set, where we eliminate noise, outliers and treat missing data, for example.

4 Methodology

Initially, we construct clusters (dictionaries) with the states of operation of the devices and the total signal. We expect to obtain a dictionary of less complexity and preserve the representativeness of the data set [10]. In the second step, we elaborate the dictionaries. Finally, we look in REDD by the solution of the disaggregation. We use Waikato Environment for Knowledge Analysis (WEKA), an open source tool widely used in data mining [8]. We initially identify and attenuate noises and outliers in REDD, keeping the redundancies. For data transformation, we analyze the attributes of the database and apply transformations that allow us to obtain more knowledge than to analyze the raw database. In the end, we use MATLAB to find the solution of the disaggregation.

We analyze the results of this work in terms of complexity in obtaining the groupings (dictionaries), number of iterations of the chosen algorithms, processing time and accuracy of disaggregation. The measure of performance of the disaggregation adopted will be the accuracy. This is defined, for a quantity of n instances, as the relation between the total electric energy consumption $\bar{x}(t)$ and the sum of the estimated electric energy consumption signals $\hat{x}_i(t)$ of the L devices, represented by Eq. (1).

$$\text{Accuracy} = 1 - \frac{\sum_{n,L} \left| x_i(t) - \hat{x}_i(t) \right|}{2 \sum_n \bar{x}(t)} \tag{1}$$

5 Experiments

We used a desktop computer 1.7 GHz Intel Core i5 processor, 2 cores, 4 GB memory, 1600 MHz DDR3, MacOS Sierra version 10.12.5. We use WEKA version 3.8.1 and MATLAB version R2016b (9.1.0.441655), 64 bits.

We use the public database of energy consumption data Reference Energy Disaggregation Data Set (REDD) [7]. This database contains records of the total (aggregated) electrical consumption of 6 residences and the devices that compose them. We will use data from residence 2 that has 9 devices.

After importing the REDD database into WEKA, we apply a filter called **unsupervised.instance.Resample** with **noReplacement = true** and **sampleSizePercent = 10.0**. This means that we re-sampled the database at 10% without replacement. The purpose of this procedure is to minimize noises and outliers, in addition to lowering the database to 10% of the original, making the cost of processing less. After pre-processing, we apply the normalization transformation, standardizing them in the [0, 1] range. We do not standardize or reduce the dimensionality of the data. In WEKA, we perform this procedure by applying **unsupervised.attribute.Normalize** filter, with default parameters. We use two methods of data mining to get the clusters (dictionaries), k-Means and Expectation and Maximization (EM). We used cross-validation holdout with 66% of the database for training and 34% for test. In the training, we execute the two configured methods in order to obtain 10, 20 and 30 clusters. In addition, we performed the EM method with the number of groups calculated automatically by WEKA.

Finally, we designate a disaggregation solution for each sample from the test set. We import the results from WEKA into MATLAB, and then, for each sample of the test set, we look for the closest clustering, using total consumption as the search index. This instance will be referred as disaggregation solution.

Table 1 Result of the execution of the methods k-Means and EM by WEKA

Execution	Method	Groups declared	Groups obtained	Processing time (s)	Iterations
1	k-Means	10	10	0.24	6
2	k-Means	20	20	1.09	17
3	k-Means	30	24	0.90	12
4	EM	10	10	3.88	0
5	EM	20	19	8.21	0
6	EM	30	28	16.16	0
7	EM	–	9	322.99	0

Table 2 Results of the disaggregation accuracy

Execution	Method	Accuracy (%)
1	k-Means	76.47
2	**k-Means**	**84.34**
3	k-Means	50.12
4	EM	80.73
5	EM	82.67
6	EM	78.03
7	EM	77.71

6 Results

In the data mining stage, we applied the k-Means and EM methods separately to the REDD re-standardized and normalized database. In Table 1 we present the number of parameterized groups in WEKA, the generated groups, the processing time and the number of iterations for the algorithm k-Means. In Table 2, we present the results of the disaggregation accuracy for each case.

7 Conclusions

Initially, even though we created a new dataset with 10% of the raw dataset size we have noted that the statistical information of minimum value (Min), maximum value (Max), number of unique samples, number of distinct samples, (Unique), mean (Mean) and standard deviation (StdDev) were very similar. We conclude that we can, through resampling, select a reduced subset of the raw data set and at the same time maintain its statistical properties. On normalization, we find that all instances have been standardized in the range [0, 1].

The transformation step of the data, we observe in Table 1 that the number of clusters generated by WEKA are close to the parameterized quantities. The processing time of the EM method is greater than that of the k-Means, which was already expected. Note that the EM algorithm, once parameterized to self-determine the number of clusters, was the most time-consuming execution and generated fewer clusters.

In Table 2, we checked the results of the disaggregation accuracy for each execution. Note, in the k-Means method, that the accuracy does not necessarily increase with the number of generated clusters (dictionary size). For 20 clusters, the value of accuracy was the highest observed. For the EM method, observe that the accuracy showed smaller variation due to the modification of the number of clusters formed. Also note that execution 7, containing 9 clusters (or a dictionary of size 9), showed accuracy of disaggregation close to that of the other executions. In Table 3 we present the disaggregation accuracy obtained by other methodologies. Note that we have outperformed these, using the k-Means and EM methods.

In general, we observe that the disaggregation of a signal of energy consumption by data mining, as presented in this paper, depends on several parameters. One of them is a priori knowledge about the nature of the data. This allows you to choose

Table 3 Results of accuracy of others disaggregation methods

Method	Accuracy (%)
Simple Mean [11]	39.00
FHMM [7]	59.60
Powerlets [11]	79.00

more appropriate approaches which leads to better results. The parametrization of the methods is also another relevant point. As seen, the accuracy may vary depending on the number of groups chosen, especially in the k-Means method.

To conclude, we observe that through data mining techniques, we can obtain better accuracy than those obtained by other complex techniques. Notice in Table 1 that both the processing times and the number of iterations of the algorithms can be considered low.

For future work, we suggest exploring other data mining techniques. We also suggest using other databases. We suggest applying other techniques of pre-processing and transformation of energy consumption signals, verifying to what extent we can reduce the database without reducing the accuracy of the disaggregation. Other methods of data mining can also be investigated.

References

1. Hart, G.W.: Nonintrusive appliance load monitoring. Proc. IEEE **80**(12), 1870–1891 (1992)
2. Chan, W.L., So, A.T., Lai, L.L.: Harmonics load signature recognition by wavelets transforms. In: Proceedings of International Conference on Electric Utility Deregulation and Restructuring and Power Technologies, drpt 2000, pp. 666–671. IEEE (2000)
3. Rodrigues, P.C.: Principal component analysis of dependent data. In: 15th European Young Statisticians Meeting. Castro Urdiales/Spain (2007)
4. Kolter, J.Z., Jaakkola, T.: Approximate inference in additive factorial hmms with application to energy disaggregation. In Artificial Intelligence and Statistics, pp. 1472–1482, March 2012
5. Kelly, J., Knottenbelt, W.: Neural nilm: deep neural networks applied to energy disaggregation. In Proceedings of the 2nd ACM International Conference on Embedded Systems for Energy-Efficient Built Environments, pp. 55–64. ACM, November 2015
6. Deshmukh, A., Lohan, D.: CS446 Project: Electric Load Identification using Machine Learning (2015)
7. Kolter, J.Z., Johnson, M.J.: REDD: a public data set for energy disaggregation research. In Workshop on Data Mining Applications in Sustainability (SIGKDD), vol. 25, pp. 59–62. San Diego, CA, August 2011
8. Garner, S.R.: Weka: the waikato environment for knowledge analysis. In: Proceedings of the New Zealand Computer Science Research Students Conference, pp. 57–64, April 1995
9. Fayyad, U., Piatetsky-Shapiro, G., Smyth, P.: From data mining to knowledge discovery in databases. AI Mag. **17**(3), 37 (1996)
10. Elhamifar, E., Sapiro, G., Sastry, S.S.: Dissimilarity-based sparse subset selection. IEEE Trans. Pattern Anal. Mach. Intell. **38**(11), 2182–2197 (2016)
11. Elhamifar, E., Sastry, S.: Energy disaggregation via learning powerlets and sparse coding. In: AAAI, pp. 629–635, January 2015

Bamboo: A Great Ally of the Civil Construction

Caroline Mayumi Yamaguchi, Márcia Milena Pivatto Serra and Adriana Volpon Diogo Righetto

Abstract Sustainability adoption at the civil construction has been increasing over the years in a search for sustainable alternatives to reduce pollution and waste. One of those sustainable practices which has received great attention in the last few years is the application of bamboo as a raw material. In fact, bamboo has been employed in different areas, such as in the food industry, handicrafts, housewares, and furniture. Its main features are flexibility, durability, lightweight, strength, low cost and versatility, which enable its application at the construction industry. However, its use requires specific treatments, such as fungicide and insecticide, along with careful procedures for harvesting, curing and drying. Therefore, the aim of this research is to evaluate the potential use of bamboo as a building material for the construction industry by means of mechanical tests. Measurements were performed at the Materials Laboratory, School of Engineering from the Mackenzie Presbyterian University, Campus Higienópolis. The analysis was based on two bamboo species, which were chosen after interviewing experts in the field and searching the literature. For comparison purposes, two bamboo species were studied: *Dendrocalamus aesper* and *Phyllostachys pubescens*, which were chosen according to suggestions from experts in the field and by searching the literature, especially those contributions based on physico-chemical tests with other bamboo species. Based on the analyzed experimental data, the results evidenced the bamboo species studied herein could be applied to civil contruction. Both species chosen for this work provided suitable mechanical characteristics, since the average compressive strength for the *Dendrocalamus aesper* species was 83.14 MPa and for the *Phyllostachys pubescens* species was 97.67 MPa. The tensile strength for the *Dendrocalamus aesper* was 180.71 Mpa, whereas for *Phyllostachys pubescens* was 121.88 MPa. These results are in good agreement to the literature, which makes these bamboo species suitable as a sustainable alternative for the construction industry.

C. M. Yamaguchi · A. V. D. Righetto (✉)
Center of Science and Technology, Mackenzie Presbyterian University,
Brazil Avenue 1220, Campinas, SP 13073-148, Brazil
e-mail: adriana.righetto@mackenzie.br

M. M. P. Serra
Consultoria em Estatística Pivatto Serra, Campinas, SP, Brazil

Y. Iano et al. (eds.), *Proceedings of the 4th Brazilian Technology Symposium (BTSym'18)*, Smart Innovation, Systems and Technologies 140,
https://doi.org/10.1007/978-3-030-16053-1_54

Keywords Civil construction · Bamboo · Mechanical properties

1 Introduction

Bamboo comes from the family *Poaceae* (formerly called *Graminae*), which has over 1250 species. They are adaptable to tropical, sub-tropical and mild climates [1]. Bamboo may be found in diverse regions of the world, such as in the American, African and Asian continents [2].

Even though the biggest native bamboo forest is in the Brazilian territory, particularly at the State of Acre [3], there is some preconception about the use of bamboo. It was formerly employed as a building material by indigenous communities, followed by the slaves, which conceived the meaning of poverty and misery by the society. Moreover, bamboo may be associated to the lack of salubrity and subhuman living conditions. The image of social status, from the antiquity up to nowadays, is taken according to the type of housing. This type of construction was called wattle and daub (pau a pique, in Portuguese) [4].

On the other hand, bamboo has a broad range of applications in some countries from Asia, such as Japan and China, which comes from spiritual/religious reasons up to a myriad of others. Over 5000 uses for bamboo have been found all over the world [2]. Its versatility ranges from handicrafts, landscape design, paper fabrication, fabrics, musical instruments, reforestation of Riverside vegetation, recovery of eroded areas, control of steep slopes and a raw material for civil construction [5].

Some countries next to Brazil, such as Costa Rica, Ecuador and Colombia, have a higher acceptance to the use of bamboo as the raw material for construction, wherein it is employed for low-cost housing, as well as in large construction projects, such as bridges and pavilions. Therefore, it is well known that bamboo is not only a potential building material, but also a low-cost, eco-friendly and easy to handle alternative [6].

Even though the use of bamboo as a raw material in construction dates back to ancient times, it has become a sustainable alternative when the use of natural resources tend to decrease environmental pollution by giving priority to the natural environment, especially at the civil construction industry, since its residues are amongst the main sources of pollution to the environment. Moreover, bamboo has suitable physico-chemical properties, great durability, workability and versatility. It also serves as a tool for land restoration and erosion control [3].

Based on the suitable features cited above, the study of bamboo was proposed by considering its plantation, harvesting, drying, treatment and application in the housing industry, as well as the investigation of its mechanical properties.

A stable bamboo production takes about five to seven years, whereas its ripening takes place within three to four years. During this period culms and shoots are harvested to continue the cultivation and production. Bamboo irrigation is required, whereas no pesticide is needed. Manual harvesting may be carried out by using axes or jigsaw, which might help fortify the bamboo crop [7].

During bamboo growth, its fibers may become more resistant and harder. Its application in housing is recommended when it has the age of three to five years, since it acquires the maximum resistance [8]. Harvesting depends on the location: in tropical regions it is recommended to be carried out at the end of the dry season. In subtropical regions it should be performed during the winter season due to the smaller amount of water and starch in bamboo culms, since they store its reserves within the rhizome, which decreases the chances of fungi and insect attacks [9]. Moon phase and harvest time are also taken into account. Optimum harvest time would be on waning moon and before sunrise [7].

In order to increase durability and decrease the chance of fungi and insect attacks, harvested bamboo must undergo curing and treatment processes. Curing is highly recommended to be carried out inside the forest, since bamboo poles will still have its leaves and branches. They should be kept in a vertical position at the harvest site for four weeks over neighboring bamboo poles and rocks to avoid humidity. It should be done at the end of the dry season, but in the time, local pressure and humidity it was developed [7]. The treatment process occurs mostly by substituting starch by chemical substances. In Brazil, the most frequently used method is treating with boric acid and borax. The treatment occurs by dipping bamboo poles into a solution, which lowers starch concentration by trapping it into the tank water. Cleaning the tank from time to time helps avoid accumulation of reagents. Salts are absorbed and adhered to the bamboo walls, which hinders the growth of xylophagous organisms. The use of varnish and sealants is recommended for preserving bamboo [10].

Bamboo drying is an important step to provide suitable resistance and durability. If it were not done properly, bamboo pieces may suffer shrinkage, which will cause structural failure. It is highly recommended to dry bamboo inside greenhouses, since they are more efficient by harvesting sunlight during the day, without direct light illumination, and yet keeping it warm at night [7].

Storage should be done indoors, protected from the rain and sunlight, with the poles arranged in layers with some room for air ventilation. Bamboo poles should be kept 15 cm off the ground to avoid moisture. [9].

Bamboo has a high structural performance in terms of compression, torsion and bending. It is known to have good bending properties in light of its tubular volumetry and the longitudinally-oriented fibers, which create micro-tube bundles [10]. Nevertheless, the mechanical properties of bamboo are mainly influenced by its age, weather conditions, harvest time, species and moisture content [10].

Bamboo is a sustainable, low-cost, eco-friendly material with a large scale production. It provides less damage to the environment, both in its production and application in the housing industry, which decreases the amount of polluting residues generated there from. It is a environmentally friendly material with a myriad of applications in construction [11]. It is an efficient substitute for certain wood species in many construction types [12], besides being affordable to most social classes [13]. Therefore, the aim of this work is to show the mechanical properties of bamboo by carrying out tests for two bamboo species: *Dendrocalamus aesper* and *Phyllostachys pubescens*. Compressive and tensile strength were measured for both bamboo species and its technical viability studied for the housing industry.

2 Methodology

This research was carried out at the Materials Laboratory, School of Engineering from Mackenzie Presbyterian University—Campus Higienópolis, Brazil. The aim of this work was the study of mechanical and structural properties from bamboo culm from two species: *Dendrocalamus aesper* and *Phyllostachys pubescens*, which are native bamboo species from Atibaia, State of São Paulo, Brazil.

Due to the lack of specific standardization, the tests carried out in this work were based on the usually adopted standards and processes from the literature for the same analyses [14, 15].

2.1 Compressive Strength

According to suggestions from the literature, each sample was cut at a height of approximately two times its external diameter, as shown in Figs. 1 and 2. Those samples were polished from both sides in order to make them uniform, as well as to allow the uniform distribution of tension during compressive tests.

The height and inner diameter of each sample were measured by using a digital calliper with a resolution of 0.01 mm, which is similar to the universal calliper with the same resolution. Since throughout the bamboo sample the cross-sectional areas are not totally circular and there are changes in thickness, three measurements from the diameters were performed herein in different positions: top, center and bottom.

Tests were performed in a Universal Testing Machine, as shown in Fig. 3, with a maximum load of 60 tf and a loading rate of 10 mm/s.

The sample was placed in such a way that the movable cross head would match the cross-sectional area of the specimen. The compressive strength was calculated from

Fig. 1 Bamboo samples for compressive strength tests from the *Dendrocalamus aesper* species. *Source* Author's own figure

Fig. 2 Bamboo samples for compressive strength tests for *Phyllostachys pubescens* species. *Source* Author's own figure

Fig. 3 Universal Testing Machine for the compressive strength tests. *Source* Author's own figure

the relationship between the maximum supported load of each bamboo specimen and the external area (the average sample internal diameter was taken into account), as shown in Fig. 3.

The maximum compressive strength was achieved when bamboo failed. Figures 4 and 5 show two broken bamboo samples, just after tensile measurements. Twenty samples were measured: 10 from *Dendrocalamus aesper* and other 10 *Phyllostachys pubescens* species.

Fig. 4 Fractured bamboo samples from the *Dendrocalamus aesper* species. *Source* Author's own figure

Fig. 5 Fractured bamboo samples from the *Phyllostachys pubescens* species. *Source* Author's own figure

2.2 *Tensile Strength*

Bamboo samples for the axial tensile strength tests were molded as shown in Figs. 6, 7 and 8. Strips were taken from culms with a height ranging from 20 to 29 cm and a thickness depending on the culm wall.

Bamboo strips were labeled sequentially from the edges (approximately 5 cm from the table length) in order to hold them tightly to the press. The central part of the specimen measured approximately 5 cm in length and 5 mm in width. Within the transition zone the width ranged from 5 up to 6 mm.

After the templates were ready and bamboo samples were properly labeled, the next step was to polish them to guarantee constant uniformity within the cross-sectional area to be fractured, making sure no error could happen. It is recommended,

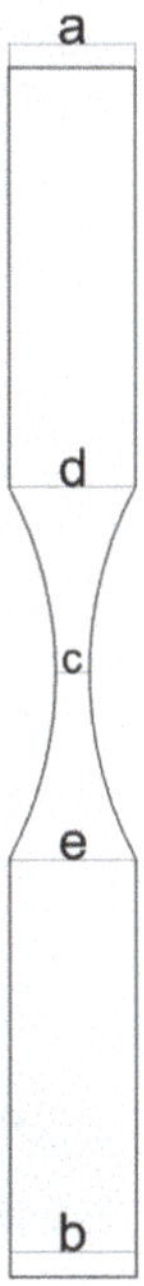

Fig. 6 Model for conducting the bamboo strip samples for tensile strength tests. *Source* Author's own figure

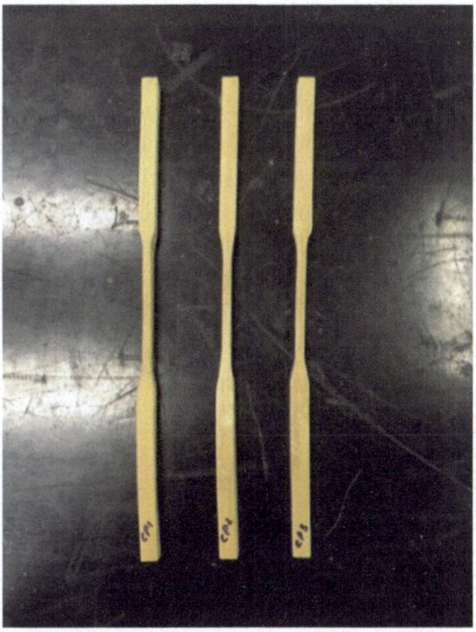

Fig. 7 Bamboo strip samples for tensile strength tests for the *Dendrocalamus aesper* species. *Source* Author's own figure

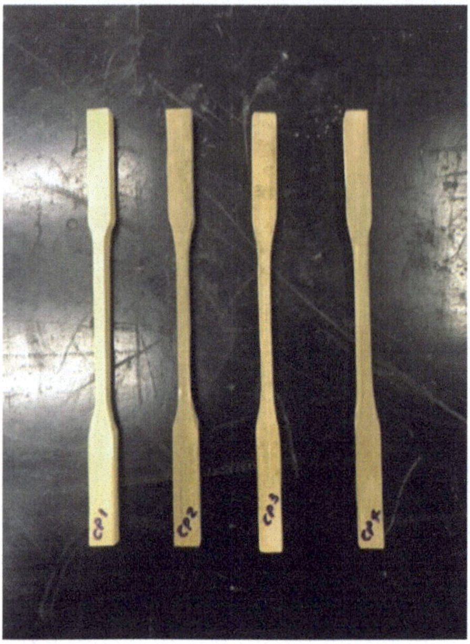

Fig. 8 Bamboo strip samples for the tensile strength tests for the *Phyllostachys pubescens*. *Source* Author's own figure

if fracture occurs away from the central region of the bamboo sample, that this result should not be analyzed. The reason for this fracture is usually shear stress.

In order to improve the adherence between the sample holder and the specimen a piece of sandpaper was inserted on each edge. This procedure aimed at minimizing sample slippage during tests, as shown in Fig. 9.

Width and thickness measurements were performed at the fracture zone of the specimen in order to determine the cross-sectional area before the tests were performed, exactly where it would occur.

The tensile strength tests were performed at a Universal Testing Machine with a load of 60 tf, as shown in Fig. 10. A total of 7 samples were analyzed: 3 from *Dendrocalamus aesper* and 4 from *Phyllostachys pubescens*. One sample from *Phyllostachys pubescens* species was used as an experimental run to specify test requirements for the other samples.

The tensile maximum load is found when there is a failure of the specimen. Figure 11 shows the failure for both bamboo species.

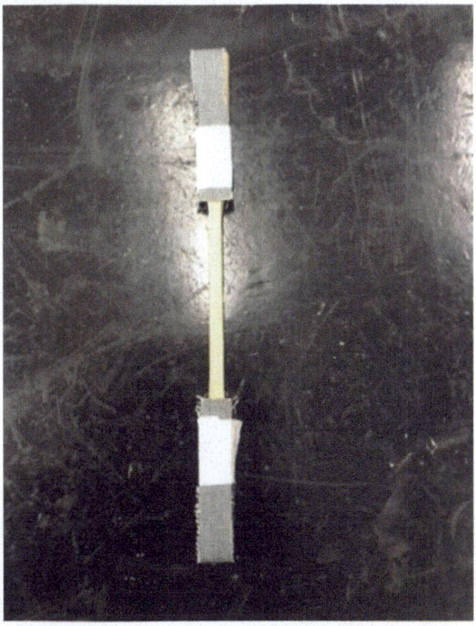

Fig. 9 Specimen for tensile strength tests. *Source* Author's own figure

Fig. 10 Universal Testing Machine used for tensile strength measurements. *Source* Author's own figure

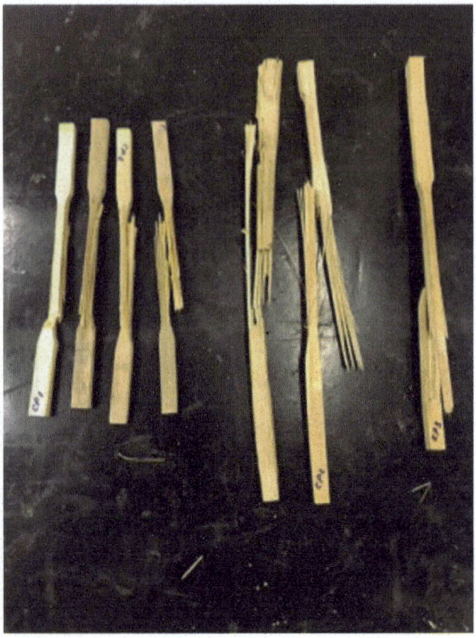

Fig. 11 Samples from both bamboo species: *Dendrocalamus aesper* and *Phyllostachys pubescens*. *Source* Author's own figure

3 Results and Discussion

Measured data was analyzed by using a statistical software (*Statistical Package for Social Sciences*—SPSS, IBM, EUA). Both descriptive (especially those measured) and inferential analyses were performed at a significance level of 5%.

The first step was to check normality from the results by means of the Shapiro Wilk statistical test. Most of the results did not show normal distribution. Both non-normal, non-parametric analyses were associated in order to compare the results from the bamboo species, as recommended by the Mann-Whitney test.

The data from the statistical analyses for compressive and tensile strength are summarized in Tables 1 and 2.

Table 1 Statistical data from the variables of compressive stress (tf) and compressive strength (MPa) from two bamboo species, 2018

Statistics	Species	Compressive stress	Area	Compressive strength
Medium	*Dendrocalamus aesper*	24.56	29.10	83.14
	Phyllostachys pubescens	15.20	15.31	97.67
Minimum	*Dendrocalamus aesper*	19.80	24.72	59.95
	Phyllostachys pubescens	13.60	13.30	88.62
Maximum	*Dendrocalamus aesper*	31.40	36.09	98.24
	Phyllostachys pubescens	18.10	17.49	124.43

Table 2 Mann-Whitney test for comparison of species, variables: length (mm), external diameter (cm), weight (g), thickness (mm), internal diameter (cm) compressive stress (tf), area (cm^2) e compressive strength (MPa) de duas espécies de bambu, 2018

Variable	Specie	Median	p-valor
Length	*Dendrocalamus aesper*	304.65	0.0001*
	Phyllostachys pubescens	237.93	
Extern Diameter	*Dendrocalamus aesper*	14.19	0.0001
	Phyllostachys pubescens	10.69	
Weight	*Dendrocalamus aesper*	1145.60	1.0000**
	Phyllostachys pubescens	1145.60	
Thickness	*Dendrocalamus aesper*	13.24	0.0001
	Phyllostachys pubescens	9.84	
Internal Diameter	*Dendrocalamus aesper*	12.90	0.0001
	Phyllostachys pubescens	9.69	
Compressive Stress	*Dendrocalamus aesper*	23.70	0.0001
	Phyllostachys pubescens	14.73	
Area	*Dendrocalamus aesper*	28.21	0.0001
	Phyllostachys pubescens	15.16	
Compressive Strength	*Dendrocalamus aesper*	83.3750	0.0111***
	Phyllostachys pubescens	94.8445	

*Significant a 1%; **Equivalent; ***Significant a 5%

3.1 Compressive Strength Tests

Statistical data from Table 1 were achieved from compressive strength tests for 20 samples, 10 from each species. The average, maximum and minimum compressive strength supported by each sample in the test was determined from a general data analysis for each species.

Tensile data, as described earlier, was measured from sample failure. The area was calculated by subtracting two area values: the external and internal areas of bamboo samples. The outer and inner diameters were measured by using a digital caliper. A circular area was considered for bamboo. The corresponding areas were determined and the final area was calculated by subtracting the two values. Then the average, minimum and maximum data were determined for each species. Compressive strength was then determined from the relationship between tension and area.

According to Table 1, both bamboo species were within the reported parameter for the uniaxial compressive strength [16]. Samples from the *Dendrocalamus aesper* species provided a compressive strength of 83.14 MPa and those from *Phyllostachys pubescens* species 97.67 MPa.

Table 1 shows that the *Dendrocalamus aesper* species is able to withstand a higher load than *Phyllostachys pubescens*, since it has a higher cross-sectional area deriving from its structure and diameter. Nevertheless, when comparing both species by means of the average statistics, it is noticed that the *Phyllostachys pubescens* has a better compressive strength compared to *Dendrocalamus aesper*, which is ca. 17.48% higher, despite having a smaller structure, i.e., a smaller cross section to withstand the load.

The average, maximum and minimum data from each species described in Table 2, were achieved by a general analysis from 6 samples, 3 from each species, starting from tensile strength tests for each sample after their failure. Then the cross-sectional area was calculated from the failure location. Data were measured by using a caliper before failure. Tensile strength was derived from the relationship between tension and area.

Although most of the variables used in the Shapiro-Wilk test could be considered normal, nonparametric tests were chosen because of the small number of samples.

For Table 2, most of the variables were significant at 1%, except the mass, between the two species studied. However, when the medians such as length, internal and external diameter, thickness, compression stress, and area and between the species *Dendrocalamus aesper* and *Phyllostachys pubescens* are different, it is observed that the one of the first species cited is superior to of the second. Excluding the mass, which has a similar behavior between both species, and the exception appears in the variable resistance to compression, where the one of the species *Phyllostachys pubescens* is larger than the one of *Dendrocalamus aesper*. Therefore, it was concluded that in the majority of the measures evaluated the species *Dendrocalamus aesper* had higher values.

3.2 Tensile Strength

Table 3 shows data from tensile tests for parallel fibres from the bamboo sample. The tensile strength from the bamboo samples, in general, ranges from 40 to 215 MPa, with or without the presence of nodes [16]. Therefore, the average of tensile strength for both bamboo sample species was within the required parameter: 180.71 Mpa for *Dendrocalamus aesper* species and 121.88 MPa for the *Phyllostachys pubescens* species.

The analysis of Table 3 shows that the tensile strength for *Dendrocalamus aesper* species is higher than for *Phyllostachys pubescens*, i.e., 87.46% higher, which is able to withstand a higher load.

When comparing both species for their tensile strength, the *Dendrocalamus aesper* species had a better performance than *Phyllostachys pubescens*, since it provided a ca. two-fold increase for this parameter.

Both bamboo species chosen in this work provided suitable mechanical properties, which serve as a motivation for the use of bamboo as a substitute for steel [16], as well as a driving force for its use in association with other building materials such as concrete, in order to increase its mechanical resistance.

Compressive strength data was, on average, three times lower than tensile strength [14]. Then it was concluded that both *Dendrocalamus aesper* and *Phyllostachys pubescens* species have a higher tensile strength than compressive strength.

Here, also, non-parametric tests were chosen because of the small number of samples. although most of the variants used in the Shapiro-Wilk test can be considered normal. According to Table 4, although no test is significant at 5%, since a small sample has been used and non-parametric tests have been used, it can be considered that at 10% there is a significant difference between the medians for the measured variables c, height, tensile stress, area and tensile strength. Another important observation in Table 4 is that for all the significant measures the species *Dendrocalamus asper* has higher values than the species *Phyllostachys pubescens*.

Table 3 Statistical data from variables of tensile stress (tf) and tensile strength (MPa) from two bamboo species, 2018

Statistics	Species	Tensile stress	Area	Tensile strength
Medium	*Dendrocalamus aesper*	1256.67	0.68	180.71
	Phyllostachys pubescens	670.00	0.50	131.88
Minimum	*Dendrocalamus aesper*	990.00	0.64	151.22
	Phyllostachys pubescens	640.00	0.48	125.88
Maximum	*Dendrocalamus aesper*	1420.00	0.71	202.52
	Phyllostachys pubescens	720.00	0.52	136.44

Table 4 Mann-Whitney test for comparison of variable species: height (mm), mesure a (mm), mesure b (mm), mesure c (mm), tensile stress (tf), área (cm^2) e tensile strength (MPa) of two species of bamboo, 2018

Variable	Specie	Median	p-valor
Mesure a and b	*Dendrocalamus aesper*	11.47	0.2291
	Phyllostachys pubescens	11.87	
Mesure d and e	*Dendrocalamus aesper*	6.12	0.6291
	Phyllostachys pubescens	5.96	
Mesure c	*Dendrocalamus aesper*	5.73	0.0571*
	Phyllostachys pubescens	5.53	
Height	*Dendrocalamus aesper*	299.10	0.0571
	Phyllostachys pubescens	200.61	
Tensile Stress	*Dendrocalamus aesper*	1360.00	0.0571
	Phyllostachys pubescens	660.00	
Area	*Dendrocalamus aesper*	0.69	0.0571
	Phyllostachys pubescens	0.50	
Tensile Strength	*Dendrocalamus aesper*	188.38	0.0571
	Phyllostachys pubescens	132.60	

*Significant a 10%

4 Conclusion

The supply chain of the construction industry generates an enormous amount of pollutants to the environment. The main reason is associated to the chosen materials, which might be rather expensive in most cases. Therefore, thinking about sustainable and low-cost materials means to search for alternatives to comply with global environmental requirements, population growth and housing shortage.

Bamboo is a versatile material, which may be found in most of the Brazilian territory. However, there is some prejudice against its usage in Brazil since it is usually associated with poor houses, whereas in some other countries, especially in Asia, its use is encouraged since it is a highly resistant material.

The world has been changing and it is open to new cultures. In Brazil, the scenario is not different. Nevertheless, there is a lack of specific technical criteria for Engineers and Architects when they want to apply bamboo in their projects. The Brazilian standard for this material is yet under discussion and research studies could contribute for broadening its applications, corroborating its relatively high compressive and tensile strength, as well as in destroying the prejudice of "poor" housing. High complex structures endowed with strong aesthetics may be created by using this material.

The data from mechanical tests from *Dendrocalamus aesper* and *Phyllostachys pubescens* species showed that bamboo is a viable raw material for application in small and big construction projects.

According to the statistical studies carried out in the research, it was observed that the species *Dendrocalamus aesper* has characteristics superior to that of the species *Phyllostachys pubescens*. Suggesting that the geometry of *Dendrocalamus aesper* is best for use in construction.

Therefore, since bamboo is a sustainable material with large scale production, a low-cost and eco-friendly material, its use in housing projects can decrease the generation of polluting residues. Since it is a renewable resource, has a broad range of applications and is affordable to most social classes, it is considered a great material candidate for use in the housing industry.

References

1. Rech, F., et al.: Use of Bamboo in structures of popular houses. In: 4th National Seminar on Sustainable Constructions, pp 1–8. 1st Forum Performance of Buildings, Passo Fundo, RS (2015)
2. Meirelles, C.R.M., Osse, V.C.: Use of Bamboo in Architecture: The Issues of Environmental Comfort and Structure (2010)
3. Dantas, T., et al.: Bamboo—Historys of Japan, p. 180 (2017)
4. da Silva, C.G.T.: Conception and Preconceptions Concerning Constructions in Crude Earth, p. 155. Dissertation (Master's Degree in Public)—National School of Public Health, Oswaldo Cruz Foundation, Subarea Sanitation and Environmental Health, Rio de Janeiro, RJ (2000)
5. Ribas, R.P.: Bamboo: Potential Plant of Sustainable Development (2008)
6. Teixeira, A.A.: Bamboo Panels for Economic Housing: Performance Evaluations of Mortar Coated Panels, p. 204 Dissertation (Master's Degree in Architecture e Urbanism)—Architecture and Urbanism College, University of Brasília, Brasília, DF (2006)
7. Neto, J.S.P., et al.: Application of Bamboo in rural constructions. Mag. Higher Agric. Educ., 67–77 (2009)
8. Ghavami, K.: Bambo … an alternative material in engineering, (1992). In: Engineering, n. 491, pp. 23–27. In: Costa, A.M.: Thermal Treatment of *Dendrocalamus aesper* for Laminated Stainless Steel Bamboo (BLC), p. 37. Department of Forestry Engineering, University of Brasilia, Brasilia, DF (2015)
9. Padovan, R.B.: The Bamboo in Architecture: Design of Structural Connections, p. 184 Dissertation (Master's Degree in Design)—College of Architecture, Arts and Communication, University State of São Paulo "Julio de Mesquita Filho", Bauru, SP (2010)
10. Marçal, V.H.S.: Use of Bamboo in Civil Construction, p. 60. Department of Civil and Environmental Engineering, University of Brasilia, Brasilia, DF (2008)
11. de Oliveira, T.F.C.S.: Sustainability and Architecture: A Reflection the Use of Bamboo in Civil Construction, p. 136 (Master's Degree in Dynamics of Inhabited Space)—Technology Center, Federal University of Alagoas, Maceió, AL (2006)
12. Paes, J.B., et al.: Physical-Mechanical characterization of the Glue Laminated Bamboo (*Dendrocalamus giganteus*), vol. 19, n. 1, pp. 41–51 (2009)
13. Debarba, A.L., et al.: Use of bamboo in interior decoration. Mag. Infinity **1**(1), 1–17 (2016)
14. Ghavami, K., Marinho, A.B.: Physical and mechanical properties of the whole culm of bamboo of the *Guadua angustifolia* species. Mag. Braz. Agric. Environ. Eng. **9**(1), 107–114 (2005)
15. Carbonari, G., et al.: Bamboo—the vegetal stell. Mix Sustentável **3**(1), 17–25 (2017)

16. Azzini, A., et al.: Bamboo: characteristics and applications. In: Freire, W.J., Beraldo, A.L. (eds.) Constructions Technologies and Alternative Materials, 4nd edn., Ch. 9, pp. 253–330. Unicamp, Campinas, SP (2017)

Fingerspelling Recognition Using Histogram of Oriented Point Cloud Vectors from Depth Data

José Elías Yauri Vidalón and José Mario De Martino

Abstract The high degree of freedom of hand movements produces a high variability of shapes and hand appearances that still challenges hand gesture recognition algorithms. This paper presents an approach to recognize sign language fingerspelling. Our approach, named histogram of oriented point cloud vectors (HOPC), is based on a new descriptor computed only from depth images. The segmented depth image is mapped into a 3D point cloud and divided into subspaces. In each subspace, 3D point vectors are mapped into their spherical coordinates around its centroid. Next, it is computed their orientations angles H_φ and H_θ onto two cumulative histograms. Normalized histograms are concatenated to form the image descriptor and used to train a Support Vector Machine classifier (SVM). To assess the feasibility of our approach, we evaluated it on a public data-set of American Sign Language (ASL) composed of more than 60,000 images. Our experiments showed a recognition accuracy average of 99.46%, achieving the state of the art.

Keywords Sign language recognition · Fingerspelling recognition · Handshape recognition

1 Introduction

Sign language is a visual-spatial language, which is used by deaf people as a natural communication system [17]. Instead of using sound patterns, sign language uses gestures to convey meaning. A gesture can be manual or non-manual, and has a conventional meaning to the deaf community. A manual gesture is performed by the

J. E. Yauri Vidalón (✉)
School of Technological Sciences and Engineering,
University of Ayacucho Federico Froebel, Ayacucho, Peru
e-mail: elias@dca.fee.unicamp.br

J. E. Yauri Vidalón · J. M. De Martino
School of Electrical and Computer Engineering,
University of Campinas, Campinas, SP, Brazil
e-mail: martino@dca.fee.unicamp.br

Y. Iano et al. (eds.), *Proceedings of the 4th Brazilian Technology Symposium (BTSym'18)*, Smart Innovation, Systems and Technologies 140,
https://doi.org/10.1007/978-3-030-16053-1_55

arms, hands and fingers, while a non-manual gesture is carried out by movements of the body, head and facial expressions.

Sign language has a rich lexicon to express concepts, ideas, feelings, etc. Thus, for spelling out names for which there are no signs, e.g., names of people, countries, places, movies, sign language uses fingerspelling. To fingerspell a word, the signer spells out the letters of the word by using its respective representation in the manual sign language alphabet. For example, the word Montreal should be fingerspelled by the sequence of signs M + O + N + T + R + E + A + L. The importance of fingerspelling lies on the regular presence of fingerspelled words during communication between deaf people. According to Padden [12], fingerspelling represents almost 18% of a typical dialog. Figure 1 illustrates the American Sign Language (ASL) alphabet which consists of 26 handshapes, two of which have motion.

Since deaf people interact daily with hearing people, misunderstandings may arise because deaf people usually do not learn to speak given their hearing limitations; moreover, hearing people usually do not learn sign language. Aiming to improve the quality of life of deaf people, sign language recognition (SLR) has gained increased attention of the research community over the past decades [1, 6, 16, 23]. Nevertheless, despite the advances, automatic recognition of sign language gestures remains a yet unsolved problem, being a reason for much effort in the computer vision community.

This work presents an approach to recognize fingerspelling of the American sign language alphabet. Inspired by recent studies showing that depth images lead to more promising results than traditional RGB image approaches [18, 22], the new approach is based on a new descriptor named Histogram of Oriented Point Cloud Vectors (HOPC) computed only from depth images. In our approach, assuming that the hand is the most foreground object in the image, the hand region is first cropped

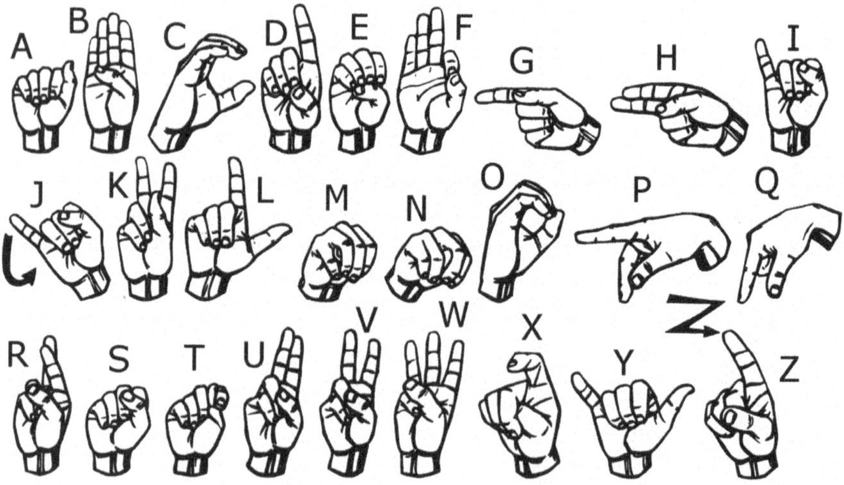

Fig. 1 American sign language alphabet. Adapted from Peach [13]

and segmented. Next, to be robust against rotation variances, the hand is aligned vertically with the Y vertical axis using the major axis of the hand region as reference.

After, the hand region is mapped into a 3D point cloud and divided into subspaces. In each subspace, 3D points are converted to vectors around the centroid of the cloud and projected to their spherical coordinates. Then, two cumulative histograms H_φ and H_θ are computed with each point contributing to a histogram bin according to its orientation angles in the 3D spherical coordinate system. Both histograms are concatenated and normalized. The final descriptor consists of the concatenation of all the subspace histograms in which the image was divided. For recognition, we use a Support Vector Machine (SVM) classifier. Our proposal was assessed on a public available data-set of 24 signs—each sign has 500 samples, and altogether 60,000 depth images—of the ASL alphabet [14]. The remainder of this paper is organized as follows: Sect. 2 provides a brief overview of related work; Sect. 3 details our proposal; Sect. 4 presents the experimental results; and finally, Sect. 5 outlines the conclusions and future work.

2 Related Work

2.1 Depth Sensing

Recent advances in depth sensing technologies have made possible consumer sensors at low cost with capabilities to capture depth images in real time. The depth sensor improvements have motivated real-world applications beyond traditional entertainment, and have boosted research on gesture recognition for human computer interactions and well as sign language recognition [3, 22].

Although depth data generated by the current low cost sensors still have low resolution and are noisy, they have the advantage of being robust against illumination and signer appearance (e.g., lighting conditions, skin color and signer's clothing). Therefore, methods have been developed to segment objects, background subtraction and to calculate image features directly from depth images [18].

2.2 Hand Gesture Recognition Methods

Many approaches in gesture recognition extract features from RGB images. For instance, Isaacs and Foo [5] recognized 24 static gestures of the ASL alphabet by using wavelet features from edge detected images and Artificial Neural Network (ANN), achieving a recognition rate of 99.90%. In this work, the user's hand is the only foreground object against a color flat background.

Otiniano-Rodríguez et al. [11] proposed the use of invariant geometric moment features to train a Support Vector Machine (SVM) classifier in order to recognize

24 static gestures of the ASL alphabet, achieving a recognition rate of 96%. In this work, the user's hand is the only object in the controlled scene, presenting only slight scale variations.

Approaches that use RGB cameras are strongly dependent on illumination and background conditions and require setup conditions and assumptions to reduce the complexity of the hand location and segmentation, which ultimately restrict their application to real life gesture recognition systems.

Recently, Pugeault and Bowden [14] used both RGB and depth images to recognize the 24 static gestures of the ASL alphabet. They used Gabor filters to extract features from both images and trained with a Random Forest (RF) classifier. In the case of using half of the data for testing and half for training, they reported recognition rates of 73, 69, and 75% using only color features, depth features, and combined features, respectively. They also presented a public dataset where five subjects perform the gesture while moving the hand in front of a Kinect camera.

Xiaolong and Wong [21] used both RGB and depth images to recognize the 24 static gestures of ASL [14]. They used Scale Invariant Feature Transform (SIFT) on color images and Local Binary Pattern (LBP) on depth images, then fused these features by means of a kernel descriptors, and finally, constructed a bag-of-visual-words model which was trained using SVM. They achieved a recognition rate of 88.94%. SIFT points vary widely between frames due to lighting changes, and not addressing the LPB scale variations reduces performance.

Keskin et al. [7] presented a Shape Classification Forest aiming to recognize hand poses from depth images. They achieved a recognition rate of 97.8% in the dataset [14]. Hand poses are 3D models which are matched with depth shapes during classification. Since different 3D hand configurations can produce the same 2D appearance, it has a high computational cost for training and testing.

Recently, Rioux-Maldague and Giguere [15] explored the Deep Belief Networks to recognize the static signs of the ASL alphabet [14], achieving a rate of 99.00%. Also, Li et al. [9] combined a feature learning approach based on Sparse Auto-Encoder (SAE) with Convolutional Neural Network (CNN), achieving a rate of 99.05% in the dataset [14]. These approaches use both color and depth images, and due to the nature of the feature learning approach presented in [9, 15], they are high dimensional and computationally expensive when compared to our proposal.

Our proposal aims to recognize static hand gestures of the ASL alphabet by using only depth images classified using SVM. The proposed descriptor, histogram of oriented point cloud vectors (HOPC), is less complex than other methods, whilst preserving the spatial and geometric information of the hand that allow to discriminate between hand shapes of similar 2D appearances but with different 3D configuration.

It is worth mentioning that a similar descriptor was proposed by Tang et al. [19], but the source and meaning of the vectors are different. They computed normal vectors from the gradients of the depth image in the spatial domain. In that domain, the image only contains the relative distance to the sensor. Unlike them, working in the point cloud space, our approach takes into account the true 3D real world structure of the object in the scene, which is a great advantage for modern computer vision applications.

3 Our Proposal

This section describes our approach to recognize hand gestures of the ASL alphabet (Fig. 1). The proposal is shown in Fig. 2, which consists of three stages: depth image processing, feature extraction, and classification.

3.1 Dataset

We used the public dataset of ASL presented by Pugeault and Bowden [14]. The dataset was recorded with a Kinect v1 and comprises 24 static signs of the ASL alphabet (excluding letters J and Z because they involve motion). It contains 500 samples per sign performed by five signers, and altogether 60,000 images. A illustrations of samples from the dataset is shown in Fig. 3.

Since signer changes freely their hand location while recording, the shape of the handshape changes and even may be overlapped with body parts. To add complexity, some signs of ASL alphabet share similarities. For example, signs for A, E, M, N, S, and T have almost the same handshape, which can lead to misrecognition. Figure 4 presents some instances of such similarities between signs.

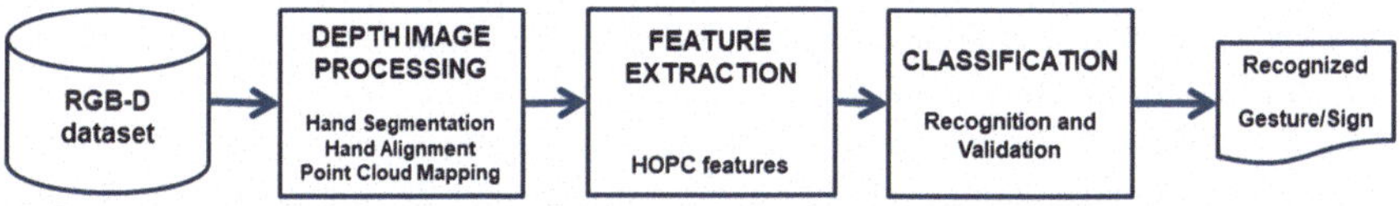

Fig. 2 The proposed recognition system

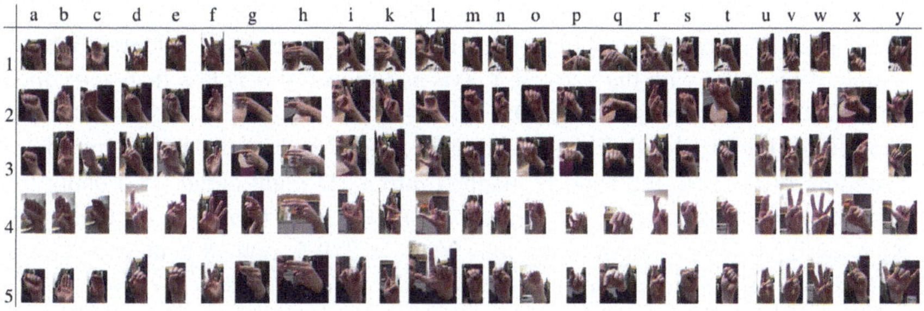

Fig. 3 Illustration of samples from the dataset [14]

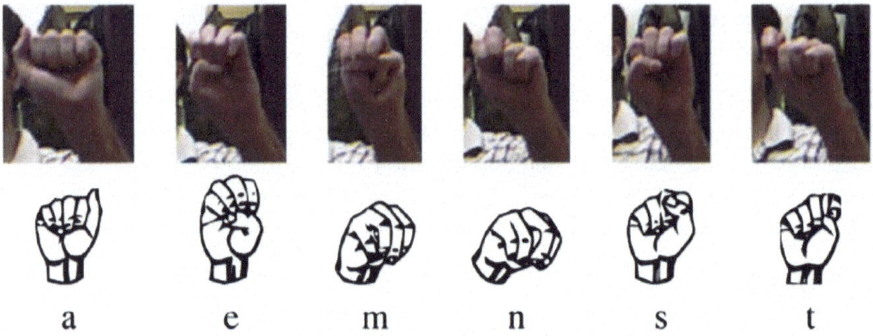

Fig. 4 Illustration of similarities between signs in the ASL alphabet [14]. Signs are configured by a fist with a subtle change of the thumb finger

3.2 *Image Processing*

This stage aims to segment the hand from the depth image, to align the hand, and to map it into a 3D point cloud using the intrinsic parameters of the Kinect sensor.

Being $depth = I(x, y)$ the depth image of the hand, we proceed as follows:

- **Hand segmentation**: Assuming that the hand is the most foreground object in *depth*, we segment the hand region using a depth threshold. Pixels with depth greater than the threshold are zeroed. The obtained depth mask is used to extract the pixels of the hand.
- **Hand alignment**: To be robust against rotation changes, we compute the angle α between the horizontal X axis and the major axis of the segmented hand—the major axis is computed from the statistical second moment property of the depth mask [4].

  Next, a rotation is performed in a counterclockwise direction in order to align the region of the hand with the vertical Y axis as follows: if α is less than zero, then rotate $270 - \alpha$, otherwise rotate $90 - \alpha$.
- **Mapping to point cloud**: The aligned hand shape is cropped using its minimum bounding box. Then, the hand is mapped into a 3D point cloud (PC) using the intrinsic camera parameters of Kinect v1: $fx = 525.0$, focal length along axis X; $fy = 525.0$, focal length along axis Y; $cx = 319.5$, optical center in X; and $cy = 239.5$, optical center in Y.

$$PC(x, y, z) = mappingToPointCloud(depth(x, y))$$

Figure 5 illustrates the main steps of the image processing to extract the shape of the hand.

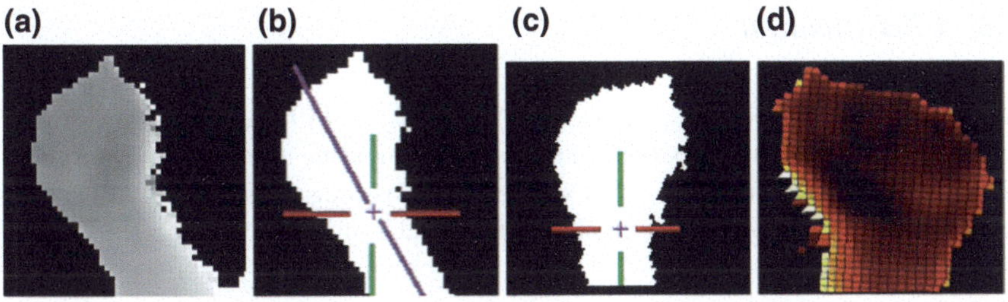

Fig. 5 Depth image processing steps in **a** depth image, **b** depth mask of the hand and its major axis, **c** aligned depth image , and **d** depth image mapped into 3D point cloud

3.3 Feature Extraction

To extract feature descriptors, we proceed as follows:

- Along the viewing direction of the camera, divide the 3D point cloud into subspaces and compute for each one their feature descriptors. For instance, the point cloud is subdivided into $k \times k$ subspaces S_i.

  Each subspace is processed as follows:

  - Find the central point CP of the subspace, $C_{S_i} = (x_c, y_c, z_c)$.
  - Generate a vector v_i for each point p_i of the subspace as $v_i = p_i - C_{S_i}$
  - Map vectors $v_i = (x, y, z)$ to spherical coordinates $v_i = (\varphi, \theta, r)$ using Eq. 1, where φ is the azimuthal angle formed with the X axis, θ is the polar angle formed with the Z axis, and r is the radial. Notice that $\varphi \in [0, \pi]$, $\theta \in [0, 2\pi)$, and $r \in [0, \infty)$.

$$\begin{aligned} \varphi &= \tan^{-1} \frac{y}{x} \\ \theta &= \cos^{-1} \frac{z}{r} \\ r &= \sqrt{x^2 + y^2 + z^2} \end{aligned} \tag{1}$$

  - Compute the cumulative histogram of orientation angles of each vector v_i. We proceed in an analogous way as Histogram of Oriented Gradients (HOG) presented in [2]. Each vector contributes to the bins of the two histograms H_φ and H_θ according to their orientation angles φ and θ. Histograms comprise [0, 180] degrees subdivided into nine bins. The angle θ is treated as an unsigned angle.
  - The subspace histograms are concatenated and normalized between [0, 1]

- The final feature descriptor consists of the concatenation of all histograms from all subspaces in which the point cloud was subdivided.
- The feature descriptor dimension is $D_{dim} = k \times k$ *subspaces* $\times$ 2 *histograms* $\times$ 9 *bins*.

3.4 *Classification*

To perform recognition tasks, we use an SVM classifier [10]. A support vector machine classifier performs supervised learning by computing a separating hyperplane between classes. We use *libsvm* to execute multiclass classification, and it is setup in one-against-one mode to operate in multiclass classification.

4 Results

To assess our approach, experiments were conducted with training and testing sets with different partition proportion, namely 80–20%, 60–40% and 50–50%. Additionally, the training set was five-fold cross validated.

For the depth image processing stage, we use a threshold of $TH = 15$ cm. In the feature extraction step, the point cloud was divided into 7×7 subspaces along the viewing direction of the camera. For classification, the SVM was configured using the radial basis function, with cost $C = 100$ and gamma $g = 0.1$. Each experiment was repeated ten times.

Table 1 presents the achieved recognition accuracy in the ASL dataset. This database has balanced data for each class [14].

Table 2 presents the confusion matrix of the classification of 24 signs of the ASL alphabet under the 80–20% partition scheme.

Because the high degree of freedom of hand movement while the gesture is performed, some alphabet signs have similar appearances due to the occlusion of the fingers and changes in the viewing direction. Therefore, it was expected that some signs were misclassified. For instance, M and N, and S and T, and P and Q, and R, U and V have been little confused. Other signs were well classified.

For further validation of our proposal, Table 3 compares our approach against other techniques under the 50–50% partition dataset scheme. Our proposal achieved 99.17% and outperforms other approaches [8, 14, 15, 20]; even, it performs better than [9], the state of the art method.

Approaches presented in [9, 15] are heavily dependent on the feature learning method applied. They are complex, time consuming and demand high performance computing (e.g., Convolution Neural Network (CNN) has high computational costs and has to setup many hyper-parameters) to produce accurate results. On the other

Table 1 Recognition accuracies in the ASL data-set using 7×7 subspaces in the point cloud

Dataset partition scheme (%)	Accuracy (%)	Standard deviation
80–20	99.46	0.0012
60–40	99.33	0.0010
50–50	99.17	0.0006

Table 2 Confusion matrix of the classification of 24 signs of the ASL alphabet (80–20% partition scheme)

	A	B	C	D	E	F	G	H	I	K	L	M	N	O	P	Q	R	S	T	U	V	W	X	Y
A	100	0	0	0	0	0	0	0	0	0	0	0	0	0	0	0	0	0	0	0	0	0	0	0
B	0	100	0	0	0	0	0	0	0	0	0	0	0	0	0	0	0	0	0	0	0	0	0	0
C	0	0	99.78	0	0	0	0	0	0	0	0	0	0	0	0.22	0	0	0	0	0	0	0	0	0
D	0	0	0	100	0	0	0	0	0	0	0	0	0	0	0	0	0	0	0	0	0	0	0	0
E	0	0	0	0	99.78	0	0	0	0	0	0	0	0	0	0	0	0	0.22	0	0	0	0	0	0
F	0	0	0	0	0	100	0	0	0	0	0	0	0	0	0	0	0	0	0	0	0	0	0	0
G	0	0	0	0	0	0	100	0	0	0	0	0	0	0	0	0	0	0	0	0	0	0	0	0
H	0	0	0	0	0	0	0.22	99.78	0	0	0	0	0	0	0	0	0	0	0	0	0	0	0	0
I	0	0	0	0	0	0	0	0	100	0	0	0	0	0	0	0	0	0	0	0	0	0	0	0
K	0	0	0	0.22	0	0	0	0	0	99.56	0	0	0	0	0.22	0	0	0	0	0	0	0	0	0
L	0	0	0	0	0	0	0	0	0	0	99.78	0	0	0	0.22	0	0	0	0	0	0	0	0	0
M	0	0	0	0	0	0	0	0	0	0	0	100	0	0	0	0	0	0	0	0	0	0	0	0
N	0.22	0	0	0	0	0	0	0	0	0	0	0.22	98.89	0	0	0.22	0	0	0.44	0	0	0	0	0
O	0	0	0	0	0	0	0	0	0	0	0	0	0	99.78	0.22	0	0	0	0	0	0	0	0	0
P	0	0	0	0	0	0	0	0	0	0	0	0	0	0	98.89	1.11	0	0	0	0	0	0	0	0
Q	0	0.22	0	0	0	0	0.22	0	0	0	0	0	0	0	1.33	98.22	0	0	0	0	0	0	0	0
R	0	0	0	0	0	0	0	0	0	0	0	0	0	0	0	0	100	0	0	0	0	0	0	0
S	0	0	0	0	0.44	0	0	0	0	0	0	0.22	0	0	0	0	0	98.89	0.44	0	0	0	0	0
T	0	0	0	0	0	0	0	0	0	0	0	0	0	0	0.22	0.22	0	0.22	99.33	0	0	0	0	0
U	0	0	0	0	0	0	0	0	0	0	0	0	0	0	0	0	0.44	0	0	99.56	0	0	0	0
V	0	0	0	0	0	0	0	0	0	0	0	0	0	0	0	0	0.22	0	0	0.22	99.56	0	0	0
W	0	0	0	0	0	0	0	0	0	0	0	0	0	0	0	0	0	0	0	0	0	100	0	0
X	0	0	0	0	0	0	0	0	0	0	0	0	0	0	0	0	0	0	0	0	0	0	99.78	0.22
Y	0	0	0	0	0	0	0.22	0	0	0	0	0	0	0	0	0	0	0	0	0	0	0	0	99.78

Table 3 Recognition accuracies in the ASL alphabet dataset [14]

Method	Accuracy (%)
Pugeault and Bowden [14]	75.00
Kuznetsova et al. [8]	87.00
Weerasekera et al. [20]	92.14
Rioux-Maldague and Giguere [15]	99.00
Li et al. [9]	99.05
Proposed method	**99.17**

hand, our approach is simple, fast, and based on human understandable features computed from a single depth image. The proposed histogram of oriented point cloud vectors (HOPC) computed from depth values encodes both the shape and the appearance of the object from the 3D real world, and also is tolerant to small viewing direction and scale changes, which is useful for real-time sign language recognizers.

5 Conclusions and Future Work

This paper presented an approach to recognize figerspelling of sign language alphabet based on a new descriptor named histogram of oriented point cloud vectors, which captures both the shape and the appearance of hand gestures from only depth images. We validated the proposal on a publicly available data-set, achieving a recognition rate of 99.46% in more than 60,000 real depth images.

As for future work, we intend to extend these features to recognize dynamic gestures from sign language and test them both in real life scenarios and in large datasets.

References

1. Chanu, O.R., Pillai, A., Sinha, S., Das, P.: Comparative study for vision based and data based hand gesture recognition technique. In: 2017 International Conference on Intelligent Communication and Computational Techniques (ICCT), pp. 26–31. IEEE, Dec 2017
2. Dalal, N., Triggs, B.: Histograms of oriented gradients for human detection. In: IEEE Computer Society Conference on Computer Vision and Pattern Recognition, 2005. CVPR 2005, vol. 1, pp. 886–893 (2005)
3. Fossati, A., Gall, J., Grabner, H., Ren, X., Konolige, K.: Consumer Depth Cameras for Computer Vision: Research Topics and Applications. Springer Publishing Company, Incorporated (2013)
4. Gonzalez, R.C., Woods, R.E.: Digital Image Processing. Prentice-Hall, Inc. (2008)
5. Isaacs, J., Foo, S.: Hand pose estimation for American sign language recognition. In: Proceedings of the Thirty-Sixth Southeastern Symposium on System Theory, 2004, pp. 132–136. IEEE (2004)
6. Joudaki, S., Mohamad, D.B., Saba, T., Rehman, A., Al-Rodhaan, M., Al-Dhelaan, A.: Vision-based sign language classification: a directional review. IETE Tech Rev **31**(5), 383–391 (2014)

7. Keskin, C., Kirac, F., Kara, Y.E., Akarun, L.: Hand pose estimation and hand shape classification using multi-layered randomized decision forests. In: Proceedings of the 12th European Conference on Computer Vision—Volume Part VI. ECCV'12, pp. 852–863. Springer, Berlin, Heidelberg (2012)
8. Kuznetsova, A., Leal-Taixe, L., Rosenhahn, B.: Real-time sign language recognition using a consumer depth camera. In: 2013 IEEE International Conference on Computer Vision Workshops, pp. 83–90. IEEE, Dec 2013
9. Li, S.Z., Yu, B., Wu, W., Su, S.Z., Ji, R.R.: Feature learning based on SAE-PCA network for human gesture recognition in RGBD images. Neurocomputing **151**, 565–573 (2015)
10. Murphy, K.P.: Machine Learning: A Probabilistic Perspective. The MIT Press (2012)
11. Otiniano-Rodríguez, K.C., Cámara-Chávez, G., Menotti, D.: Hu and Zernike moments for sign language recognition. In: Proceedings of the 2012 International Conference on Image Processing, Computer Vision, and Pattern Recognition. IPCV 2012, vol. 2, pp. 918–922 (2012)
12. Padden, C.A.: Learning fingerspelling twice: young signing children's acquisition of fingerspelling. In: Schick, B., Marschark, M., Spencer, P.E. (eds.) Advances in the Sign Language Development of Deaf Children, pp. 189–201. Oxford University Press (2005)
13. Peach, D.: David Peach's LearnSigns.com. http://www.learnsigns.com/sign-language-alphabet-asl/ (2018). Accessed on 24 June 2018
14. Pugeault, N., Bowden, R.: Spelling it out: real-time ASL fingerspelling recognition. In: 2011 IEEE International Conference on Computer Vision Workshops (ICCV Workshops), pp. 1114–1119, Nov 2011
15. Rioux-Maldague, L., Giguere, P.: Sign language fingerspelling classification from depth and color images using a deep belief network. In: 2014 Canadian Conference on Computer and Robot Vision, pp. 92–97. IEEE, May 2014
16. Sahoo, A.K., Mishra, G.S., Ravulakollu, K.K.: Sign language recognition: state of the art. ARPN J. Eng. Appl. Sci. **9**(2), 116–134 (2014)
17. Sandler, W., Lillo-Martin, D.: Sign Language and Linguistic Universals. Cambridge University Press (2006)
18. Suarez, J., Murphy, R.: Hand gesture recognition with depth images: a review. In: 2012 IEEE RO-MAN. pp. 411–417 (2012)
19. Tang, S., Wang, X., Lv, X., Han, T.X., Keller, J., He, Z., Skubic, M., Lao, S.: Histogram of oriented normal vectors for object recognition with a depth sensor. In: Proceedings of the 11th Asian Conference on Computer Vision—Volume Part II, pp. 525–538. Springer (2013)
20. Weerasekera, C.S., Jaward, M.H., Kamrani, N.: Robust ASL fingerspelling recognition using local binary patterns and geometric features. In: 2013 International Conference on Digital Image Computing: Techniques and Applications (DICTA), pp. 1–8. IEEE, Nov 2013
21. Xiaolong, Z., Wong, K.: Single-frame hand gesture recognition using color and depth kernel descriptors. In: 2012 21st International Conference on Pattern Recognition (ICPR), pp. 2989–2992, Nov 2012
22. Zanuttigh, P., Marin, G., Dal Mutto, C., Dominio, F., Minto, L., Cortelazzo, G.M.: Time-of-Flight and Structured Light Depth Cameras. Springer International Publishing (2016)
23. Zheng, L., Liang, B., Jiang, A.: Recent advances of deep learning for sign language recognition. In: 2017 International Conference on Digital Image Computing: Techniques and Applications (DICTA), pp. 1–7. IEEE, Nov 2017

Methodology for Reducing Staff Turnover in Service Companies Based on Employer Branding and Talent Management

Amy Espinoza, Estefania Rojas, Jose Rojas and Carlos Raymundo

Abstract The turnover rate of personnel in companies in the service sector is approximately 10%. In Peru, this figure is above 15%, which generates high cost overruns for organizations. Companies with the highest turnover are the service companies, and the operational staff is regarded the key factor of these organizations. Previous research has been carried out on this issue, and the solutions are models with the objective of retaining employees, using Employer Branding, Endomarketing, and Talent Management techniques. However, for this study, a model based on the combination of these techniques was designed and applied. Thus, it is intended that the model managed to meet the concerns raised and to reduce costs in companies. The model was applied in a company in the fast food sector, aiming the frontline/operational personnel of the organization. The results showed that an agile recruitment process and development of personnel training significantly reduces high rates of personnel turnover and the costs presented by this. Additionally, it is proven that the leadership of store managers plays an important role in increasing employee satisfaction and commitment within the organization.

Keywords Employer Branding · Staff turnover · Talent Management · Productivity · Fast food

A. Espinoza (✉) · E. Rojas · J. Rojas
Escuela de Ingeniería de Gestión Empresarial, Universidad Peruana de Ciencias Aplicadas (UPC), Lima, Perú
e-mail: U201310458@upc.edu.pe

E. Rojas
e-mail: U201317285@upc.edu.pe

J. Rojas
e-mail: jose.rojas@upc.pe

C. Raymundo
Dirección de Investigación, Universidad Peruana de Ciencas Aplicadas (UPC), Lima, Perú
e-mail: Carlos.raymundo@upc.edu.pe

Y. Iano et al. (eds.), *Proceedings of the 4th Brazilian Technology Symposium (BTSym'18)*, Smart Innovation, Systems and Technologies 140,
https://doi.org/10.1007/978-3-030-16053-1_56

1 Introduction

As global economies grow, organizations face a constant war for talent. This growth opens up new opportunities for personnel, and thus, new challenges for companies in attracting top talent. Voluntary turnover occurs when employees join and leave a company in a certain period [1].

High turnover rates are detrimental to organizations because they generate cost overruns, such as recruitment and training costs, payment of overtime to fill vacancies, low employee morale, time taken to hire new employees, and training time required, depending on the job specialization. Moreover, high turnover affects employee productivity until the new employees become fully competent in the job. Furthermore, due to operational disruptions triggered by this issue, the time and talent that should be spent on ongoing improvement of the process is used to combat high turnover rates.

Latin America is not immune to this problem. According to a study conducted by APERHU (Peruvian Association of Human Resources), Peru is the third country with the highest turnover rate, standing at an average 18%, compared to rates ranging 5–10% in the remaining countries [2]. Moreover, overruns generated by turnover account for about 43% in any type of company, according to Teresa Morales, a talent management specialist [3].

Studies conducted by the Ministry of Labor and Promotion of Employment (MINTRA, 2017) report a total of 10,725 employees joining and leaving companies, out of which 62% belong to the service industry. This industry comprises service provisions, restaurants, hotels, financial institutions, education, social services, gas, and water. To this extent, the restaurant sector, subdivided into the fast food sector, is constantly growing in Peru; in 2016 alone, it went up by 53% and reached 45,582 stores by the end of the year.

This study aims to reduce employee turnover and align employees with company goals.

2 State of the Art

2.1 Endomarketing

There are several models to reduce high turnover rates across companies. Based on prior research [4, 5] related to endomarketing, the authors adopt this strategy to solve the issue in fast food restaurants. The models uphold that organizational support, supervisor support, and organizational commitment are related to the employees' willingness to stay in the current organization. Both authors concur that about 25.5% of employees leave a company because there is no real organizational commitment. However, these studies do not consider that the creation of training programs boost commitment within the company.

2.2 Leadership

Another key element of endomarketing is leadership, which affects employee turnover. A total of 16.5% of employees quit because manager expectations are unclear or because their direct bosses lack sufficient leadership. Additionally, an employee under somebody's leadership performs better, feels more motivated to work, and maximizes productivity. Both authors agree that a leadership program is not only essential for employees but also empowers bosses. Studies validate the importance of leadership but do not outline a resulting comparison or the implementation plan [6, 7].

2.3 Employer Branding

The study on employer branding includes in its models the seven dimensions that comprise this technique; namely, social value, interest value, economic value, application value, development value, management value, and the value of a work–life balance [8]. However, the application of these components varies across sectors and organizational goals. The study found that, at present, 39% of companies are willing to invest in this strategy. The authors also agree that the adoption of solid employer branding exerts a powerful impact on personnel satisfaction and organizational goal alignment [9].

2.4 Talent Management

Talent management is the baseline of all research on human resources (HR) within organizations [10]. Majority of the models designed by using talent management practices are aimed at achieving better results and gaining a competitive advantage. One of the main findings of the model is the estimation that 43% of the leaders argue that inadequate human capital management prevents organizational growth.

The strategies and models used to reduce employee turnover were previously examined. However, validations of the proposals fail to completely solve the aforementioned issues. Although there are service-related studies, there is a scarcity of conclusive studies concerning the fast food industry that comprises such a sector.

3 Development of the Model to Decrease Turnover Across Service Companies

To conceptualize this proposal, a comparative chart (see Fig. 1) of the applied models was prepared on the basis of the literature reviewed, to examine each methodology in the sectors examined.

The first model consists of a case in the United States, which intends to reduce employee turnover in fast food restaurants. The research focuses on decreasing turnover by implementing internal marketing techniques and analyzing how they enhance job satisfaction and commitment to retain employees. The study seeks to show that a lack of satisfaction and commitment affects turnover.

The second model is a case from India, which is based on employer branding and focuses on the technique's dimensions. The study concludes that employer branding plays a crucial role for organizations.

The third model was also designed in India, using secondary data to prove its theories. It is based on four steps: (1) attraction proposition, (2) capacities, (3) experiences, and (4) the commitment to achieve a long-term employer–employee relationship.

Using a combination of these techniques is recommended to reduce employee turnover in any service company.

Upon applying the techniques in the fast food industry—a neglected sector—the purpose is to create a model that promotes personnel retention, improves organizational climate, cuts costs, enhances productivity, and boosts organizational commitment.

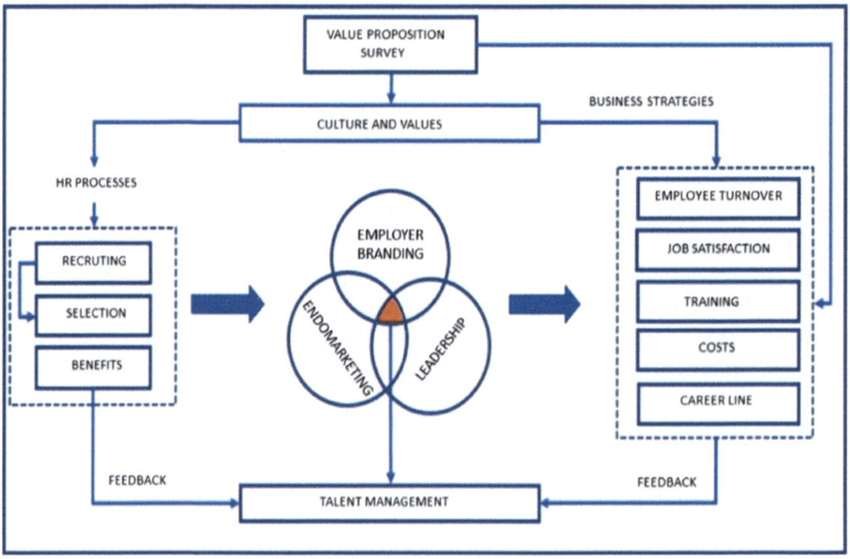

Fig. 1 Models to reduce employee turnover

Table 1 Value proposition

Recognition
I was granted an award for good performance
My bosses recognize my work and praise me for a job well done
The company fosters fairness in the workplace
The company implements a recognition program
The company uses distinguished employees as role models

Section 3.1 analyzes each component of the model, and Sect. 3.2 deals with the indicators for the proposed model.

3.1 Input for Value Proposition

The model is built from a survey based on employer branding, which used variables such as recognition, communication, leadership, career line, motivation, and interpersonal relationships.

To perform this survey, the Likert scale was adopted (not at all important, slightly important, moderately important, important, and very important), and each variable comprised five statements that were randomly distributed within the survey.

Table 1 presents the five statements concerning "Recognition"; the same method was adopted for the other variables.

The results reveal the employee value proposition to be applied according to the type of organization and processes. However, not all the variables may be considered to obtain the value proposition, as the model would be highly biased.

3.2 Input of Human Resources Processes

Every company in the service industry should enhance its HR management, as employees advance the value proposition to the end customer; "satisfied employees lead to satisfied customers," as many HR leaders assert.

Therefore, an HR management process flowchart should be used to identify corporate areas that need improvement to achieve the baseline results, design a development plan according to the areas detected and compare the baseline results with the indicators implemented post-implementation.

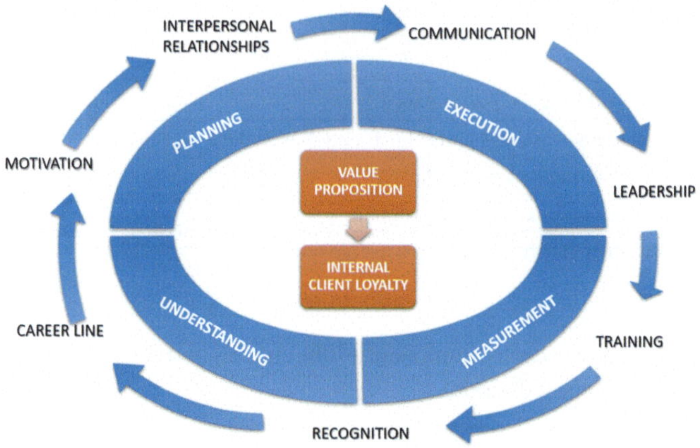

Fig. 2 Adapted model employer branding

3.3 *Methodologies to Be Implemented*

Following strategies were implemented in the aforementioned processes to obtain the proposed model.

Employer Branding: This is used to find the company's value proposition; it is the first step to drive change in an organization. It may also be used to streamline the recruiting and selection sub-processes. The goal is to win the external client through value proposition and reduce costs with the new selection processes (Fig. 2).

These practices are aimed at increasing personnel satisfaction and, consequently, the external client's satisfaction. These practices are coupled with leadership in training operators to boost organizational commitment and empower chain store managers so that, in turn, they motivate their employees.

Talent Management: This strategy encompasses the model, as it facilitates retention efforts such as training, compensation, and organizational climate. It also includes improvement initiatives in the area sub-processes according to the diagnostics within the company.

3.4 *Indicators*

Table 2 presents the indicators that are recommended to measure the model across the implementation stages.

Table 2 Model indicators

HR processes	**KPI**
Selection	Employee turnover rate
	Average time to fill
	Average cost per hire
Training	Effectiveness of the inductions attended by new employees/store managers
	Number of audit findings
	Satisfaction with the program undergone
Career line	Number of personnel promoted

4 Validity

As mentioned, the model was applied in a fast food restaurant in Lima, Peru. The same model has been applied in the other three scenarios to multiply the number of indicators and conduct a thorough analysis of the case study.

4.1 Model Index Comparative Analysis

Below is a comparison of the four stores chosen to validate the model. The primary model indexes used as base were employee turnover, organizational climate, inductions attended, and productivity.

Different scenarios were compared for assistants and delivery persons.

Table 3 presents the comparison of the previous scenarios, which matches the improvement achieved in the four cases with the baseline situation.

For assistants, the average turnover rate and organizational climate increased by 61.91 and 77.12%, respectively. Only assistants underwent inductions; hence, the average effectiveness of the inductions attended stood at 77.24%, raising the desired productivity by 68.94%. Delivery persons improved the average turnover, as it fell by

Table 3 Comparison of the Scenarios with the Baseline

	Assistants		Delivery persons	
Index (%)	Scenarios proposed	Baseline	Scenarios proposed	Baseline
Employee turnover	61.91	76.26	67.21	82.25
Organizational climate	77.12	75.00	75.90	75.00
Inductions attended	77.24	–	–	–
Productivity	68.94	62.50	67.97	62.50

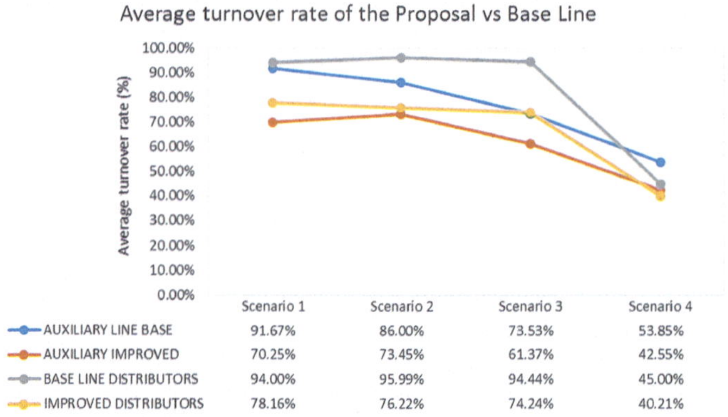

Fig. 3 Turnover Rate per position v. Baseline

15.04%, and the organizational climate for them improved by 75.90%. No inductions were conducted for this position, so no index is available. Productivity grew by 5.47%.

To facilitate the comparison of the turnover rates—the subject matter of this study (Fig. 3)—the different baseline indexes per position were compared with the average indexes after implementing the model in the four cases.

The savings earned from implementing the model for assistants and delivery persons amount to PEN 6,782.73 and PEN 13,624.10, respectively.

5 Conclusions

Upon performing a comparative analysis of the four scenarios chosen to validate the model with respect to the baseline, it may be concluded that, through statistical tests, there is an overall improvement in employee turnover, organizational climate, the number of inductions attended, and productivity. The model's implementation also allows the organization to cut internal costs from the high turnover experienced in 2016 and 2017.

This methodology combines employer branding, endomarketing, and leadership, thereby providing talent management with strategies to create a value proposition suitable for an organization. Throughout applying the techniques and the case study, the recruiting process was streamlined; a trainer position was created; and training programs were optimized. The devised methodology may be applied to any organization.

The final model indicators are right on course and are optimum to monitor improvement on an ongoing basis. Store managers and new hires should continue to undergo training to foster a favorable organizational climate.

We recommend implementing the model in future engagements involving administrative personnel and companies with high turnover rates in a sector other than the service industry.

References

1. Goleman, D.: La Inteligencia Emocional en la Empresa. Grupo Zeta, Argentina, p. 460 (1999)
2. Batt, R., Eun L. J., Lakhani, T.: A national study of human resource practices, turnover and customer service in the restaurant industry, **24** (2014)
3. Morales, T.: 43% de sobrecostos sería por mala selección de personal. La República [En línea] Disponible en: https://larepublica.pe/economia/788014-empresas-43-de-sobrecostos-seria-por-mala-seleccion-de-personal, last accessed 2018/05/03 (2016)
4. Gestión: El pollo se impone: Norky´s, Roky´s y KFC lideran ventas de restaurantes y fast foodl Empleo y Management. Gestión.pe [En línea] Disponible en: https://gestion.pe/economia/empresas/pollo-impone-norky-s-roky-s-kfc-lideran-ventas-restaurantes-fast-food-140556, last accessed 2018/05/03 (2017)
5. Hyun-Woo, J., Goh, B., Huffman, L.: Investigating relationships between internal marketing practices and employee organizational commitment in the food services industry. Int. J. Contemp. Hosp. Manage. (2015)
6. Nasyira, M., Othman, M. Y., Ghazali, H.: Predictors of intention to stay for employees of casual dining restaurant in Klang Valley area. Int. Food Res. J. 863–871 (2014)
7. Ariyabuddhiphongs, V., Kahn, S.I.: Transformational leadership and turnover intention: The mediating effects of trust and job performance on cafe employees in Thailand. J. Human Res. Hosp. Tourism, 215–233 (2015)
8. Bde W., Yau Y., Yang, C.: How appealing are monetary rewards in the workplace? A study of Ethical leadership love of money, happiness and turnover intention. Springer Science + Business Media, 1–14 (2015)
9. Dabirian, A., Kietzmann, J.: A great place to work? Understanding crowdsourced employer branding. Business Horizons Elsevier, 197–205 (2016)
10. Tanwar, K., Prasad, A.: Exploring the relationship between employer branding and employee retention. Glob. Bus. Rev. (2016) https://doi.org/10.1177/0972150916631214

Technological Model for the Exchange of Goods in the Peruvian Agricultural Business Sector Using the Smart Contracts and Blockchain

Christopher Cortez-Zaga, Víctor Casas-Llamacponcca and Pedro Shiguihara

Abstract To make the purchase or sale of a product through the internet, an intermediary is needed to validate certain operations, which make the products more expensive; according to INEI (National Institute of Statistics and Informatics), 68% of Peruvian agricultural sector distrusts the use of the internet for the sale of their products, so it is unlikely that they will carry out certain operations over the internet. As a result, two emerging technologies were created that offer new opportunities for the design of decentralized markets, allowing to increase the security of transactions, saving money by not depending on an intermediary, increasing the confidence of farmers and ensuring transparency and immutability of each transaction; these technologies are the Smart Contracts and Blockchain. They are technologies store information in a shared and transparent system for all the members; likewise, that the records of transactions cannot be altered, thus ensuring the transparency and immutability of the information. This document proposes a model that uses the Smart Contract and Blockchain technologies applied in the Peruvian agricultural business sector, which allows both the buyer and seller to commercialize agricultural products.

Keywords Smart contract · Blockchain · Farming · Commercialization

1 Introduction

The Peruvian agricultural sector is one of the engines of vital importance for the economy of the country; not only provides food and raw materials, but also employment

C. Cortez-Zaga (✉) · V. Casas-Llamacponcca · P. Shiguihara
Universidad Peruana de Ciencias Aplicadas, Lima, Peru
e-mail: u201112762@upc.edu.pe

V. Casas-Llamacponcca
e-mail: u201317002@upc.edu.pe

P. Shiguihara
e-mail: pedro.shiguihara@upc.edu.pe

Y. Iano et al. (eds.), *Proceedings of the 4th Brazilian Technology Symposium (BTSym'18)*, Smart Innovation, Systems and Technologies 140,
https://doi.org/10.1007/978-3-030-16053-1_57

opportunities to a significant amount of population. Likewise, it also contributes to the growth of the Peruvian economy by having a growth in agricultural production. This sector has been improving and this is demonstrated in the report of MINAGRI (Ministry of Agriculture and Irrigation) [1], which indicates that in 2017 it was consolidated as the second economic activity generating the largest foreign currency for the country. This result was due to the growth of our agroexports, mainly of non-traditional products, which they registered an increase of 11% in their export value.

On the other hand, the current infrastructure of the Peruvian agricultural business sector limits the commercialization of goods among farmers due to the dependence of intermediaries, this is due-to-the-fact that there is no efficient system of exchange of goods that represents one of the keys to favor a correct formation of prices according to the market forces [1] and a greater security and speed in the transactions.

Under this reality, there are technologies that have been implemented in similar situations in other countries, which are Smart Contracts and Blockchain. These are emerging technologies that offer new opportunities for decentralized market designs [2]. At present, the Peruvian agricultural business sector does not have a model that allows the exchange of goods using the Smart Contracts and Blockchain technologies, in order to reduce the aforementioned problems.

That is why, to improve the disconnection in the information regarding the offer and demand of the product, ensure the traceability of transactions and reduce the dependence of a third party to make sales or purchases of a product [8], has developed a marketing model applied in the business sector, where a simulation will be presented in a company of the Peruvian agricultural sector, which allow both the buyer and the seller to be able to exchange agricultural products bilaterally. In Sect. 2, the related work is presented. In Sect. 3, the smart contract implementation model is described. In Sect. 4, the methodology of the survey is presented. In Sect. 5, a discussion of the results is stated. Finally, in Sect. 6, the conclusions of the work are described.

2 Related Work

Blockchain can be defined as a large accounting book or a public accounting record that is increasing as all the entries and exits of money are recorded. It is a technology that allows the transfer of digital data with a very sophisticated coding and in a completely secure way. In addition, it contributes with a tremendous novelty: transfers do not require a centralized intermediary that identifies and certifies the information but is distributed in multiple independent nodes that register and validate it without the need for trust between them [3]. This technology also has other uses such as the one that will be introduced in this article: Smart Contracts, which can have the ability to modulate transactions according to the levels of trust of the parties involved. An inventory of benefits that Blockchain can offer in Table 1 has been prepared.

In [4] they propose to establish an electricity market (M2M) in the context of the chemical industry through IoT, which is to present a scenario that includes two

Table 1 Benefits offered by Blockchain when used in projects with different scenarios

Benefit	Description	References
Eliminate dependence on a central entity	All product information can be stored in a shared and transparent system for all members throughout the supply chain	[7]
Integrity of process	Users can rest assured that their transactions will be executed exactly as they mark the protocol, without the need to supervise third parties	[7]
Irrevocable transactions	The transaction log cannot be modified	[5]
Security and privacy	A level of privacy is introduced as each participant can use one or more anonymous portfolios	[5]
Transparency and immutability	Any modification to public Blockchain can be publicly seen by each party, ensuring transparency	[4]

electricity producers and one electricity consumer who trade with each other through of a Blockchain. Producers publish energy exchange offers (in kWh) per currency (in USD) in a data stream. The consumer reads the offers, analyzes them and tries to satisfy their energy demand at a minimal cost. When an offer is accepted, it is executed as an exchange. On the other hand, [5] it provides energy consumers and consumers with a decentralized market platform to market local energy generation without the need for a central intermediary, this is done through an initial concept test of a LEM scenario. (market platforms that offer agents the opportunity to virtually exchange energy within their community) simple with artificial agents implemented in a private Blockchain.

The market mechanism is implemented through an Smart Contract stored in a private Blockchain. Therefore, a central entity is not needed once the market is implemented. In addition to the orders, the payment is made through the Blockchain as well. The integration of these issues allows to establish a safe and effective market, so they propose a trade applying Blockchain. In such a way [7] provides a detailed description of how Blockchains and Smart Contracts work, to identify the advantages and disadvantages that your introduction to a system brings, and to highlight the ways in how the Blockchains and the IoT can be used together. Finally, in [6] it is discussed how the Blockchain technology can contribute to the interruption and innovation in business models, this study shows that the Blockchain technology can affect diverse dimensions of the business models in different industries. It is recommended that managers should follow developments in this field in order to prepare for possible disruptions in their industries. In addition, it is proposed that there are three crucial ways in which the Blockchain technology can affect and alter business models: through the authentication of traded goods, through disintermediation and through the reduction of transaction costs.

3 Smart Contract Implementation Model

The proposed model seeks to ensure that the commercialization of agricultural products is carried out successfully ensuring the traceability of information throughout the process through the Smart Contracts, that is why we define two types of Smart Contracts for their implementation and integration with the web tool.

3.1 Smart Contract Model

For the elaboration of the technological model, a work divided into stages has been carried out, which have been adapted for the development process of a web tool that allows us to validate the usability and functionality of the Smart Contracts. The proposal is divided into three stages for its correct implementation, which will be detailed below.

Analysis of need. This first stage includes the analysis of the results of the needs of the agricultural companies, in which the basic functionalities are detailed so that the users can simulate a flow commercialization in a successful way; these functionalities are transformed into user stories, where all the requirements are specified and what users expect when accessing a specific option in the web tool.

Solution Design. This second stage involves the design and development of the solution, where the user stories previously developed in the analysis stage were taken as a basis to be able to elaborate a prototype that serves as the basic scheme when implementing the web tool. As a second part of the solution design, a database diagram was developed detailing the relationship between the tables to be used in the model, as well as the type of data contained in each of them. In addition, a component architecture was carried out, specifying all the functionalities of the web tool, the programming language and the libraries that will be used for its development, as well as the design of the Smart Contracts and the necessary tools to be able to carry out the integration of the web tool with the Smart Contracts.

Implementation of tool. Finally, in this stage, the implementation of the web tool was carried out based on the analysis and previous design of the solution, for the development of the web tool HTML and CSS were used for the visual development of the tool (Frontend) and for the development of the functionalities (Backend) JavaScript was used as the programming language and the Angular framework. The development of this web tool serves as a basis to implement the fundamental part of the proposed model, which is the development of Smart Contracts.

3.2 Smart Contract Design

For the development of the model, it is proposed to carry out the definition and implementation of two Smart Contracts, which are "Product Contract" and "Transaction Contract".

Product Contract. The first Smart Contract to be developed comprises a link between the vendor and the web tool, since at the time of registering a new product a unique record of the product is generated, where the detailed information of the product is stored in the contract which is stored in the Blockchain and cannot be altered by any user. The fields that are used to create the Product Contract are the following: ID Product, Product Name, Unit Price, Amount, Delivery Time, Product Description, ID Seller.

Transaction Contract. The second Smart Contract to be developed is generated once the buyer makes the purchase of a specific product, that's where the product validation is done to verify if the product actually exists and has an available stock, a query is made to the Product Contract to verify this information and be able to proceed with the creation of the new Transaction Contract where the information is likewise stored in the Blockchain to ensure the integrity and security of the transaction. The fields to be used for the creation of the Transaction Contract are the following: ID Buyer, ID Seller, Product Contract Hash, Quantity to buy, Total Price, Additional clauses.

3.3 Tools for Smart Contract Implementation

For the implementation phase of the Smart Contracts, the Ethereum Blockchain was used to store the Smart Contracts. Ethereum is a decentralized platform that allows us to perform the execution of Smart Contracts, these applications are executed in a customized Blockchain with a shared global infrastructure throughout the network of nodes.

Additionally, in order to develop Smart Contracts in the Blockchain of Ethereum it is necessary to use its programming language called Solidity, which is a high-level language oriented to contracts to implement Smart Contracts, this language is mainly influenced by JavaScript language.

In order to carry out the development of Smart Contracts using the Solidity programming language, it is necessary to use the Truffle framework, Truffle is a framework that allows us to work in a development environment to connect with the Ethereum Blockchain and be able to carry out the development of applications based on Smart Contracts.

3.4 Smart Contract Use Case

In order to carry out tests of the web tool integrated with the Smart Contracts, a flow was simulated with the companies that were selected to validate the model, where they made a user registration to access the web tool, later they were able to visualize the products previously registered where it's show in detail the characteristics of each one of them, we also simulated a new product registration flow where we proceed with the generation of a new Smart Contract between the seller and the web tool, in addition they could observe their user profile where their personal information is displayed. Finally, the simulation of a purchase was made by selecting a product from among all the categories and subsequently generating a new Smart Contract between the buyer and the seller.

4 Methodology of the Survey

A survey methodology was carried out to validate the model, this included the analysis of a survey carried out to a group of companies related to the agricultural environment in which they sought to measure the level of interest they have regarding the main characteristics of the tool using smart contracts and blockchain technologies which are:

- The information security
- Data integrity
- Process traceability
- Variety of products.

4.1 Case Study with Businesses

To validate the case in study, a survey is prepared for the companies. The survey carried out on the companies had five characteristics to evaluate which will be detailed in Table 2.

Table 2 Main characteristics of the survey to be evaluated

Code	Characteristics
EC1	Keep users' information confidentially
EC2	Keep an immutable products registration
EC3	Ensure transactions' traceability
EC4	Keep a historical record of transactions
EC5	Make contracts by transactions

Table 3 Questions used to citizens about the characteristics mentioned above

Code	Characteristics
CP1	Have you ever purchased products online?
CP2	Do you know web pages that sell agricultural products?
CP3	Did you buy a product?
CP4	What has motivated you or motivated you to buy online?

Table 4 These are some similarities and differences that our proposal shows with the works explained in the literature review

References	Similarities	Differences
[4]	Establish a market using Blockchain	The established market is related to electricity
[5]	Proposes a decentralized market platform to market without the need for an intermediary	The platform is focused on local energy generation
[6]	Demonstrate that Blockchain technology can affect various dimensions of business models in various industries	It shows through reliable articles
[7]	Eliminate trust issues, such as fraud, corruption and information falsification	They develop a traceability system of the food supply chain for the tracking of food in real time based on HACCP (Analysis of hazards and critical control points), blockchain and the internet of things

As can be seen, in characteristic EC1, 100% of companies consider that the information of users must be kept confidential. In the EC2 characteristic, 81.8% consider that an immutable record of the products must be maintained, in the EC3 characteristic 81.8% and in the EC4 90.9% contemplate that the traceability of the transactions must be ensured when making a purchase or sale and a historical record of them must be maintained. Finally, in characteristic EC5, 100% agree with the conclusion of contracts for transactions.

On the other hand, some questions were asked to the citizens, mostly to find out their opinion on the use of technologies for the purchase of agricultural products. This survey can be seen in Table 3.

From the citizen survey we obtained as a result, that a large part of them do not know web pages that commercialize agricultural products in Peru. Observing this result, we can say that the solution would produce great benefits for agricultural companies as well as for Peruvian citizens, since they will have a tool where they can buy and sell purely agricultural products (Table 4).

5 Discussion

The Blockchain technology is emerging in a disruptive manner in different areas of application, so much that it is changing the way in how companies want to keep their information stored and that is how the use of Smart Contracts also comes into, which allows us to have a more autonomous and secure control of how to handle information. Knowing this, the use of these two technologies Blockchain and Smart Contracts in the e-commerce gives us a greater ease of how to manage customers (buyers and sellers), have greater control of products and maintain a historical record of transactions performed in a more modern and secure environment than would be done like using a conventional database. Based on this research, we can have a change in how e-commerce platforms perform transactions through payment gateways such as Visa or MasterCard, since the blockchain technology can replace them and not only that, but also, we can include the payments with cryptocurrencies on these platforms.

6 Conclusions

In this work an in-depth study has been carried out to identify the level of acceptance of the new technological model of exchange of goods based on Smart Contract and Blockchain for the Peruvian agricultural business sector, with which people or companies (buyer or seller) can perform the purchase or sale of your products in a more secure and integrated way. The proposal provides a set of deliverables with which you can move to the development stage and implement agricultural trade using Smart Contract and Blockchain.

When carrying out the expert judgment evaluation, it is concluded that its use would allow a better commercialization of agricultural products, since 50% of them considered as "Very acceptable" and the other 50% as "Acceptable". In addition, in the evaluation of the marketing proposal that was built based on new technologies, there was great acceptance by the companies of the agricultural sector that participated in the process, 60% as "Very acceptable" and the 40% as "Acceptable", which indicates that they look promising this way to market agricultural products through Smart Contracts.

References

1. Oficina de comunicaciones e imagen institucional ministerio de agricultura y riego.: Sector Agricultura se consolidó como el segundo generador de mayores divisas para el Perú. http://minagri.gob.pe/portal/publicaciones-y-prensa/noticias-2018/20660-sector-agricultura-se-consolido-el-2017-como-el-segundo-generador-de-mayores-divisas-para-el-peru-2 (2018)
2. Swan, M.: Blockchain: Blueprint for a New Economy. O'Reilly Media, Inc. (2015)

3. Zhang, Y., Wen, J.: The IoT electric business model: using blockchain technology for the internet of things. Peer-to-Peer Netw. Appl. **10**(4), 983–994
4. Sikorski, J.J., Haughton, J., Kraft, M.: Blockchain technology in the chemical industry: machine-to-machine electricity market. Appl. Energy **195**, 234–246 (2017)
5. Mengelkamp, E., Notheisen, B., Beer, C., Dauer, D., Weinhardt, C.: A blockchain-based smart grid: towards sustainable local energy markets. Comput. Sci.-Res. Dev. **33**(1–2), 207–214 (2018)
6. Nowiński, W., Kozma, M.: How can blockchain technology disrupt the existing business models? Entrepreneurial Bus. Econ. Rev. **5**(3), 173–188 (2017)
7. Tian, F.: A supply chain traceability system for food safety based on HACCP, blockchain & Internet of things. In: 2017 International Conference on Service Systems and Service Management (ICSSSM), (pp. 1–6). IEEE (June 2017)
8. Papadopoulos, G.: Blockchain and Digital Payments: An Institutionalist Analysis of Cryptocurrencies, pp. 153–172. ResearchGate (2015)

Technological Model of Facial Recognition for the Identification of Patients in the Health Sector

Diego La Madrid, Martín Barriga and Pedro Shiguihara

Abstract The identification of patients within medical institutions is an important issue to provide better care in health centers and avoid identity personifications. The risk of medical identity theft is one important factor for patient safety. Technologies are improving, such as fingerprints, atrial biometry or electrocardiograms to improve safety measures. However, biometric counterfeiting methods have increased and violated the security of these technological models. This article proposes a technological model of facial recognition to efficiently identify patients according to cognitive services in medical centers. The technological model was implemented in the UROGINEC clinic for the proof of concept. The results of the identification of the patient were successful with a precision percentage of 95.82 in an average of 3 s. This allowed the clinic to prevent identity theft with alert messages and improved the user experience within the medical institution.

Keywords Information systems · Patient identification · Biometrics · Facial recognition · Health sector · Cloud computing

1 Introduction

Medical identity impersonation is the fraudulent use of the identification information of another person without permission. This is done with the purpose of assuming their identity in a medical institution, to be able to obtain the services of insurers or beneficiaries who access health benefits in the mentioned entity. The management of the risk of medical identity theft is one very important factors for patient safety.

D. La Madrid (✉) · M. Barriga · P. Shiguihara
Facultad de Ingeniería, Universidad Peruana de Ciencias Aplicadas, Lima, Peru
e-mail: u201012614@upc.edu.pe

M. Barriga
e-mail: u201211625@upc.edu.pe

P. Shiguihara
e-mail: pedro.shiguihara@upc.edu.pe

Y. Iano et al. (eds.), *Proceedings of the 4th Brazilian Technology Symposium (BTSym'18)*, Smart Innovation, Systems and Technologies 140,
https://doi.org/10.1007/978-3-030-16053-1_58

In general, medical institutions request an identity document for the identification of the patient, being a National Document or an immigration document. According to the Federal Trade Commission (FTC), it was reported that 9.9 million Americans suffered some type of identity theft in 2017. This affects consumers an equivalent of $50 billion annually [1]. Medical identity theft accounted for 249,000 people. More recently, the Ponemon Institute estimated that there were 1.84 million victims of medical identity in 2013 [2]. The World Privacy Forum (WPF) report counted claims of medical identity theft between 2013 and 2017 in the United States. Due to this, medical institutions have chosen to use different technological models in order to efficiently identify patients in a hospital environment. Over the years, the number of biometric modalities has grown significantly to include, among others, fingerprints, atrial biometry, electrocardiograms, etc. However, biometric counterfeiting methods have increased and have violated the security of these technological models. In addition, the response times and the use of the media are not the most efficient for this type of process [10]. On the other hand, dependent objects have been used for the identification of patients such as bracelets with barcodes and RFID tags. However, these solutions have not been sufficient and errors in patient identification are common [3, 4]. Before these efforts, this work proposes to implement a technological model of facial recognition based on "Deep Learning" technology for the process of patient identification. The structure of the paper is organized as follows: Section 2 describes the related studies. Section 3 provides the description of the proposed method. Section 4 describes the experiment and implementation carried out. Then, the results are reported in Sect. 5 and in Sect. 6 the conclusions of the paper.

2 Review of Literature

The technological models of identification of patients already carried out have as their main function to collect the biometric characteristics of people, transform them into digital information and store them in a database. This will allow a comparison of these characteristics when identifying a patient. The challenge lies in methods designed to obtain high identification accuracy.

Accuracy is an important factor in providing the efficiency and ability to identify patients. In [4] an identification system is proposed among patients who receive treatment in hospitals using identification bracelets. The accuracy was found to be 83.9% of correctly identified patients. In contrast, in [5] a method of atrial biometry is proposed for the identification of patients in health. A unique ear identification algorithm (left and right) was developed and the study was conducted with 25 adults. The identification rate obtained was 88.3% accuracy in only one ear. In [6] they investigated the design of automated patient verification using three-dimensional magnetic resonance (MR) plates of the brain. The study was conducted with 730 MR pairs of patients' brains and the efficiency performance was established at 98.6%.

The time of identification of a patient is a fundamental part for an optimal quality of care. Situations that require help and are considered emergencies could occur. Therefore, it must be ensured that the demand of the population is met in a reasonable and determined time. In [7] they propose combining common physiological signals (vital signs). These physiological signals are the electrocardiogram (ECG) and blood pressure (ABP) to identify authenticate a patient in an average of 3 s, which makes it practical in a hospital setting. However, patients cannot be expected to have normative cardiac signals, since many of them may be suffering from ailments that affect the cardiac process. In [8] the authors attempt to consider individuals with normative and non-normative physiological signals (ECG) to evaluate their identification approach. The results regarding the time of identification of the patient is 2 s. However, this approach produces relatively low accuracy and high false negative rates. Demonstrating the transparency of deep learning in the given information is fundamental for the performance of the systems. The way to train automatically must be efficient in terms of resources and time of execution. In [9] the authors perform an Electrocardiogram Fusion and Blood Pressure Signals for the identification of patients. To learn the physiological signals on the electrocardiogram it takes an average of 20 min. In contrast to these models, our project uses cognitive facial recognition service to effectively identify patients. The learning method is based on machine learning services in the cloud.

3 Method

In this section, the approach used to identify patients by facial recognition was presented. The goal is to develop a model that completes patient identification solutions reliably and efficiently. Figure 1 shows the structure of the model. It is divided into 3 phases: entry, process and exit. Under the method used, the technology is used machine learning in cloud services. It is based on a discipline in the field of artificial intelligence. System developing machine learning to identify complex patterns required data are created.

3.1 Input

In the entry phase, the basic patient data of an existing database and an image of his face are required. This information is within the process of admission of a patient in the medical center. The image is sent to the process phase by means of the cloud to perform the cognitive analyzes for face detection and store in the database. The patient's information is stored in the database.

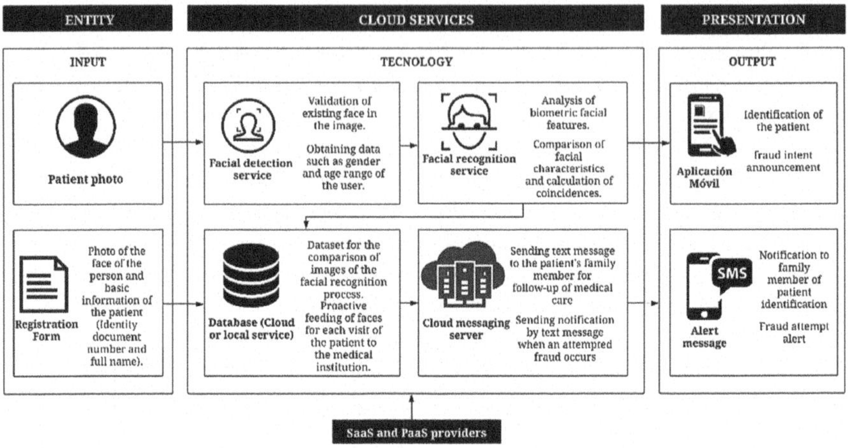

Fig. 1 Technological model of patient identification by facial recognition

3.2 *Process*

This phase is developed in the cloud environment. It is divided into 4 sub-processes: face detection, facial identification, database management and messaging notification. Under the cloud platform PaaS and SaaS, cognitive services of face detection and facial identification are used through Machine Learning. The facial detection service uses a series of algorithms to detect the face in a low image at a pre-definition level. Then the facial recognition service performs automatic learning under the neural network system to interpret the image and detect a single patient.

3.2.1 Face Detection

The image loaded to the PaaS cloud goes through the process of facial analysis. Its function of the service is to detect the faces of an image. Performs different biometric algorithms to determine the facial feature. For the model, only the patient's face is needed. The selection of the required face will be requested based on the detected ones. Once the face is selected, the service transforms the photo into a json file and performs face analysis to detect the age and sex attributes. Once the process is completed, the face information is sent to the facial identification process to learn the face or identify it in the database. If a face is not identified in the image, the process does not allow it to continue with the flow of the model and another image will be requested.

3.2.2 Facial Identification

The process collects face information to run the facial identification service in Cloud PaaS. The analysis performs the cognitive algorithms of facial recognition of deep learning. The service will learn the biometric points, along with their attributes, so that the patient can be identified when searching for faces. The information learned is registered with an ID and stored in the database. If the patient needs to be identified, the process performs the comparison of learned facial features. When finding the largest percentage of face matching, the database ID is called to obtain the patient information. Finally, a deep learning to the face that has been compared and a new json file that will be stored in the database is generated. This last function is given to prevent reduction the percentage of authenticity.

3.2.3 Storage Management

Storage management is responsible for storing patient files. You save the json files for comparison of facial recognition. Perform a proactive feeding at each patient visit within the medical center. It assigns the access in the entrance and exit of the information to the requests requested in the process of the facial identification. The distribution of the database is divided into information and pictures of the patient. Only patient information can be developed locally.

3.2.4 Messaging Notification

When the patient is identified, the notification process sends a text message to the relative's number registered in the database. The process is performed for the user experience around the model. If the user is not identified, the National Identification will be requested to identify the patient in the database and consult their telephone number to notify by a text message the intent of the patient's care in the health center.

3.3 Output

The exit phase shows the identification of the patient with the main information. Simultaneously, a notification is sent by SMS to the patient's or family member's mobile phone with the notification of the attempt to attend the medical center.

4 Experiments

To perform the proof of concept, a mobile application has been developed in Android Studio 2.3. The minimum level of the chosen API is 23 (Android 6.0)—Marshmallow, because it can use cloud computing services. The cognitive service that was used to perform face detection and identification was "Face API" from Microsoft Azure. To start using the Face API service, it is necessary to create an instance in the IA + Machine Learning category of the Azure portal. Therefore, we will obtain an Api key, which will be the only parameter that the provider requests to be able to identify us when the facial recognition service is used. For security measures, the Api key was encrypted and stored in a variable within the mobile application so that the service can be used without problems. The mobile application has a module that will allow identifying patients already registered in the database. Allowing to take a photo of the face of the patient who visits the medical institution to send it to the cloud service. The Face Api cognitive service will return the information and accuracy percentage of the candidate found in the database and it will be displayed in a new form. In addition, the "Simple Notification Service" messaging service of Amazon web service will send a text message to the relative of the patient who visits the medical institution notifying him of the care provided. Either for emergency situations, consultations or medical appointments. In case the patient to be identified is not in the database and is presenting an identity document that does not belong to him, an alert will be opened indicating an attempted fraud. This ad will receive the identity document number as the entry data to obtain the telephone number of the person belonging to the document and the alert will be sent by means of a text message. In this way, the corresponding complaint can be made. The tests were performed at the UROGINEC clinic.

5 Results

When we finished the tests in the clinical center we found the following results: We managed to register 142 patients throughout the validation of the technological model. We obtained the images of the faces and basic information of the patient. In addition, the patient identification and model training times were stored in the database to calculate our metrics. A comparison in identification response time was made in patients who have from one to three photos stored in the data set. Figure 2 shows that patients with a greater number of photos, the response time is faster than with people who only have a photo. This shows that the method of storing face images per patient visit is effective. The cognitive facial recognition service responds in a more efficient way to the identification of patients when training with more facial features per person in the database. Figure 3 shows the precision analysis of the identified patients. For this, the following equation is used:

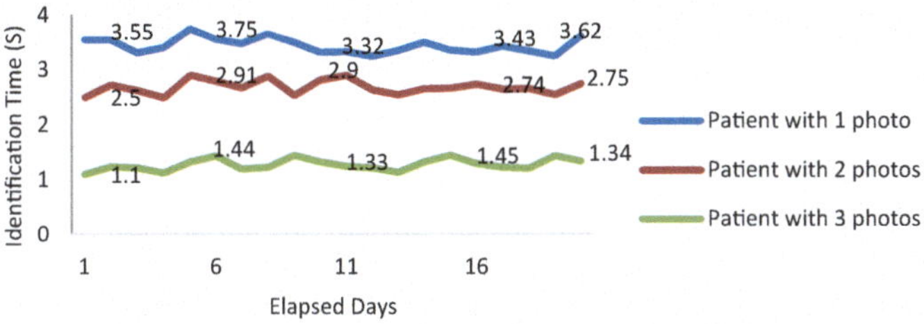

Fig. 2 Comparison of the time of identification of patients with respect to the number of photos stored in the database

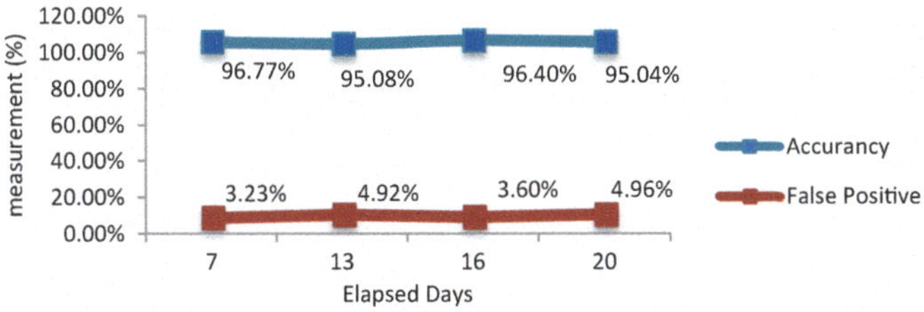

Fig. 3 Patient identification accuracy

$$\mathrm{TP} + \mathrm{TN}/(\mathrm{TP} + \mathrm{TN} + \mathrm{FP} + \mathrm{FN}) \tag{1}$$

where TP (true positive) are existing patients who were successfully detected, FP (false positive) are existing patients who were not identified and FN (false negative) are non-existing patients who were not detected.

The concurrence of identification of our work is 95.82% of 142 authenticated patients. On the other hand, the clinic had reports of identity thefts with an average of 11% of the total number of patients per year that have only been identified by the claims made. Regarding the validation carried out with a sample of 142 patients, false positive identification of patients was identified. That is, a patient has not been identified. These cases obtained an average of 4.18% and were for various reasons such as: lighting problems, poor camera focus, new patients and identity theft.

Table 1 shows that the technological model has more efficient results compared to other jobs. Finally, a survey was conducted with the evaluation scores: excellent, good, fair and bad. To determine the level of satisfaction of user experience regarding the notification of medical care to their relatives and the alerts of attempted fraud. An approval on the part of the patients was obtained an average of 92% with the scores of excellent and good. With this we can show that the user experience increases with the proposed technological model in contrast to other technological models that do not show this metric.

Table 1 Performance comparison with other works

Approach	Accuracy	False positive	Enrollment time	Identification time
Evaluation of electrocardiogram for biometric authentication [7]	82	7	15 s	15 s
ECG authentication for mobile devices [8]	84.93	1.29	30 s	4 s
Fusion of electrocardiogram and arterial blood pressure signals for authentication in wearable medical systems [9]	97.43	1.2	20 min	3 s
Our current approach	95.82	4.18	7 s	2.75 s (with 2 photos)

6 Conclusions

The identification of patients is a crucial process to provide an adequate care in medical institutions, both public and private. In this paper we present a more efficient facial recognition technology model to existing proposals. Because these solutions have patient identification errors, the authentication and/or training times of the system are not acceptable for the care of a patient in emergency situations although the validation of these works gives a sample of no more than 40 people. Our work achieved 95.82% accuracy of patients' identification; this percentage increased as the patient visited the medical institution, as we stored the patient's most recent face image. In this way, the facial recognition service was constantly trained and obtained more biometric characteristics of the person. In addition, the maximum training time was 15 s despite the number of people registered in the database and the time of identification of the patient occurred in only 3 s. In comparison with other proposals, the complexity and implementation of our model is minimal, since we have designed a mobile application that integrates two cloud computing services: facial recognition and text messaging. The workload is processed in the cloud and not locally. This will allow patient identification, notifications and alerts to be given in the shortest possible time. Previously, the UROGINEC clinic did not have methods to verify an identity theft. These cases were observed months later through patient complaints. This had a direct impact with the costs assumed by the insurance companies to users who did not correspond receive any payment for a health service. Now, with the technological model implemented, they receive alerts of identity impersonation attempts at moment and are automatically notified to the real patient. In the future, we plan to improve some aspects of our work. For example, the position of the mobile device will be evaluated so that it has a fixed location and is not manipulated by the nurse. In addition, we plan to completely supply the database with the images of the faces of all the patients of the medical institution to obtain an extensive validation of the idea in this work.

References

1. Lacey, C.: ITRC 2017 identity theft and fraud predictions—Identity Theft Resource Center. Identity Theft Resource Center, 2018 [Online]. Available: http://www.idtheftcenter.org/Identity-Theft/the-2017-identity-theftand-fraud-predictions.html. Accessed: 15 June 2018
2. Medidfraud.org: 2018 [Online]. Available: http://medidfraud.org/2013-survey-on-medical-identity-theft. Accessed: 16 April 2018
3. Rosenbaum, B.: Radio frequency identification (RFID) in health care: privacy and security concerns limiting adoption. J. Med. Syst. **38**(3) (2014)
4. Hoffmeister, L., Moura, G.: Use of identification wristbands among patients receiving inpatient treatment in a teaching hospital. Rev. Latino.-Am. Enfermagem **23**(1), 36–43 (2015)
5. Ragan, E., Johnson, C., Milton, J., Gill, C.: Ear biometrics for patient identification in global health: a cross-sectional study to test the feasibility of a simplified algorithm. BMC Res. Notes **9**(1) (2016)
6. Ueda, Y., Morishita, J., Kudomi, S., Ueda, K.: Usefulness of biological fingerprint in magnetic resonance imaging for patient verification. Med. Biol. Eng. Compu. **54**(9), 1341–1351 (2015)
7. Singh, Y., Singh, S.: Evaluation of electrocardiogram for biometric authentication. J. Inf. Secur. **03**(01), 39–48 (2012)
8. Arteaga-Falconi, J., Al Osman, H., El Saddik, A.: ECG authentication for mobile devices. IEEE Trans. Instrum. Measur. **65**(3), 591–600 (2016)
9. Cai, H., Venkatasubramanian, K.: Fusion of electrocardiogram and arterial blood pressure signals for authentication in wearable medical systems. In: IEEE Conference on Computer and Network Systems CNS Workshop, pp. 1–5 (2016)
10. Cai, H., Venkatasubramanian, K.: Patient identity verification based on physiological signal fusion. In: 2017 IEEE/ACM International Conference on Connected Health: Applications, Systems and Engineering Technologies (CHASE) (2017)

Predictive Model for the Evaluation of Credit Risk in Banking Entities Based on Machine Learning

Brenda Haro, Cesar Ortiz and Jimmy Armas

Abstract In this paper, we propose a technology model of predictive analysis based on machine learning for the evaluation of credit risk. The model allows predicting the credit risk of a person based on the information held by an institution or non-traditional sources when deciding whether to grant a loan. In this context, the financial situation of borrowers and financial institutions is compromised. The complexity of this problem can be simplified using new technologies such as Machine Learning in a Cloud Computing platform. Azure was used as a tool to validate the technological model of predictive analysis and determine the credit risk of a client. The proposed model used the Two-Class Boosted Decision Tree algorithm that gave us a greater AUC of 93% accuracy, this indicator was taken as having greater repercussion in the proof of concept developed because it is wanted to predict more urgently the number of possible applicants who do not comply with the payment of debits.

Keywords Credit risk · Machine learning azure · Cloud azure · Credit risk analysis · Machine learning studio

1 Introduction

Major changes are occurring because of the emergence of new technologies and their adoption in different business processes, and industries such as banking and finance are not alien to it. By the year 2020, the rate of digital transformation is accelerating with trends in technology such as the Internet of Things, the ingrained

B. Haro (✉) · C. Ortiz · J. Armas
Escuela de Ingeniería de Sistemas de Información, Universidad Peruana de Ciencas Aplicadas (UPC), Lima, Peru
e-mail: u201214147@upc.edu.pe

C. Ortiz
e-mail: u201212957@upc.edu.pe

J. Armas
e-mail: jimmy.armas@upc.pe

Y. Iano et al. (eds.), *Proceedings of the 4th Brazilian Technology Symposium (BTSym'18)*, Smart Innovation, Systems and Technologies 140,
https://doi.org/10.1007/978-3-030-16053-1_59

use of smartphones and tables, as well as convergence and the explosion of Cloud services [1]. In this article, the case study focuses on the assessment of credit risk, where a key point to consider is the increase in data and the sources that provide it. Non-traditional sources such as social media, online transactions and more help to approximate more accurately the profile of a person applying to a loan. These new sources to consider also achieve to increase the threshold of clients evaluated and therefore of loans granted and the respective profits. For example, there is a range of potential clients that do not have enough historical data related to traditional data sources and are therefore discarded from credit assessments in most entities. This represents a significant loss on the radar of financial institutions. This distinguishes the need to adapt to these new features that offer opportunities for those who are prepared and weaknesses for those who do not align on time. On the other hand, to process and analyze all this data the machine learning techniques arise to help recognize the subtle relationships of the variables in the datasets, these analytical systems allow to quickly deliver the output, presented in the desired form in a way that is easily interpretable by the end users or interested parties [2] As more precision is reached in predicting the credit risk of a client, the probability of granting loans to people who fail to repay the entire loan will be lower.

This paper presents a model that adapts to the needs of adapting to the presented context and takes technologies in the Cloud as an opportunity and an effective means to achieve the necessary flexibility, integration and security.

In this case, the model is applied to the credit risk assessment process from start to finish.

This paper is structured as follows. We will begin with the explanation of the different solutions that exist for the predictive model and, later, we will describe the detail of the components of the proposed model. Finally, the conclusions about the results obtained in the case study will be addressed.

2 Literature Review on Predictive Model

There are machine learning techniques that are applied to solve problems of different nature, but none have been identified as the best over the others since it is a very complex task and many of the existing studies have not been carried out with the care that is required. [3]. Jelte, focusing on the problem he had to solve, selected the AdaBoost algorithm as the one that fitted the best. Other investigations identify LR (Logistic Regression) and SVM (Support Vector Machine) as the techniques that result in better performance in their experiments [4]. For the risk rating when granting a credit, techniques such as Naive Bayes and ID3 have been used in the reviewed literature [5] with an accuracy of up to 82 and 76% respectively. In general, the solution to this kind of problem focuses on identifying the eligibility of those who apply for a loan. And data mining has been a proven value tool for this evaluation.

On the other hand, for the evaluation and application of Machine Learning techniques it is necessary to have a considerable amount of data. The greater the amount of data, the accuracy of predictions increases. The information is the driver for making corporate decisions at strategic, tactical and operational levels. Fortunately, the amount of data available to be collected by companies is increasing at an accelerated rate [6]. This is how even not being able to extract the available data and not having access to real, precise and meaningful information represents a risk for organizations. Therefore, the leaders of these organizations must be aligned to these contexts and new technologies, identifying opportunities in them when applying them and not risks by ignoring them.

3 Proposed Predictive Model

The proposed model allows to improve the quality in the evaluation of credit risk carried out by banking entities with the support and benefits of different technological services granted by a Cloud provider. Three important factors were identified for the design of this model when having a focus on the cloud: The flow and processing of data through the services acquired, security for handling sensitive data and predictive analysis experiment modeling. Each of these factors and the elements that compose them are related to each other to produce value indicators that contribute to better decision making when granting credits to applicants. See Fig. 1.

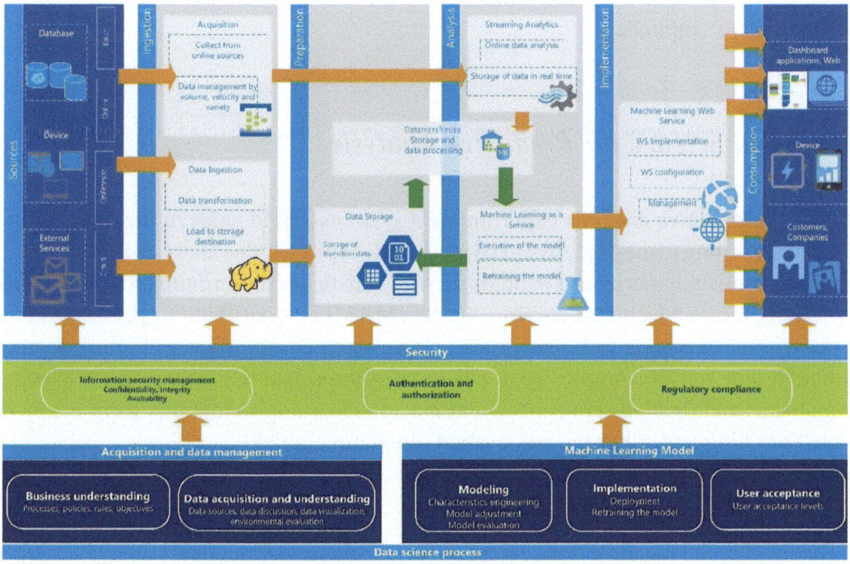

Fig. 1 Predictive model for credit risk evaluation

3.1 Flow and Processing of Data

The elements of this section are the stages through which the data transits through Cloud services. Each of the stages adds value from start to end so that finally the model, through the selected Machine Learning algorithms, throws the predictions based on the data that is ingested.

- Sources: Databases, dispositive, external services.
- Ingestion: Collection of data towards repositories.
- Preparation: Data cleaning.
- Analysis: Predictive model based on Machine Learning.
- Implementation: Deployment of the model in production.
- Consumption: Consumption of the model through web services, web applications, mobile applications.

3.2 Information Security

When dealing with the prediction of the credit risk assessment of multiple clients, the necessary data must maintain:

- Confidentiality, Integrity and Availability
- Authentication and authorization
- Regularity compliance.

3.3 Data Science for Predictive Analysis

To create predictive analysis experiments it is necessary to analyze the data that is needed according to the objective to be predicted. Typically, this analysis is performed by data scientists, experienced in fields such as statistics or data mining. Solutions like Microsoft Azure Machine Learning Studio (MAMLS) contribute to perform these tasks without the need to be an expert. However, it is important to highlight that, although they reduce complexity in design, it is important to have basic knowledge in statistics and previous concepts about machine learning so that this work is much more enriching and solutions are reached much more quickly.

For this factor, the following elements are considered:

- Acquisition and Management of Data
  Business understanding: Processes, policies, rules and objective.
  Acquisition and understanding of the data: Data sources, data discussion, data display, environment evaluation.
- Machine Learning Model
  Modeling: Characteristics engineering, fit of the model, evaluation of the model.

Implementation: Deployment and retraining of the model User acceptance: Levels of user acceptance.

4 Validation

For the validation of the proposed model, concept tests were carried out to evaluate the credit risk of clients for a banking entity.

Due to the nature of the tests to be performed, the database to be used is public access with 150,000 records of a German banking entity, which information was altered to protect confidentiality, but allows testing the data source is a representation of data from a client of a bank [7]. The objective column is SeriousDlqin2yrs, with which we seek to build a model that borrowers can use to help make the best financial decisions.

The following Table 1, shows the variables that the dataset contends. With the help of the IDEAR tool, the data in question was analyzed interactively. This tool allows us to explore, visualize and analyze data [8] with features such as: Automatic Variable Type Detection, Classification of the variable and identification of the objective variable and Visualization of high- dimensional data.

When evaluating the trained models, the following results were obtained with the Two-Class Boosted Decision Tree algorithm shown in the following Table 2.

Table 3, shows the values of the statistical result of the experiment with the Two-Class Support Vector Machine algorithm.

Figure 2, shows the receiver operating characteristics curves (ROC), this graph represents the ratio of true positives to false positives, a value by which a case is defined as positive. Among the tests what is compared is the area under the curve, where 1 represents a perfect diagnostic value and 0.5 is a test without discriminatory capacity, for our case we have a value of 0.933 which means that there is a 93% probability that the evaluation carried out is of a customer with credit risk. For this reason, it is advisable to always choose the test that presents the greatest area under the curve.

Finally, the consumption of the predictive model was implemented through Web Service and the endpoint test was successful in less than 410 ms on a cloud platform.

5 Results

The three graphs of Fig. 2, originated with the module Evaluate Model are the ROC curve, Precision/Recall and Lift Cueva ROC: The closer this curve approaches the upper left corner, the better the classifier performance will be. The truly positive rate is the proportion of positive results that are predicted correctly, also known as sensitivity or probability of detection. The false positive rate is the proportion of negative results that are falsely predicted to be positive [9]. Then, in MAMLS the

Table 1 Dataset variables

Variable name	Description
SeriousDlqin2yrs	90 days after default or worse
Revolving Utilization of Unsecured Lines	Total balance on credit cards and personal lines of credit except real estate and without installment debt such as car loans divided by the sum of the credit limits
Years	Age of the borrower in years
Number of times 30–59 passed date no worse	Number of times the borrower has been between 30 and 59 days past due, but has not worsened in the last 2 years
Debt ratio	Number of times the borrower has been between 30 and 59 days past due, but has not worsened in the last 2 years
Monthly income	Monthly income of the borrower
Number of lines of open credit or loans	Number of open loans (fees like car loan or mortgage) and Lines of credit (for example, credit cards)
Number of times 90 days late	Number of times the borrower has spent 90 days or more
Amount of loans or mortgage loans	Number of mortgage and real estate loans including mortgage lines of credit
Number of times 60–89 past no worse date	Number of times the borrower has gone from 60 to 89 days but has not worsened in the last 2 years
Number of dependents	Number of dependents in the family excluding them (spouse, children, etc.)

Table 2 Evaluation of results two-class boosted decision tree

True positive	False negative
11,502	**3502**
False positive	True negative
3152	**38,875**
Accuracy	Precision
0.883	**0.785**
Recall	F1 score
0.767	**0.776**
Threshold	AUC
0.5	**0.933**
Positive label	Negative label
1	**0**

Table 3 Evaluation of results two-class support vector machine

True positive	False negative
2908	**12,096**
False positive	True negative
490	**41,537**
Accuracy	Precision
0.779	**0.856**
Recall	F1 Score
0.194	**0.316**
Threshold	AUC
0.5	**0.812**
Positive label	Negative label
1	**0**

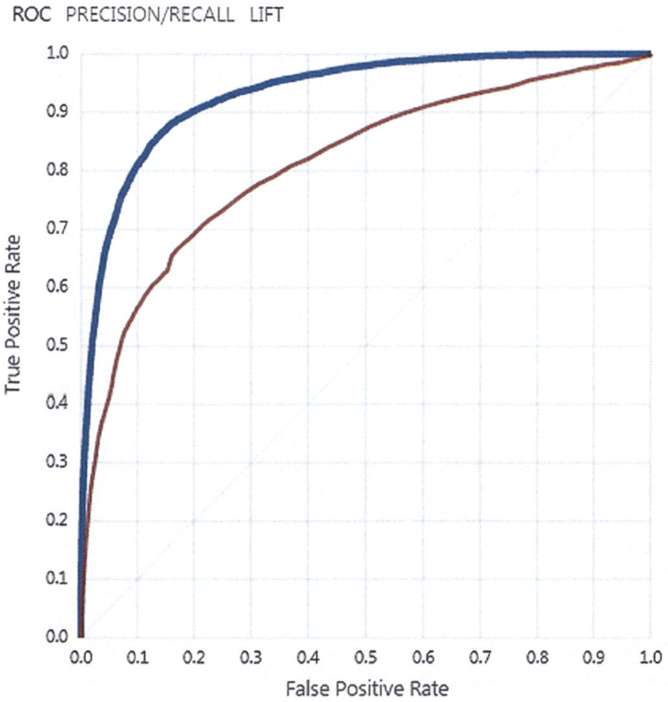

Fig. 2 Comparative graph ROC

algorithm of the evaluated ones that achieves a greater AUC, in evaluation of credit risk, is: Two-Class Boosted Decision Tree con 0.933. This indicator was taken as the one with the greatest impact on the proof of concept developed since we want to predict more urgently the number of possible applicants that do not comply with the payment of debts, represented as the positive variable in this experiment.

6 Conclusions

The proof of concept performed in the scenario of a bank entity executed in MAMLS allows us to understand the opportunity represented by technologies such as Cloud and Machine Learning as enablers of integral solutions for the evaluation of credit risk through the proposed model.

From the loading of data to the tool, its ingestion, preparation, analysis with predictive models, implementation and consumption with optimal results compared with classic solutions of this type. The base indicator was AUC which gives us a clear idea about the accuracy of up to 0.933 that the designed model can reach through each of its layers.

References

1. Efma, A.T., Kearney.: Going Digital—The Banking Transformation Road Map. 9 (2014)
2. Chen, C.P., Zhang, C.Y.: Data-intensive applications, challenges, techniques and technologies: a survey on Big Data. Science direct, 314–347 (2014)
3. Vink, J.P., de Haan, G.: Comparison of machine learning techniques for target detection. Springer Science + Business Media Dordrecht 1–2 (2015)
4. Guo, Y., Wei, Z., Keating, B.J., Hakonarson, H.: Machine learning derived risk prediction of anorexia nervosa. BMC Medical Genomics, 6 (2016)
5. Madyatmadja, E.D., Aryuni, M.: Comparative study of data mining model for credit card application scoring in bank. J. Theor. Appl. Inf. Technol. **1**, 5 (2014)
6. Florian Kache, S.S.: Challenges and opportunities of digital information at the intersection of big data analytics and supply chain management. Int. J. Oper. Production Manag. **1**, 6 (2017)
7. Kaggle.: Kaggle. Obtenido de Gime me some credit. https://www.kaggle.com/c/GiveMeSomeCredit/data (2011)
8. Cortana Intelligence and ML Blog Team. (18 de Octubre de 2016).: Cortana Intelligence and Machine Learning Blog. Obtenido de Two New Utilities to Boost Your Data Science Productivity. https://blogs.technet.microsoft.com/machinelearning/2016/10/18/two-new-utilities-to-boost-your-data-science-productivity/
9. Microsoft (10 de Agosto de 2016).: Microsoft Azure. Obtenido de. https://msdn.microsoft.com/en-us/library/azure/dn906025.aspx

Analysis of the Energy Quality of the Electronic Laboratory of the Federal University of Tocantins Using Artificial Immunological Systems

Jadiel Caparrós da Silva, Luiz Felipe R. Moreira and Stefani Carolline L. de Freitas

Abstract This article will bring an analysis of the electric power quality of the Electronic Laboratory of the Federal University of Tocantins. The proposed methodology was developed in two stages. The first one consists of obtaining the database through the acquisition of the oscillography images of the electrical network, in which the Laboratory is inserted using Fluke Power Log equipment. These images were recorded in the 34-h period, where there were moments of use and disuse of the Laboratory. In the second stage of the methodology, an algorithm was developed to read the input data (oscillography of the created database), to detect possible anomalies, to classify them, to generate a report showing the events occurred and to propose specific solutions. This algorithm is one of the most recent Artificial Intelligence in the bibliography, called the Artificial Immunological System. In order to perform the detection of the disturbances, the input signals were scanned, and through this process, these data are separated in periods, compared to a set of detectors previously established to finally assess the affinity between they. Therefore, this research aims to present a method based on Artificial Intelligence that helps in the correction of voltage events, providing greater reliability and quality of electric energy.

Keywords Energy quality · Fluke power log · Artificial immune system

1 Introduction

Generation, Transmission and Distribution of electrical energy make up the Electric Power System (SEP), which has the function of providing electrical energy with reli-

J. C. da Silva (✉) · L. F. R. Moreira · S. C. L. de Freitas
Federal University of Tocantins, Block 109 North Avenue NS 15, Palmas, TO, Brazil
e-mail: jadiel@uft.edu.br

L. F. R. Moreira
e-mail: luizks@uft.edu.br

S. C. L. de Freitas
e-mail: stefaniclf@uft.edu.br

Y. Iano et al. (eds.), *Proceedings of the 4th Brazilian Technology Symposium (BTSym'18)*, Smart Innovation, Systems and Technologies 140,
https://doi.org/10.1007/978-3-030-16053-1_60

ability and quality. In order to guarantee the quality and safety of the generation and supply of energy, the SEP must have highly reliable systems that allow the reduction of the execution time of the protections, triggered by atmospheric discharges, equipment failure, collision, among others [1, 2].

In recent years, due to the growth of electricity demand, SEP operations have become very complex [2]. The large number of buses implies an increase in the impedance matrix, and it is necessary to use optimization techniques to reduce the execution time of the program that will perform network load flow analysis in the disturbance detection procedure. This increase brought about a greater reflection in the distribution systems, since these are the terminal circuits of energy supply [3]. Therefore, it is now perceived that the great challenge imposed on distribution concessionaires is to provide electricity to their consumers in a sustainable and reliable way [1, 2]. In this context, the growing use of efficient technologies in electrical system analysis, applied to the challenge of granting self-recovery to distribution networks, through the development of new techniques for the detection and diagnosis of electrical power quality disturbances (QEE).

In this way, artificial systems arise as a bioinspired methodology to propose resolutions of complex engineering problems, for example to diagnose faults in electric power distribution systems. Among Artificial Intelligence (AI), the most relevant ones in the specialized literature are: Genetic Algorithms, Wave-let Transform, Fuzzy Logic, Artificial Neural Networks (RNA), and more recently Immune Systems Artificial Intelligence (AIS). These methodologies are the basis of the self-healing capacity, which among other resources guarantee the continuity of the service to users with quality and for the longest time possible [4].

Na literatura não foi encontrada nenhuma referência baseada em SIA para a analise da QEE em sistema de distribuição real, o que valoriza ainda mais esta pesquisa, proporcionando uma inédita e importante contribuição.

2 Methodology

The methodology proposed in this work is based on two stages. The first one consists of the construction of a data bank with voltage oscillography images of the electrical network, in which the Electronic Laboratory of the Electrical Engineering course of the Federal University of Tocantins—UFT is inserted. In order to construct this database, we used the Fluke VR1710 device that uses the Po-wer Log software. According to the user's manual, [5, 6], the registration parameters should be defined using Power Log software, so that the register accurately collects all the parameters and then this data is transferred to the computer and later the analyzes are carried out. All logging and storage settings are available in the user manual [5, 6].

The second step is the use of SIA based on the ASN Negative Selection Algorithm. The SIAs propose a new computational paradigm, with the objective of computationally reproducing the main characteristics, properties and abilities of the principles of Biological Immunological Systems (SIB) [7, 8].

Therefore, this article presents an automatic method of QEE analysis using SIA. From the reading and processing of the oscillographs images, the ASN [9] is applied to recognize and classify the images through the fineness-of-patterns method. This is the process that represents the discrimination that SIB makes with cells, between own and non-own. The algorithm is executed in two phases, the Censoring phase, where it is defined the set of own chains to be protected, and the second phase is called Monitoring, where the affinity evaluation between the chains and the generated detectors will occur in the censoring phase.

The data set can be used both in the monitoring and in the monitoring of the system, respecting a maximum limit of 30% of the data to generate the set of detectors [9]. This limit is due to the fact that in a SIB the number of detectors distributed by the organism represents 30% of all the infectious agents present in the body. Consequently, in order to make the proposed process dynamic and real, the use of a maximum of 30% of the data as sensors in the censorship phase was adopted.

3 Electrical Power Quality Disturbance

Equipment malfunctions, fires, animal contacts to energized parts, natural phenomena, are some examples of problems that make the operation of distribution systems susceptible to failures. These events can cause power interruption, introduce harmonics, and degrade QEE rates provided by power utilities, thereby generating high operating costs.

There are three main classes of disturbances that affect the electric power distribution system, being, short-circuit fault, lack of high impedance and the Voltage Disturbances that are also known as QEE disorders, related to voltage problems (and may be instantaneous, momentary or temporary) in magnitude from a nominal value [10–12]. The main effects are interruption of power supply, interference in communication systems, overheating of conductors, inaccurate measurements and undue relaying [3].

Therefore, the analysis of the QEE is made more and more necessary in order to solve the disturbances before they take on greater proportions. Some of the most common solutions are: harmonic filters, line reactors, improvements in wiring, grounding and insulation transformers, surge suppressors, voltage regulators, installation of static reactive compensators, among other measures [10–12]. However, most of these solutions already exist in distribution systems. The big problem lies in optimizing the detection of disturbances, causing them to be eliminated quickly and efficiently.

4 Artificial Immunological Systems

In the last decades, researchers have begun new studies in the field of natural phenomena, which, in an efficient way, act in the resolution of problems in various

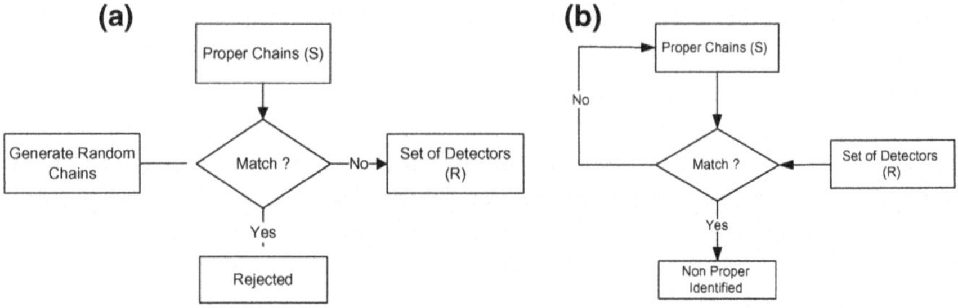

Fig. 1 ASN flowcharts

regions of the human body, going beyond any existing technological strategy [7]. These natural systems are composed of several distinct elements that have limited capacity and specific actions. However, when they function in a collaborative way, they acquire complex and robust behaviors, which is one of the main characteristics that make them attractive to study [8].

The ASN was proposed by [9] to detect changes in systems and is based on the negative selection of T lymphocytes within the thymus in order to classify the own cells in the Censoring phase and non-own in the monitoring phase 3.

Figure 1 shows the flowcharts of the Censoring and Monitoring phases, respectively [13].

4.1 Marriage and Affinity Criteria

It will be used the criterion called marriage, also known as Match, which has a positive characteristic that it is not necessary to transform the problem into binary, thus, the nominal values are used, obtaining a reduction of the response time. Marriage can be divided into perfect where it is necessary that the two chains being analyzed are perfectly equal or partial where the two strands being analyzed need not be the same. Thus, the affinity ratio [14] is defined as representing the degree of similarity necessary for marriage to occur between the two chains under analysis:

$$TAf = (An/At) * 100 \tag{1}$$

being:

TAf Affinity ratio;
An Number of own chains;
At Total number of own and non-own chains;

Also used in this work, the concept of marriage and affinity presented in [14], thus will be used the definition presented in (1), and described in the methodology, a

TAf of 70%, that is, to confirm a marriage between two signals requires at least 70% of the detector points to be equal.

5 Results and Discussions

This chapter presents the results obtained through the application of the methodology. The characteristics of the machine used to perform the simulations are: Intel Core i5 1.8 GHz processor, 8 Gb RAM and MacOS High Sierra operating system.

The oscillographs were collected for 34 h using the Fluke VR1710 equipment. As already mentioned, the equipment was plugged into an outlet at the Electronics Laboratory of the Electrical Engineering course at UFT. During these 34 h, there were different types of load, such as moments of disuse and moments in which the Laboratory was being used by groups of 20 students arranged in 5 benches conducting electronic experiments.

The set of detectors is generated in offline mode in the sensing phase. In this specific work, three types of detectors were generated: normal operation, voltage sag and voltage rise. These detectors were generated because from a previous study, they are the most recurrent in a distribution system. In order to perform the detection of the disturbances, the windowing of the input signals was performed and through this process the data are separated in periods and compared with a set of detectors, thus evaluating the affinity between them. It is important to emphasize that the choice of three images is done in a totally random way, making the system dynamic and real.

5.1 Results of the Application of the AIS to the Analysis of the QEE

In the simulation of the SIA for the detection of disturbances were made test, where for each one was presented in the Tables the set of detectors (images used) and different censoring and monitoring, besides the Hit Rate. This shows that the SIA randomness criterion was met perfectly. Among the simulations, five were selected for exemplification. It is important to note that the results and the processing time depend directly on which images are selected to compose the detector sets, and the image number 01, used in all tests, represents the normal operation of the system.

In Table 1, 112 images of events of tension disturbances occurred from 8 am to 6 pm on the first day (Laboratory in full use) were used.

It could be observed that an error occurred in the monitoring phase in Test 3. This problem in the classification of a voltage sink occurred because, as shown in Fig. 2, there were two voltage sags in a short period of time, approximately less than 2 ms.

Table 1 Results for the first period

	Image used	Census time (s)	Monitoring time (ms)	Hit rate (%)
Test 1	01-58-101	1.5543	0.3019	100
Test 2	01-78-95	1.2394	0.2581	100
Test 3	01-61-111	1.9274	0.2395	99.1071
Test 4	01-94-107	1.0147	0.2011	100
Test 5	01-87-93	1.1024	0.1487	100

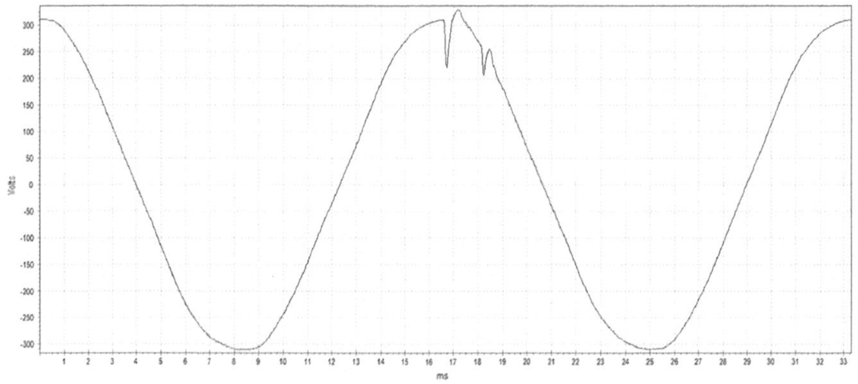

Fig. 2 Example of sinking voltage in a short period

Table 2 Results for the second period

	Image used	Census time (s)	Monitoring time (ms)	Hit rate (%)
Test 1	01-11-26	1.3093	0.1409	92.8571
Test 2	01-18-28	1.3001	0.1398	100
Test 3	01-21-20	1.8510	0.1321	100
Test 4	01-17-23	1.1042	0.1201	100
Test 5	01-20-25	1.1049	0.1013	100

In Table 2, 28 images were used for the period of 18 h from the first day to 8 h of the second day, that is, the Laboratory and external loads in disuse, which justifies the least amount of disturbance events, the least amount of images.

Again, in Table 2, it is observed that an error occurred in the monitoring phase in Test 1 due to the same problem discussed in Table 1.

Finally, in Table 3, 84 images were used for the period from 8 o'clock in the morning to 6 o'clock on the second day (Laboratory in disuse but with external loads).

In the tests presented in Table 3, the monitoring process obtained 100% accuracy in the classification of the images.

Table 3 Results for the third period

	Image used	Census time (s)	Monitoring time (ms)	Hit rate (%)
Test 1	01-67-83	1.3632	0.2056	100
Test 2	01-34-76	1.3021	0.2001	100
Test 3	01-49-73	1.4295	0.1249	100
Test 4	01-54-84	1.2194	0.1231	100
Test 5	01-40-93	1.1821	0.1037	100

The ASN concepts were tested and excellent results were achieved. The emphasis given to the criteria of an AIS (TAf and detector set) established by the literature is justified by its importance throughout the process so that the optimal performance could be achieved. In addition to defining the appropriate TAf for test sets, these results are due to the fact that all images have been correctly processed by the program. Each of them has 2 wave cycles, as can be seen from Fig. 2, and the algorithm performs image slicing to extract the characteristic of all 2 cycles in each image. Thus, minute adjustments were made so that all cuts were made without the desired stitches.

6 Conclusion

This work presents a method for classification and detection of voltage disturbances in real electric power distribution systems using ASN based SIA. The first step is to obtain the database through the installation of the Fluke VR 1710 and Power Log software. The second step consists in the development of a algo-pace capable of reading the input data, to detect possible voltage disturbances and to generate a report showing the events that occurred to propose timely and immediate solutions.

The TAf and the number of the detector set are fundamental parameters for the success of the classification process due to noise and loss of quality when the images are processed by the software. In this way, the value chosen for the TAf should be sufficient to have the marriage, even if partial. It should be noted that the set of detectors is generated off-line in the sensing phase and its time, when presented in the Tables, is on the second scale. Once the detectors are generated, monitoring is performed, which presents a considerably shorter time, in the order of milliseconds. These extremely small times accredit the program to be a tool used in real time, where the response provided by monitoring the system should be immediate for the decision to be made and no problems occur in the equipment.

The results obtained were promising, showing that the implementation of SIA is quite reliable and safe, and is therefore a valid computational tool within the QEE scenario. In the light of the tests carried out during this work, some observations should be made to highlight the advantages of applying the CIS:

- Once the detector set is generated, the system is able to perform the moni-toring several times, without having to restart to perform new tests;
- AIS is capable of performing the detection and classification of disorders from a limited number of information;
- SIA has the characteristic of being a system with continuous learning, which allows new elements to be added for detection without the need to erase the data from memory.

In addition to the results already presented in this paper, proposals for future work will be presented:

- The Fluke VR1710 Recorder equipment has a huge potential for data to be explored. Following the methodology of this work, one can use the Equipment Voltage Guide that displays a temporal graph with the voltages recorded.
- Another suggestion is the use of techniques to treat input data from AIS, in order to increase the differentiation of images and raise the affinity rate, generating even more promising, accurate and less time consuming results.
- Expand the use of the AI equipment and techniques used in this work in order to create a comparative basis of which are the best options. Expand the study here explored to other locations, such as residences and businesses.
- Finally, to develop a low-cost QEE analyzer, based on the tools explored in this work. Thus, to propose an Integrated System capable of providing response and restoration to power grid faults.

As noted, the application of SIAs is not limited, so it is important to expand their study in all directions, allowing great contributions to society, a fundamental role of the scientific community.

References

1. Dongli, J., Meng, X., Song, X.: Study on technology system of self-healing controlin smart distribution grid. In: International Conference on Advanced PowerSystem Automation and Protection, pp. 1–5 (2011)
2. Gungor, V.C., Sahin, D., Kocak, T., Ergut, S., Buccella, C., Cecati, C., Hancke, G.P.: Smart-grid technologies: communication technologies and standards. IEEE Trans. Industr. Inf. **7**(4), 529–540 (2011)
3. Lima, F.P.A., Lotufo, A.D.P., Minussi, C.R.: Artificial immune systems applied to voltage disturbance diagnosis in distribution electrical systems. In: Proceedings of the IEEE PowerTech, Grenoble, Rance, pp. 1–6 (2013)
4. Alvarenga, R.F., Sales, A.G.B., Folha, A.A., Silva, J.C.: Classificação de Distúrbios de Tensão em Sistemas de Distribuição de Energia Elétrica Utilizando Sistemas Imunológicos Artificiais, in: Simpósio de Estudos em Engenharia Elétrica do Tocantins, pp. 1–7 (2017)
5. Corporation, Fluke: Manual do Usuário, VR1710 Voltage Quality Recorder, p. 18. WA, EUA, Everett (2007)
6. Fluke Corporation.: Manual do Usuário, Power Log PC Application Software, p. 34, Everett, WA, EUA (2007)
7. de Castro, L.N., Timmis, J.: Artificial Immune Systems: A New Computational Intelligence Approach, 1st edition, Springer (2002)

8. Dasgupta, D.: Artificial Immune Systems and Their Applications. Springer-Verlag, New York Inc, Secaucus, NJ, USA (1998)
9. Forrest, S., Perelson, A., Cherukuri, A.L.R.: Self–nonself discrimination in a computer. In: Proceeding of do IEEE Symposium on Research in Security and Privacy, pp. 202–212 (1994)
10. Barros, A.C., Tonelli-Neto, M.S., Decanini, J.G.M.S., Minussi, C.R.: Detection and classification of voltage disturbances in electrical power systems using a modified euclidean ARTMAP neural network with continuous training. Electr. Power Compon. Syst. **43**(19), 1–11 (2015)
11. Dugan, R.C., McGranaghan, M.F., Beaty, H.W.: Electrical power systems quality. p. 265, McGraw-Hill, New York (1996)
12. Bollen, M.H.J.: Understanding power quality problems, New York, (2000). Disponível em: https://docs.google.com/file/d/0B5vXY4-Kg5GeY3ltZ041VkZDVW8/edit. Acesso em: 30 set. 2018
13. Silva, J.C., Lima, F.P.A., Lotufo, A.D.P., Batista, J.M.: Identity recognition using an artificial intelligence based on artificial immune system. In: 3rd Asia-Pacific Conference on Computer Aided System Engineering (APCASE) vol. 1, p. 1–5 (2015)
14. Bradley, D.W., Tyrrell, A.M.: Immunotronics—novel finite-state-machinearchitectures with built-in self-test using self-nonself differentiation. IEEE Trans. Evol. Comput. **6**, 227–238 (2002)

Prototype of a Single-Phase and Two-Phase Electric Power Meter for Residential Consumption Monitoring Using an ESP32

Rachel Batalha de Lima and Edgard Luciano Oliveira da Silva

Abstract This article describes a prototype of an electric power meter in the monophasic and biphasic levels. Its purpose is to monitor residential consumption, based on a low-cost microcontroller, which through sensors can obtain the current and voltage values to find the variants of power and cost. The communication is established through a web server that is developed in the C language in conjunction with the Arduino libraries.

Keywords ESP32 · Electric power · Internet of things

1 Introduction

The electricity wasting in Brazil reaches annually the quota of 50 thousand gigawatts/hour that, estimating from the tariff charged in 2014, are equivalent to 3.2 billion dollars approximately [1]. Because of these worrying data, it's important to start with public policies about energy conscious usage, both to consumers, with projects to allow people to know their average consumption, and researchers, providing budget to energy efficiency researches, for example.

The number of consumers in the residential electrical sector has been growing every year [1], as shown in Table 1, which consequently impacts on the increase of electrical consumption in this sector. Consumers are separated according to the voltage level they use. The rule says that the higher the consumption, the higher the voltage level: Group A—Class A1 (230 kV or more), Class A2 (88–138 kV), Class A3 (69 kV) and Class A4 (2.3–25 kV); Group B—Low Voltage (Less than 2.3 kV) and Residential (110–220 V) [2].

R. B. de Lima (✉)
SIDIA Instituto de Ciência e Tecnologia, Manaus, AM 69075-830, Brazil
e-mail: rachel.l@samsung.com

E. L. O. da Silva
Universidade do Estado do Amazonas, Manaus, AM 69050-010, Brazil
e-mail: elsilva@uea.edu.br

Y. Iano et al. (eds.), *Proceedings of the 4th Brazilian Technology Symposium (BTSym'18)*, Smart Innovation, Systems and Technologies 140,
https://doi.org/10.1007/978-3-030-16053-1_61

Table 1 Number of consumers per year (millions)

	2012	2013	2014	2015	2016
Residential	61.697	63.862	66.007	67.746	69.277

The development of the electric energy meter prototype at the monophasic and biphasic levels is important because of several factors. First, it is known that the metering of the electrical energy consumption is done by a reader using a specific form or an electronic apparatus called a data collector. After collecting the readings, the reader returns to the supplier, to proceed with next bill's calculation, which shall be delivered at the customer address. This process is done within 30 days [3].

Due to this process, the prototype of the low-cost electric meter is important because it will allow the consumer to control electric power consumption of his residence in real time and compare with the service charged by the concessionaire. Thus, avoiding undue charges and reducing energy waste.

In this context, the main contributions of this work are: to develop an intelligent meter that is efficient and low-cost, using IoT concepts; compare and analyze the work developed with preexisting solutions in the market, presenting the advantages and disadvantages of the methodology used.

The paper is organized as follows: Sect. 2 presents the necessary theoretical basis for the comprehension of this paper, Sect. 3 points out some of the main related projects about the subject, Sect. 4 describes the approach used in the proposed meter development, Sect. 5 exposes the experiments and results obtained, and finally Sect. 6 presents the conclusions and suggestions for further projects.

2 Theoretical Basis

2.1 Monophasic and Biphasic Systems

In monophasic systems, the power grid is constructed with two wires: one phase and one neutral. The maximum electrical voltage that can be supplied by this system is 127 V at 60 Hz, a frequency suitable for short-range distribution in household appliances [4].

The monophasic networks need to respect a limit that comprises the sum of powers from all equipment in a residence and cannot exceed 8000 W, if in the residence there is a quantity of equipment that exceeds this value, the user can request the increase of load, transforming in a biphasic network that provide powers in the values of 12,000 W up to 25,000 W [5].

In biphasic systems, there is usually a primary voltage phase and a transformer with a central grounded outlet. In a biphasic network with neutral it is always possible to have two voltages: a measurement between the phase and the neutral, in this case it is like a measured voltage in a monophasic network; And another voltage measured between two phases: in this case the voltage is always twice the voltage measured between phase and neutral, since the phase difference between the two phases is 120° [6].

2.2 Embedded Systems and ESP32

An embedded system is a fully encapsulated microprocessor system where its architecture is dedicated to the device or system it controls. Unlike general-purpose computers, such as the personal computer, an embedded system performs a set of predefined tasks, usually with specific requirements [7]. The microcontroller, present in this system, is a semiconductor device in the form of an integrated circuit on a single chip that integrates all the basic units of a microcomputer [8].

An example of an embedded systems development platform is Arduino, which has become popular because it is open source, low-cost, with very complete documentation having contributions from various developers around the world and a vast collection of sensors, actuators and modules of communication available, making it accessible to all inquisitive and tech savvy. The Raspberry Pi project stands out as an example of more powerful embedded systems, using an ARM family processor, including Wi-Fi, Bluetooth, four USB ports, 40 input and output pins and various interfaces such as camera and external memory [9].

The architecture used in the prototype was the ESP 32, that allows to be programmed independently, without the need for other microcontroller boards such as the Arduino, for example. The main advantages of using are: low power consumption, high power performance, low noise amplifier, robustness, versatility and reliability [10].

2.3 IoT and Client-Server Communication

Internet of Things (IoT) is not only to interconnect "things" through the Internet, but also, to make them intelligent, capable of receiving and processing from the environment in which they are inserted or from the networks that are connected. The growing evolution of technology and the decrease in the price of solutions that use IoT have made this communication feasible and very attractive for use in many different areas such as: Smart Grid, Smart Home, Smart Health, among others [9].

The TCP/IP (Transmission Control Protocol) communication model is divided into layers: interface to network, network, transport and application. The lower layers have the function of transmitting the data sent by the application layer reliably, but do not provide direct services to the users. In the TCP/IP model there are no sectioning and presentation layers, which in most applications are little used. These two layers are included in the application layer.

The client with its specific application software enables communication with the server. The server with its specific software allows to receive the message, interpret it and return the response to the client and the network [11].

For an internet page to be initialized, the user uses a browser (for example, Google Chrome, Internet Explorer, Netscape, among others) that works as a client in relation to a web server, (TCP/IP), where it uses a high-level protocol called HTTP (Hypertext Transfer Protocol), acting as a request-response protocol [12].

3 Related Projects

In the work of Lima, Rachel Batalha [12], presents a meter composed of a current sensor in conjunction with the Arduino, which sends data to a web platform, and the total cost of this project was R$ 180.00.

As exemplified in the work of Pereira, Luiz Henrique Junior [13], entitled Monitoring of Electrical Energy Consumption and Equipment Control Via Application, an ASC12 current sensor and a relay were used, with the aid From ESP8266 by sending the data to a spreadsheet in Google Docs, and implementing the IoT MQTT Dash protocol.

Seeking to develop a prototype for electric energy measurement Silva, Igor Alexandre Dutra e [14], presents a prototype with a cost of R$ 108.00. Formed by a non-invasive current sensor and a voltage filter, a Bluetooth sensor was used to send data to the computer.

The present work differs from the previous ones due to the use of a voltage sensor to improve the accuracy of the variants and the substitution of the Arduino Uno with the ESP32, which has a superior processing power, also in a prototype with the most accessible price and makes use of IoT concepts so that the user obtains a better experience with the use of the software.

4 Proposed Approach

The proposal of developing the low-cost power meter prototype consists of two modes:

1. Monophasic Mode:
    - Composed by a current sensor, a voltage sensor and the ESP32.
2. Biphasic Mode:
    - Composed of two current sensors and the ESP32.

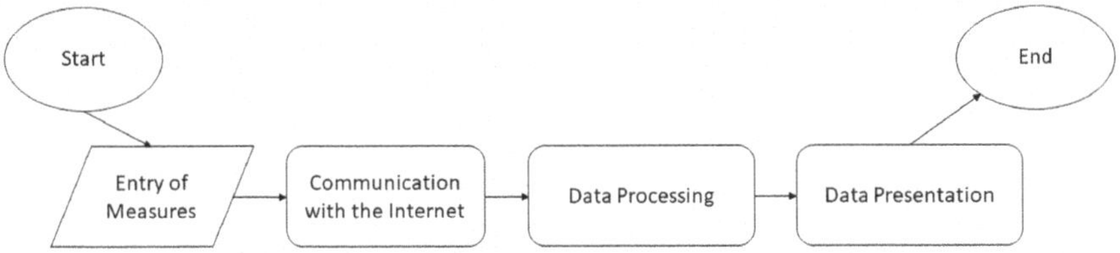

Fig. 1 Measurements on a lamp with the smart meter

The idea of the monophasic mode is to be able to measure the consumption in the equipment connected to the residence jacks whereas in the biphasic mode the measurement occurs in the general energy picture.

The flowchart of the meter operation proposed in the work can be seen in Fig. 1. Its contents are: (i) first, the data input is obtained through the sensors. (ii) The data are sent via the internet through client-server communication via the HTTP protocol with the help of ESP 32. (iii) Processing is done by ESP 32 and (iv) with some Arduino, the meter will send the data to a Web platform, for a better control of user data visualization.

The microcontroller chosen was ESP 32 from Espressif has a processing power three times higher compared to the Arduino one, with more memory, Bluetooth, capacitive sensors, integrated Wi-Fi module and low power consumption energy [9]. Main features:

- Main processor: LX6 32-bit Dual-core, operating 2–240 MHz
- Secondary processor: ULM (Ultra Low Power coprocessor) 8 MHz and consumes 150 uA.
- FLASH: 4 MB
- RAM: 520kB
- GPIO: 34, with 3.3 V and 12 mA among others.

5 Experiments and Results

The experiments carried out in this work had the purpose of validating the prototype measurements in monophasic mode with the aid of the PZEM-061 m, which has a four-parameter measurement function: voltage, current, active power and energy, having an energy metering range of 100 A/22,000 W and accuracy class equal to 1.0, each prototype would be allocated in a residence outlet. The comparisons of the prototype measurements with the PZEM-061 can be seen in the Table 2.

In biphasic mode evaluation, the prototype would be allocated in the general picture of the residence and the validations were carried out with the aid of DT266 digital ammeter. The comparisons of the prototype measurements with the DT266 can be seen in the Table 3.

Table 2 Comparison of measurements between prototype and Pzem-061

Variables	Prototype	PZEM-061
Tension (V)	124 V	125 V
Current (A)	0.35 A	0.36 A
Power (W)	41.29 W	42.25 W
Cost (R$)	R$ 124.71	–

Table 3 Comparison of measurements between prototype and DT266

Variables	Prototype	DT266
Tension (V)	127 V	128 V
Current 1 (A)	0.32 A	0.33 A
Current 2 (A)	0.27 A	0.28 A

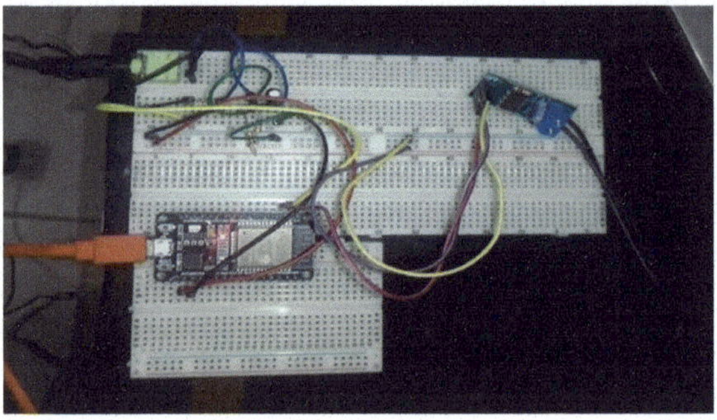

Fig. 2 Monophasic prototype: measurements and communication

The current prototype can be seen in Fig. 2 and this hardware represents the first phase of the process, the measurement phase. This hardware is composed of an ESP-WROOM32 microcontroller, a non-invasive current sensor SCT-013, 100 A, this sensor uses the magnetic properties to generate an induced alternating current that is proportional to the total alternating current of the circuit.

Another component is the voltage sensor P8, which detects the AC voltage (alternating current) and has an optocoupler, that ensures the insulation of the network, and can withstand an input voltage of 127/220 V. It is also composed of resistors, a 10 μF capacitor and a p2 connector for better sensor coupling.

The second phase is the communication phase, which comprises the web server, developed in the C language in conjunction with the Arduino libraries, responsible to accept client's requests through the HTTP protocol and to answer by sending the data to the client web platform, as can be seen in Fig. 3.

According to Aneel—National Electric Energy Agency, linked to the Ministry of Mines and Energy, was created to regulate the Brazilian electricity sector, through Law N° 9.427/1996 and Decree N° 2.335/1997. The main requirements of the monophasic metering systems for group B are: class of meter A, which has an error tolerance of at most more or less 2.5, having as a mandatory parameter presenta-

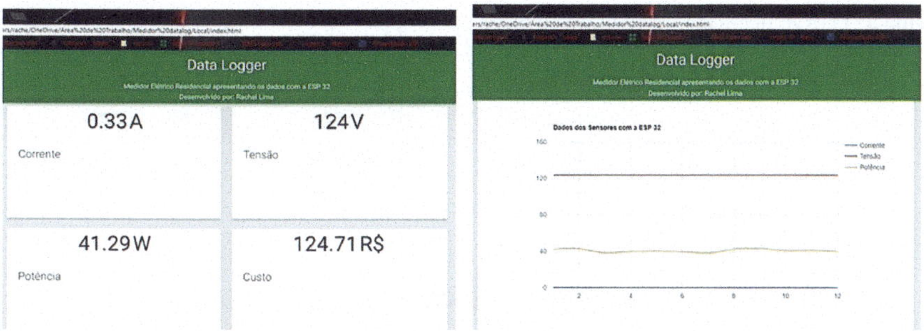

Fig. 3 Web platform

Fig. 4 Biphasic prototype

tion of the active power. Therefore, the prototype of the meter meets the proposed requirements [15].

The biphasic prototype, as can be seen in Fig. 4 is coupled in the phases located in the general energy picture of the residence. It consists of the ESP-WROOM32 microcontroller, two current sensors SCT-013, one capacitor 10 μF, resistors and two connectors P2 for better coupling. To validate the results, the DT266 ammeter was used.

According to Aneel, the main requirements of the biphasic medical systems for group B are: class of meter B, which has an error tolerance of at most ±1.5%. Therefore, some refinements are still needed to obtain a better measurement, such as the implementation of the voltage calculation [15].

The cost of making the monophasic and biphasic prototypes is in line with the market values of microprocessor systems in Manaus. If we compare with the version presented in the first stage of the conclusion work, we can see that the current prototype has a better cost benefit with the use of the ESP32 microcontroller, the values can be observed in the following Table 4:

Table 4 Cost of the main components of prototypes

Component	Monophasic	Biphasic
Tension	R$ 50.00	R$ 50.00
Current	R$ 75.00	R$ 150.00
Power	R$ 20.00	–
Cost	R$ 145.00	R$ 200.00

6 Conclusion and Further Projects

In this work, the prototype of a monophasic and biphasic electrical energy meter for residential consumption monitoring, based on a low-cost microcontroller, was presented.

As a result of the monophasic mode of measurement experiments, all the requirements were met in relation to the class of precision and parameters that it needed to have [15]. With respect to the biphasic prototype, some adjustments are necessary mainly in relation to the voltage measurement [15]. In general, the prototype of the monophasic and biphasic meter corresponded to the expectations, achieving the communication with a web platform and being consolidated with a low-cost IoT platform. The use of the ESP 32 microcontroller added considerably so that all communication functions could be implemented more easily at a lower cost.

As future works, improvements will be made to the measurement, as well as the study and implementation of a protocol of the internet paradigm of things, in order to acquire control of the data to the user to obtain the history of residential consumption and the development of the module for three-phase systems.

References

1. Brazilian Association of Conservation Services Companies of Energy—ABESCO: Waste of generated losses of 12.6 billion 2017 [Online]. Available: http://www.abesco.com.br/en/novidade/desperdicio-de-energia-geraperdas-of-r-126-billion (15 Sep 2018)
2. Energy Research Company—EPE: Statistical Yearbook of Electrical Energy 2017 [Online]. Available: http://www.epe.gov.br/AnuarioEstatisticodeEnergiaEletrica/Forms/Anurio.aspx (23 Mar 2018)
3. Brazilian Association of Large Industrial Energy Consumers and Free Consumers—ABRACE: An Introduction to the Sector Electric 2017 [Online]. Available: http://www.abrace.org.br/wpcontent/uploads/2015/12/manualelectricpower.pdf (24 April 2018)
4. Agostini, N.: Electricity and Energy. Brazilian System of Technology—SIBRATEC (2016)
5. Architects, E.: What is an embedded system? 2018 [Online]. Available: http://www.embarc.com.br/p1600.aspx (28 June 2018)
6. What's New?: How to make your reading, 2017 [Online]. Available: https://goo.gl/NYP5BW JUNIOR, J. S. d. S. (2018, May 30). Single systems basic, two-phase and three-phase 2018 [Online]. Available: https://mundoeducacao.bol.uol.com.br/fisica/sistemas-monofasicos-bifasicos-trifasicos.htm (12 May 2017)
7. Gimenez, S.P.: Microcontrollers 8051: Theory of Hardware and Software/Applications in Digital Control/Laboratory and Simulation. Pearson Education of Brazil Ltd, São Paulo (2002)

8. De Oliveira, S.: Internet of Things with ESP8266, Arduino and Raspberry Pi. Novatec Editora (2017)
9. EXPRESSIF: ESP32 Datasheet 2018 [Online]. Available: https://www.espressif.com/sites/default/files/documentation/esp32datasheeten.pdf (30 April 2018)
10. Tanenbaum, A.S.: Computer networks. Pearson Education (2003)
11. Silva, I.A.D.: Intelligent Voltage-Current Sensor with Monitoring and Control by Smartphone (2016)
12. de Lima, R.B., da Silva, E.L.: Prototype of a Single-Phase Power Meter for Consumption Monitoring Based, Low-Cost Microcontroller, ERIN (2017)
13. Pereira, L.H., Jr.: Monitoring of Energy Consumption Electric a and Control of Equipment Via Application (2018)
14. Silva, Igor Alexandre Dutra e. Intelligent Voltage-Current Sensor with Smartphone Monitoring and Control Online, UFRS (2016)
15. ANEEL: Public Consultations 2017 [Online]. http://www2.aneel.gov.br/aplicacoes/Consulta/publica/. Access date: 23/10/2017 (10 Dec 2017)

A Model for Medical Staff Idleness Minimization

C. E. V. Marinho, Vânia Vieira Estrela, Hermes José Loschi, Navid Razmjooy, Albany E. Herrmann, Y. Thiagarajan, Mikhail P. Vishnevski, Ana Carolina Borges Monteiro, Reinaldo Padilha França and Yuzo Iano

Abstract Operation research concepts can help healthcare facilities with unplanned situations, crisis, and handling of supplies, utilities, and strategies. A methodology for dealing with the Man-Hour (MH) distribution in a cellular type Healthcare Organization (HCO) is introduced, and its goal is to maximize the use of resources and workforce. The problem is modeled via Linear Programming (LP), which results in

---

C. E. V. Marinho
FACITEC, Distrito Federal, Brasília, Brazil
e-mail: carlos.marinho@estacio.br

V. Vieira Estrela (✉)
Department of Telecommunications, Fluminense Federal University, Niterói, RJ, Brazil
e-mail: vania.estrela.phd@ieee.org

H. J. Loschi · A. C. B. Monteiro · R. P. França · Y. Iano
FEEC, University of Campinas—UNICAMP, Campinas, SP, Brazil
e-mail: eng.hermes.loschi@ieee.org

A. C. B. Monteiro
e-mail: monteiro@decom.fee.unicamp.com

R. P. França
e-mail: padilha@decom.fee.unicamp.com

Y. Iano
e-mail: yuzo@decom.fee.unicamp.com

N. Razmjooy
Department of Electrical Engineering, Tafresh University, Tafresh, Iran
e-mail: xnavid@gmail.com

A. E. Herrmann
Department of Romanistic, University of Tuebingen, Tuebingen, Baden-Wuerttemberg, Germany
e-mail: albany.estrela@gmail.com

Brasilien-Zentrum—Uni Tuebingen, Tuebingen, Baden-Wuerttemberg, Germany

Y. Thiagarajan
Sri Venkateshwaraa College of Eng. and Technology, Pondicherry, India
e-mail: thiagu2517@gmail.com

Y. Iano et al. (eds.), *Proceedings of the 4th Brazilian Technology Symposium (BTSym'18)*, Smart Innovation, Systems and Technologies 140,
https://doi.org/10.1007/978-3-030-16053-1_62

a minimal cost flow problem with the simplex algorithm. The proposed framework (MCFP) can motivate individuals, reward ability and individual knowledge (not only moneywise) plus improving patient care. Such model can help to devise a software tool for decision making with performance and efficiency; it brings direct profit for the HCO and its staff due to the superior management of their MHs, and it increases the benefits to the community they serve. The methodology models MHs in an HCO to optimize management, decision-making tasks, and resource distribution.

Keywords Idleness minimization · Healthcare system model · Minimal cost flow problem · Motivation rewards · Linear programming · Graph theory applications · Simplex algorithm · Healthcare work assignment

1 Introduction

This work delineates key resource strategies and concepts to manage human resources in Healthcare Organizations (HCOs) better using applied operations research concepts [1, 2]. Any HCO requires a complicated process for allocating resources and utilities/services into some modules. Money, workforce, buildings, and equipment must be allocated to various administrative levels, locations, client groups, and institutions while responding to (1) the accessibility and transferability of the resources, (2) the demands or needs, and (3) the provision standards. Healthcare planning tries mostly to balance these factors, and with the exercise of priority decisions in those situations where needs cannot be met at acceptable standards with the present resources. Neglecting any element increases the demands for the other two [3–5].

A patient is an inpatient when staying hospitalized or an outpatient when going to the clinic for diagnosis and/or treatment without occupying a bed.

The inefficiencies drain the restricted resources allotted for healthcare in developing countries. Hospital managers and other agents want to maximize quality (in the absence of a profit motive) where the inpatient mortality rate is the leading indicator of quality.

Decision-makers must use ethics to conciliate the maximum benefit per patient with the maximum benefit for the entire community according to the following principles:

(a) They require being essentially just and caring for all individuals likewise.
(b) To treat all victims in the best possible manner.
(c) The best result for the most people with the resources on hand.
(d) The decision-making procedure and criteria should be clear.
(e) Consistency in their way to handle patients experiencing a crisis.

M. P. Vishnevski
LCMAT, UENF, Campos de Goytacazes, RJ, Brazil
e-mail: mikhail@uenf.br

(f) The amount of resource limitation should be proportional to the demands.
(g) Triage staff should be capable of securing their decisions via documented actions and estimated potential of them.

Managing expectations is vital for planning and executing reactions. Simulations before a real event make the staff calmer with resource provision plans and decisions in an incident, which yields the following benefits:

- Community members understand the notion of resource shortage;
- Better standards for triage of patients, supplies, and medicines; and
- The public develops confidence in medical providers.

Every day, administrators face decisions about what should go where and how. The decision-making problem resembles a transportation problem because it can be seen as an exchange process where resources flow from one working cell to another. A working cell can be a laboratory, department or other sector or some other location of production, consumption or consolidation in an HCO. Network flows help scheduling, personnel assignment, and DNA sequencing to name a few possibilities [6, 7].

An HCO corresponds to a network consisting of nodes and edges. Every node corresponds to a working cell. Each edge joins a pair of nodes, which correspond to a duo of working cells. Each edge has a capacity that restricts the amount it can transport. The goal is to balance the resources transported through the edges of the network, which lowers costs for service providers and improves service to customers.

Network flow problems generate more efficient solutions computationally speaking. The Shortest Path (SP) and Minimum Cost Flow (MCF) are fundamental operations research problems. Algorithms for these problems also function as building blocks to create healthcare management software for complex situations encountered in HCOs. Therefore, extensive literature on several features of these two problems to unravel them in polynomial time. Nevertheless, adding one or more extra constraints makes them intractable. This paper concentrates on the SP problem, which entails the network path estimation with minimum cost from an initial node to an end node so that the total delay sum of the paths becomes smaller than or up to a threshold.

The MCF Problem (MCFP) consists of transmitting flow from a given set of supply nodes, over the network arcs to a group of demand vertices, at a minimum total cost, and without infringing the lower and upper bounds on flows from each side of the arcs. The MCF structure is wide-ranging and can model some more specific network problems, like Assignment, Transport and Transshipment. The Network Simplex Method (NSM) corresponds to a variant of the bounded variable primal simplex procedure, especially for the MCF problem. The primary network contains a rooted spanning tree, in which arcs represent variables. The process iterates by interchanging basic and non-basic arcs until it finds an optimal solution. Some pricing strategy chooses an incoming arc while establishing an arc to leave the basis at each iteration. There are many policies for picking out the entering arc, and they influence the convergence to the solution.

This paper intends to introduce a model to handle workforce imbalances in the context of HCO. Section 2 sets scenarios where the proposed model can be used.

Section 3 contextualizes the allocation of resources as an optimization problem. The specific model is presented in Sect. 4. Finally, the conclusions and future work appear in Sect. 5.

2 Problem Characterization

Throughout the crises, a calamity or disease occurrence, healthcare providers may be forced to make complicated decisions concerning the allocation of resources and medications. However, epidemics and natural crises are not the only times when doctors may have to make tough calls about who gets what. The lists of scarce medicines in hospitals and clinics bring to mind the real possibility of not having access to crucial medical supplies and medicines when they are required. Planning for shortages is not only sensible but also essential.

Calls for cost restraint, legislation for managing competition plans, and the financial demands of customers increased financial responsibility. These issues suggest a greater need for sound administration principles, including purposeful planning and resource allocation models in HCOs.

During a tragedy, medical care occurs along a range that goes from usual to the crisis. The objective is to stay in either usual or emergency care mode while feasible, as this presents the least risk to patients. HCO with the best expansion capacity and vigilance plans should be able to run longer in these modes than ill-prepared facilities. Throughout significant disasters, appropriate strategies must be implemented. With these appear a heightened risk of poor outcomes for individuals, as existing resources must be applied strategically to benefit the population as a whole, instead of some patients. When feasible, suppliers should exploit a hands-on triage procedure (one with formal decisiveness standards) rather than determine reactively according to their will. Inducing a proactive triage attack, having selected triage staffs, using decision instruments, and integrating with the event command system all will develop the value of the decisions reached.

The most severe dilemma confronting hospital administrators nowadays is linked to business and economic management, above all the motive to control spending and preserve resources [6]. Still, commercial and monetary issues are among the most challenging and sensitive areas in analyzes. Issues on administrative efficiency and managerial execution have been regarded as problems for the past decades in healthcare management.

Even though outer investigation and internal appraisal matters are becoming the dominant worry of healthcare executives, the managerial tools employed by these professionals have not kept cadence. Few hospitals, e.g., are presently using any form of environmental analysis, user surveys, and financial plans and resource distribution procedures in their strategic preparation practices [8].

The collapse to a slot in resource allocation procedures in strategic provision may come from being concentrated on revenue augmentation and economic processes. Upcoming tactical preparation methods and models utilized by hospitals must integrate resource use models, which keep resources, get rid of duplication of practices and equipment within departments, and permit efficient price control.

HCOs need formal strategic administration tools, which offer direction for internal supply allotment decisions as well as external awareness and promotional expense allocations. HCOs need more sophisticated data, furnished by clients, which should be employed to get better the medical organization's aptitude to please its clientele by exceeding their service prospects.

The usage of client surveys assessing satisfaction in medical care can ameliorate the effect of healthcare administration. Data from client surveys can help executives create better internal and external supply allocation determinations [1].

2.1 Methodology

The initial analysis step creates full documentation of medical services (e.g., inpatient tasks like obstetric or pediatric treatment, cardiology, and respiratory rehabilitation; and outpatient services like cancer treatment, and radiology). Surveys assess these issues:

(a) the residents' awareness degree about the health services on hand by the neighborhood hospital, (b) the significance they attach to having each of these services accessible within the society, and (c) their hospital's operation opinion when it comes to delivering services based on their individual experience or on what they had discovered from others. HCOs must deliver services according to the target community to thrive among the competition. Emergency aid generally is the essential outpatient service while nutritional assistance and prenatal instruction are less critical.

2.2 Strategic Implications

The bulk of the services labeled as very important and extremely pleasing made up such as intensive-care nursing, cardiology, surgical accompaniment, laboratory practices, mammography, and respiratory treatment, radiology, and mammography, which are tactical for HCOs and demand functionality safeguarding.

To conclude, a mixture of inpatient and outpatient routines above all, vital services, health nursing, outpatient surgical procedures, and maybe oncology—can be regarded services that echo the HCOs domains of strategic limitations. Therefore, they have a propensity to provide the best occasion for bettering performance and client happiness. To achieve the utmost advantage from the costs of resources and/or managerial deliberation, these services should be given exceptional priority in any intervention/change effort.

Such inpatient-customized services require a policy concerning marketing/promotional actions. The bulk of resources and campaigns for advertising the HCO services should be spent on more essential areas for the clientele and turn them into opportunity market areas. Other less apparent but still potentially significant service contributions that may deserve more attention include inpatient nuclear medicine and ultrasound, physiotherapy, outpatient urology, and mammography. Advertisement of services should be closely supervised and providentially restructured before primary extra resources or efforts are spent. These may entail support services necessary for a full-service healthcare center, but without central importance to the community being served. Other promotional supervising should entail a periodic reassessment of the public necessities to resolve if services are rising in value to patients and/or such non-patient hospital components as regional physicians or insurance institutions [5, 9].

3 Mathematical Formulation of the MH Flow Problem

Network flow inquires the time to proceed somewhere subsequently and how [10–12]. Assuming that labor is the answer to individual development and progress of the person and as a result, the idleness is not something desired, this text reorganizes the workforce in the organizations to reduce the idleness of its employees and to increase its production of it. Additionally to causing financial losses to the HCO proprietors, the idleness dilemma in the workspace interferes subjectively on the worker, and lack of motivation is its most significant drawback [8, 10].

In this work, the idleness problem for an HCO is treated as a network flow problem, and this flow is expressed in Man-Hours (MHs). When estimating the amount of MHs to be consumed by a productive system, there is a corresponding fixed MH amount per month. Thus, the proposed production model has to be adjusted to real or predictable demand. The workforce apportioned to a specified cell can become unused all through some periods, while other cells can become overloaded due to scarce supplies. For the last case, instead of seeking human resources outside the HCO, workforce migration is allowed from one cell to another, and this process is called MH flow.

It is indispensable to identify the price involved in this procedure so that the MH replacement can be done in the HCO. By cell, it is meant a room, sector, department or another type of unit that gathers staff, utilities, and resources.

To supply the cost of the transport of expertise within the hospital requires organization and financial mapping of that business and information on the healthcare workers of each cell and also knowledge and documentation of all the productive processes the group.

A directed graph $G = (V, E)$ consisting of working cells (aka nodes or vertices) connected or not by edges (i, j) with n nodes and m arcs can characterize an HCO. The directed network has edge costs c_{ij} for each edge $(i, j) \in E$, denoting the cost to transport one MH flow unit between nodes over that edge. n designates the number V of vertices, and m the number of edges in the graph.

To typify the problem as a MCFP, one may think that cells with idleness create a supply of MHs. Then again, cells whose requirements are not fitted to attract this stream of MHs according to a linear programming problem with nodes in place of organization cells as follows:

$$\begin{aligned} &\text{Minimize } z = \sum_{(i,j)\in E} c_{ij}x_{ij} \\ &\text{subject to: } \sum_{(i,j)\in E} x_{ij} - \sum_{(j,i)\in E} x_{ji} = d_i, \quad i = 1,\ldots,n \\ &\qquad\qquad x_{ij} > 0, \quad \forall (i,j) \in E \end{aligned} \tag{1}$$

where z is called the minimization functional;
x_{ij} is the MH flow between cells; and
d_i denotes the current state of a cell in terms of MH.

If $d_i > 0$, then a cell can supply MHs. When $d_i < 0$, then it is an attractor or demanding cell. $d_i = 0$ means that the cell is just part of a transmitting path (transshipment node). A minimum cost flow problem that has only transshipment nodes is also called a minimum cost circulation problem. The variables x_{ij} indicates how many units of flow are sent over the edge (i, j).

It is important that transportation costs also carry information on the skills of the workers from the demanding cell. The production cells possess different values added to the production costs and by the degree of workforce specialization. E.g., if a cell contains highly informed individuals, then MH cannot migrate to a node requiring basic knowledge for a given operation. In this case, the MH flow between these cells must be penalized. By definition, the graph nodes are numbered according to the importance of each cell in the business scenario. The more important a cell is, the higher the number generated. It is also assumed that cell (node) 1 will supply MH because it has the smallest importance index (α).

4 Optimizing the MH Flow Between Cells

The MCFP optimizes the MH flow between the cells of the model [6, 7, 10]. Clarify what the following cells define based on [6, 7] and explain their relationship. In the first step, the application of the algorithm makes MH available, so that its origin cells have no loss since there is a cost to maintain a worker in a cell even if he is inactive. The simplex algorithm that resolves the MCFP is founded on the exploration of the best possible Generating Tree (GT), which represents the optimum allotment of the MH flows between nodes, with smallest cost, after some iterations (Fig. 1).

The numbers above the graph nodes show the supplies and demands of MHs. For this model, cells 1 and 2 have positive numbers; hence, they contain rations of MH whereas the others whose cells have negative numbers have demands. The number related to each arc explains the transfer cost. If an arc has an associated negative cost, then this means that there is an affinity or a partnership between the two cells required. All supplies in the network have to be exhausted within the network. The summation of supplies is equal to the sum of requirements.

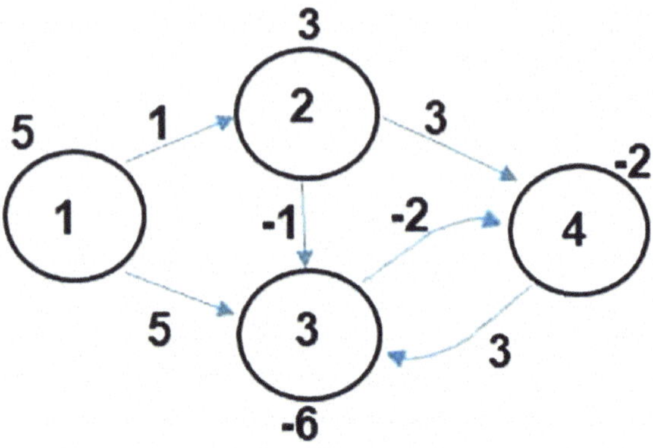

Fig. 1 State flow diagram—Graph of the MCFP model

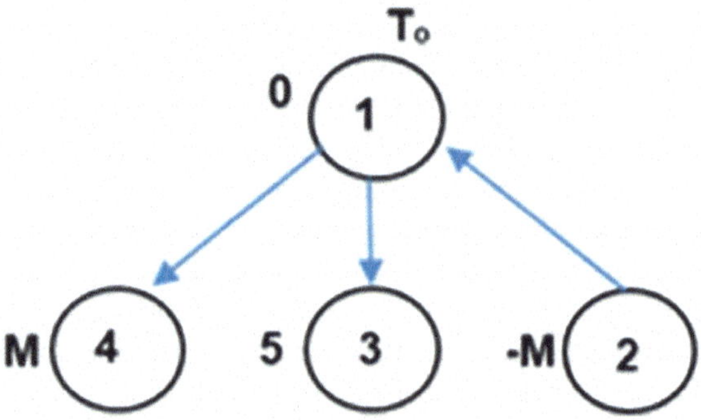

Fig. 2 Initial basic and optimum solution of the MCFP model

The MCFP algorithm determines a fundamental GT for each iteration where a solution can be represented by the arcs belonging to the GT (set of basic arcs), and by the set of non-basic arcs, which are the ones that belong to the original graph and are not part of the tree. The GT represents an optimum solution for the MCFP when the reduced costs of the non-basic arc satisfy the optimality conditions. By definition, node 1 (cell 1) will always be the root of the GT. Try to give a better understanding of the next generated tree explain what is T_O, M, and—M (Fig. 2 depicts the initial basic solution).

In the first fundamental tree, all the cells must be linked to the root node (cell 1). If there is no original relationship between the root and the nodes of the graph, then artificial arcs must be introduced in the tree. Hence, there is a sure path between the root and a node, with the matching costs sanctioned with large values M. The direction of the arrows connecting nodes hinges on the fact that a given node is related to demand or supply. As there is demand, the orientation is from the root to the cell, if not, it is from the node to the root. If an unreal arc joins a provision node to the root node, this arc must be leaning from node to root. Namely, the root draws the flow of this cell. While the procedure develops, all the artificial arcs must be taken from the tree by the technique. When the optimal answer contains one or two artificial arcs, the problem is not workable.

For each interaction, the algorithm reviews all the non-basic arcs employing the computation of their individual reduced costs (C_{ij}^{w}). The arcs meeting the optimality stipulations (that is, $C_{ij}^{w} \geq 0$) stay classified as non-basic arcs. The only candidates to get into the tree are the ones going against the optimality conditions ($C_{ij}^{w} \geq 0$). If there is more than one contender with negative cut down costs, the most negative one is chosen to quicken the optimization [7]. The reduced arc costs are calculated as

$$C_{ij}^{w} = w(i) - w(j) + C_{ij}, \tag{2}$$

where i is a source node, j is a target cell, C_{ij} is the transport expenditure for arc (i, j) $\in$ A and w(i) and w(j) are dual variables. This process of transforming a certain tree T into another one is identified as a pivoting procedure.

Adding an arc to the tree produces a cycle that must be cracked throughout pivoting to create a tree. By definition, a tree gives all the graph cells with only $n-1$ arcs relating these nodes, where n is the number of cells. When this cycle sets up a descendant relation for the nodes, there will be the so-called negative cycle. For the negative cycle, the betterment is limited to the cycle, and all the nodes that do not belong to this cycle are lost. In the absence of descendants (no arcs opposing the direction of the entrance arc), the entering arc flow will be the smallest one among the arc flows with contrary directions.

While updating the arc flows of the cycle, the arcs whose orientation are contrary to the entering arc will have their flows decreased by the amount equivalent to the flow assumed for the entering arc or increased by the same amount in case they have the same orientation. An arc whose flow is zero will be removed from the GT. The dual values for the new GT will be updated according to the expression $w_j = w_i + C_{ij}$ when j is a descendant of i. Otherwise, $w_j = w_i - C_{ij}$, that is, i descends from j. To check the new tree for optimality, one must compute the reduced costs of the non-basic arcs. Once the optimality conditions are verified, then we can say the tree is optimal. If not, an optimal solution must be sought, which means the search process must go on. Figure 3 represents an HCO floor. The nodes correspond to production cells, and the arcs are equivalent to the MH flow possibilities between them. If the node number is positive, then there is a supply of MH. Negative ones stand for demands. The number on top of the arcs represents the cost to transport one MH unit.

Arc (5, 4) has the most negative (<<) reduced rate. Hence, it must be added to the tree. The resulting tree with new updated flows and prices is presented in Fig. 4. The computation of the reduced costs for the non-basic arcs indicates the optimal conditioning of T_2. In the optimized tree (Fig. 4), node 1 (node 1) meets the MH demand of cell 2 entirely while supplying MHs to fulfill cell 3 demands and, additionally, providing 50 MHs to be consumed by the next nodes. In this case, the supply is entirely consumed by cell 4, which also absorbs all the supply from cell 5.

The relationship between supply and demand must go from a cell with less importance to a busier one. This extra value is characterized by the cell factor α [8].

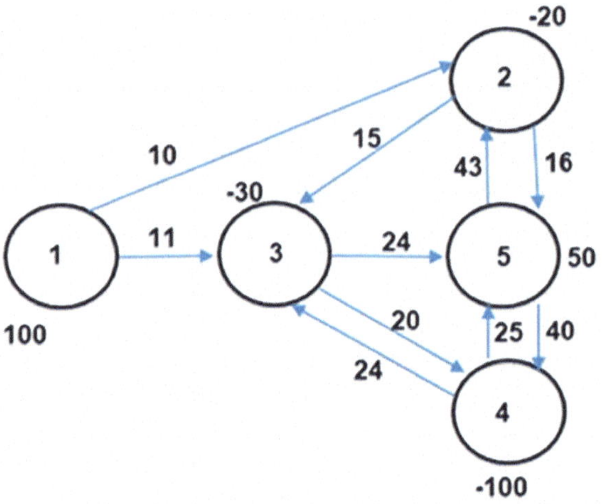

Fig. 3 State flow diagram—Graph representation of the problem (T_1)

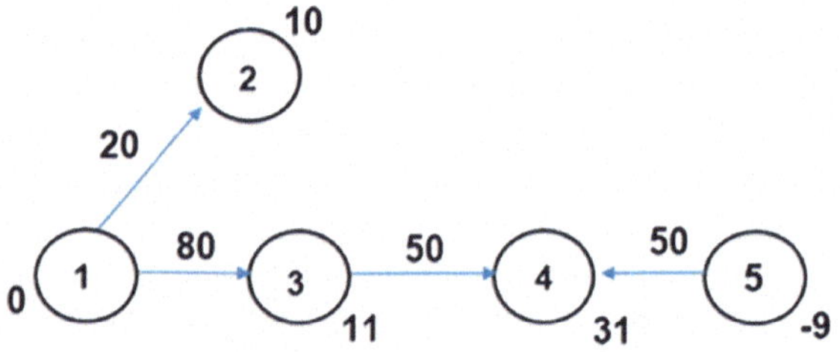

Fig. 4 Improved solution (tree T_2)

Nodes are numbered in crescent order of their αs. The node increase ($Profit_i$), which will be required to estimate the individual gain for this cell given by:

$$Profit_i = X_{ij}{}^*CHHM_j - X_{ij}{}^*C_{ij}, \text{ where} \tag{3}$$

where X_{ij} is the MH path transported from node i to cell j; $CHHM_j$ is the average cost of MHs in cell j, and C_{iJ} is the transportation expenditure of a unit of MH from cell i to cell j, that is, the cost of node i to keep a worker in this cell. A worker's weight estimation (PF) will be given to each staff element, and his/her gain will be proportional to it. The worker's weight becomes

$$\mathrm{PF}_{ij} = \sum \mathrm{TEC}_{ij} \cdot \alpha_k, \tag{4}$$

where i is the node index related to the worker origin; j is the employee index in cell i; k is the node index where the member of staff j was transmitted; TEC is the quantity of MHs of staff member j in cell k; and α is the importance factor of cell k, in the HCO scenery. The individual gain per cell is proportional to the weight assigned to each healthcare professional as:

$$\frac{Profit_i}{\sum_{j=1}^{m} PF_{ij}} = \frac{Gain_1}{PF_{i1}} = \frac{Gain_2}{PF_{i2}} = \cdots = \frac{Gain_m}{PF_{im}}, i = 1,2,\ldots,n \quad (5)$$

where $Gain_j$ is the gain of employee j; m is the entirety of workers in a node i, and PF_{ij} is the weight assigned to worker j in cell i. The MH allocation problem turns into a minimum cost flow one, where the cells providing and demanding MHs form a graph and the simplex procedure will determine the lowest cost flow between graph nodes.

The MCFP algorithm applies non-binary trees as data structures to represent the minor cost flow in the network; yet, there are practical issues, which produce pathologic or degenerated trees. So, it will be indispensable to include in the system some data handling routines to solve this kind of problem. To compute the negative flow when an arc with negative cost is added to the graph calls for a modified simplex algorithm.

5 Conclusion

As HCOs need to optimize the use of their internal resources, utilities, and staff reallocations to fund strategical programs, these assets may be shifted from some services to others. Lastly, service combination involves data from the organizational, therapeutic, and technical staffs [12–14].

This paper introduces a workforce allocation policy based on the MCFP algorithm to better handle patient care, medical data and the administrative process [2]. It performs analyses and follows the results over time to control the success level achieved while supporting HCOs to stay receptive to their constituencies and adapt to changing surrounds. The analyses result help to (a) evaluate the importance employees give to their jobs and outcomes, (b) identify areas in need of interference, and (c) evade unproductive workforce policies. These decision-making and organizational gains incorporate:

(i) well-defined info on the organization's strategic effectiveness and handicaps;
(ii) a greater discernment of the alteration types that can advance performance; and
(iii) explicit characterize where to focus managerial resources and decision-making.

The idleness and shortage of perspective concerning salary payoffs often frustrate people in the HCOs. Thanks to the current practice of selecting and hiring workers with multi-abilities, it is implicit that an employee can work for another cell, even though he/she may be allotted to another one [15]. This work suggests: (i) to study the redistribution the unused MHs on some cells to redirect them to other nodes, in an optimized manner, and (ii) to divide the gain while dispensing it fairly. The proposed idleness conduct is not an innovation as far as management policies function, but it models the idleness problem while encouraging a fair remediation scheme [8]. The MCFP helps the distribution of WHs so that the HCOs organizes processes and its workforce, which is already exceptionally valuable while documenting and normalizing of the procedures to obtain quality.

An HCO should plan for supply apportioning during crises and with additional discussions about allotment of restricted resources so that all actors can attack situations with self-assurance, knowing they are well prepared for decisions using the best existing data and in a way that fosters trust.

An administrator faces the dispute for resources to conform to the goals with economy and efficiency. The connection between the range of productive inputs and the corresponding outputs results in an output function, which depicts the maximum quantity of production that can be obtained from any specific combination of resources (or inputs) used in creating a product (or output) or in other words the connection between output and inputs that maximizes economic return.

1. Suppliers can limit waste and needless care.
2. Dispense resources fairly based on value [16–18].
3. Take on supervised care plans to disseminate healthcare resources more ably and cleverly where physicians may review policies to equilibrate the healthcare of a particular patient.
4. Ethical matters related to healthcare resource distribution.

This research opens new paths concerning the way WHs and resources are deployed [16–18] even in diverse contexts such as palliative care [19, 20], HCO management simulation [14], healthcare system design and resiliency [21], home care [20, 22, 23], and staff fatigue mitigation [24].

After the system via simulations, studies employing computational intelligence techniques [25, 26] to circumvent big data problems arising from the medical applications will be done [27–29]. The effects of cloud-available distributed computational resources [13, 25, 29] and different types of medical databases [27, 28] must also be considered since medical systems are cyber-physical systems [26, 30].

References

1. Dal Poz, M.R., Varella, T.: Methodology guide for the analysis of remuneration systems and human resource incentives in the health sector. UERJ, RJ, Brazil (1999)
2. Dorfman, R., Samuelson, P., Solow, R.M.: Linear Programming and Economic Analysis. McGraw-Hill Book Company (1958)
3. Agho, A.O.: Problem areas faced by hospital administrators. Hosp. H Serv. Adm **37**, 131–135 (1992)
4. Bergthold, L.A.: Purchasing Power in Health. Rutgers Univ. Press, New Brunswick (1990)
5. Hemmasi, M., Graf, L.A., Nielsen, W.: Triggering organizational change: application of importance satisfaction analysis. Creat. Inn. Man **1** (240–244) (1992)
6. Ahuja, R.K, Magnanti, T.L, Orlin, J.B.: Network Flows, Theory, Algorithms and Applications. Prentice-Hall, Inc (1993)
7. Cunningham, W.: A Network Simplex Method. Math Prog 11
8. Santos Neto, A.J.: Gain distribution methodology in production cells and idleness treatment as a minimum cost flow problem. M. Sc. thesis, UENF, Brazil (in Portuguese)

9. Estrela, V.V., Monteiro, A.C.B., França, R.P., Iano, Y., Khelassi, A., Razmjooy, N.: Health 4.0: applications, management, technologies and review. Med. Technol. J. 2(4) 262–276 (2019) https://doi.org/10.26415/2572-004x-vol2iss1p262-27
10. Marinho, C.E.V., Santos Neto, A.J., Paula Júnior, G.G.: Uma proposta de ganho individual a partir da minimização da ociosidade no espaço de trabalho. Vértices **5**(3), 109–133 (2003)
11. Aidman, E., Galanis, G., Manton, J., Vozzo, A., Bonner, M.: Evaluating human systems in military training. Aust. J. Psychol. **54**(3), 168 (2002)
12. Brandeau, M., Sainfort, F., Pierskalla, W.: (2004) Operations research and health care: a handbook of methods and applications. Kluwer Int'l Series, ISBN: 1 -4020-7629-0
13. Bhardwaj, T., Sharma, S.C.: Cloud-WBAN: an experimental framework for cloud- enabled wireless body area network with efficient virtual resource utilization. Sust. Comp. Inf. Syst. **20**, 14–33 (2018)
14. Monks, T., Robinson, S., Kotiadis, K.: Can involving clients in simulation studies help them solve their future problems? Transfer of learning experiment. Eur. J. Oper. Res. **249**(3), 919–930 (2016)
15. Brozino, J.D., Gover, J.E.: Medical technology: a solution to the healthcare cost problem. IEEE Eng. Med. B Mag. **13**(3), 313–315 (1994). https://doi.org/10.1109/51.293995
16. Hemmasi, M., Strong, K.C., Taylor, S.A.: Measuring service quality for strategic planning and analysis in service firms. J. Appl. Bus. Res. 10(24–33) (1994)
17. Zallocco, R.L., Joseph, W.B.: Strategic market planning in hospitals: Is it done? Does it work?. J. Health Care M. 11(5–11) (1991)
18. Bond, C.A., Raehl, C.L., Pitterle, M.E., Franke, T.: Health care professional staffing, hospital characteristics, and hospital mortality rates. Pharmacotherapy 19(130–138) (1999)
19. Clark, J., Barnes, A., Gardiner, C.: Reframing global palliative care advocacy for the sustainable development goal era: a qualitative study of the views of international palliative care. Exp. J. Pain S Man **56**(3), 363–370 (2018)
20. Adhvaryu, A., Nyshadham, A.: Health, enterprise, and labor complementarity in the household. J. Develop. Econ. **126**, 91–111 (2017)
21. Rosso, C.B., Saurin, T.A.: The joint use of resilience engineering and lean production for work system design: a study in healthcare. A Ergon **71**, 45–56 (2018)
22. Napier, P., Norris, P., Braund, R.: Introducing a checking technician allows pharmacists to spend more time on patient-focused activities. Res. Soc. Adm. Pharm. **14**(4), 382–386 (2018)
23. Gallipoli, G., Makridis, C.A.: Work and tax evasion incentive effects of social insurance programs: evidence from an employment-based benefit extension. J. Pub. E **117**, 211–228 (2014)
24. Givi, Z.S., Jaber, M.Y., Neumann, W.P.: Modelling worker reliability with learning and fatigue. Appl. Math. Model. **39**(17), 5186–5199 (2015)
25. Hemanth, D.J., Estrela, V.V.: Deep learning for image processing applications. Adv. Par. Comp. IOS Press. ISBN978-1-61499-821-1 (print) 978-1-61499-822-8 (online) (2017)
26. Wu, Z., Guo, Y., Lin, W., Yu, S., Ji, Y.: A weighted deep representation learning model for imbalanced fault diagnosis in cyber-physical systems. Sensors **18**, 1096 (2018)
27. de Jesus, M.A., Estrela, V.V.: An introduction to data mining applied to health oriented databases. OJCST **9**(3) (2017). https://doi.org/10.13005/ojcst/09.03.03
28. Herrmann, A.E., Estrela, V.V.: Content-based image retrieval (CBIR) in remote clinical diagnosis and healthcare. In: Encyclopedia of E-health and Telemedicine. IGI Global, pp. 495–520 (2016). https://doi.org/10.4018/978-1-4666-9978-6.ch039
29. Aviso, K.B., Mayol, A.P., Promentilla, M.A.B., Santos, J.R., Tan, R.R., Ubando, A.T., Yu, K.D.S.: Allocating human resources in organizations operating under crisis conditions: a fuzzy input-output optimization modeling framework. Resour. Conserv. Recycl. **128**, 250–258 (2018)
30. Estrela, V.V., Saotome, O., Loschi, H.J., Hemanth, J., Farfan, W.S., Aroma, J., Saravanan. C., Grata, E.G.H. Emergency response cyber-physical framework for landslide avoidance with sustainable electronics. Technologies **6**(42) (2018)

Environmental Education and Its Influence on Childhood Learning for the Formation an Aware Citizen

Ana Claudia Mendes de Seixas and Giovanna Ramos Maccari

Abstract Environmental education is becoming a topic increasingly addressed in society, being of extreme importance for the participation of the individual in the midst of a community. Since the 1760s, with the Industrial Revolution, man began the degradation of the environment as a source of natural resources for the rapid production of the industrial means of the time, generating a great problem with different types of pollution in the world, among them the management of solid waste. It was only in the mid-1970s that this vision began to be changed. In Brazil, after many years, the National Policy on Solid Waste (Law 12,305 of 2010) was instituted aiming at the non-generation, reduction, reuse and adequate final destination in a way that does not harm the environment. The problem of waste only grows each year, generating concern about the future. n this way, environmental education enters as an instrument of awareness of people from the influence of individuals with society. Therefore, this work was developed for children, in the case students of the Pio XII Application School, aiming at the reduction and adequate management of solid waste, applying a suitable environmental education for a significant change of behavior from the creation of materials (primer and games easy to understand issues related to solid waste management.

Keywords Environmental education · Solid waste · University extension

1 Introduction

Since the Industrial Revolution in the eighteenth century, there has been an accelerated industrial, technological and population development throughout the world. This historical period initiated the exacerbated use of natural resources, these resources for the most part, inexhaustible and nonrenewable, seriously damaging the environment

A. C. M. de Seixas (✉) · G. R. Maccari
Pontificia Universidade Católica de Campinas, São Paulo, SP 08544, Brazil
e-mail: acmseixas@gmail.com

Y. Iano et al. (eds.), *Proceedings of the 4th Brazilian Technology Symposium (BTSym'18)*, Smart Innovation, Systems and Technologies 140,
https://doi.org/10.1007/978-3-030-16053-1_63

in which man lives. The combination of these factors led to an increased generation of solid waste without adequate management and disposal.

Due to the lack of knowledge added to the paradigm of consumption that the capitalist system brings with it, a new global perspective on the environment became necessary, starting in 1970, a point exposed in the Universal Declaration of Human Rights where it is said that every human being has the right to a quality of life [1].

In Brazil, several themes related to the environment were addressed over time. However, in 2010 alone the country started to worry about solid waste, with the publication of the National Policy on Solid Waste (PNRS—Law 12305/2010). The National Policy on Solid Waste (PNRS) defines itself as: "an economic and social development instrument characterized by a set of actions, procedures, and means to enable the collection and restitution of solid waste to the business sector, for reuse, in its cycle or in other cycles productive, or other environmentally appropriate final destination" [2].

According to data from the Ministry of the Environment, Brazil produces about 90 million tons of garbage per year and each Brazilian generates approximately 1 kg, depending on the region in which it resides and its purchasing power [3, 4]. In this context, as cities develop, it becomes necessary to sensitize the population about solid waste management, seeking alternatives to mitigate problems related to this.

In this way, according to the approach of Ref [5], environmental education is a topic of extreme importance for the formation of a conscious individual. It favors the formation of opinions related to life on the planet, dissemination of more sustainable practices between society and nature, seeking solutions to environmental problems, reinforcing the role of the school since this work is carried out in an educational environment [6].

Environmental Education applied in the city of Campinas-SP, Brazil, is an educational concept that proposes that the schools have actions regarding the environment, involving social actors in the school community, from the students to the relatives and employees collaborating with the development of new citizens [7, 8].

Therefore, the school is an ideal place for the production and dissemination of knowledge for a large number of people and at a stage where it is possible to form critical thinking about the environment. For this reason, this work was developed for children, in this case, students of the Pio XII Application College aiming at the reduction and proper management of solid waste, applying a suitable environmental education for a significant change on individual behavior.

2 Objectives

The objective of the present work was to create materials (booklet and educational games) of easy understanding with themes related to solid waste management in order to minimize the generation of waste with educational actions through awareness, developing critical thinking using 'learning through play' approaches and forming conscientious and respectful citizens in relation to the environment and the community in which they live.

3 Methodologic

Socio-educational workshops were carried out and aimed at attracting and encouraging students about the environment. The workshops began with the creation of a booklet on solid waste separation (Fig. 1). Selective waste management and program awareness help on creating a better understanding of its importance and encourages practices in homes.

This was done from parallel workshops for waste reuse and research in the school's computer lab. Its content refers to the explanation of the colors for waste separation, how to identify the materials and tips for efficient separation, as well as to inform the days of the week that the selective waste truck goes in the districts of the city of Campinas, so that the students and parents do the disposal of these wastes on the correct days.

On the other hand, the educational games were designed so that the students understood the importance of the environment in which they live in an educational way, from the re-reading of four games created by them and those responsible for the work.

Monopoly (Fig. 2) was made from the original game, modifying the lands by states in Brazil that could possibly be degraded and in need of recovery and for that reason, purchase of the lands was necessary for planting trees and recovering the lands biodiversity. This was fundamental for teaching about degraded areas, making this game so sought after by students.

The "Who Am I" game was adapted to solid waste learning, in which students found out the object by asking questions about waste to identify it and learned about characteristic of each residue (Fig. 3).

The elaboration of the games was also based on the research done in the computer lab and the re-reading of the original games. The "Conscious Memory Game" was adapted with the content from selective waste management, referring to the separation colors of the solid and electronic residues and helping to identify them, with the purpose of memorizing the colors of the selective waste management (Fig. 4).

The game "Environment detective" (Fig. 5) was adapted from the concept of areas degraded by electronic waste, in which the objective is to discover by means of "clues" what crime occurred, who committed it and where it happened and, finally, find one or more solutions to this environmental crime. The player that finds out committed crime, the suspect and the place, according to the letters of the confidential envelope, wins the game and should seek one or more solutions to it.

The booklet and games about students' daily themes were adapted to environmental themes. Both were elaborated in an educational and creative way with the help of the students themselves, so that they could understand the concepts about the themes and disseminate this idea to people living close to them.

Fig. 1 Selective waste management booklet. *Photos* Seixas 2018

Fig. 2 Game about monopoly. *Photos* Seixas 2018

Fig. 3 Game about "Who Am I". *Photos* Seixas 2018

Fig. 4 Conscious memory game with students. *Photos* Seixas 2018

Fig. 5 Environment detective. *Photos* Seixas 2018

4 Results and Discussion

With the execution of the workshops throughout the year and elaboration of the informative materials it was possible to identify students' interest in environmental issues as they prepared the informational materials. This interest in the workshops made it possible to change the student´s behavior and the involvement of the community surrounding them, once the parents and staff of the school also got involved, causing an increase in the reach of information and population about the environment. The booklet was given to students and games borrowed so they could play with other classmates from the school.

The information communicated about waste will directly interfere with landfills, as with the content learned about waste separation and management, it is possible to apply the selective separation and management and waste reduction, thus allocating less and less waste to landfills, causing a reduction in deposition area and thus increasing its lifetime.

5 Final Considerations

At the end of the workshops and elaboration of the information materials and games, it was possible to identify the behavioral change of some students who were part of the work, verifying the students' interest in separating the waste and carrying out processes for their reduction and reuse. For this reason, Environmental Education is fundamental for the formation of the individual.

The workshops related to the environment have fostered the awareness of those involved since childhood, being extremely important for their socialization process and for the formation of critical thinking about the environment, consumption and the importance of preserving it for collective good. This is the principle for spreading the idea to friends, family, teachers, or school staff.

Acknowledgements The Pro-Rectory of Extension and Community Affairs of the Pontifical Catholic University of Campinas for all support, which from the first moment made feasible and contributed to make Extension Work a reality. The Direction of the College of Application Pio XII of Campinas (SP) that make possible and support the Extension Project.

References

1. ONU.: Organização das Nações Unidas. Assembleia Geral da ONU. Declaração Universal dos Direitos Humanos. 217 (III) A. Paris (1948)
2. BRASIL.: Política Nacional dos Resíduos Sólidos- PNRS. Decreto-lei nº 12.305, de 2 de agosto de 2010. Disponível em: <http://www.planalto.gov.br/ccivil_03/_ato2007-2010/2010/lei/l12305.htm>. Acesso em 10 julho de 2017
3. IBGE: Instituto Brasileiro de Geografia e Estatística. Cuidando do lixo. Disponível em: ‹www.ibge.gov.br›. Acesso em: 11 de maio de 2018
4. Nunesmaia, M.F.S.: Lixo: soluções alternativas, p. 152. UFES, Feira de Santana (1997)
5. Pereira, C.G., Silva, L.F., Souza, L.O., Souza, R.F.: Um estudo diagnóstico sobre educação ambiental nas escolas do ensino fundamental e médio de São Miguel do Guamá-Pará. Enciclopédia Biosfera, **10** (2014)
6. BRASIL.: Política Nacional de Educação Ambiental - Lei nº 9795/1999. 1999. Acesso em 27 de maio de 2015. Disponível em http://www.planalto.gov.br/ccivil_03/leis/l6938.htm. Acesso em 11 de maio de 2018
7. Alencar, M.M.M.: Reciclagem de lixo numa escola pública do município de Salvador. Candombá –Revista Virtual 2005 jul/Dez; **1**(2), 96 –113
8. CAMPINAS.: Prefeitura Municipal de Campinas: Departamento de Limpeza Urbana (DLU). Campinas-SP (2018)
9. BRASIL.: Ministério do Meio Ambiente. Política Nacional dos Resíduos Sólidos- PNRS. Decreto-lei nº 12.305, de 2 de agosto de 2010. Disponível em: <http://www.planalto.gov.br/ccivil_03/_ato20072010/2010/lei/l12305.htm>. Acessado em 31 março de 2018

Poemathics

Leandro Sopeletto Carreiro, Vânia Vieira Estrela, Mikhail P. Vishnevski, Wilma D. Huacasi, Albany E. Herrmann, Hermes José Loschi, Yuzo Iano, Navid Razmjooy and Y. Thiagarajan

Abstract Mathematics is a discipline much feared by most students in schools. Handing out strategies to demystify this discipline through interdisciplinarity, that is, to show how mathematics can be used in other branches of knowledge is our goal. Thinking about this issue, we developed the POEMATHICS Project as an alternative

---

L. S. Carreiro (✉)
Mathematics, IFF, Campos de Goytacazes, RJ, Brazil
e-mail: leandro.carreiro@iff.edu.br

V. Vieira Estrela
Department of Telecommunications, Fluminense Federal University (UFF), Rio de Janeiro, Brazil
e-mail: vania.estrela.phd@ieee.org

M. P. Vishnevski · W. D. Huacasi
LCMAT, UENF, Campos de Goytacazes, RJ, Brazil
e-mail: mikhail@uenf.br

W. D. Huacasi
e-mail: wilma@uenf.br

A. E. Herrmann
Department of Romance Languages, University of Tübingen, Tübingen, Baden-Württemberg, Germany
e-mail: albany.estrela@gmail.com

Brasilien-Zentrum–Eberhard-Karls Universität Tübingen, Tübingen, Germany

H. J. Loschi · Y. Iano
LCV, Faculty of Electrical and Computer Engineering, UNICAMP, Campinas, SP, Brazil
e-mail: eng.hermes.loschi@ieee.org

Y. Iano
e-mail: yuzo@decom.fee.unicamp.com

N. Razmjooy
Department of Electrical Engineering, Tafresh University, Tafresh, Iran
e-mail: navid.razmjooy@ieee.org

Y. Thiagarajan
Sri Venkateshwaraa College of Engineering and Technology, Pondicherry, India
e-mail: thiagu2517@gmail.com

Y. Iano et al. (eds.), *Proceedings of the 4th Brazilian Technology Symposium (BTSym'18)*, Smart Innovation, Systems and Technologies 140,
https://doi.org/10.1007/978-3-030-16053-1_64

activity for the teaching of mathematics allied to the Portuguese language teaching. This initiative consists of relating the mathematics with the written language. Students should develop poems, poetry, texts, stories, music, etc., using mathematical terms and ideas. To promote this activity is necessary the participation of teachers of a language (this work deals with English examples and a workshop done in Portuguese) and Mathematics so that students can be accompanied and together awaken the pleasure of learning as well as improving math and linguistic skills.

Keywords Playful games · Mathematical education · STEM education · Cognition · Multiple intelligence types · Semantics · Semiotics · Neurolinguistics

1 Introduction

Intelligence, according to Sternberg is "the ability to solve abstract problems." For Terman, an individual is intelligent as far as he/she can think in abstract terms [1, 2]. "One can also conceptualize intelligence as the ability of a human being to create and/or modify things; The ability to extract the essence of certain content and still reflect and discuss it. For [2], "an intelligence implies the ability to solve problems or create products that are important in a particular environment or cultural community." Until very recently intelligence was measured through a series of psychometric tests, such as the IQ test. Most tests focus on logical-mathematical reasoning, and individuals who have good memorization and quick-thinking skills are better qualified. A person is considered intelligent when he/she has quick reasoning, good memorization capacity and takes good grades at school.

The intelligence is much more multifaceted than tests can measure, even because most tests measure only the logical-mathematical knowledge of individuals, not being able to evaluate, for example, their creation capacities [1]. Jean Piaget's studies emerged as a fashion counterpoint established by intelligence tests. For Piaget, the knowledge is not in the subject or the object, but it builds the interaction of the subject with the object. When the subject interacts with the object produces the ability to know and produce the knowledge itself [3, 4].

Another question about intelligence is the possibility that there are mechanisms that can increase or improve it. For [5], it is possible to stimulate intelligence through the use of vitamins, physical exercises (according to this author, some aerobic exercises when performed moderately allow better retention of information) and mental exercises such as reading books and crossword puzzles among others showing that the environment directly influences the intellectual capacity.

Different intelligence types can interact with each other, and an individual can use two or more types of intelligence [1, 2]. By viewing this spectrum, activities can develop, not only one but several intelligence forms simultaneously.

According to Gardner [1, 2], the brain has eight intelligence forms: linguistic or verbal, logical-mathematical, spatial, sound or musical, kinesthetic-corporal, naturalistic, biological or ecological, intrapersonal and interpersonal. This work considers logical-mathematical intelligence that involves the ability to recognize patterns, to work with abstract symbols (such as geometric numbers and shapes), as well as discern relations and/or see connections between separate parts and distinct them.

It is up to the school to stimulate these intelligence types by using pedagogical strategies (games, toys, and pranks) [3, 6, 7]. Thus, the teacher who teaches mathematics in the early series of elementary school should always act as a facilitator, the one who helps the student to overcome their limits—using creative activities and evaluations that allow the student to build meaningful learning and knowledge from the environment.

2 Importance of the Games Using Words and Ideas

Semiotics deals with the exploitation of all human senses, using them as antennas of receiving verbal and non-verbal messages, visible and invisible in the structure of the texts with which humans interact regularly.

Associating the semiotic contents with the training of mental discipline, the organization of scientific reasoning, thus contributing to the efficiency of the production of reading and, above all, the learning process of the Portuguese language. It is then seen that the goal is fundamentally pedagogical because it is intended to understand the pathways of semiosis and significant production better, focusing on the written, verbal representation as an eminently symbolic process [8, 9]. Semiotics studies the signs in general. So, the sound, the image, the ballet, the painting, the drawing, the writing, the spoken, etc., everything is semiotic matter including all school disciplines and thus can be used as a theoretical-methodological frame, establishing the transitional nexus between the discipline in focus and those intentions to relate to it.

2.1 Case Studies

Some examples of activities involving word games aiming at mental distraction, but also reinforce various aspects of the various intelligence types, besides contributing to the improvement of the capacity of expression. The examples below are from Karen Beatty [10, 11], who performed experiments with children of 4th and 5th series of the first degree of the Cadarackque Public School (Figs. 1, 2, 3 and 4).

2.2 The Evaluation of Mathematical Learning

Nowadays, schools, in the vast majority, have a policy of evaluation of school-based income assessment, as it were, in the approval/reproach dichotomy. In this context, there is no room for a practice of evaluation, which helps in the identification of

Fig. 1 How to write a math poem [10, 18]

Fig. 2 How to write a math poem [19]

Fig. 3 How to write a math poem: the slang problem [20–22]

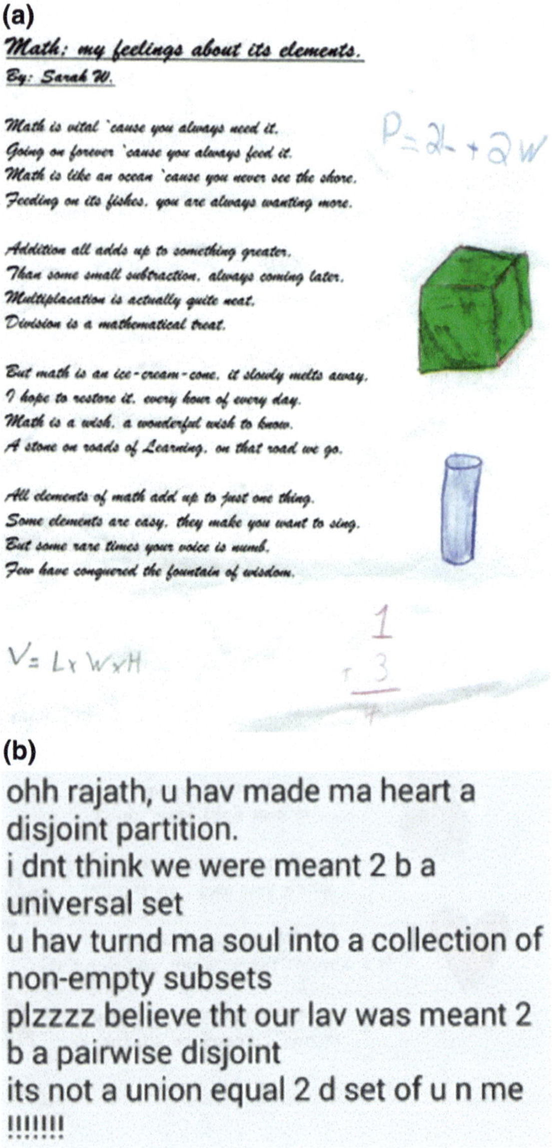

Fig. 4 Matharena: Linear equations lyrics by Vicki Young. Melody: "Macarena" [22]

We're solving equations; we want to find the answer.
It has to fit just right; that's the number we are after.
At first we'll take our time, but then we'll travel faster!
Do the matharena!

Solving equations, I want to be emphatic!
You must decide: is it linear or quadratic?
Then check your plan - plan to check, while you're at it!
Do the matharena!

What is the plan when "x" is only first degree?
Clear - that means distribute - and collect like terms you see!
Last (but not least) you must cancel so that "x" is free!
Do the matharena!

What do you do if certain terms just disappear?
There are 3 cases - we need to make them crystal clear!
One solution, no solution, or any number anywhere!
Do the matharena!

As long as an "x" is left in your equation,
You'll have one solution, you'll find it if you're patient!
That answer will be zero if the number side is vacant!
Do the matharena!

When both sides are zero, your answer's any number.
When only letters drop out, what's left may make you wonder:
The answer's no solution - you no longer have to ponder!
Do the matharena!

overcome difficulties in the teaching and learning process, both sides: of the pupil and the teacher.

The evaluation in most of our schools, public or not, is eminently summed up, always concerned about the final results that lead to irreversible situations concerning the students' performance, without being taken into account to many implications including social, a fatal decision-making process from an educational standpoint.

The idea of combining mathematical concepts with literature, in particular with poetry, is not new. Concerning Brazil, one can mention the "Mathematical Poem" by Millôr Fernandes and the song "Math Class" by Tom Jobim.

3 The POEMATHICS Project

Some research in first and second-degree schools revealed that many students present difficulties in mathematics, in general, and in activities requiring writing, where students need to express their ideas in written form. They present several problems related to the organization and chain of ideas, as well as vocabulary deficiencies [12, 13].

Based on these shortcomings, the development of activities that stimulate both logical-mathematical and linguistic intelligence are sought. Mathematics national curricular parameters consider the discussions and reflections in the context of contemporary mathematical education; however, about the development and stimulation, of the intelligence forms, this was not enough, perhaps because of the shortage of adequate literature. These parameters should emphasize more general issues since this topic was addressed superficially in Brazilian curricular proposals, and without much attention during the previous training of the teachers.

3.1 Objectives

It is with the intention of demystifying the mathematics, by learning while associating it to a gratifying and kind act, developing the logical reasoning and the ability to use it together with other disciplines, that we elaborate such activity. Students are expected, at the end of the activity, to be able to write texts using technical writing techniques, seeking to use mathematical terms coherently in their work.

The primary objective is to achieve that students use all their knowledge of mathematics and Portuguese to draw up their texts. This activity will stimulate research, both in mathematics books and in Portuguese books and dictionaries. The teacher will be able to emphasize the content that is being taught in the period in which the activity is carried out, making the learning more enjoyable. These activities require students to exercise their logical reasoning, among other requirements. Students will have to relate mathematical terms to insert them into text (prose or verse). There will arise the need to choose the words and terms so that the text can acquire meaning.

3.2 Methods

The activity requires the partnership between Portuguese and Math teachers besides the players. Students should write texts, poetry, prose, stories, using mathematical terms always seeking to express ideas and coherently compose sentences or verses. It is necessary, first, to develop the prerequisites. One of them belongs to the Portuguese teacher, who should work on the writing and spelling techniques with the students. The other is up to the math teacher, who should make a small mathematical dictionary along with the students, seeking meanings for the terms used in this discipline and what other meanings could have in our daily lives. After the two procedures have been done, students should now write texts using the terms they learned in math classes. It is advisable to be presented with texts that can serve as a guide for students, as this will probably be an entirely new activity for them.

It is interesting that the activity is carried out in the classroom because it is a new proposal, unknown to most students, there will undoubtedly be many difficulties for the elaboration of the texts. Therefore, it is indispensable to monitor teachers for better student performance. It will be necessary to take a class that will serve to expose to students what is to be developed, the goals and how much fun it is to merge disciplines. In an introductory class, it is interesting that the teacher read some texts as an example for the students and so that they can have a better view of what is being proposed. As a suggestion, the teacher with the class can formulate a bank of mathematical terms and words with their meanings to elaborate the text.

To stimulate the students more, rewards to the best works such as watching an animation or seeing the poetry made into an animation can be awarded, but always with the care of not devaluing the other works did not come to achieve the goals. Four group activities or meetings can serve to carry out the project. Of course, this time is flexible and depending on both the student as well as the teacher can lengthen or compress the suggested time.

4 Discussion and Future Work

The association between art and mathematics considers the existing Mathematics in artistic productions as architecture, literature [12], sculpture, music [13] and painting. Highlighting that associations seem to be mathematics secret and art in itself, this text discusses the role of artistic gratification, imagination, identification and originality in mathematics, and vice versa.

One aspect that called the authors' attention is the fact that seniors feel more at ease when it comes to writing. When seniors and children are paired in an activity involving math and literature, in the long run, better results appeared, especially when there are computers involved. Seniors tend to have a better vocabulary while the children feel more comfortable with technology. Since both groups are not in general found of math, once they establish a bond, they can cooperate towards better texts with an increase in computer literacy by the seniors. The experiments gave some interesting spinoffs. Given that the population is aging fast, games and transmedia activities can further explore new possibilities with different age groups [14–17].

5 Conclusions

Despite the focus of this work, which is, stimulating the taste for reading and fixing mathematical concepts, it is worth to point out that this idea can also be used in order to involve other areas of knowledge.

This activity is conventional outside standards. One may find diverse reactions and behaviors in students. At first sight, it is impossible to do a job involving two different disciplines, Portuguese and mathematics. Some students may present resistance to develop the activity. However, the vast majority are enthusiastic about the proposal and face the activity as a challenge. It is necessary to have patience and convincing arguments to show the students more closed to new proposals that this activity in addition to improving their writing techniques, exploring the grammar and the mathematical terms and concepts, it develops the power of abstraction and logical reasoning by mixing the richness of words in many ways in one job.

Many students have difficulties in elaborating their texts for various reasons, so it is necessary to show texts to inspire students and to clear the real sense of what should be done. The whole process should be followed closely and with quantitative methodologies so that the students produce texts with the greatest authenticity possible. It is clear that our goal is not to form poets, chroniclers or any writer, but rather that the works are done with the most significant commitment possible exposing what our students can accomplish.

References

1. Gardner, H.: Inteligências múltiplas: a teoria na prática. Artes Médicas (1995)
2. Greenberg, D., Silbert, J.: Max's Math Adventures (2004). http://teacher.scholastic.com/max/tguide.htm
3. Brenelli, R.P.: O jogo como espaço para pensar: a construção de noções lógicas e aritméticas. Papirus, Campinas, SP (1996)
4. Gardner, H.: Estruturas da mente: a teoria das inteligências múltiplas. Artes Médicas Sul (1994)
5. Drummond, T.: Elixirs for Your Memory. Time (1999)
6. Carretero, M.: Construtivismo e Educação. Artes Médicas, Porto Alegre (1997)
7. Franco, S.R.K.: O construtivismo e a educação, 4th edn. Mediação (1995)
8. Nöth, W.: Handbook of Semiotics. Indiana University Press (1995)
9. Nöth, W.: Panorama da semiótica: de Platão a Pierce, ANNABLUME (1995)
10. Gadanidis, G., Simmt, E., Sterenberg, G., Tumanov, V.: Literature. In: Proceedings of Mathematics as Story: A Symposium on Mathematics Through the Lenses of Art & Technology, pp. 62–65. Faculty of Education, University of Western Ontario (2004)
11. Beatty, K.: Math poems (2004). http://www.durham.edu.on.ca/programmath2/Math_Poems/math_poems.html
12. Carreiro, L.S., Estrela, V.V., Vishnevskii, M.P., Huacasi, W.D.: Poematica. In: XXVIII CNMAC, Sao Paulo, SP (2005)
13. dos Santos, T.F.: Matematica, musica e educação, Dissertation, UENF, RJ (2004). https://doi.org/10.13140/2.1.2849.5360
14. Estrela, V.V., Monteiro, A.C.B., França, R.P., Iano, Y., Khelassi, A., Razmjooy, N.: Health 4.0: applications, management, technologies and review. Med. Techn. J. **2**(4), 262–276 (2019). https://doi.org/10.26415/2572-004x-vol2iss1p262-276
15. Herrmann, A.E., Estrela, V.V., Loschi, H.J., Vishnevski, M.P.: Some thoughts on transmedia communication. OJCST **11**(4) (2018). https://doi.org/10.13005/ojcst11.04.01
16. Ahmed Zakariya, N., Rachid, B.: Health e-learning using virtual-reality technology in Algerian universities. Med. Techn. J. **1**(3), 50–51 (2017). https://doi.org/10.26415/2572-004X-vol1iss3p50-51
17. Tabari, P., Montaseri, N.: Investigating the remote monitoring usage for home health care. Med. Techn. J. **1**(4), 136 (2017). https://doi.org/10.26415/2572-004X-vol1iss4p136-136
18. https://www.edu.uwo.ca/mpc/files/mm-sept1999-poems.pdf
19. Latterell, C.M., Wilson, J.L.: Math is like a lion hinting a sleeping gazelle: pre-service elementary teachers' metaphors of mathematics. Eur. J. Sci. Math. Ed. **4**(3), 283–292 (2016)
20. https://files.eric.ed.gov/fulltext/EJ1107832.pdf
21. https://www.quora.com/What-happened-when-you-told-your-crush-about-your-feelings-How-did-he-she-react

22. http://mathsongs.info/matharena_le.htm
23. https://owl.uwo.ca/access/content/group/d9c3b137-1d5d-4026-9794-2079b0d9f6a8/Mathematics%20As%20Story.pdf

The manufacturer's authorised representative in the EU is Springer Nature Customer Service Centre GmbH, Europaplatz 3, 69115 Heidelberg, Germany. If you have any concerns regarding our products, please contact ProductSafety@springernature.com

Printed and bound by CPI Group (UK) Ltd, Croydon, CR0 4YY
30/09/2026
02225440-0017